Medical Genetics and Genomics

Fundamentals of Biomedical Science

Medical Genetics and Genomics

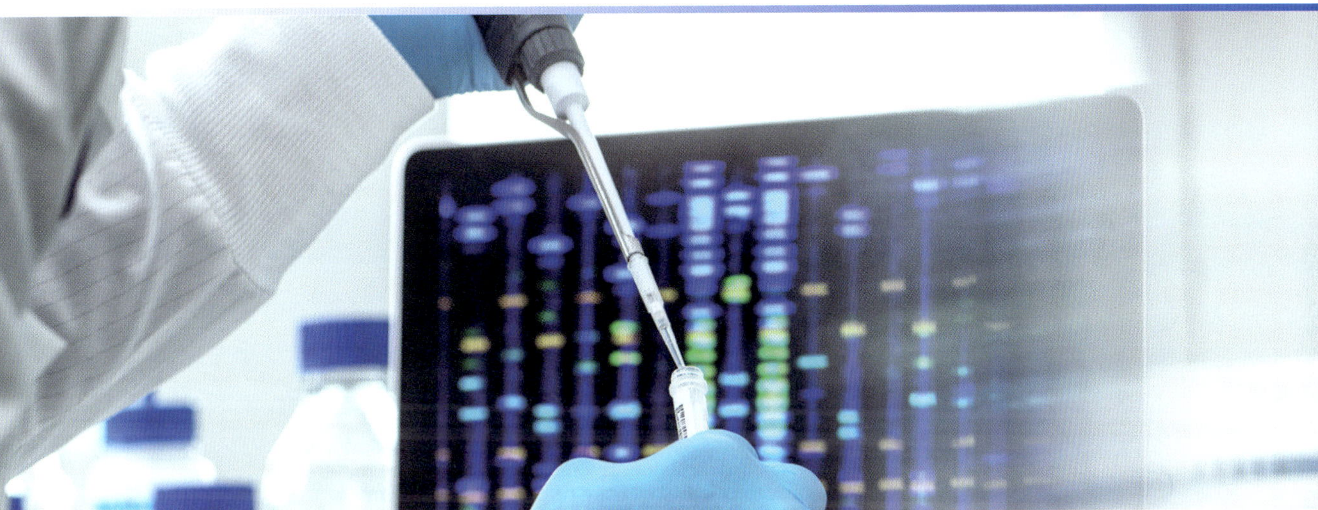

Edited by

Emanuela Volpi
Professor, University of Westminster

Lorna Tinworth
Senior Lecturer, University of Westminster

Great Clarendon Street, Oxford, OX2 6DP,
United Kingdom

Oxford University Press is a department of the University of Oxford.
It furthers the University's objective of excellence in research, scholarship,
and education by publishing worldwide. Oxford is a registered trade mark of
Oxford University Press in the UK and in certain other countries

Published in the United States of America by Oxford University Press
198 Madison Avenue, New York, NY 10016, United States of America

British Library Cataloguing in Publication Data
Data available

Library of Congress Control Number: 2024952658

ISBN 978–0–19–883472–4

Printed and bound in Great Britain by Bell and Bain Ltd, Glasgow
Ashford Colour Ltd., Gosport, Hampshire

The manufacturer's authorised representative in the EU for product safety is Oxford University Press
España S.A. of El Parque Empresarial San Fernando de Henares, Avenida de Castilla, 2 – 28830 Madrid
(www.oup.es/en or product.safety@oup.com). OUP España S.A. also acts as importer into
Spain of products made by the manufacturer.

An introduction to the Fundamentals of Biomedical Science series

Clinical scientists and biomedical scientists form the foundation of modern healthcare, from cancer screening to diagnosing HIV, from blood transfusion to food poisoning, and infection control.

Without scientists, the diagnosis of disease, the evaluation of the effectiveness of treatment, and research into the causes and cures of disease would not be possible.

However, the path to becoming a scientist is a challenging one: trainees must not only assimilate knowledge from a range of disciplines but must understand—and demonstrate—how to apply this knowledge in a practical, hands-on environment. The UK's Institute of Biomedical Science (IBMS) is the leading professional body for clinical and biomedical scientists. One of its key roles is the provision of training material, guidelines, and, in collaboration with Oxford University Press, a comprehensive series of textbooks aimed at those taking IBMS examinations.

The *Fundamentals of Biomedical Science* series is written to reflect the challenges of biomedical science education and training today. It blends essential basic science with insights into laboratory practice to show how an understanding of the biology of disease is coupled with the analytical approaches that lead to diagnosis.

The series provides coverage of the full range of disciplines to which a clinical or a biomedical scientist may be exposed—from microbiology to cytopathology to transfusion science. Alongside volumes exploring specific biomedical themes and related laboratory diagnosis, an overarching *Biomedical Science Practice* volume provides a grounding in the general professional and experimental skills with which every laboratory scientist should be equipped.

Produced in collaboration with the Institute of Biomedical Science, the series:

- understands the complex roles of scientists in the modern practice of medicine;
- understands the development needs of employers and the profession;
- places the theoretical aspects of biomedical science in their practical context.

The website for each title in the series also features figures from the book for registered adopters to download for use in lecture presentations and resources.

Learning from this series

The Fundamentals of Biomedical Science series draws on a range of learning features to help readers master both biomedical science theory, and biomedical science practice.

CASE STUDY 3.1: Swyer syndrome

A 30-year-old patient with a female external phenotype presented to the clinic with complaints of primary amenorrhea. There was no similar family history of infertility, amenorrhea, abnormal external genitalia development, or cryptorchidism. on physical examination, the patient presented with normal female external genitalia and normal breast abdomen. Hormonal assays further confir nadal dysgenesis. The patient was referre testing. The cytogenetic laboratory report karyotype, confirming the Swyer syndrome sis (**Figure 3.13**). The clinician advised the p prophylactic gonadectomy of the left dyspl

Case studies illustrate how the biomedical science theory and practice presented throughout the series relates to situations and experiences that are likely to be encountered routinely in the biomedical science laboratory.

BOX 3.1 Fluorescence *in situ* Hybridizati

fluorescence *in situ* hybridization (fiSH) is a combined molecular and cytological technique that can be applied to evaluate the numerical and structural integrity of genetic sequences on cellular specimens. for its ability to conceptually and experimentally link 'cell and sequence', the introduction of FISH many decades ago prompted the reinvention of cytogenet- and the slides are p that varies accordin The hybridization te ficity and sensitivity cubation, the excess with phosphate bu

Additional information to augment the main text appears in **boxes**.

Key Points

G-banding is a Giemsa-based staining technique that generates a reprodu of alternating dark and pale chromatin bands along the chromosome leng

Key Points reinforce the key concepts that the reader should master from having read the material presented, while **Summary** points act as an end-of-chapter checklist for readers to verify that they have remembered correctly the principal themes and ideas presented within each chapter.

nucleolus

The primary site of production and assembly of ribosomal units within the nucleus of a eukaryotic cell.

Before applying fi SH and other immunofluoresc cleolus was the only example of a visually recog with a critical functional significance. The **nucleol** sis, which occupies within the nucleus an area cr genes-harbouring chromosomes and their coord that the nucleolar aspect (number of nucleoli pe lished biomarker in histopathology as nucleolar somal synthesis, which is necessary to sustain acc in other words, the nucleolus provides a paradi

Key terms provide on-the-page explanations of terms that the reader may not be familiar with; in addition, each title in the series features a glossary, in which the key terms featured in that title are collated.

SELF-CHECK 3.1

Which modern cytomolecular techniques have progressed our understanding of chromosom organization within the cell nucleus?

3.1.2 Why and how does the appearance of chromosomes change during the cell cycle?

Self-check questions throughout each chapter and extended **discussion questions** at the end of each chapter provide the reader with a ready means of checking that they have understood the material they have just encountered. Answers to these questions are provided in the e-book's end-of-chapter pedagogy.

Cross reference

Please refer to **Chapters 5** and **6** to learn more extensively about PCR, microarrays analysis, and whole genome sequencing.

ically associating domains (TADs boundaries characterized by loca interactive as DnA sequences wit sequences outside the TAD. The d highly conserved across species, genomes (Dixon et al. 2016). Alth functions, understanding their fu Costa-nunes and nordermeer 202

Cross references help the reader to see biomedical science as a unified discipline, making connections between topics presented within each volume, and across all volumes in the series.

Preface

Understanding how genes influence disease and individual responses to treatment, and how genetic knowledge can be applied for disease diagnosis, prognosis, and prediction is key in biomedical science education and training. The past two decades have seen significant transformation of clinical practice brought about by advances in the application of genomic technologies in healthcare. As academics teaching within Institute of Biomedical Science accredited courses at both undergraduate and postgraduate levels, throughout this transformation we have been in a constant process of modernizing our course content. We work to seamlessly bridge foundational learning in genetics and genomics with education and skills around the integration of cutting-edge knowledge and techniques in the field into modern healthcare.

Over the past few years, it became clear to us that in order to meet the challenge of teaching this subject and bringing it alive for our learners we needed a handbook of modern medical genetics for biomedical science students and healthcare professionals. Of course, careers for biomedical science graduates are not limited to employment within the healthcare system. Given the wide relevance of genetics and genomics beyond the clinic, it is our intention that new graduates who have studied this book will be equipped with transferable knowledge applicable not only in healthcare settings, but also in many different related professional spheres. As scientists and educators, we are naturally inclined towards orderly conceptual progression when presenting processes and discoveries. Accordingly, we have organized the book contents to enable learners to follow, almost chronologically, the evolution of thought and competencies that have led to the progressive integration of genetics into healthcare.

Chapter 1 is an introductory chapter that sets the scene. It gives the reader a retrospective of the most important discoveries and strategic initiatives that have led the United Kingdom to become a world leader in genomic medicine. Chapters 2 to 4 cover essential genetic concepts around the functional organization of the human genome, single gene disorders, chromosomes in health and disease, and polygenic inheritance. Chapter 5 presents a comprehensive overview of the evolution of DNA sequencing technologies and their applications. Chapter 6 gives the reader an up-to-date understanding of the diagnostic laboratory workflow. Chapter 7 will give the reader knowledge and understanding of the key concepts in cancer genetics, taking the learner from foundational knowledge through to applications in clinical practice, and providing examples of translational research. With the clear understanding that genomics technologies now bridge many disciplines and that an individual and their microbiome may be viewed as an holobiont, Chapter 8 focuses the reader's attention on the genomics of microbiology. It explains to the reader how modern genetics and genomics techniques can be used to study both commensals and pathogens. Offering detailed case studies, Chapter 9 provides students scope for learning and reflection around the legal and ethical challenges surrounding the application of genomics in clinical practice and beyond. Finally, with the understanding that the global demand for precision medicine is increasing exponentially, Chapter 10 provides our readers with a strong and detailed introduction to the area.

The content of each chapter is comprehensive but by no means exhaustive. Partly because it would be impossible in a book to cover real-time developments in a field that advances at breakneck speed, and partly because encyclopaedic coverage is beyond the design of this series. At the end of each chapter the learner will find suggestions for further reading and exploration designed to broaden their horizons still further and pique their curiosity.

In order to challenge students' overreliance on well-trodden learning paradigms and conceptual frameworks, we have, within some of the chapters, experimented with unconventional approaches to content organization. Essentially, encouraging students to approach their learning from different perspectives. For instance, in Chapter 3, more advanced concepts such as the organization of chromosomes in the interphase nucleus come before the traditional topics, such as mitosis and meiosis, which would conventionally have been given first. Or in Chapter 5, a reference to sequence motifs placed at the beginning of the chapter draws the readers' attention towards the functional meaning of DNA sequence patterns, down to a single nucleotide, and hence to the relevance of sequencing.

Some content repetition across different chapters is given to support conceptual amalgamation and synoptic learning throughout the book. This repetition also ensures that each chapter stands alone as an independent and comprehensive portion of information on a well-defined topic of interest, to be studied either within the broad context of the book or in isolation.

This book is principally for students taking an undergraduate course in Biomedical Science, or for graduate students and medical sciences students seeking to consolidate previous learning. However, the book could also be consulted by healthcare workers and biomedical science professionals who, having possibly not received much formal training in medical genetics and genomics, seek reference material to help them navigate and build confidence in this rapidly developing field of growing relevance for clinical practice.

—Emanuela Volpi and Lorna Tinworth

Acknowledgements

We thank the following organizations and individuals without whom this book would never have been published:

- IBMS former Executive Head of Education, Alan Wainwright, and IBMS former Deputy Chief Executive, Sarah May, for endorsing the idea of this book and initial brainstorming around content.
- All those who have contributed chapters, for their sustained commitment and enduring patience.
- Oxford University Press for editorial support. In particular, we would like to thank Lauren Wing for giving us expert guidance and steadfast encouragement.
- Our families for their understanding, forbearance, and support while this book was being written.
- The University of Westminster, our professional home, for providing us the opportunity to grow as educators.

We also extend our thanks to the following external reviewers, along with those who wished to remain anonymous:

- William Edward Allen, Queen's University, Belfast
- Jim Boyne, Leeds Beckett University
- Aparna Duggirala, University of Derby
- Antonio Marco, University of Essex
- Michael Randles, University of Chester
- Kevin Smith, Abertay University
- Jim Taylor, York St. John University
- Sarah Westbury, University of Bristol
- Kenneth White, London Metropolitan University

We dedicate this book to our students, whose curiosity, imagination, tenacity, and limitless ambition are our inspiration.

—Emanuela Volpi and Lorna Tinworth (2025)

Contributors

Angus Clarke
Cardiff University

Drew Ellershaw
NE Thames Regional Genetics Service

Zoe Hatt
Synnovis UK

Janine Lamb
University of Manchester

Hussein Sheikh Ali Mohamoud
University of London

Jamal Nasir
University of Northampton

Dianne Newbury
Oxford Brookes University

Shivani Shah
University College London Hospital

Nicola Taverner
Cardiff University

Lorna Tinworth
University of Westminster

Vincenzo Torraca
King's College London

Nirmal Vadgama
Stanford University

Emanuela Volpi
University of Westminster

Heather Williams
Columbia University Medical Center

Abbreviations

Chapter 1

CMO	Chief Medical Officer
DNA	Deoxyribonucleic Acid
HGP	Human Genome Project
NGS	Next Generation Sequencing
NHS	National Health Service
NIH	National Institutes of Health
SNP	Single Nucleotide Polymorphism
RNA	Ribonucleic Acid

Chapter 2

ATP	Adenosine triphosphate
BRE	TFIIB recognition element
DNA	Deoxyribonucleic Acid
DPE	Downstream Promoter Element
dsDNA	Double-stranded DNA
GWAS	Genome-Wide Association Study
INR	Initiator
LCR	Locus Control Region
MND	Motor Neurone Disease
mtDNA	Mitochondrial DNA
MTE	Motif Ten Element
PIC	RNA polymerase preinitiation complex
RNA	Ribonucleic Acid
rRNA	Ribosomal RNA
SNP	Single Nucleotide Polymorphism
TFBS	Transcription Factor Binding Site
TFIIB	Transcription factor II B
TSS	Transcription Start Site

Chapter 3

3C	Chromosome Capture Conformation
3D-FISH	Three-dimensional FISH
CMA	Chromosomal Analysis by Microarrays
Cryo-FISH	FISH on cryosections
DAPI	4′, 6-diamidino-2-phenylindole
DNA	Deoxyribonucleic Acid
FISH	Fluorescence *in situ* Hybridization
Hi-FISH	High-throughput FISH
M-FISH	24-colour FISH
qPCR	Quantitative Polymerase Chain Reaction
RNA	Ribonucleic Acid
TADs	Topologically Associated Domains

Chapter 4

aCGH	array Comparative Genomic Hybridization
CFTR	Cystic Fibrosis Transmembrane Conductance Regulator
CNV	Copy Number Variant
ddPCR	droplet digital Polymerase Chain Reaction
ENCODE	ENCyclopedia of DNA Elements
FISH	Fluorescence *in situ* Hybridization
GA1	Glutaric Aciduria type 1
GDP	Gross Domestic Product
GERP	Genomic Evolutionary Rate Profiling
gnomAD	Genome Aggregation Database
GWAS	Genome-Wide Association Study
HCU	Homocystinuria (pyridoxine unresponsive)
HI	Haploinsufficiency
Indel	Insertion/deletion mutation
IVA	Isovaleric Acidaemia
LoF	Loss-of-Function
MCADD	Medium-Chain Acyl-CoA Dehydrogenase Deficiency
miRNA	Micro Ribonucleic Acid
MSUD	Maple Syrup Urine Disease
mtDNA	Mitochondrial DNA
NGS	Next-Generation Sequencing
NumtS	Nuclear mitochondrial sequences
PCR	Polymerase Chain Reaction
PKU	Phenylketonuria
pLI	probability of being Loss-of-function Intolerant
qPCR	quantitative Polymerase Chain Reaction
RVIS	Residual Variation Intolerance Score
SNP	Single Nucleotide Polymorphism
SNV	Single Nucleotide Variant
WGS	Whole Genome Sequencing

Chapter 5

ACMG	American College of Medical Genetics
ALT	Alternative allele
BAM	Binary Alignment Map
bp	base pairs
BRE	B-Recognition Element
CML	Chronic Myeloid Leukaemia
CNV	Copy Number Variant
ddNTP	Dideoxynucleoside Triphosphate
dNTP	Deoxynucleoside Triphosphate
DPE	Downstream Promoter Element
ENCODE	Encyclopaedia of DNA Elements
gnomAD	Genome Aggregation Database
IGSR	International Genome Sample Resource
IRUD	Initiative on Rare and Undiagnosed Diseases
LoF	Loss of Function
MeDIP-seq	Methylated DNA Immunoprecipitation sequencing
NGS	Next Generation Sequencing
OR	Olfactory Receptor
PCR	Polymerase Chain Reaction
REF	Reference allele
RNA-seq	RNA sequencing
SAM	Sequence Alignment Map
sc-RNA-seq	Single-cell RNA sequencing
SNV	Single Nucleotide Variant
VCF	Variant Call File
VUS	Variant of Uncertain Significance

Chapter 6

aCGH	Array Comparative Genomic Hybridization
ART	Assisted Reproductive Technology
BAC	Bacterial Artificial Chromosome
BPG	Best Practice Guidelines
BrdU	Bromodeoxyuridine
CBAVD	Congenital Bilateral Absence of the Vas Deferens
CF	Cystic Fibrosis
CffDNA	Cell-Free Foetal DNA
CMA	Chromosomal Microarray
CNV	Copy Number Variant
DECIPHER	Database Of Genomic Variation and Phenotype in Humans Using Ensembl Resources
DMR	Differentially Methylated Region
DNA	Deoxyribonucleic Acid
DSD	Disorder of Sexual Differentiation
DTC	Direct To Consumer
EDTA	Ethylenediaminetetraacetic Acid
EQA	External Quality Assessment
ETS	Extension to Scope
FFPE	Formalin-Fixed Paraffin-Embedded
FISH	Fluorescence *in situ* Hybridization
FMU	Foetal Medicine Unit
GA1	Glutaric Aciduria type 1
GLH	Genomic Laboratory Hub
gnomAD	Genome Aggregation Database
HCPC	Health and Care Professions Council
HSC	Haematopoietic Stem Cell
HCU	Homocystinuria
HSP	Healthcare Science Practitioner
HSST	Higher Specialist Scientific Training
IRT	Immunoreactive Trypsinogen
ISCN	International System for Human Cytogenetic Nomenclature
IVA	Isovaleric Acidaemia
LH	Lithium Heparin
LIMS	Laboratory Information Management System
MCADD	Medium-Chain Acyl-CoA Dehydrogenase Deficiency
MCC	Maternal Cell Contamination
MDT	Multi-Disciplinary Team Meeting
MLPA	Multiplex Ligation-dependent Probe Amplification
MRD	Monitoring Residual Disease
msMLPA	Methylation-Specific MLPA
msPCR	Methylation-Specific PCR
MSUD	Maple Syrup Urine Disease
NGS	Next Generation Sequencing
NHS	National Health Service
NIPD	Non-Invasive Prenatal Diagnosis
NIPS	Non-Invasive Prenatal Screening
NIPT	Non-Invasive Prenatal Testing
NT	Nuchal Translucency
OMIM	Online Mendelian Inheritance in Man
PCR	Polymerase Chain Reaction
PGD	Preimplantation Genetic Diagnosis
PHA	Phytohaemagglutinin

PKU	Phenylketonuria
PND	Prenatal Diagnosis
PTP	Practitioner Training Programme
QF-PCR	Quantitative Fluorescent PCR
PWM	Pokeweed Mitogen
QMS	Quality Management System
qPCR	Quantitative PCR
RHDO	Relative Haplotype Dosage
RNA	Ribonucleic Acid
rt-PCR	Reverse Transcriptase PCR
SPRT	Sequential Probability Ratio Test
SNP	Single Nucleotide Polymorphism
SOP	Standard Operating Procedure
STP	Scientist Training Programme
UKAS	UK Accreditation Service
UKFASP	UK Foetal Anomaly Screening Programme
UPD	Uniparental Disomy
WES	Whole Exome Sequencing
WGS	Whole Genome Sequencing

Chapter 7

aCGH	array Comparative Genomic Hybridization
ACT	Paediatric Adrenocortical Tumour
ALCL	Anaplastic Large Cell Lymphoma
CLL	Chronic Lymphocytic Leukaemia
CML	Chronic Myelogenous Leukaemia
Cn-LOH	Copy-neutral Loss of Heterozygosity
CNV	Copy Number Variations
CPGs	Cancer Predisposition Genes
DAPI	4′,6-diamidino-2-phenylindole
DCIS	Ductal carcinoma *in situ*
FAP	Familial Adenomatous Polyposis
FISH	Fluorescence *in situ* Hybridization
GoF	Gain of Function
HBOC	Hereditary Breast and Ovarian Cancer
HGNC	Human Gene Nomenclature Committee
HGVS	Human Genome Variation Society
HUGO	Human Genome Organization
IARC	International Agency on Cancer
IDC	Invasive ductal carcinoma
ISCN	International System for Human Cytogenetic Nomenclature
LFS	Li–Fraumeni Syndrome
LoF	Loss of Function
MDS	Myelodysplastic Syndrome
NGS	Next Generation Sequencing
PCR	Polymerase Chain Reaction
RSV	Rous Sarcoma Virus
SEER	US National Cancer Centre Surveillance, Epidemiology, and End Results Programme
SNP	Single Nucleotide Polymorphism
SNP-A	Single Nucleotide Polymorphism Array
TCGA	The Cancer Genome Atlas
vHL	von Hippel–Lindau syndrome
WES	Whole Exome Sequencing
WGS	Whole Genome Sequencing

Chapter 8

ADRS	Acute Respiratory Distress Syndrome
AFLP	Amplified Fragment Length Polymorphism
AIDS	Acquired Immunodeficiency Syndrome
AP	Alkaline Phosphatase
ART	Antiretroviral Therapy
AST	Antimicrobial Susceptibility Testing
bDNA	Branched DNA
cDNA	Complementary DNA
Ch	*Chlamydia* species
Chr	Chromosome
CMV	Cytomegalovirus
COVID-19	Coronavirus Disease 2019
CSF	Cerebrospinal Fluid
DDH	DNA–DNA hybridization
DNA	Deoxyribose Nucleic Acid
dNTPs	Deoxyribonucleotide triphosphates
dsDNA	Double-stranded Deoxyribonucleic Acid
dsRNA	Double-stranded Ribonucleic Acid
EBV	Epstein–Barr Virus
EUCAST	European Committee on Antimicrobial Susceptibility Testing
EV	Enterovirus
fg	Femtogram
flu A	Influenza A Virus
flu B	Influenza B Virus
Gbp	Giga base pairs
gDNA	Genomic DNA
HAdV	Human Adenovirus
HBoV	Human Bocavirus
HBV	Hepatitis B Virus

HcoV	Human Coronaviruses
HCV	Hepatitis C Virus
HDV	Hepatitis D Virus
HERVs	Human Endogenous Retroviruses
HEV	Hepatitis E Virus
HGT	Horizontal Gene Transfer
HIV	Human Immunodeficiency Virus
HMPV	Human Metapneumovirus
HPIV	Human Parainfluenza Viruses
HPV	Human Papillomavirus
HRC	C-terminal heptad repeat
HRSV	Human Respiratory Syncytial Virus
HRV	Human Rhinovirus
HSV	Herpes Simplex Virus
huDNA	Human DNA
huRNA	human RNA control
IC	Internal control
Indel	Insertion/deletion mutation
kbp	Kilo base pairs
LAMP	Loop-Mediated Isothermal Amplification
LCR	Ligase Chain Reaction
Mbp	Mega base pairs
MDR	Multidrug-resistant
MLST	Multilocus sequence typing
Mp	Mycoplasma pneumoniae
mRNA	Messenger RNA
MRSA	Methicillin-resistant Staphylococcus aureus
NASBA	Nucleic Acid Sequence-Based Amplification
NGS	Next Generation Sequencing
nm	Nanometre
PCR	Polymerase Chain Reaction
PFGE	Pulse-Field Gel Electrophoresis
pl	Plasmid
RFLP	Restriction Fragment Length Polymorphism
RNA	Ribonucleic Acid
rRNA	Ribosomal RNA
RSV	Respiratory Syncytial Virus
RT	Reverse Transcriptase
RT-PCR	Reverse Transcriptase Polymerase Chain Reaction
RT-qPCR	Real-Time quantitative PCR
SARS-COV-2	Severe Acute Respiratory Syndrome Coronavirus 2
SDA	Strand Displacement Amplification
SNP	Single Nucleotide Polymorphism
ssRNA	Single-stranded RNA
STI	Sexually Transmitted Infection
TB	Tuberculosis
TMA	Transcription Mediated Amplification
tRNA	Transfer RNA
UTI	Urinary tract infection
VZV	Varicella Zoster Virus
WGS	Whole Genome Sequencing
XDR	Extremely Drug-Resistant

Chapter 9

ABI	Association of British Insurers
ACMG	American College of Medical Genetics & Genomics
BRCA	BReast CAncer gene
DMD	Duchenne Muscular Dystrophy
DNA	Deoxyribonucleic Acid
GCP	Good Clinical Practice
GDPR	General Data Protection Regulation
GINA	Genetic Information Non-discrimination Act (USA)
GMP	Good Manufacturing Practice
GNDA	Genetic Non-discrimination Act (Canada).
hGGE	Human Germline Gene Editing
HIV	Human Immunodeficiency Virus
IEC	Independent Ethics Committee
IgG	Immunoglobulin G
IRB	Institutional Review Board
MYLK	Myosin Light Chain Kinase
NHS	National Health Service (UK)
NIPT	Non-Invasive Prenatal Testing
UK	United Kingdom
VUS	Variants of Uncertain Significance
WGS	Whole Genome Sequencing

Chapter 10

ACGS	Association for Clinical Genomic Science
ACMG	American College of Medical Genetics and Genomics
ADA-SCID	Severe Combined Immunodeficiency due to Adenosine Deaminase deficiency
ADH	Alcohol Dehydrogenase
ADME	(drug) Absorption, Distribution, Metabolism, and Excretion

ALDH	Aldehyde Dehydrogenase enzymes
AMD	Age-related Macular Degeneration
BGI	Beijing Genomics Institute
CAR-T	Chimeric Antigen Receptor T-cell
CPIC	Clinical Pharmacogenetics Implementation Consortium
CRISPR	Clustered Regularly Interspersed Short Palindromic Repeats
EGFR	Epidermal Growth Factor Receptor
EMA	European Medicines Agency
FDA	Federal Drug Administration
gnomAD	Genome Aggregation Database
HAART	Highly Active Antiretroviral Therapy
HER2	Human Epidermal Growth Factor Receptor 2
HIV	Human Immunodeficiency Virus
HLA	Human Leukocyte Antigen
MHRA	Medicines and Healthcare products Regulatory Agency
NGS	Next Generation Sequencing
NHS	National Health Service
NICE	National Institute for Health and Care Excellence
NSCLC	Non-Small Cell Lung Cancer
RCT	Randomized Controlled Trials
TPMT	Thiopurine Methyl Transferase
ZFN	Zinc Finger Nucleases

Contents

1

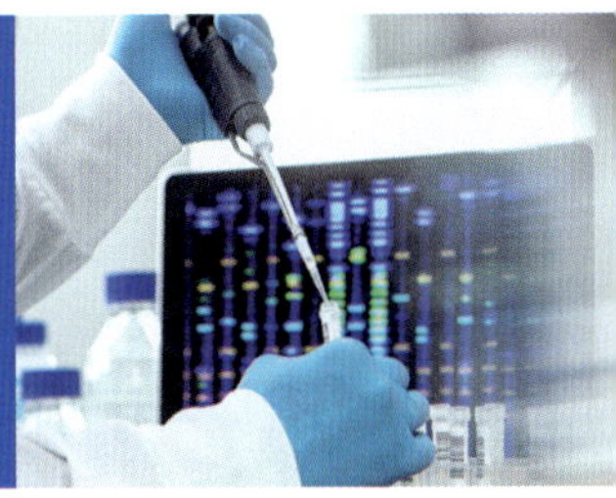

Human Genetics in the Post-Genomic Era

A brief history of discoveries, innovations, and integration in healthcare

Emanuela Volpi

Learning Objectives

After reading this chapter, you should be able to:

- Appreciate the remarkable feat of scientific endeavour that has allowed our current understanding of human genetics
- Describe the key milestones which have supported the progressive integration of genetics into medical practice in the post-genomic era
- Discuss the challenges and opportunities that arise when translating research discoveries in genetics into clinical applications for society.

When approaching the study of a field such as human genetics, which is advancing at a very fast pace and increasingly impacting our daily lives, especially in connection to medical applications such as the diagnosis, treatment, and prevention of disease, it might be tempting to focus on the most recent and exciting developments, and perhaps disregard as no longer relevant what would have preceded the latest breakthrough. However, approaching our learning of human genetics by considering the scope and scale of previous efforts and achievements is crucial to gaining a balanced perspective. Using the past as a frame of reference, we can gain valuable insights into how we got where we are and, most importantly, infer what could be realistically expected in the near future. 'If I have seen further it is by standing on the shoulders of giants', a famous metaphor used by Sir Isaac Newton in a letter he wrote to Robert Hooke in 1675, exemplifies how scientific progress builds on the ideas of those who have preceded us. In this chapter, and indeed in the following chapters of this book, we will do precisely that. We will be ***standing on the shoulders of giants*** to see further. We will reflect on how foundational research discoveries and ambitious technological innovations made, particularly around and after the completion of the Human Genome Project, have gradually paved the way for our current understanding of human genetics and, most crucially, are enabling the integration of human molecular genetics into modern clinical practice. The aim of this short introductory chapter is not to give an exhaustive coverage the history of modern human genetics and its medical applications but to recognize and illustrate of some of the most important developments. It is an important acknowledgement of the incredible scientific efforts and collective strategic thinking that have led us to where we are today.

1.1 What have we learned from the Human Genome Project?

The Human Genome Project (HGP), or the accomplishment of the first (almost) complete sequence of a human genome in 2003, has been befittingly described as one of the most incredible feats of exploration in history. Rather than an outward journey of exploration, we should consider it as an inward journey of discovery. Exploring our genome has undoubtedly allowed us to discover unexpected aspects of being human, providing fascinating insights into our individuality and unicity. It has also crucially informed our collective sense of identity, fuelling some of the most prominent societal discourses surrounding many aspects of our everyday lives.

It is hard to believe the HGP was completed only fifty years after the famous 1953 *Nature* paper in which Jimmy Watson and Francis Crick described for the first time the deoxyribonucleic acid (DNA) double helix, a molecular structure they somewhat understatedly anticipated would be of 'considerable biological interest', the discovery of which later led them to win the Nobel Prize for Medicine (Watson and Crick 1953). The publication of Watson and Crick's explanation of the DNA molecule ushered in the era of modern genetics. The connection between the DNA molecule and the transmission of genetic information had previously been proposed following several independent experiments carried out by molecular biologists on viruses and bacteria in the early/mid-twentieth century. These experiments enabled the identification and naming of the DNA molecule as the 'transforming principle' (McCarthy 2003). However, Watson and Crick's description of the double-helical structure, which was primarily based on X-ray diffraction data captured by Rosalind Franklin and Maurice Wilkins, provided the key to interpreting previous research findings around some of the critical properties of the DNA molecule, such as its chemical stability. It also provided a means to comprehend the significance of some of the characteristics and functionalities of DNA that make it uniquely suitable as the carrier of genetic information, such as the complementarity of the nucleotides (adenine/thymine, cytosine/guanine), which ensures accuracy during the replication and transcription processes, as well as its mutability, which is essential for adaptation and evolution.

As in space exploration, discoveries leading to conceptual disruption in modern biomedical research are often driven by ground-breaking technological developments. The invention of Sanger sequencing, a pioneering method to 'read' the DNA molecule at single nucleotide resolution (Sanger et al. 1977), can be singled out as the methodological development in DNA analysis that most significantly contributed to the accomplishment of the HGP and the completion of other foundational genomic studies independently undertaken by many research laboratories towards the end of the twentieth century. This world-famous type of sequencing procedure, based on the random incorporation of chain-terminating dideoxynucleotides during *in vitro* DNA replication followed by separation of the different DNA fragments by capillary electrophoresis, is still ubiquitously applied in diagnostic settings in the present day. Unsurprisingly, Fredrick Sanger, the Cambridge-based researcher who pioneered this revolutionary method and subsequently was awarded two Nobel Prizes for his work, is considered by many the 'father of genomics'.

Initiated in 1990 and completed after thirteen years, the HGP was a highly collaborative undertaking of rare and significant scale. The project was delivered by a consortium of twenty universities and research centres from the United States, the United Kingdom, France, Germany, Japan, and China. More than 2,800 researchers were named authors in the first human genome draft (Lander et al. 2001), with even more investigators contributing to the completion of the project in 2003 (Human Genome Project Sequencing Consortium 2004). The staggering final cost of the HGP was around $300 million worldwide, 60% of which was covered by the National Institutes of Health (NIH).

The completion of the HGP broadened and deepened our understanding of human genetics. For instance, establishing the precise order of almost all three billion DNA base pairs of the whole genomic sequence revealed the physical maps of our chromosomes. These maps underpin our understanding of genetic linkage and the non-random co-segregation of traits. When newly acquired, this sequence information on physical linkage enabled researchers to piece together some of the unresolved aspects of the inheritance traits puzzle, such as the observed exceptions to Mendel's law of independent assortment. To revise Mendel's laws of genetic inheritance, refer to **Box 1.1**.

It is important to mention that the first iteration of the human genome sequence assembled upon completion of the HGP presented significant gaps, mainly in non-gene-containing regions comprising highly repetitive DNA. Those gaps in sequence (equivalent to less than 10% of the entire human genome) have only recently been bridged. In 2022, the NIH-funded Telomere-to-Telomere (T2T) research programme led to the publication of the first truly complete, gapless sequence of a human genome (Nurk et al. 2022). Whilst T2T is important for factual accuracy, we shall leave it aside for the moment and instead focus on the excitement that the ability to read, for the first time, nature's almost complete genetic blueprint of a human being generated in 2003. This huge excitement was not only felt by those directly involved in scientific research. Indeed, the completion of the first human genome sequence captured the imagination of society as a whole, as even for those not formally trained in biology, it was easy to start envisaging—at times, overenthusiastically—the potential collective implications, particularly in connection with health, longevity, and family planning.

BOX 1.1 Mendel's Laws of Inheritance

Based on observations made while experimenting with plant breeding in a monastery garden in the second half of the nineteenth century, Gregor Mendel formulated three laws to explain the inheritance of traits through generations. What is remarkable about Mendel's work is how prescient the interpretation of his findings around 'particulate inheritance' was when there was no notion of our traits being encoded in our DNA, and nobody knew that genes existed. Mendel's discoveries marked the birth of classical genetics, and the principles of inheritance he described in his laws have supported our understanding of biology and evolution over the last two centuries. The modern formulation of Mendel's laws based on our current understanding of genes, alleles, meiosis, and inheritance of genetic traits states that (1) during gamete formation, the two alleles at a gene locus segregate from each other, with each gamete having an equal probability of containing either allele (first law or law of segregation) and (2) pairs of alleles of two (or more) genes are sorted into gametes independently (second law or law of independent assortment), allowing for different combinations of traits in the offspring. Mendel also developed the law of dominance, according to which one allele exerts a more significant influence than the other on the same inherited character.

Once the first genome had been sequenced and the 'reference genome' or master sequence established, the step that followed was the sequencing of other human genomes. Comparison of these later human genomes with the reference genome started to reveal the unexpected scale of human genetic variation, intended as the collection of differences or variants that exist between different human individuals in the DNA sequence of their genomes. Whilst 99.9% of DNA is identical across individuals, the remaining 0.1% that accounts for variation is potentially millions of genetic variants across the three billion DNA base pairs of our genome. For example, compared to the HGP sequence, the genome draft competitively generated by the privately owned biotech company Celera Genomics revealed 2.1 million differences at the single nucleotide level, also known as single nucleotide polymorphisms (SNP) (Venter et al. 2001). When the genome of a specific, single individual human was first sequenced and compared to the HGP reference genome (which had instead been assembled by combining DNA sequence information from many different individual genomes), more than 4.1 million different types of DNA variants were revealed. Collectively, these differences accounted for 12.3 million DNA base pairs (Levy et al. 2007). Famously, the first individual genome to be sequenced was that of Dr Craig Venter, the scientist and entrepreneur at the head of Celera Genomics. This made Dr Venter the first human ever to be able to read their own genome. These and other discoveries around inter-individual genomic variability marked the beginning of the era of individualized genomic information. This important step laid the conceptual foundations for modern personalized medicine wherein a patient's genomic profile informs disease management.

As genomes of other organisms started to be sequenced and explored, and evidence of intra- and inter-species genomic similarities and differences accumulated, some perplexing aspects emerged, such as the incongruity of genome sizes across the evolutionary scale, with no apparent correspondence between genome size and organismic complexity (known as the C-value paradox) as well as the unexpected similarity in gene content in genomes of different organisms (known as the G-value paradox). For example, the not-too-dissimilar number of genes found in the genomes of mice and humans (~25,000 vs ~26,000). Whilst subsequent research on the functional organization of different genomes and, more specifically, on the regulation of gene expression in eukaryotic cells has provided clues to interpreting some of the most intriguing biological and evolutionary puzzles, the full complexity of the human genome organization has not yet been fully elucidated and is still the focus of intense investigation.

1.2 Strategic milestones in the post-genomic era and future directions

Once the HGP was completed and the complexity and scale of human genetic variation started to emerge, it became clear that only by scaling up the sequencing efforts would it be possible to gather sufficient information to start *translating* research findings by genomics into practical benefits for society, particularly in the medical field. To avoid confusion, it must be specified that in this context, the word 'translation' and its derivations have nothing to do with the molecular process of protein chain assembly that follows RNA transcription or the copying of RNA from the DNA template. Translational research refers instead to a systematic approach to convert research discoveries into societal benefits. In this case, translating genomic discoveries into healthcare improvements. The most pressing 'knowledge gap' hindering translation was around the unexpected scale of genetic variation and the challenges in ascertaining the potential pathogenic significance of those genetic variants. It became clear that many replicates would be required to facilitate robust statistical analysis to establish an incontrovertible association of rare genetic variants with specific diseases or associations that would potentially be worthwhile in clinical settings.

Similarly to what was described before, with the development of the Sanger sequencing pushing forward genomic discoveries at the time of the HGP, the introduction of a new type of high-throughput DNA sequencing method and its commercialization in the early years of the new millennium provided a timely technological breakthrough to escalate genome analysis to its next phase. The new sequencing method or next-generation sequencing (NGS) was initially called 'massive parallel sequencing' as it essentially enabled the simultaneous sequencing of many DNA strands instead of one at a time, as in traditional sequencing. All of a sudden, albeit initially at a relatively high cost, it was possible to sequence an entire genome in a single run and in a relatively short time (especially when compared to the thirteen years that took the HGP consortium to sequence the first genome!) and with increased precision, crucial to allow confidence in variant calling. The field of modern Genomics was born, a novel methodological approach to the study of Genetics, enabled by high-throughput sequencing technologies for the base-by-base decoding of whole genomes, the suffix 'omics' indicating the involvement of large-scale data analysis.

In current scientific literature and everyday discourse, the words Genetics and Genomics are often juxtaposed to suggest similar yet conceptually different disciplines, with Genetics described as the study of single genes and inheritance of traits and Genomics as the study of genomes. This is, of course, an oversimplistic and potentially misleading distinction, as long before the advent of high-throughput whole genome sequencing technologies, through the study of karyotypes, position effects, epistasis, penetrance, variable expression, genomes' three-dimensionality, inheritance of complex traits, complexity of cancer genomes, and many other cornerstones in Genetics, it had become evident that genes do not exist and function in a molecular and biological vacuum. Indeed, studying whole genomic contexts as the physical and functional 'homes' of genes is, in most cases, essential to understanding genetic inheritance and investigating the genetic basis of disease. So, perhaps, the distinction between Genetics and Genomics, rather than theoretical and purpose-related, should be redefined around differences in their analytical properties, with Genomics as a powerful method conferring Genetics as a field of study with unprecedented resolution and latitude.

With the increased accessibility and rapidly lowering cost of NGS technologies, the circumstances became favourable for the first significant investment in translational genomic research. An investment of £100 million was announced in 2012 by the British Government to launch the '100,000 Genomes Project', a large-scale programme of genome sequencing aimed at linking genomic and clinical data through new scientific discoveries to enable medical insights. The investment was also anticipated to kick-start the development of a 'genomics industry' in the UK, with the ambitious scope of improving health whilst at the same time generating wealth. Genomics England, wholly owned and funded by the Department of Health & Social Care, was set up to deliver this flagship project and sequence 100,000 whole genomes from National Health Service (NHS) patients affected by rare disorders and cancer. This programme was the first of this kind in the world. The project was initiated in 2013 and reached its patient recruitment target in 2018. Examples of significant outcomes so far include a report on rare disease diagnosis published by the project consortium in the *New England Journal of Medicine* in 2021, which showed a crucial increase in diagnostic yield across a range of rare diseases, with immediate clinical actionability in 25% of those who received a clinical diagnosis (The 100,000 Genomes Project Pilot Investigators 2021). Also, the findings of the Cancer Programme, published in *Nature Medicine* in 2024, demonstrated how linking genomic and real-world clinical data enables the identification of gene variants with prognostic and predictive implications in cancer (Sosinsky et al. 2024).

The 100,000 Genomes Project was the first strategic step towards creating an NHS Genomic Medicine service, with the UK aiming to be the first national healthcare system to offer whole genome sequencing as part of routine care for the benefit of patients. Since its launch, the 100,000 Genomes Project has attracted huge investments by the Wellcome Trust, the US-based genomics company Illumina, the Medical Research Council, and the NHS towards the development of sequencing infrastructures, including in 2017 the reorganization of genetic laboratories and services from one hundred hospitals into thirteen new genomic medicine centres. Sequencing plans have since escalated, and the NHS is in the process of completing the sequence of 500,000 whole genomes in rare disease and cancer cases.

Gaps in infrastructure and service provision were highlighted in 2016 in the Chief Medical Officer's annual report. The Chief Medical Officer (CMO) is a governmental figure and a senior physician who leads a team of medical experts on public health matters and provides clinical advice to ministers in the Department of Health. While the recruitment of consented patient participants in the 100,000 Genomes Project was still ongoing, as part of their statutory role, the then CMO submitted to the Government a 256-page report entitled 'Generation Genome', which explored the potential of genomics to improve health and prevent illness. The report presented evidence of the 'fantastic work' already happening in three specific areas: genetic screening, diagnosing rare diseases, and genomic applications for personalized medicine or the potential to customize through genetic information medical treatments for individuals, leading to improved health outcomes. Crucially, at that time, the report highlighted gaps around infrastructures for genomic service provision, particularly concerning computational capabilities and bioinformatic know-how necessary to process the vast amount of data generated by high-throughput sequencing technologies. The report also underlined the need for workflow standardization at the national level. The report recommended addressing those limitations to develop the necessary resources for implementing genomic medicine for primary care. The report also included ethical considerations regarding privacy, consent, and data protection and the need to ensure equitability of access to genomic medicine to benefit all socio-economic groups and prevent health inequalities. This anticipated the scale of discussion and public debate we have seen lately on these very topics. Ultimately, the report sent an unequivocal and timely message about embracing genomics as a critical component of future healthcare. It advocated that policymakers, healthcare providers, and researchers work together to make this happen whilst managing the associated ethical and practical challenges.

Following the CMO Generation Genome report, the Secretary of State for Health and Social Care commissioned an independent review to advise on preparing the healthcare force to deliver the digital future. The report that summarized the review outcomes—named after Doctor Eric Topol, a Professor of Molecular Medicine at the Scripps Research Institute in California, who was asked to lead the study—was published in 2019. The Topol Review provided comprehensive advice on how, in the foreseeable future, technological developments, including genomics, digital medicine, robotics, and artificial intelligence, will likely change the roles and functions of clinical staff. It also covered the implications of these changes regarding the skills required for those professionals and the consequences for the selection, education, training, and lifelong development of current and future NHS staff. In the report, 'reading the genome', or the ability to capture genetic variants in individual genomes through high-throughput sequencing, and 'writing the genome', through genome-editing technologies and synthetic biology, were highlighted as two of the top ten technological advances expected to impact healthcare significantly over the next few decades. As part of the review, Dr Topol appointed an expert advisory panel on genomics to contribute recommendations based on the anticipated impact that the macro-scale digital implementation of genomics in primary healthcare could have on patients, the workforce, and the whole healthcare system. Given the complex ethical issues that genomic data can raise around privacy, insurance, and employment, the panel recommended that, in partnership with relevant regulatory bodies, the NHS should establish a clear, robust framework by which healthcare professionals use genomic data. This framework would safeguard patients' confidentiality and inspire the support and confidence of citizens and the wider community. Regarding the workforce, the panel advised that healthcare professionals should receive core training in genomic literacy to help them understand the broad principles of genomics and its benefits and ethical considerations. The availability of life-long training and continued support, including access to dynamic digital updates and online genomic information resources, was highlighted as essential. The panel also recommended that healthcare professionals receive accredited genomic training in vital clinical specialities to incorporate genomic testing and counselling into their practice. Regarding required actions for systemic changes, the panel advised building capacity in the genomic medicine sector, including introducing attractive career pathways for clinical bioinformaticians and a framework for genomic leadership. Finally, the role of academic institutions in this transformational process is discussed, with the recommendation that genomics and data analytics be given prominence in undergraduate curricula for healthcare professionals and that the undergraduate capacity in genomics, bioinformatics, and data science be expanded.

Not long after the Topol Review, in September 2020, the Government published a ten-year strategy entitled 'Genome UK: The future of healthcare', which maintained and extended the ambition of the United Kingdom to 'create the most advanced genomic healthcare system in the world, underpinned by the latest scientific advances, to deliver better health outcomes at lower cost'. The document focuses on three areas identified as pillars of the vision outlined in the strategy: Diagnosis and personalized medicine, Prevention, and Research. Incorporating the latest genomic advances into routine healthcare for the diagnosis, stratification, and treatment of disease is core to the strategy, with specific emphasis on the introduction of whole genome sequencing in routine healthcare, tailored drug treatments, and new operating models for cancer that draw on multidisciplinary expertise to improve patient outcomes. As an example of the successful application of pharmacogenetics or the study of how genes affect a person's response to drugs in clinical practice, the report showcases the introduction of genetic testing to identify cancer patients with mutations in the *DPYD* gene that could lead them to experience severe adverse reactions to fluoropyrimidines (drugs used in colorectal cancer chemotherapy and also commonly used in other types of cancer). Genomics is also seen as an enabler of preventative care through early life and targeted screening. In the strategy, investment in translational research is asserted to be fundamental to ensuring a seamless transition of discoveries and innovations from the laboratory bench to the hospital bedside. In the wake of the first pandemic wave, the strategy also articulates the intention of using genomics to deepen our understanding of how and why different people's immune systems respond differently to pathogens. It makes specific reference to the Genetics of Mortality in Critical Care Consortium's (GenOMICC) COVID-19 whole genome sequencing programme, which compared genetic variants of people who required intensive care because of COVID-19 with people who had mild symptoms (Pairo-Castineira et al. 2020; 2023). In the strategy, there is also a specific reference to the expanding need for functional genomics research (or the study of how genes and other genomic regions contribute to biological processes) to facilitate better understanding and diagnosis of disease and support drug discovery. Bio-sampling capabilities, such as the UK Biobank, are crucial resources in functional genomics, allowing the prospective collection of bio-samples for clinical research. To read more about the UK Biobank and its role in genomic medicine, please refer to **Box 1.2**. The strategy also reiterates and amplifies the relevance of certain practice aspects and partnerships previously identified by the CMO Report and the Topol Review as critical actors in the cultural transformation required for society to embrace genomic medicine successfully. These are: public engagement, workforce development, support for industrial growth, robust ethical frameworks, and nationally coordinated approaches to data and analytics. Specific mention of scalable and secure informatics systems to link the

BOX 1.2 The UK Biobank, an unprecedented resource of health and genomic data

The UK Biobank is a world-leading biomedical research resource offering non-preferential access to data and bio-samples to approved researchers undertaking health-related studies for the public good. The resource follows the health and well-being of 500,000 volunteer participants in the UK since 2006 as part of a large-scale prospective study. The consented participants—aged 49–60 at the time of recruitment—undergo anthropometric measurements, share detailed information about their lifestyle and health records, and regularly provide bio-samples, such as blood, urine, and saliva. This makes for unprecedented biological and medical data generated by a single research initiative. In 2019, a £200 million investment from the government, industry, and charity secured all participants' whole genome sequencing, making the UK Biobank the world's most genetically characterized medical research resource. The addition of genomic information is expected to provide valuable insights into how genetics combines with the environment and lifestyle to cause disease. So far, UK Biobank has led to over 12,500 published scientific studies that describe opportunities for new diagnoses and treatments for everything from cancer to diabetes, heart disease, and depression. Since its inception, the UK Biobank has adopted pioneering technology to protect its biomedical assets, including automated freezers (to store the samples of the 500,000 participants in 10 million separate tubes) and security management systems to keep the resource contents secure.

NHS with the research community for clinical decision support and large-scale data processing and analytics is indicated as one of the next decade's critical objectives.

Another significant milestone in the chronicle of strategic initiatives aimed at progressively integrating genomics in primary healthcare was the announcement in 2022 of the Newborn Genomes Programme by Genomics England in collaboration with the NHS. The Generation Study, launched as part of this programme the following year, proposes to explore the benefits and challenges of sequencing the genomes of 100,000 newborns. Results from this study, which is due to be completed in 2025, will help decide whether whole genome sequencing should become part of mainstream NHS practice for newborn screening and eventually replace the current newborn blood spot screening programme. The blood spot programme, also known as the heel prick test or Guthrie test, helps identify newborns with any of nine severe yet actionable genetic disorders, including Phenylketonuria, Sickle Cell Anaemia and Cystic Fibrosis, before babies become ill. This type of screening has been embedded in the NHS and has been successfully used for fifty years. However, consideration has been recently given to its limited scope (nine diseases only) when compared to a potentially new screening programme based on whole genome sequencing through which so much more could be revealed, such as the presence of 200 or more genetic conditions, including sporadic disorders, and genetic predisposition to pathological conditions. There is currently extensive expert and societal discussion around the potential advantages and drawbacks of whole genome sequencing for newborn screening programmes, including concerns around the ethicality and risks of collecting genomic data in excess of what would be necessary for diagnosing a list of specific diseases (Horton et al. 2024; Lucassen and Horton 2024; Salisbury 2024).

Given the vast amount of data that genomic activities produce and work with, the application of artificial intelligence (AI) in genomic research and healthcare provides significant opportunities for system-wide improvements. However, it also magnifies ethical and legal challenges related to the sensitive nature of genomic data, such as patient safety, data governance, and respect for individuality and human dignity. As stated in the Topol Review, care must be taken to ensure that healthcare technologies are used in ways that remain faithful to the core ethical principle of medical care: 'do no harm'. Unquestioning overreliance on technology and lack of validation might present serious patient risks. Researchers and practitioners in genomic medicine must give themselves time and scope to explore the safe use of AI in genomics. It is reassuring to see the recent establishment of a new NHS Genomic AI Network, a national community made of clinicians, AI experts, and patients working together to discuss, explore, promote, test, and ultimately shape in a participatory fashion how AI should be used in genomic medicine.

Lack of diversity and under-representation is often highlighted as a weakness in past and present genomic datasets. Considerations of what might have historically, geographically, culturally, and socio-economically caused limited participation and lack of inclusivity in medical research studies and sequencing initiatives are complex and beyond the scope of this chapter. However, it is now accepted that because of the lack of diversity in genomic datasets, some of the diagnostic, prognostic, and therapeutic innovations increasingly applied to personalized medicine might not be applicable as widely as anticipated, with potential health inequalities implications. A significant development in this respect was the creation of the Our Future Health partnership in 2022, a collaboration between the public, charity, and private sectors to build the UK's largest health programme. The partnership, supported by UK Research and Innovation (UKRI), aims to create an extensive and comprehensive health research database by collecting and analysing data, including genetic data, of up to 5 million participants across the UK from different demographics and all age groups. The initiative is focused on addressing the lack of representation by recruiting a diverse cohort of participants to ensure that health discoveries are relevant and applicable across different ethnicities and socio-economic backgrounds, thus reducing health disparities.

1.3 Concluding remarks

The last twenty years have seen a surge of public and private investments, research, and entrepreneurial activities collectively aimed at leveraging our growing knowledge of human genetics and the latest innovations in genomic technology to improve healthcare. Whilst the push towards personalized medicine through genomics can be considered a global phenomenon, several high-level strategic initiatives have enabled the UK to become a world-leading example of this cultural shift in clinical practice. Notwithstanding the incredible conceptual and technological progress so far, several challenges around infrastructures, workforce capabilities, and ethical and legal aspects will need to be addressed before genomics can be safely and equitably embedded into routine healthcare. The application of AI to genomics is presenting incredible opportunities but also magnifying concerns around respect for autonomy and justice, and action needs to be taken to seek reassurance for both practitioners and members of the public. We are witnessing a promising trend towards a more open and participatory type of research practice in biomedicine, with an increasingly prominent place for research participants' and patients' perspectives. Also, more inclusive approaches to recruitment in genomic research will allow more extensive strata of society to benefit from converting those discoveries into healthcare improvements. We must keep reminding ourselves as a society that the Universal Declaration of Human Rights asserts that we all have a right to participate in and benefit from scientific advancement.

Chapter summary

- The study of the human genome provides us with the opportunity to learn more about our individuality and unicity but also enables us to develop a sense of a collective identity.
- From a biomedical perspective, the completion of the Human Genome Project has marked the beginning of a new era in modern human genetics, characterized by distinctive translational efforts at the global level to convert genomic discoveries into improved health.
- The UK is at the forefront of genomic medicine implementation on account of high-level strategic initiatives and significant public and private investment in research and infrastructures.
- Widening research participation and ensuring robust ethical frameworks are essential aspects to be considered to take genomic healthcare forward and benefit society as a whole.

Discussion questions

1.1 What does human genetic variation tell us about our being human?

1.2 What key strategic initiatives have enabled the UK to become a world leader in genomic medicine?

1.3 What are the potential reasons behind the under-representation of specific demographics in genomic research?

Useful websites

- **https://digitallibrary.hsp.org/index.php/Detail/objects/9792** A digital copy of the original letter by Sir Isaac Newton to Robert Hooke in 1675, containing his famous pronouncement about 'standing on the shoulders of giants'
- **https://www.genome.gov/human-genome-project** A comprehensive repository of information on the human genome project curated by National Human Genome Research Institute, NIH
- **https://www.genome.gov/about-genomics/telomere-to-telomere** A webpage dedicated to the Telomere-to-Telomere (T2T) sequencing project that led to the first complete, gapless sequence of a human genome
- **https://www.genomicsengland.co.uk/initiatives/100000-genomes-project https://www.qmul.ac.uk/whri/research/featured-research/the-100000-genomes-project-holding-the-key-to-21st-century-healthcare/**
- NHS England » NHS Genomic Medicine Service >> NHS Genomic Medicine Service
- **https://www.gov.uk/government/publications/chief-medical-officer-annual-report-2016-generation-genome** The CMO 2016 Annual Report 'Generation Genome'

- **https://topol.hee.nhs.uk/** The Topol Review
- **https://www.gov.uk/government/publications/genome-uk-the-future-of-healthcare** Genome UK the future of healthcare strategy
- **https://www.ukbiobank.ac.uk/** UK Biobank
- **https://www.generationstudy.co.uk/** The Generation Study
- **https://ourfuturehealth.org.uk/** The Our Future Health research programme
- **https://genomicainetwork.nhs.uk/** The Genomic AI Network

2

Nucleic Acids and Human Inheritance Patterns

Lorna Tinworth

By studying this chapter, you will learn key facts and explore core concepts that you will need to be familiar with if you are to work in any job related to medical genetics and genomics. Firstly, we will look at the basic structure of nucleic acids (DNA and RNA) then study gene structure and consider briefly how the different areas of our genome function. We will then consider how DNA is passed down generations by looking at simple patterns of inheritance. You will find that many places within this chapter link out to other areas of the book. It is recommended that you read this whole chapter first, before exploring these cross references.

Before we begin to consider human medical genetics and genomics, we must be sure to have basic knowledge of the molecules and processes that make complex life possible. Even if you have already completed high-level biology courses and understand molecular biology well, you are still encouraged to go slowly through this chapter. Please pay specific attention to the tables and figures and absorb the information. Ask yourself the 'self-check' questions as you progress to be sure that you have enough foundational knowledge to benefit from the other chapters in this book. Please note that this chapter is intended only as a broad overview of the concepts covered in the learning objectives for this chapter and to serve as a basis for the rest of the book. If you wish to better acquaint yourself with the molecular structures and mechanisms mentioned and/or inheritance patterns, please refer to the recommended reading given at the end of the chapter.

Learning Objectives

After studying this chapter, you should be able to:

- Describe the basic structure of nucleic acids, DNA and RNA, and explain how their molecular structure supports their functions
- Describe the basic structure of a human gene

- Describe, in simple terms, the basic processes of DNA replication, transcription, and translation
- Describe several types of small genetic variations and their impact on protein coding, if relevant
- Describe human mitochondrial DNA and explain why it is of special interest
- Discuss the distinction between somatic and germline DNA and the importance of timely regulation of gene expression
- Describe basic inheritance patterns and interpret simple Punnett squares and pedigree diagrams.

2.1 Deoxyribonucleic acid (DNA)

Deoxyribonucleic acid (DNA) and ribonucleic acid (RNA) are both nucleic acids. Cast your mind back to what you already know about cell biology, and think about the double membrane-bound nucleus that can be found in eukaryotic cells. The nucleus is where most of the DNA in human cells (chromosomal DNA) is to be found; hence 'nucleic acids'. Do not confuse nucleic acids with amino acids. Amino acids are the primary building blocks of proteins—they do not carry genetic information.

leukocyte
White blood cell: any of the colourless blood cells of the immune system including; basophils. eosinophils, lymphocytes, monocytes, and neutrophils.

Let us begin by focusing on deoxyribonucleic acid (DNA) and keep in mind its only function, which is to carry information. In the late 1860s, Friedrich Miescher first identified DNA from **leukocyte** nuclei. In 1919, Phoebus Levene gave us the fact that DNA contained adenine, guanine, thymine, cytosine, deoxyribose, and a phosphate group. In the late 1940s Erwin Chargaff figured out the proportionality of these components within nucleic acids and gave us Chargaff's rules (see **Box 2.1**).

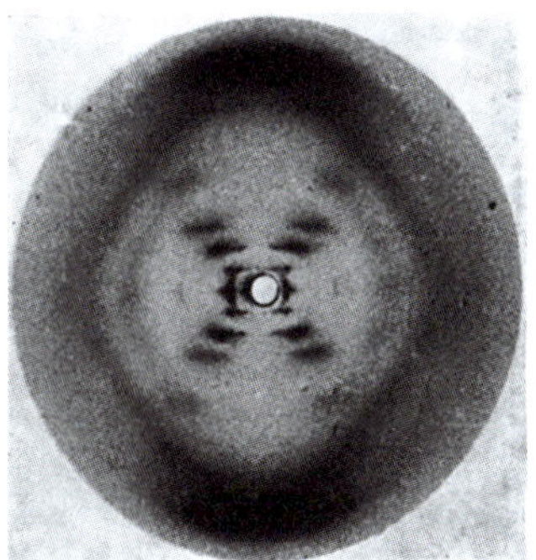

FIGURE 2.1
Rosalind Franklin's X-ray crystallography image of DNA. Evagorou M, Erduran S, Mäntylä T (2015) The role of visual representations in scientific practices: from conceptual understanding and knowledge generation to 'seeing' how science works. *IJ STEM* Ed 2, 11. https://doi.org/10.1186/s40594-015-0024-x

In 1952, Rosalind Franklin published **Figure 2.1**. It is from X-ray crystallography of DNA in a right-handed helical formation.

As you view **Figure 2.1**, imagine looking down the end of a long DNA molecule as it twists, tunnel-like, away from you. The dot in the centre of the image represents the 'light at the end of the tunnel'. The dark patches are where electrons have passed through the structure, and the pale areas are where DNA bases have blocked the path of electrons. This image enabled the chemists Watson and Crick to elucidate the structure of DNA and bring their ideas to the world.

In aqueous environments with lots of water and low salt, as is the situation inside our cells, DNA is found in the well-recognized 'B' formation. **Figure 2.2** represents the conformation of B-DNA. Notice the major and minor grooves, the two sugar–phosphate 'backbones' spiralling the length of the molecule, and the neat base pairing in the centre. It is in this conformation that DNA best performs its coding functions. If placed in other conditions—low humidity, high salt, urea, or formamide for example—its conformation will change, and its function will be impaired.

BOX 2.1 Chargaff's Parity Rules

Rule 1: In double-stranded DNA, the quantity of A (adenine) equals the quantity of T (thymine), and the quantity of G (guanine) equals the quantity of C (cytosine).

Rule 2: In a single DNA strand the percentage of adenine is approximately equal to the percentage of thymine and the percentage of guanine is approximately equal to the percentage of cytosine.

It was these rules that enabled scientists to elucidate the structure of DNA, because the equal quantities of complementary bases in the first rule supports the base-pairing model. Also, because of the different percentages found in different organisms for rule two, they were able to recognize DNA as the molecule carrying genetic information, varying in its composition between species.

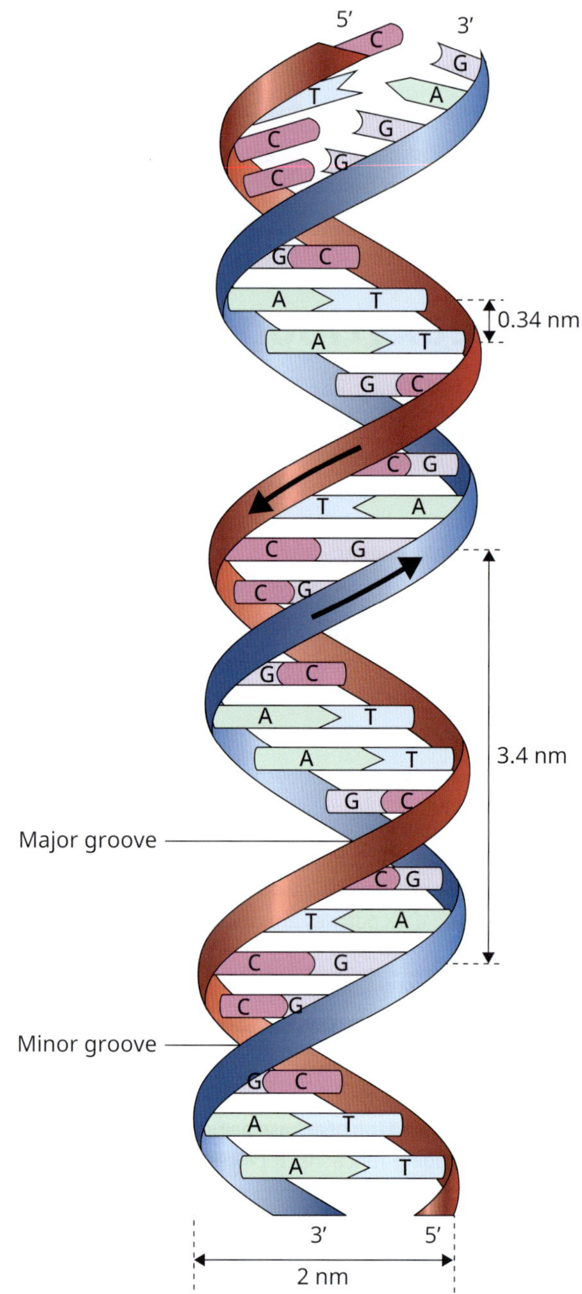

FIGURE 2.2
Canonical image of DNA double helix in B formation with sizes.

To be able to carry information DNA must:

- be chemically stable,
- be capable of mutational change to ensure evolution,
- be repairable should damage occur,
- be able to carry information from parent cell to daughter cell, and
- contain information for its own replication.

Let's consider just the first three of these as we look more closely at the structure of DNA.

See how one larger, double-ringed 'purine' is paired with one smaller, single-ringed 'pyrimidine'. Specifically, and because of the availability of hydrogen, nitrogen, and oxygen atoms, adenine pairs with thymine, and cytosine pairs with guanine.

Notice how there are two hydrogen bonds (shown by dotted lines) between adenine and thymine and three hydrogen bonds between cytosine and guanine. This makes the cytosine-to-guanine bonds stronger. Areas of the genome that are rich in C–G bonds require more energy to open up. These hydrogen bonds are vital in enabling DNA to fulfil its functions. In aggregate, they provide immense stability, holding the two parts of the molecule together. Imagine trying to pull a zip in your clothes apart by pulling the fabric on either side in opposite directions. Because hydrogen bonds are weak in comparison to covalent bonds, the molecule can be easily 'unzipped' and will spontaneously 'zip' back together because of the attraction between the partially charged areas.

As you continue to consider the chemical stability of DNA, notice the alternating sugar–phosphate–sugar–phosphate pattern (labelled P and D) on both sides of **Figure 2.3**. These are referred to as the backbones of the molecule because they provide its structural strength whilst also being able to move and give the molecule flexibility, because the molecule is a coil. Notice how the sugar–phosphate backbone runs up and down either side. The partial negative charges are depicted as being on one of every oxygen within each phosphate group. In an aqueous environment, these would cause water molecules to line up around the backbone in an orderly fashion.

Notice the five prime (5′) and three prime (3′) ends of the backbones at the top and bottom of the image. Where do these numbers come from? Point to the oxygen within the five-sided (orange, D-labelled) pentose sugar at the top left of the image and count the carbons clockwise around the pentagram, as represented by each angled joint, then out to land on the first carbon that points towards the phosphate group. Stop there. This carbon that sits between the pentose sugar, shown in orange, and the oxygen of the phosphate group will be the fifth or five prime (5′) end. Do the same at the top right of the figure. Start at the oxygen of the pentose sugar and count the carbons clockwise up and over to the carbon before the OH group: this will be the third carbon. Have a look at the small depiction of the pentose ribose sugar in the top left of **Figure 2.3** and see the numbering of the carbons (in red). DNA strands run in a reverse complementary direction. Notice how one backbone runs from five prime to three prime as the other runs three prime to five prime. When reading DNA code, it is vital that we know which end is five prime and which is three prime, so that the order of the bases is correct. Look at **Figure 2.3** again and notice how the bases are all attached to carbon number one of the pentose sugar.

It is useful at this point to consider the modular structure of DNA, which must be capable of mutational change and repairable. When a base (in DNA, adenine (A), cytosine (C), thymine (T), or guanine (G)) is attached to a pentose sugar the resulting molecule is referred to as a nucleoside. The pentose sugar in DNA is deoxyribose (in RNA it is a ribose sugar). When one phosphate group is added, it is then referred to as a nucleotide monophosphate. Nucleotide monophosphates can be considered subunits of the linear polymer that is a DNA molecule. If another phosphate is added the structure is referred to as a nucleotide diphosphate, and adding a third gives us a nucleotide triphosphate. Look at the labelling on the left-hand side of **Figure 2.4**. Triphosphate molecules carry with them a considerable amount of energy within their phosphoanhydride bonds because of the proximity of the partial negative charges on the oxygens. Now compare the general structure of a nucleotide triphosphate with the structure of the well-known store of energy at the cellular level, adenosine triphosphate (ATP), which is shown on the right-hand side of **Figure 2.4.** Note how the depiction of the bonds between the

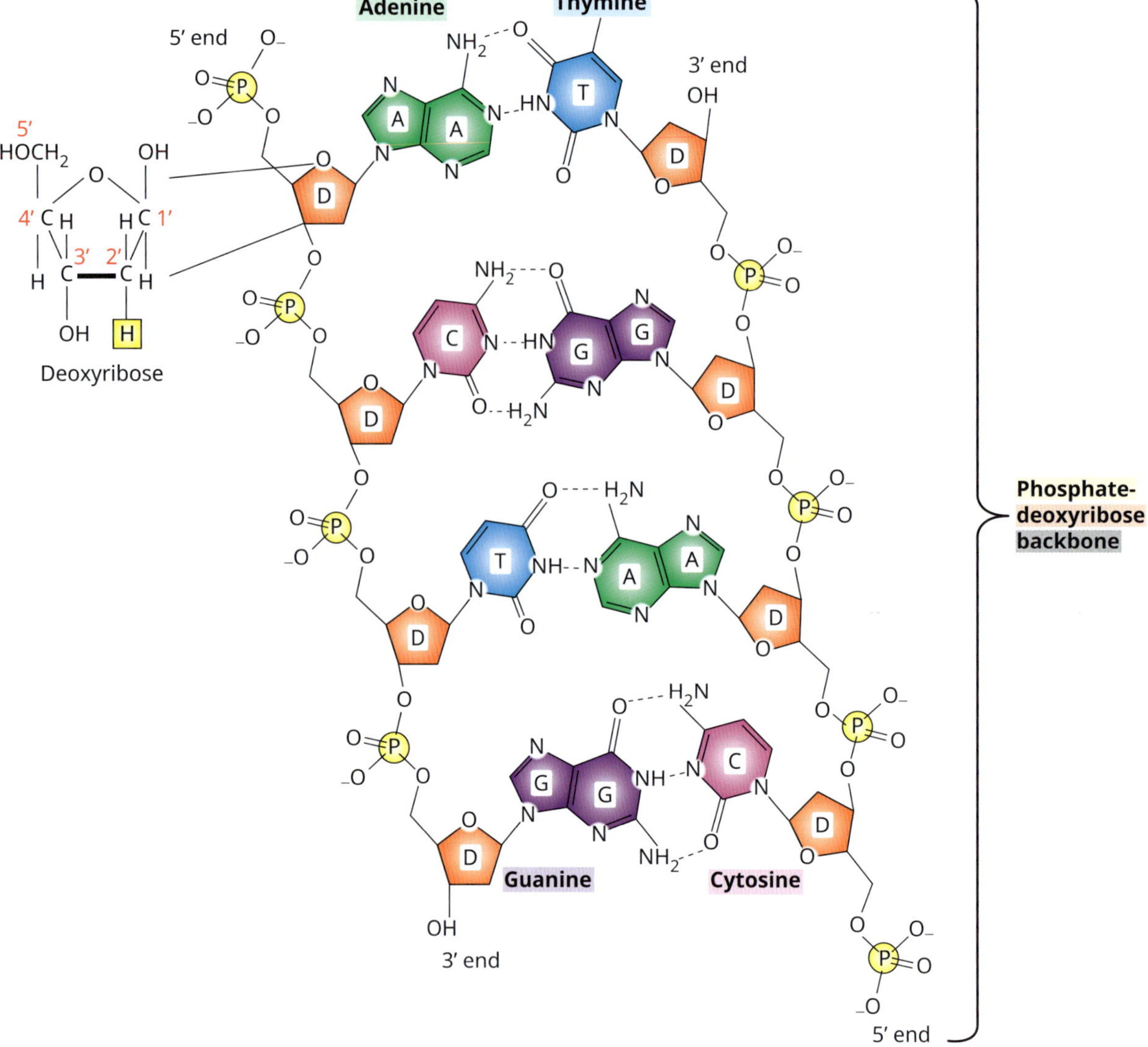

FIGURE 2.3

A two-dimensional, flattened 'ladder' view of DNA. This is not how DNA appears in nature, but it is useful to help us consider the structure. Notice the four differently shaped bases:

- Adenine (A) in green,
- Cytosine (C) in pink,
- Guanine (G) in purple, and
- Thymine (T) in b`lue.

phosphate and oxygen molecules is different. The image to the left simply shows more detail about the angles of the bonds to remind you of the three-dimensional nature of the molecule.

By convention we always give a DNA sequence in the five prime to three prime orientation (5′-3′). However, if given a sequence, please always check! Keeping your focus on **Figure 2.3**, consider the base sequence. The right-hand strand reads down: adenine, cytosine, thymine,

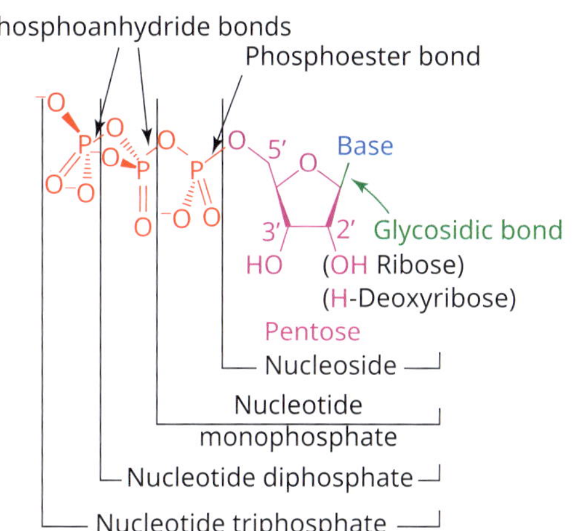

FIGURE 2.4

Comparison of a nucleotide triphosphate with adenosine triphosphate (ATP), the familiar molecule that acts as an energy store for many cellular processes. ATP is shown on the right and consists of a pyrimidine base, adenine (shown in purple); a ribose sugar (shown in pink); and a triphosphate group (shown in red). The general structure for nucleotide triphosphates is shown to the left. In DNA the base can be either one of adenine (A), cytosine (C), thymine (T), or guanine (G). The base is joined to the pentose sugar with a glycosidic bond. The pentose sugar in DNA is deoxyribose and in RNA is ribose (see the label underneath the second carbon of the pentose ring). A nucleotide monophosphate has one phosphate group; a diphosphate has two phosphate groups; and a triphosphate has three phosphate groups.

guanine (ACTG). The left-hand side reads up: TGAC. If given only one of these, knowing the base pairing rules (A–T, C–G), you can work out what the other will be. . .

5′ ACTG 3′ gives 3′ TGAC 5′, which is 5′ CAGT 3′

SELF-CHECK 2.1

What is the sequence of the complementary strand for the following short single-stranded DNA (ssDNA) molecule? Please format your answer in the conventional 5′ to 3′ orientation.

5′ TGATTGGAAGCCGCGATATGCATAT 3′

Key Points

DNA has a specific chemical structure—its double-stranded, modular, reverse self-complementarity is critical to enabling both its coding and replicative functions.

As we think about DNA structure, base pairing, and how the molecule carries genetic information, it is important not to forget that sequence-dependent modulations of DNA structure are just as important to its function. Depending on the ratio and pattern of the bases, and which RNA and protein molecules are interacting with the DNA, its overall shape will change. Groove width, helical twist, curvature, mechanical rigidity, and resistance to bending can all alter.

If the overall shape of the DNA does not enable it to interact with the myriad of other cellular molecules correctly, then the information contained with the base pairing will not be accessible to the cell. Although the focus is often on the 'coding sections' of DNA, by which we mean the genes that encode RNA molecules some of which go on to code for proteins, much of the DNA in our genomes has structural purpose. In particular, the regions around the **centromeres** and at the ends of chromosomes (**telomeres**) both of which contain large areas of repeating sequences. These provide stability and attachment points for elements of the cytoskeleton during cell division and DNA replication.

centromeres
The area of a chromosome to which the spindle attaches during mitosis and meiosis.

telomeres
Structures made from specific DNA sequences and proteins found at the ends of most eukaryotic chromosomes.

DNA is not found naked within a cell; it is always in association with complex mixtures of proteins and RNA molecules which is collectively referred to as **chromatin**. **Heterochromatin** specifically refers to DNA that is compacted. **Euchromatin** specifically refers to DNA that is more open and active. Runs of cytosine residues in DNA are often found to have methyl groups attached to them in areas referred to as CpG islands. This methylation does not interfere with base pairing, but it stops the cellular machinery from accessing the methylated areas, thereby switching any coding regions in them off. At the same time, the histone proteins with which the DNA is closely associated are found in a de-acetylated state which increases their attraction for one another, further ensuring that those areas remain inactive. In **Chapter 3** you will learn about the exquisite architecture of chromosomes and their packaging at different times during the cell cycle. However, for now, bring your thoughts back to base pairing and the DNA sequence. Having looked briefly at the chemical stability of DNA, keep in mind how DNA is capable of mutational change and is repairable, as we study the process of DNA replication.

Cross reference
To learn more about the conformation of chromatin during different phases of cell cycle please refer to **Chapter 3, Section 3.1**.

chromatin
A complex in eukaryotic cells mostly composed of DNA, histone and other proteins involved in the regulation of gene expression, that is usually found dispersed in the interphase nucleus and is condensed during mitosis and meiosis.

heterochromatin
A condensed form of DNA that is transcriptionally inactive. Heterochromatin is tightly packed and inaccessible to transcription factors. Compare with euchromatin.

euchromatin
The part of chromatin that is transcriptionally active.

2.2 DNA replication

The key enzymes in the process of DNA replication are **DNA polymerases**. They work to catalyse the formation of phosphodiester bridges as they add new nucleotides to a growing DNA strand. DNA polymerase 'clips' around a single strand of DNA which it then uses as a 'template'. Nucleoside triphosphates from the cytoplasm are incorporated into the new strand of DNA within the enzyme. This incorporation only happens if the template base forces the correct shape to allow base pairing to occur. **Figure 2.5** represents this molecular shape as pointed for A–T bonds and rounded for G–C bonds. In reality, their three-dimensional shapes are much more complex. Think about a key fitting a lock. If a fit is achieved then two inorganic phosphate groups are released from the nucleotide triphosphate, providing energy to complete the reaction, as the phosphodiester bridge is formed between the two bases. Once the bond is complete, the polymerase moves one base pair towards the 3′ end of the DNA and hopefully goes on to incorporate another correct base.

DNA polymerase
An enzyme that synthesizes DNA molecules from nucleotide building blocks.

DNA replication is conceptually divided into three phases:

- initiation
- elongation
- termination.

Initiation occurs late in Gap Phase 1 (G1) of the cell cycle, and the main activity occurs during the synthesis (S) phase.

Cross reference
To study the phases of DNA replication, please refer to **Chapter 3, Section 3.1**.

DNA molecules are replicated in a semi-conservative manner. This means that one strand is kept (conserved) and used as a template strand from which a new one is built. The process is carried out by DNA polymerase enzymes within a complex aggregation of proteins and RNA

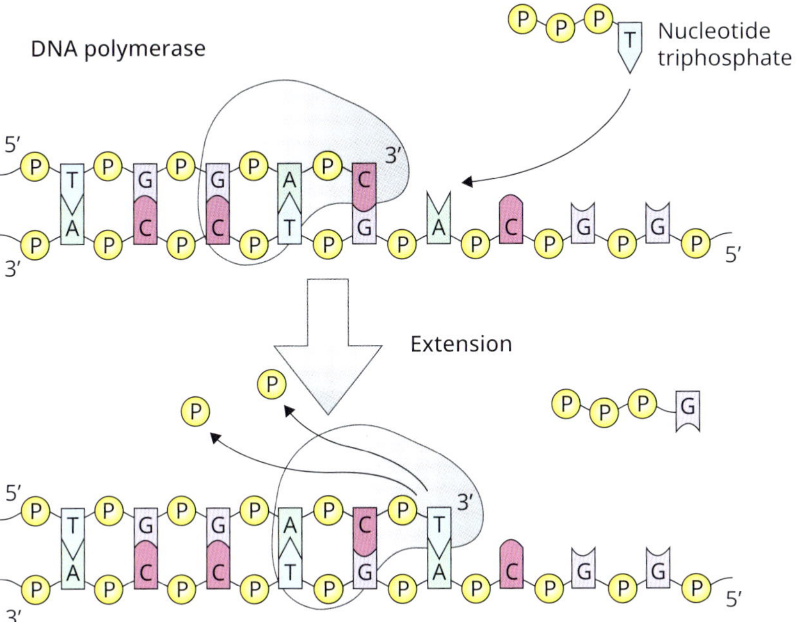

FIGURE 2.5
Basic representation of DNA replication.

molecules referred to as a **replisome**. There are genes in the human genome for at least fourteen different DNA polymerases. The five first-identified DNA polymerases (α, β, δ, ε, and γ) are essential to the reproduction of nuclear and/or mitochondrial DNA. The others are specialized DNA polymerases and work in conjunction with the first five or are key to DNA repair. These molecular machines synthesize new DNA at approximately two thousand base pairs (bp) per minute. At this rate, if each human chromosome was duplicated beginning at one end and proceeding to the other, it would take over a month to replicate the whole human genome. Instead, replication 'bubbles' form at multiple sites along the DNA, referred to as origins of replication. This enables the entire nuclear genome to be copied within about nine hours.

replisome
Multiprotein molecular machinery responsible for the replication of DNA.

Row A at the top of **Figure 2.6** shows a simple depiction of a replication bubble. Note the site of the origin of replication indicated by two black dots, one on each DNA strand. See the semi-conservative manner of replication indicated by the parental (original) DNA strand in dark blue and the newly synthesized (daughter) DNA strand in pale blue. Recognize a replication 'fork' as delimited by the black rectangle. Identify the six replication forks in the diagram. Look at row B of **Figure 2.6** and imagine each pair of forks moving away from one another, expanding the bubbles horizontally as the two new double helices are built. Each new double helix is comprised of one original parent strand and one new daughter strand.

The process of replication begins late in gap phase 1 (G1) of the cell cycle, when ATPase containing origin recognition complexes bind to multiple origins of replication within the DNA. This process is referred to as 'licensing' and is under extremely tight control to prevent aberrant replication of DNA. The exact positions of these origins of replication are still a matter of debate, although it is known that they 'fire up' in a hierarchical manner throughout the synthesis ('S') phase of the cell cycle.

Once an origin of replication (refer again to the black dots in **Figure 2.6**) is licensed, two helicase enzymes bind to each strand of the DNA and they move along in a 3′–5′ direction away from each other, uncoiling the DNA and creating the replication forks.

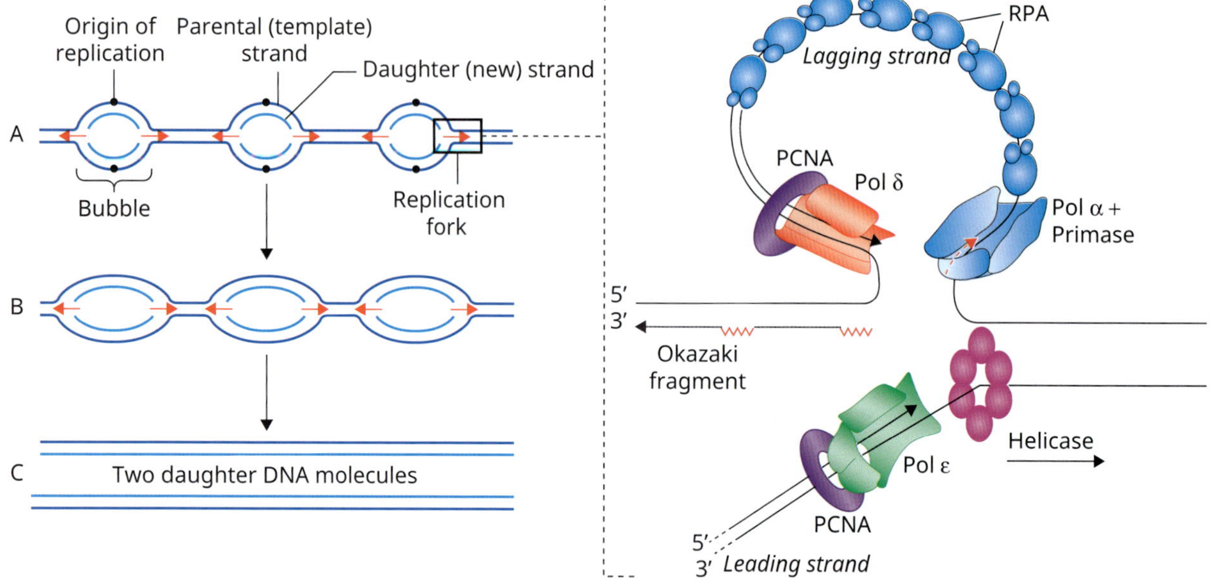

FIGURE 2.6

DNA replication bubbles and a simplified view of what happens at a replication fork. RPA, replication protein A; PCNA, proliferating cell nuclear antigen.

hexameric

A polymeric protein composed of six molecules of a monomer OR a structural subunit that is part of a viral capsid composed of six subunits of similar shape.

DNA helicases are highly sophisticated ring-shaped **hexameric** ATP-fuelled nanomachines. They go through repeated conformational changes as they hydrolyse ATP to move along and unwind double-stranded DNA (dsDNA). The strand to which a helicase is bound will be referred to as the leading strand. Look at the hexameric helicase depicted in pink, round circles in the bottom right-hand section of **Figure 2.6**. The DNA is opened up in this way in both directions and a replication fork forms at each end of the 'bubble'.

Replication complexes (replisomes), containing DNA polymerase, bind at each fork and work away from each other synthesizing new DNA. Replisomes are complex and their exact composition varies in heterochromatic and euchromatic DNA. Differing compositions have been associated with varying speeds of new DNA synthesis and in situations of various types of DNA damage. A polymerase alpha (α) holoenzyme complex binds to initiate the DNA synthesis reaction. Because DNA polymerases can only extend double-stranded DNA, the primase subunit of Polα synthesizes a short RNA primer about 10 nucleotides long and then the DNA polymerase subunit extends this primer using dNTPs for a further twenty to thirty nucleotides.

The 'zoom in' image on the right-hand side of **Figure 2.6** shows the location of Polα. Single-strand binding proteins associate with the DNA to ensure its availability to the replication machinery and Polα is replaced by either polymerase delta (Polδ) on the lagging strand, or polymerase epsilon (Polε) on the leading strand, which go on to carry out the rest of DNA replication by elongating these primers. The leading strand is built by Polε as it works towards the fork, using a template that is continually available as helicase unwinds the parent strands. However, because DNA polymerases can only synthesize DNA in a 5′–3′ direction, Polδ must wait until the template becomes available and produces new DNA in short sections of about one thousand bases, referred to as Okazaki fragments. These individually primed fragments are then joined together by DNA ligases. The work of the different DNA polymerase enzymes is coordinated so that their replication speed matches.

Replication forks work along the DNA until they meet. Once converged, a replication termination process is carried out, gaps are filled in and ligated, and the replication machinery is released.

Key Points

DNA replication occurs at a specific time in the cell cycle and begins at multiple sequence-specific sites throughout the genome. Replication forks form, and complex aggregations of molecules come together to form replisomes. These move through the genome to bring about high-fidelity replication of the entire genome within about nine hours.

2.3 Variation in DNA

You have considered the basic structure of nuclear DNA, learned how its structure enables its function, and now understand the basics of how DNA is replicated. However, molecular processes are not perfect, and neither is DNA replication. When thinking about changes to DNA, it is important to recognize two separate concepts:

1. Replication errors.
2. DNA damage, which happens *after* replication is complete.

Each of us has unique DNA code. It is a unique quarter of our total grandparental DNA code, that was recombined and independently assorted in our parents' gonadal tissue during meiosis. Then, as we grow and age, the cells within us accumulate DNA changes. Some of these changes are clinically relevant, and some are not. Clinically relevant variants (mutations) are known to be causative of a pathology. A difference in the DNA that is not pathological is simply a polymorphism. You will see these single base differences referred to as single nucleotide polymorphisms or SNPs. These variations are often found at different frequencies within healthy populations. However, things in human medical genetics and genomics are not always simple, and some polymorphic variants are known to be statistically associated with clinically relevant characteristics. You will read more about this in **Chapter 4**, **Section 4.5.2**.

Sometimes bases exist in different **tautomeric** forms, which means that hydrogen bonds can form incorrectly during the process of DNA replication and incorrect bases are incorporated. This is referred to as mispairing. Sometimes a base—or several bases—are simply left out, creating a 'deletion'. Sometimes, in areas of the genome with lots of repeating short sequences of DNA referred to as tandem repeats, DNA polymerase 'slips' and the numbers of repeats expand. Repeat expansion due to replication slippage is associated with various pathologies, which you will read about later in this chapter and in **Chapter 4**. Some are involved in a particular kind of inheritance pattern, called anticipation, that you can read about in **Section 2.11.2**. Alterations in the DNA base sequence, such as missense and nonsense mutations (that change the amino acid that the DNA codes), insertions/deletions (indels, that alter reading frames), and copy number changes cannot be referred to as a mutation or polymorphism until the change has been copied to a daughter cell. Until that time, there is always the possibility that the error will be corrected or that it will trigger the cell into apoptosis and so prevent the error being transmitted to future generations of cells (if somatic) or individuals (if germline).

DNA replication is a very high-fidelity process. It is estimated that an incorrect nucleotide is only incorporated once per hundred million nucleotides polymerized. DNA polymerases α, δ, and ε all have 3′–5′ **exonuclease** 'proofreading' activity to remove incorrect bases. This capability also plays a part in two other important fidelity-ensuring mechanisms: mismatch repair and the maturation of Okazaki fragment joining. Errors in DNA replication do occur, as does damage after replication. Look at **Figure 2.7** to see the different kinds of damage that happen.

tautomeric
A form of isomerism in which the isomers change into one another with great ease so that they ordinarily exist together in equilibrium.

Cross reference
For more information on DNA alterations, see **Section 2.10**.

exonuclease
An enzyme that breaks down RNA molecules by removing successive nucleotides from one end of the molecule. Exonucleases can have either 3′ to 5′ or 5′ to 3′ activity, and are involved in RNA processing, tRNA maturation, and RNA degradation.

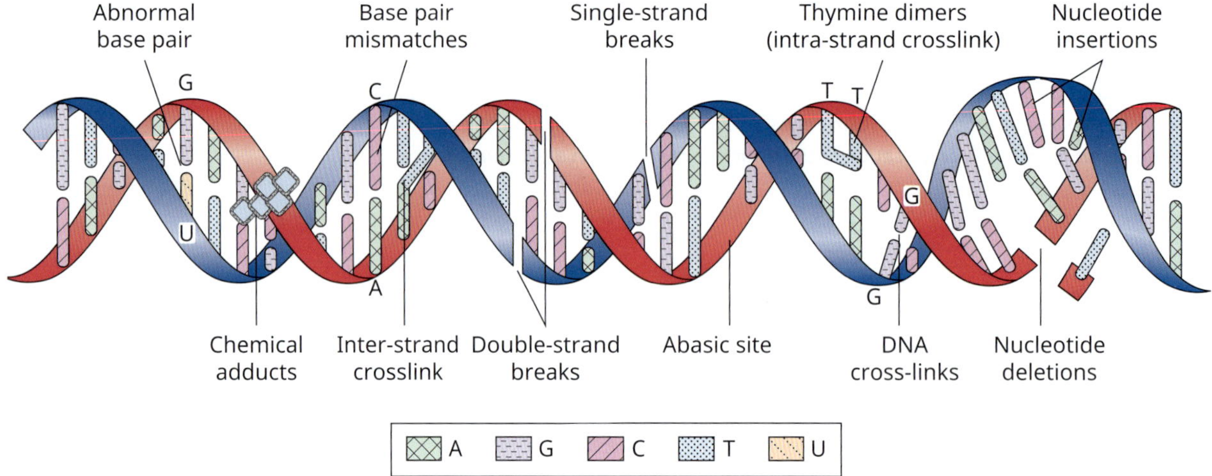

FIGURE 2.7

Eleven major types of DNA damage in eukaryotic cells.

Damaged DNA is unlikely to function correctly. However, sometimes a single base pair change will be silent (have no discernible effect). Because they have no negative impact, such changes will remain in the gene pool as single nucleotide polymorphisms (SNPs). There are about four and a half million SNPs in the human genome. To be classed as a SNP, a base pair variation must be present within about one per cent of individuals. More rare DNA changes, or those known to be causative of a pathology, are referred to as 'mutations'.

Data from Genomics England's 100,000 Genomes Project, an initiative to sequence the entire genomes from around 85,000 NHS patients affected by rare disease or cancer, now provides us all with a global reference for human genetic variation at the sequence level.

SNPs are often associated with clinically relevant phenotypes but remember, association is not causation. It may be that the polymorphism is genetically linked to another **locus** in the DNA that does play a role in the phenotype.

locus
The position in a chromosome of a particular gene or allele.

Cross reference
You will read more about linkage and association in **Chapters 3** and **4**.

The human body has various mechanisms to repair DNA. Non-homologous end joining is the primary repair mechanism for double-strand breaks. When this is not possible, they are repaired by copying the sequence from the sister chromatids in a process called homologous recombination. Inter-strand crosslink repair, nucleotide excision repair, and mismatch repair are the three other major repair mechanisms employed to keep the fidelity and function of DNA intact. Sometimes repair mechanisms themselves are compromised. Individuals with Lynch syndrome have **autosomal** dominantly inherited predispositions to cancer (you will learn about inheritance patterns later in this chapter), due to mutations in the genes that code for key proteins in mismatch repair systems.

autosomal
Belonging to, located on, or transmitted by an autosome.

Cross reference
To read about ethical issues in a case of Lynch syndrome, see **Chapter 9, Section 9.5**.

Key Points

Although DNA replication is a high-fidelity process sometimes errors do occur, and DNA can also be damaged post replication leading to sequence changes. These changes, if they do not lead to cell death, can persist in populations of cells or into future generations. Sometimes there are no discernible effects of the changes, sometimes they are positive, and sometimes negative.

2.4 Mitochondrial DNA (mtDNA)

Having studied a little about the structure of DNA and its replication, we must remember that not all our DNA is found in the nucleus. Important DNA is also found in our mitochondria. The origin of our mitochondria (and chloroplasts in plants) is thought to be symbiotic endocytosis. Multiple times, in early evolutionary history, one prokaryotic organism entered another, surrendered its autonomy, and became the energy-harnessing unit of the host cell while retaining some of its genome. Therefore, we recognize that the genome of our mitochondria (and those of chloroplasts in plants) resembles very closely the genomes of bacteria. These relatively small circles of DNA are only 16.6 kilobases (as compared to the 6.6 billion base pairs in the human **diploid** nuclear genome) and yet they play major roles in human health and disease. In particular, changes in mitochondrial DNA have been associated with many neurodegenerative disorders, such as ataxia, chorea, dystonia, motor neuron disease (MND), and parkinsonism. **Figure 2.8** shows a diagrammatic representation of our double-stranded mitochondrial DNA.

diploid
Having two sets of chromosomes, one from each parent, typical of most somatic cells.

There are thirty-seven known genes in our mtDNA. They encode twenty-two transfer RNA (tRNA) molecules, thirteen subunits of the oxidative phosphorylation (OXPHOS) complexes, and 2 ribosomal RNA (rRNA) molecules. Given that we can now sequence DNA with ease and that the thirty-seven human mitochondrial genes are simple and well understood, it would be tempting to assume that genetic disorders arising from alterations in the mitochondrial genome were

Cross reference
Please refer to **Section 2.8.3** to read about tRNAs.

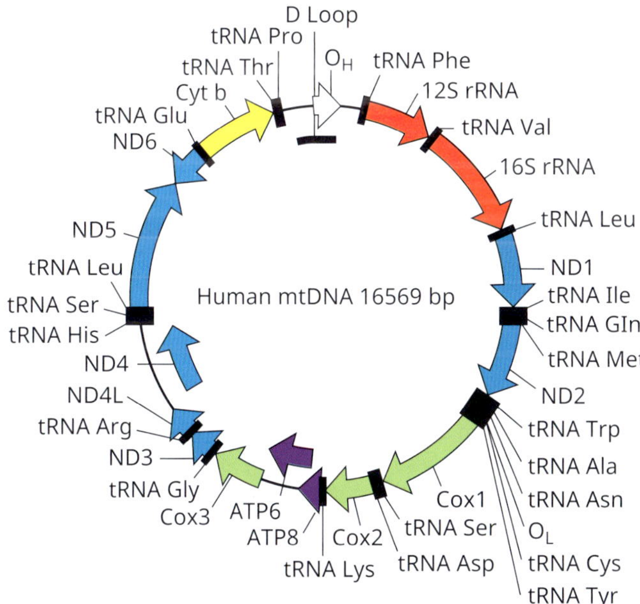

OH and OL: Origins of heavy and light strand replication, respectively.
ND1-ND6: Subunits of NADH dehydrogenase (ETC complex I) subunits 1 through 6.
COX1-COX3: Subunits of cytochrome oxidase subunits 1 through 3 (ETC complex IV).
ATP6 and ATP8: Subunits 6 and 8 of mitochondrial ATPase (complex V).
Cyt b: Cytochrome b (complex III).
ETC: Electron transport chain.
tRNAs shown for various amino acids.

FIGURE 2.8
A map of human mitochondrial DNA.

easy to characterize and followed simple patterns. However, this is not true. Although clinically relevant variants in mitochondrial DNA do often lead, as one might expect, to disruption of **oxidative phosphorylation (OXPHOS)**, they also have other wide-ranging and seemingly less likely phenotypic effects. This is because, as well as the accepted thirty-seven genes described in **Figure 2.8**, mtDNA also encodes short open reading frames (ORFs), which are translated into peptides with many important and as yet poorly understood biological functions.

oxidative phosphorylation (OXPHOS)
The metabolic pathway in which cells use enzymes to oxidize nutrients, thereby releasing chemical energy to produce adenosine triphosphate (ATP).

Mitochondrial DNA is almost exclusively inherited from the mother, via the oocyte. One important concept to keep in mind when considering the phenotypic implications of changes in mitochondrial DNA is copy number. We are diploid organisms and, as such, our **somatic** cells (which are all our body's cells except the **haploid** eggs or sperm cells that we produce) each have two copies of a genome that is a combination of maternally and paternally derived code. However, the number of copies of mtDNA within any given cell (mtDNAcn) can vary greatly. Consequently, the presence within a single cell of both slightly different mitochondrial DNA (mtDNA) sequences (referred to as **heteroplasmy**) and the proportion of **wildtype** to mutated mtDNA (i.e., the mutant load) might vary in different tissues. This ratio will change over time and is a critical factor in the expression and severity of any pathology caused by mutations of mtDNA.

Cross reference
We will consider the implications of this upon inheritance patterns later, in **Section 2.11**.

somatic
Refers to the cells of the body, in contrast to the germline cells.

haploid
Having a single set of chromosomes, typically found in gametes (sperm and egg cells).

heteroplasmy
A condition in which a cell or individual has more than one type of organellar genome. Typically used to describe the situation in which two or more mtDNA variants exist within the same cell.

wildtype
A phenotype, genotype, or gene that predominates in a natural population of organisms or a particular strain of organisms, in contrast to that of natural or laboratory mutant or variant forms. Adjectival form is wild-type.

Key Points

Mitochondrial DNA is not replicated or inherited in the same way as nuclear DNA and therefore inheritance patterns for mitochondrial disorders are different from conditions caused by variations in nuclear DNA.

2.5 What is a gene?

DNA is passed from parent to child via the **gametes** and from parent cell to daughter cell within the somatic tissue of a complex organism. It is capable of mutational change, ensuring adaptation and evolution as change is selected for or against. However, DNA change brings with it enormous risk. As you read earlier, within the DNA there are coding regions and non-coding regions. Coding regions are those areas of the DNA proven to carry the information needed to produce an RNA product, more commonly referred to as *genes*, but what exactly is a gene?

A gene is best described as a set of segments of nucleic acid that contains the information necessary to produce a functional RNA product in a controlled manner. Note that the term 'nucleic acids' and not 'DNA' was used here. Deliberately so, because many viruses have RNA-based genomes which contain genes.

Cross reference
See **Section 2.11.5** for discussion on mitochondrial inheritance.

The segment(s) of nucleic acid that make up any given gene will be found at a particular position within the genome of an organism; this location is referred to as its locus. So, for example, in all of us, our Cystic Fibrosis Transmembrane Conductance Regulator gene (CFTR) is found on chromosome seven and we have two different but very, very similar copies (one maternal and one paternal). These different versions are called **alleles**. Imagine that your parents each gave you a recipe for chocolate chip cookies. Both would be slightly different, but both would 'code' for chocolate chip cookies. For a molecular geneticist, the specific molecular locus for this gene in humans is Chromosome 7: 117,287,120 to 117,715,971, with the numbers denoting DNA base pairs along the chromosome. For a cytogeneticist, the same locus is given as 7q31.2.

gametes
Reproductive cells of a plant or animal.

alleles
Alternative forms of a gene that might be found at a given locus.

Now that we have considered the positioning of genes, let's go on to look at the architecture of a typical human gene. The purpose of a gene is to encode an RNA molecule. This RNA molecule will either have a function as it is, or it will be a messenger RNA (mRNA) that encodes a protein molecule. The section of DNA that codes for RNA (the coding sequence) consists of **exons**, which are retained in the mature RNA product, and **introns**, which are spliced out (you will read more about this in **Section 2.7**). However, a gene consists of far more than just the coding region. **Figure 2.9** is a diagrammatic representation of a typical eukaryotic gene; please study it as we consider the relative positions of specific functional areas in the DNA.

RNA transcription (see **Section 2.7**) begins within what is known as the core promoter region, at the Transcription Start Site (TSS). Look at the arrow labelled TSS in **Figure 2.9**—this marks the point in the gene where transcription begins, and it sits within a specifically sequenced section of DNA called the transcription initiator (INR). Detail of the coding region (consisting of exonic and intronic regions) is not shown in **Figure 2.9**, but it runs in the direction indicated by the arrow until a stop codon is reached (you will read more about codons later, in **Section 2.8**). After the end of the coding region typically come two distinct areas; the Motif Ten Element (MTE) and then, finally, the Downstream Promoter Element (DPE)—look at **Figure 2.9** to see their positions relative to the TSS. Also note the sections labelled within both the proximal and core promoter regions and then the labelled sections outside of these regions.

We will now consider some of these gene features in turn, as we think about the full functional architecture of a human gene, starting within the promoter at the TATA box. The TATA box has an evolutionarily conserved **consensus sequence** located twenty-five base pairs upstream of the start site; the consensus sequence in humans is TATA(A or T)A(A or T). This is seen within the promoter region of about a quarter of human genes. About half of known human genes instead contain the consensus initiator (INR) sequence (C or T)(C or T)A(any base)(C or T)(C or T). Not all promoters have a TATA box or INR sequence present. If they don't, then transcriptional

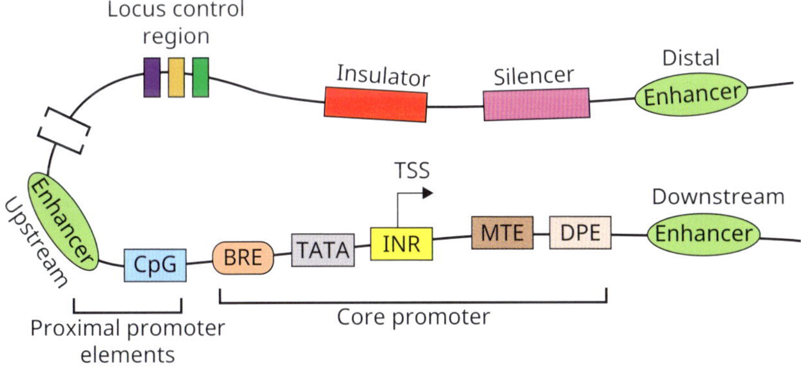

FIGURE 2.9

Diagrammatic representation of a human gene showing key features. Relative positioning of the various **promoter** elements is shown in relation to the Transcriptional Start Site (TSS) or +1 nucleotide. Distally, at the top of the diagram starting at the left, the **locus control region** (a tissue-specific, copy number-dependent long-range cis-regulatory element that enhances expression of linked genes at distal chromatin sites), the **insulator** (a long-range cis-regulatory element that can act as an enhancer blocker), the **silencer** (a DNA sequence capable of binding **repressor** proteins), and the distal **enhancer** sequences are represented. Proximally, at the bottom of the diagram starting at the left, an upstream enhancer can be seen, next a CpG island, and to the right of that the core promoter region which contains the BRE (Transcription factor II B [TFIIB] recognition element) and the Initiator sequence (INR), within which is the Transcription Start Site (TSS), After the coding region comes the Motif Ten Element (MTE) and the Downstream Promoter Element (DPE).

Cross reference

You will learn about chromosome structure and cytogenetic loci in **Chapter 3, Section 3.2**.

exon(s) (exonic)

A polynucleotide sequence in a nucleic acid that encodes information for protein synthesis and that is copied and spliced together with other such sequences to form messenger RNA.

intron(s) (intronic)

A non-coding region of a gene that is removed during RNA splicing.

consensus sequence

A common nucleotide sequence or amino acid sequence found in highly conserved regions of DNA or RNA or proteins.

promoter

A region of DNA upstream of a gene where relevant proteins (such as RNA polymerase and transcription factors) bind to initiate transcription.

locus control region

A tissue-specific, copy number-dependent long-range cis-regulatory element that enhances expression of linked genes at distal chromatin sites.

insulator

A long-range cis-regulatory element that can act as an enhancer blocker.

silencer

A specific DNA sequence capable of binding repressor proteins.

repressor

A small protein that can bind to specific DNA sequences to prevent the transcription of genes and operons.

enhancer

A nucleotide sequence that increases the rate of genetic transcription by preferentially increasing the activity of the nearest promoter on the same DNA molecule.

CpG island

Short stretches of DNA (typically 500–1500 bp long) with a CG:GC ratio of more than 0.6. Represented as 'CpG'. The 'p' indicates the phosphate linking the two bases to distinguish it from the hydrogen bonding of the GC base pairs in different DNA strands. CpG islands are frequently associated with vertebrate gene promoters.

transcription factor

A protein that regulates the transcription of a genes or set of genes.

enhancer

A nucleotide sequence that increases the rate of genetic transcription by preferentially increasing the activity of the nearest promoter on the same DNA molecule.

machinery is attracted to the area by the next feature to consider: the CG pair-rich region that is the '**CpG island**'.

CpG islands are typically from about five hundred to fifteen hundred base pairs long and have been identified within or near promoters in about forty percent of human genes. CpG islands are important sites for epigenetic silencing as, when methylated, they prevent transcription of the gene. Housekeeping genes, and many of the genes that are master regulators in early development, are known to have TATA-less promoters containing CpG islands.

Within the promoter region is a conserved transcription factor binding site (TFBS), in **Figure 2.9** this is shown as BRE (Transcription factor II B [TFIIB] recognition element). **Transcription factor** proteins are vital components of the RNA polymerase pre-initiation complex (PIC) which must form correctly before transcription can begin. All these conserved regions, where consensus sequences are recognizable and known to have coding related functions, are used when we search DNA sequences for possible coding regions.

Another area of the DNA code important to the function of many genes are **enhancers**. These are specific regions of DNA known to bind 'activator' proteins and attract non-coding RNA molecules. The proximity of these specific RNA and protein molecules increases the possibility of—or speed of transcription for—a particular gene or group of genes. Whereas promoters act to initiate transcription of the gene, are physically close to the gene in question, and always upstream before the RNA polymerase binding site, enhancers can be up or downstream and close to (proximal) or very distant (distal) from the coding region in question. They might even be found on a different chromosome. Carefully read over the caption for **Figure 2.9** and identify each element mentioned.

Do you recall back in **Section 2.1** that you learned about DNA methylation? Areas of DNA at the 5′ ends of constitutively regulated or 'housekeeping' (which are always 'on' and vital for the cell's survival or core phenotype) genes are protected from DNA methylation, so that they always remain open and active.

2.6 Ribonucleic acid (RNA)

Cross reference

You will learn about transcription in **Section 2.7**.

ribonucleic acid (RNA)

A nucleic acid present in all living cells that plays a role in coding, decoding, regulation, and expression of genes.

Having studied a little about the structure of DNA and the coding regions (genes) within our genome, let us go on to look at another very important nucleic acid: RNA. In contrast to DNA, **ribonucleic acid (RNA)** consists of one sugar phosphate backbone and these bases: adenine, cytosine, guanine, and uracil. By convention, RNA sequences are also written down in the 5′–3′ direction. Read through **Table 2.1** carefully, row by row, and think about the similarities and differences between these two important molecules.

In the rows entitled 'Bases' and 'Base Pairs', you can see that in RNA molecules, thymine is replaced by uracil. However, nucleic acid codes are not only written as strings of ACG and T/U. The International Union of Pure and Applied Chemistry (IUPAC) issued a series of letters to help us code for ambiguity in sequences, shown in **Table 2.2**.

SELF-CHECK 2.2

Now that you have studied the IUPAC code for nucleic acid bases think back to what you learned in **Section 2.3** about the TATA box consensus sequence within the promoter section of coding genes. If given the consensus sequence again, TATA(A/T)A(A/T), rework it to give a seven-letter string, replace the bases in brackets with one single letter to denote the fact that there might be either an Adenine or Thymine at positions five and seven.

TABLE 2.1 DNA/RNA Comparison Table

	DNA	RNA
Full Name	Deoxyribonucleic Acid	Ribonucleic Acid
Function	Replicates and stores genetic information.	Converts the genetic information contained within DNA to a format used to build proteins. Plays a key role in chromatin functioning and makes up the core structural components of ribosomal protein factories.
Structure	Consists of two strands, arranged in a double helix. These strands are made up of subunits called nucleotides. Each nucleotide contains a phosphate, a 5-carbon sugar molecule and a nitrogenous base.	Single-stranded but like DNA, is made up of nucleotides. RNA strands are shorter than DNA strands. RNA can form secondary and tertiary structures to fulfil many various functions related to gene expression and chromatin functioning.
Length	DNA is a much longer polymer than RNA. A chromosome, for example, is a single, long DNA molecule, which would be several centimetres in length if unravelled.	RNA molecules are variable in length, but much shorter than long DNA polymers. A large RNA molecule might only be a few thousand base pairs long.
Sugar	The sugar in DNA is deoxyribose, which contains one less hydroxyl group than RNA's ribose.	RNA contains ribose sugar molecules, without the hydroxyl modifications of deoxyribose.
Bases	Adenine ('A'), Thymine ('T'), Guanine ('G') and Cytosine ('C').	RNA shares Adenine ('A'), Guanine ('G') and Cytosine ('C') with DNA but contains Uracil ('U') rather than Thymine.
Base Pairs	Adenine and Thymine pair (A–T), Cytosine and Guanine pair (C–G)	Adenine and Uracil pair (A–U), Cytosine and Guanine pair (C–G).
Location	DNA is found in the nucleus, with a small amount of DNA also present in mitochondria.	RNA forms in the nucleolus, and then moves to specialized regions of the cytoplasm depending on the type of RNA and its function.
Reactivity	Due to its deoxyribose sugar, which contains one less oxygen-containing hydroxyl group, DNA is a more stable molecule than RNA, which is useful for a molecule which has the task of keeping genetic information safe.	
Ultraviolet (UV) Sensitivity	DNA is vulnerable to damage by ultraviolet light causing crosslinking.	RNA is more resistant to damage from UV light than DNA.

Source: DNA: RNA comparison table adapted from: Mackenzie, RJ (2023), *Genomics Research*, 14 July https://www.technologynetworks.com/genomics/articles/what-are-the-key-differences-between-dna-and-rna-296719

SELF-CHECK 2.3

Use the IUPAC code for nucleic acid bases given in **Table 2.2** to give a seven-letter string instead of (C/T)(C/T)A(A/C/G/T)(A/T)(C/T)(C/T).

RNA, much like DNA, is constructed as a single-stranded linear polymer (see the row on structure in **Table 2.1**). However, whereas DNA's destiny is to pair up with a complementary strand within a helix (note the word *double* has been deliberately left out here because in telomeric regions, DNA forms four-stranded G-quadruplexes and transient triplex DNA has been tenuously

TABLE 2.2 IUPAC Nucleic Acid Codes

IUPAC nucleotide code	Base
A	Adenine
C	Cytosine
G	Guanine
T (or U)	Thymine (or Uracil)
R	A or G
Y	C or T
S	G or C
W	A or T
K	G or T
M	A or C
B	C or G or T
D	A or G or T
H	A or C or T
V	A or C or G
N	any base
. or –	gap

reported to play a role in transcription), RNA can take on many different three-dimensional shapes, dictated by the sequence of its base pairs.

A classic example is the sequence and structure of Ribonuclease (RNase) P RNA from *Escherichia coli* bacteria. The sequence for this molecule shown in **Figure 2.10**.

Look carefully at **Figure 2.10** and think about how this RNA molecule will have been produced as a single-stranded polymer by an RNA polymerase, much like the DNA polymerase discussed earlier. Now recall what you know about base pairing: adenine will spontaneously form hydrogen bonds with uracil, and cytosine with guanine. At the very top right of the image, complementary bases form a short helix (P13), at the end of which is an open loop. This secondary structure is referred to as a 'stem-loop'. Look lower and notice the single loop below helices P7 and P5—it shows the sequence CGGG (highlighted in a box)—then see another loop to the right of helix P17 showing CCCG (also highlighted in a box). These are complementary sequences. The molecule will fold, like origami, so that these two runs of four bases can connect with one another and the hydrogen bonds are formed as indicated by the black connector line. At the very bottom of the image, there is a run of nine unpaired bases. Follow the line up

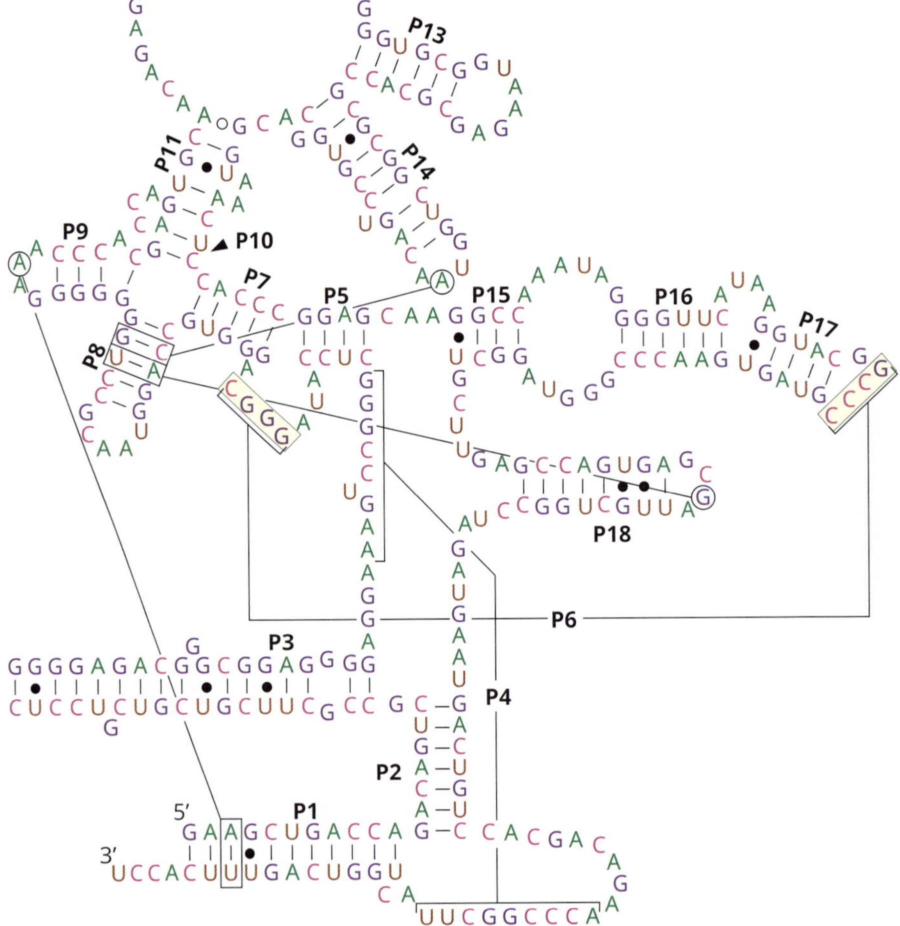

FIGURE 2.10

The nucleic acid sequence and structure of Ribonuclease (RNase) P RNA from *Escherichia coli* bacteria. Identified helices are numbered 1–18.

to the centre of the image to see which bases they will pair with. Look at the other 'connector' lines on the image and picture this RNA molecule as a very specific, base-dependent, three-dimensional shape.

RNA molecules that do not code for proteins (ncRNAs) are classified by their length. Arbitrarily, those under 200 nucleotides are referred to as small ncRNAs, and those longer than 200 nucleotides as long ncRNAs. Because of their highly labile nature we still have much to learn about ncRNAs (other than the two 'housekeeping' ncRNAs, transfer and ribosomal, that are well characterized). The structures and functions of ncRNA molecules are hugely varied, but they are known to play critical roles in gene regulation and immunity.

Now that you have captured the concept of RNA molecules capable of forming complex, sequence-dependent specific 3D shapes whose conformation enables their function, let's get back to our study of human genetics. We will consider DNA and RNA as they perform their most basic functions by studying the basic processes of transcription and translation. In the English language, the word *transcribe* means to write an exact copy, and the word *translate* means to transfer meaning from one language to another. In cell biology, during the transcription of DNA, the meaning (or 'code') is copied to a new molecule (RNA) but the 'language' (that of nucleic acids) remains the same. During translation, the meaning is kept but the language changes, from that of nucleic acids (DNA and RNA), to that of amino acids (proteins).

Key Points

RNA molecules serve many functions and have specific base sequences and three-dimensional structures which enable them to perform their functions.

2.7 Transcription

Genes are coding regions within the genome, known to produce RNA products (see **Figure 2.9**). It is important to remember that not all these RNA molecules go on to code for proteins. Many RNA molecules of various shapes and sizes perform complex roles within cell biology. Let's think about how they are formed.

Before considering the process of transcription, we must acknowledge the three different categories of RNA polymerase enzymes that work to transcribe the eukaryotic genome:

1. RNA polymerase I, which is found in the nucleolus and responsible for building ribosomal RNAs, 5.8, 18, and 28S.
2. RNA polymerase II, which is found in the nucleoplasm and responsible for building pre-messenger RNA and many types of non-coding RNA (ncRNA).
3. RNA polymerase III, which is found in the nucleoplasm and is responsible for building 5S RNA and tRNA.

All three categories use slightly different mechanisms of binding to their templates and are recruited to relevant regions along with differing aggregations of transcription factors. For brevity here we will focus on the actions of RNA polymerase II.

In **Figure 2.11**, notice the binding of a transcription factor (made of multiple subunits) to the transcription factor binding site. These proteins act as helicases and have the capacity to

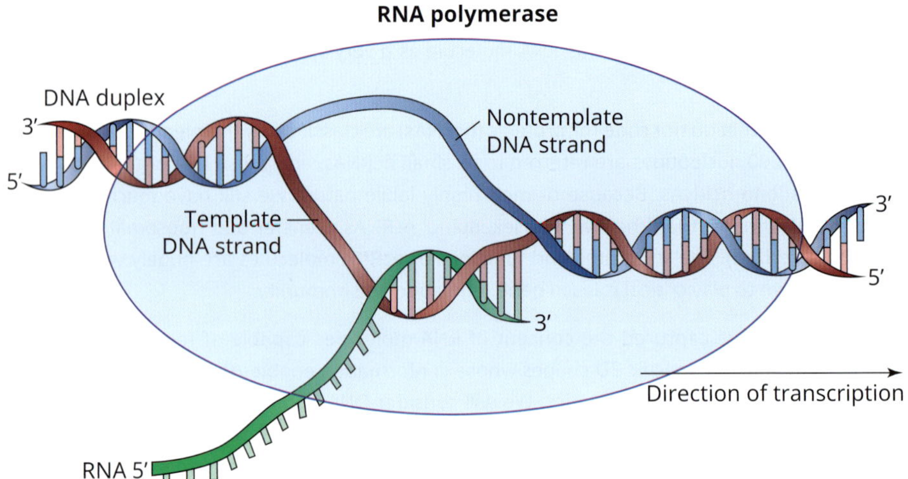

FIGURE 2.11
A diagrammatic representation of DNA transcription by RNA polymerase II.

hydrolyse ATP. Next, notice the RNA polymerase bound to the non-coding (template) strand. It will begin producing a single-stranded RNA molecule that is of complementary sequence to the DNA. Recall that in RNA, uracil replaces thymine, so where there is an adenine in the DNA template, RNA polymerase will incorporate uracil. RNA polymerase will continue to transcribe the gene, until a **terminator** sequence is reached. The recognition of polyadenylation signals (PAS) set the site for pre-mRNA cleavage and polyadenylation.

terminator
A specific section of nucleic acid sequence that marks the end of a gene or operon in genomic DNA during transcription.

monocistronic
A messenger RNA that can encode only one polypeptide per RNA molecule; compare with polycistronic.

Almost all eukaryotic messenger RNAs are **monocistronic**. This means that they encode one polypeptide. They have a short section of RNA before the coding region and a longer one after it. These are referred to as the 5′ and 3′ untranslated regions, or UTRs, respectively. The 5′ UTR begins with a 5′ terminal Cap structure, and the 3′ UTR ends with a polyadenylate (polyA) 'tail'. These polyA tails are vital to correct gene expression and have several roles. They are a signal for the molecules to be exported from the nucleus so that they can be used for transcription in the cytoplasm. They stabilize intact messenger RNA molecules, protecting them from degradation, and they are involved in the regulation of gene expression as they can be shortened or lengthened to change their trafficking and/or rate of transcription.

The precise anatomy of pre-messenger RNA (pre-mRNA) and mRNA is important because it is vital to the correct formation of proteins that mRNA is moved to the right place for processing and is correctly formatted for translation so that the right proteins are produced.

spliceosome
A large ribonucleoprotein (RNP) complex found in the nucleus of eukaryotic cells that removes introns from pre-messenger RNAs (pre-mRNAs).

Intron/exon boundary sequences are highly conserved. Mutations in them can result in aberrant splicing and consequently incorrect and non-functional protein formation. However, intron/exon boundary sequences also allow for differential recognition in different body tissues. This, very usefully, can result in **spliceosomes** producing different mature messenger RNA molecules (ergo different, very similar, functional proteins) from the same pre-mRNA. **Figure 2.12** shows how the mature RNA in the right-hand column is made up of different-sized segments because of alternate splicing. Differently spliced RNA can result in proteins that may or may not be functional dependent on the circumstances. Some exon skipping and/or intron retention occurs because of mutations or aberrant regulatory signals, some through normal, healthy regulatory routes.

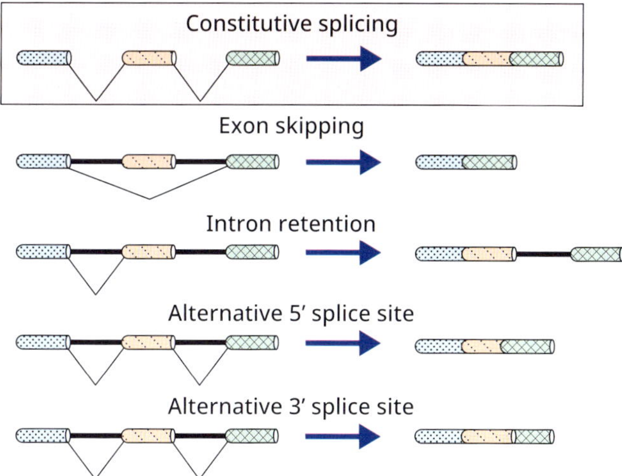

FIGURE 2.12
Diagrammatic representation of alternate splicing. Three exons are depicted, each represented by a differently coloured or patterned block. Introns are shown with lines between the exonic regions. The thin triangles show the regions of splicing activity. The top of the diagram shows constitutive splicing, where every intron is removed from the pre-mRNA, resulting in the full protein coding sequence. The second row down depicts exon skipping where both intron regions and the central exonic region are spliced out, resulting in a truncated protein. The third part of the diagram shows intron retention, where the second intronic region is retained into the mRNA and is used to code for amino acids. The final two rows show alternative splice sites, the first resulting in the central exonic region being truncated, the last one resulting in the final exonic region being truncated.

As you finish reading this section, take a moment to compare the process of DNA replication with the process of transcription. Both processes require a template. Both use fundamentally similar biochemical mechanisms, in which a polymerase constructs a new nucleic acid from nucleotide triphosphates. The direction of synthesis of the new nucleic acid strand is the same, with the addition of nucleotides to the 3′ OH. Both can be thought of as happening in three phases:

1. Initiation
2. Elongation
3. Termination.

However, during transcription only one strand of the DNA serves as the template. Note: the complementary strand might also serve as a template, but the product would be fundamentally different and not destined to pair with the product of the other template strand. Whilst DNA replication copies the entire genome, transcription only involves short sections of our DNA. Transcription uses RNA polymerase which does not need a double-stranded molecule to extend from and therefore does not require a primer.

2.8 Translation

To produce functional protein products once messenger RNA processing is complete, the process of translation must begin. There are three key molecules involved in translation:

- Messenger RNA
- Ribosomal RNA
- Transfer RNA.

Let us consider each in turn.

2.8.1 Messenger RNA (mRNA)

Messenger RNA (mRNA)
A type of RNA that carries genetic information from DNA to the ribosome, where it is used to synthesize proteins.

The first key molecule involved in translation is **messenger RNA**, a linear single-stranded nucleic acid formed as a result of transcription of a protein-coding gene, in a process that would have resulted in a pre-mRNA and then a mature mRNA once splicing had been completed (you read about splicing in **Section 2.7**—have a look back at **Figure 2.12**).

This molecule carries the code (the information) for the amino acid string that will be formed. This is possible because the molecule contains within it a 'reading frame' of sequential triplet bases, referred to as codons. This reading frame is signalled by conserved base sequences upstream of the transcription start site to which molecular machinery can bind. Imagine molecules moving along the mRNA strand and starting to translate it as a set point, much like you might read along sheet music and understand when to start singing. Or you might understand when to start jumping when you see that you are approaching the first hurdle on an athletics track.

The first of these codons is always AUG. **Figure 2.13** shows a diagrammatic representation of a messenger RNA molecule. Remember from previous sections that the helical 'ribbon' represents the sugar–phosphate backbone. Over to the right, you will see each codon represented in order. Remember that the molecule is 'read' in a 3-prime to 5-prime direction.

Cross reference
Do you remember how we also use single-letter codes to indicate base substitutions in nucleic acids? If not, look back at **Table 2.1** to remind yourself.

In **Figure 2.13**, focus on the 'code'. Codon 1 codes for the amino acid methionine. What do the others code for? Each amino acid can be represented by either a single-letter or triple-letter shortening. In **Table 2.3**, T sets out the agreed single- and triple-letter codes for each amino acid. Notice that 'M' codes stand for methionine, which can also be represented as 'Met'.

Now, look at **Table 2.4**, which is a tool that scientists use to match codons with their amino acids. The left-hand side represents the first nucleic acid in the codon (U, C, A, or G). The top of the table shows the same four options that could be the second base. The right-hand side

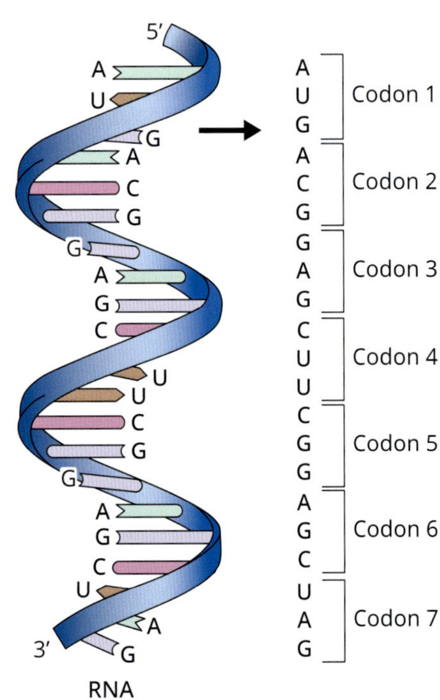

FIGURE 2.13
A diagrammatic representation of the coding section of a messenger RNA (mRNA) molecule, showing the reading frame broken down into codons, beginning with the 'start' codon, which codes for the amino acid methionine.

TABLE 2.3 International Union of Pure and Applied Chemistry (IUPAC) single and three letter codes for Amino acids

IUPAC amino acid code	Three letter code	Amino acid
A	Ala	Alanine
C	Cys	Cysteine
D	Asp	Aspartic Acid
E	Glu	Glutamic Acid
F	Phe	Phenylalanine
G	Gly	Glycine
H	His	Histidine
I	Ile	**Isoleucine**
K	Lys	Lysine
L	Leu	Leucine
M	Met	Methionine
N	Asn	Asparagine
P	Pro	Proline
Q	Gln	Glutamine
R	Arg	Arginine
S	Ser	Serine
T	Thr	Threonine
V	Val	Valine
W	Trp	Tryptophan
Y	Tyr	Tyrosine

also has the same four options that could be the third base. Now, work around the table again, selecting the first, then second, then third base to identify the amino acid coded for by the 'start' codon (AUG). If you select the first base over to the right, you will note that all codons beginning with 'A' will code for all the amino acids represented in the fourth row of the table (Ile, Thr, Asn, Ser, Met, Lys, or Arg).

The second base in the start codon (AUG) is U, so now select 'U' at the top of the table and work down. You can now discount all the amino acids to the right and find that any codon beginning with 'AU' will code for either Ile or Met, the three-letter codes in the cell that is row 4 and column 2, shown highlighted.

Look at **Table 2.3** again; what amino acid does 'Ile' code for? If you picked out Isoleucine (which is the only one in bold) then you are working well and understanding. If not, please follow the instructions for **Table 2.4** again.

In **Table 2.4**, notice that, unlike the left-hand side and top, the bases on the right-hand side are represented four times. The third base in the start codon (AUG) is G, so select this from

TABLE 2.4 A tool for identifying which amino acid is coded for by each three-letter codon

Second base in codon (columns U, C, A, G); **First base in codon** (rows U, C, A, G); **Third base in codon** (right-hand column)

	U	C	A	G	
U	Phe	Ser	Tyr	Cys	U
	Phe	Ser	Tyr	Cys	C
	Leu	Ser	STOP	STOP	A
	Leu	Ser	STOP	Trp	G
C	Leu	Pro	His	Arg	U
	Leu	Pro	His	Arg	C
	Leu	Pro	Gln	Arg	A
	Leu	Pro	Gln	Arg	G
A	Ile	Thr	Asn	Ser	U
	Ile	Thr	Asn	Ser	C
	Ile	Thr	Lys	Arg	A
	Met	Thr	Lys	Arg	G
G	Val	Ala	Asp	Gly	U
	Val	Ala	Asp	Gly	C
	Val	Ala	Glu	Gly	A
	Val	Ala	Glu	Gly	G

the right-hand side, but only within the row in which you are now working. The correct 'G' is highlighted—follow this along to the left to arrive at 'Met'. AUG codes for methionine (Met); work around the first, second, and third bases again to be sure that you can use this table. Notice how no other codon but AUG codes for methionine. Can you see another amino acid that has only one codon? Work around the table to decode UGG. Hopefully you arrived at tryptophan (Trp).

We have used **Table 2.4** to 'decode' the coding parts of a messenger RNA. However, this is not the way that our bodies do the job! The molecular biological processes of translation rely on the size, shape, and charge of molecules as they all work together in molecular machines to use the information encoded by the DNA to form the structures of functional RNA and protein molecules.

ribosomal RNA (rRNA)
Specific RNA molecules which are the major components of ribosomes.

ribosome
A cellular organelle that translates mRNA into proteins, found in both prokaryotic and eukaryotic cells.

nucleolus
A membrane-less nuclear organelle where ribosomal RNA (rRNA) is synthesized and assembled with specific proteins to form ribosomes.

epigenetic
A heritable and stable change in gene expression that occurs through alterations in the chromosome but not in the DNA sequence. An epigenetic trait is a stably heritable phenotype resulting from epigenetic changes in a chromosome.

2.8.2 Ribosomal RNA (rRNA)

The next molecule to be familiar with when seeking to understand the process of translation is **ribosomal RNA (rRNA)**. Ribosomal RNA is a crucial component of ribosomes. **Ribosomes** are the cell's translation 'machines', and they are composed of approximately 60 per cent RNA and 40 per cent protein. Ribosomal RNA molecules are coded for by clusters of genes, whose highly conserved nature can be used to study evolutionary relationships between organisms. A huge amount of ribosomal RNA is required to ensure the biogenesis of enough ribosomes to keep up with the cells' demands for protein biogenesis. So, the spatial organization of these genes within the **nucleolus** and their **epigenetic** regulation is tightly controlled. Ribosomal RNAs are clustered together to form two subunits: one large and one small.

2.8.3 Transfer RNA (tRNA)

The final type of molecule to be familiar with when seeking to understand this process of translation is **transfer RNA (tRNA)**. There are many tRNAs in eukaryote systems. Most sense codons have a matching tRNA. These small RNA molecules spontaneously adopt what is commonly referred to as a 'cloverleaf' structure. Their shape allows them to carry a specific amino acid. In **Figure 2.14**, notice the amino-acid acceptor site at the top and the anti-codon at the bottom.

transfer RNA (tRNA)
Small RNA molecule that carries amino acids to ribosomes, where they are covalently bound to each other to form proteins.

To bring about translation, the small and large ribosomal subunits lock around messenger RNA molecules and move along them. Transfer RNA molecules, loaded with specific amino acids to match their anti-codons, are attracted to the acceptor site within the large ribosomal subunit. If the mRNA codon and the tRNA anti-codon are complementary, the loaded tRNA moves along into the peptidyl site within the large ribosomal subunit, which then catalyses the formation of a peptide bond. The amino acid leaves the transfer RNA and is added to the growing polypeptide chain that is the primary amino acid product of the gene. **Figure 2.15** is a representation of all three RNA molecules working together to bring about translation of a protein-coding gene. Multiple ribosomal subunits can move along a single mRNA molecule in a structure, referred to as a polysome, ensuring rapid production of multiple identical polypeptides from a single mRNA molecule.

The products of the process of translation are not always functional. The body undertakes a multitude of post-translational modifications to ensure that the correct molecular conformations are achieved at the right time and in the right place, but describing these is beyond the remit of this book. If you wish to read more about post-transcriptional modifications, please refer to the further reading section at the end of this chapter for some recommended texts.

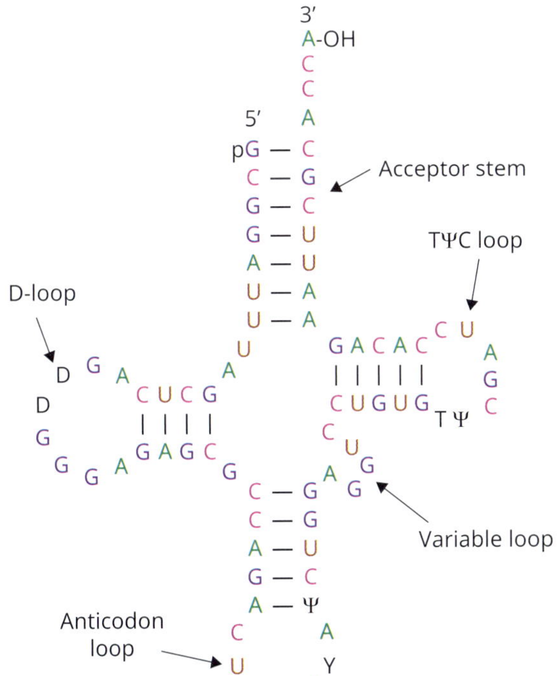

FIGURE 2.14
Canonical representation of the base sequence and secondary structure of human transfer RNA (tRNA) for the amino acid leucine.

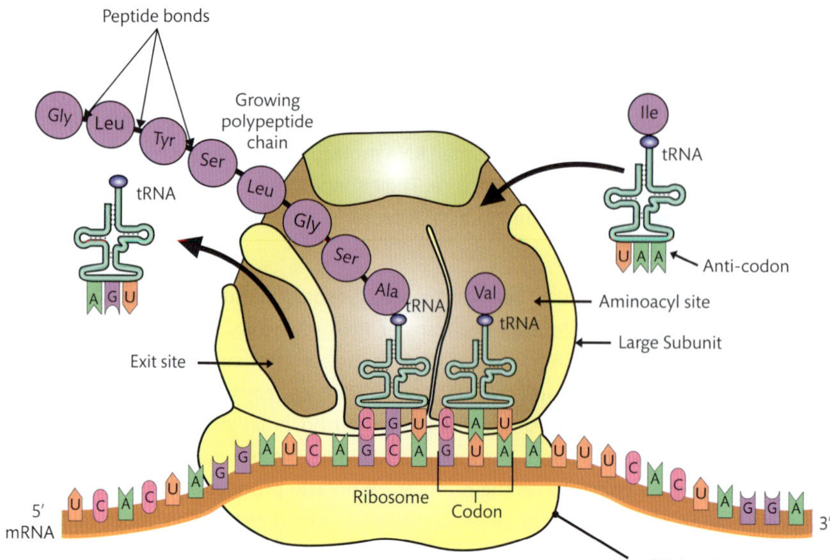

FIGURE 2.15
The large and small ribosomal subunits associated with messenger RNA (ribbon along the bottom), facilitating the building of a growing primary polypeptide chain using amino acids (circles) supplied attached to transfer RNA molecules (clover-like structures).

Key Points

Messenger RNA, ribosomal RNA, and transfer RNA molecules work together to bring about translation of specific sections of DNA code into polypeptide chains.

SELF-CHECK 2.4

What is the amino acid being added by the transfer RNA in **Figure 2.15**?

SELF-CHECK 2.5

What are the next two tRNA anti-codons that will be needed to match AUU and UUC in **Figure 2.15** and which amino acids do they code for?

2.9 Repetitive DNA

Now that you have a good understanding of the nature of DNA, and how coding regions are used to generate proteins and functional RNA products at the correct time and place to bring about the phenotype, let's go on and consider once again those regions of the genome that do not code for functional RNA or protein molecules. In **Section 2.1**, we covered repeat sequences at centromeric and telomeric regions of the genome. Repeating DNA motifs at the centromeric regions provide physical anchor points for elements within the cytoskeleton so that chromatic material can be moved correctly during cell division. Because of the inability of the replisome to cover the end portion of a DNA template, these repeats are progressively lost with each round of cell division, therefore acting as an internal molecular clock, 'counting down' cell divisions and ensuring senescence. There are many more repeating sequences in our genome. Short and long interspersed elements (SINEs and LINES) are the most well-known and are evolutionarily derived from retrotransposons. A 300 base-pair SINE known as ALU is repeated so much that it makes up approximately 17% of our genome.

2.10 Nucleic acid analysis

The focus of this book is clinical genetics for the biomedical scientist and so we are not so interested in the healthy (wild-type) functioning of the genome, but in what happens when things go wrong. Conditions caused by large structural chromosomal alterations are diagnosed by cytogeneticists, working with whole cells, mostly at metaphase, when the chromosomal material is visible in its most condensed form. Whereas genetic conditions caused by small alterations in the DNA, ranging from a single nucleotide change such as sickle cell (which is caused by a GAG to GTG mutation in the gene on the short arm of Chromosome 11) to larger repeat expansions are found using DNA sequencing techniques.

Genetic disorders are tested or screened for using many types of samples. Samples of human tissue, with associated microorganisms, are collected and sent to the laboratory for nucleic acid extraction and analysis. When taking any sample for nucleic acid analysis, it is important to be mindful of both safe and ethical working practices. Any practising biomedical scientist or trainee must be mindful of the ethical, safe, and legal working practices relevant for the country and organization within which they work. Please refer to local guidelines and, if unsure, always consult line managers and/or relevant Professional, Statutory, and Regulatory Bodies (PSRBs). In the UK, please refer to the Institute of Biomedical Science (IBMS) and the Human Tissue Authority (HTA).

It is important for the biomedical scientist to be aware that there are many complexities that influence our understanding and identification of inherited conditions. A variation in DNA, whatever its locus, may not necessarily be the cause for an observed clinically relevant phenotype. However, having considered the basic structure and function of nucleic acids, let's think now about how genetic material is passed from one individual to the next through sexual reproduction and how this looks in terms of genotype bringing about phenotype. In terms of genetic disorders, it matters very much whether a cell is **germline** or somatic. Why? Because damage to the DNA of a somatic cell, however terrible, stops with that single organism, whereas DNA damage in a germline cell can lead to inherited disorders that can be passed down generations.

Cross references

Figure 3.9 shows condensed human chromosomes at metaphase. The telomeric regions, at the very ends of each huge DNA molecule that makes up a single chromosome (fluorescing shapes) are constructed of approximately three hundred to eight thousand tandem repeats of TTAGGG.

You will read about chromosomal alterations and conditions caused by them in **Chapter 3**.

You will learn about how laboratories work to diagnose genetic conditions and sample types used in **Chapter 6**.

Because the laboratory techniques are so similar, this book also covers nucleic acid analysis of human pathogens in **Chapter 8**.

Matters of ethics are covered in **Chapter 9**, and matters of safety are briefly touched upon in **Chapter 6**.

germline

The cellular lineage of a sexually reproducing organism from which eggs and sperm are derived. The genetic material contained within this cellular lineage can be passed to the next generation.

2.11 Basic patterns of inheritance

As a biomedical scientist, the purpose of analysing nucleic acids is to support clinical staff with diagnosis, prognosis, treatment, and management for a pathology or phenotype of interest. It is therefore vital that, as a practitioner, you have a good basic understanding of inheritance patterns and their underlying genetic causes. Before we begin this section, an important distinction must be made between the terms genetic and heritable.

When we know that the cause for a particular phenotypic trait is a genetic one, then we may say that it is genetic. However, if a condition or trait appears to run in families, but we have no evidence that DNA is involved, then we must refer to it as heritable until such time as a genetic basis is evidenced.

Cross reference

To learn more about complex inheritance and its impact on how we understand the basis for genetic conditions, see **Chapter 4**.

We each have two sets of genetic material: one maternally derived (through the egg) and one paternally derived (through the sperm). Remember, though, that through the egg we also have mitochondrial DNA, but we will come back to that again later. We also have an X chromosome from our mother and, if we are female, an X from our father too. If we are male, then our father has passed on his Y chromosomal code to us instead.

When our parents made their gametes using meiosis, the DNA that they had inherited from their parents was recombined and independently assorted. This means that none of us has a chromosome that is an exact replica of that found in a parent. Rather, our chromosomal material is a complex, shuffled combination of about a quarter of the code found in each of our four grandparents. Considering inheritance, and with this in mind first, we will look at what we commonly refer to as Mendelian inheritance patterns. These patterns are named for Gregor Mendel, whose dogged analysis in the mid-1800s of inherited dichotomous phenotypic traits in pea plants gave us the data from which our understanding grew.

2.11.1 Monogenic autosomal dominant inheritance

In monogenic autosomal conditions, particular phenotypic trait(s) can be attributed to one gene (monogenic), which is known to be on an autosomal (non-sex) chromosome. An individual will inherit one copy of the gene from their mother and one from their father. This means that most of us will have two slightly different copies (alleles). In monogenic autosomal dominant conditions, the presence of one of these alleles will give rise to an observed phenotypic trait, thereby 'dominating' the presence of the other allele. In cases relevant to medical genetics, this usually happens for one of two reasons. It could happen because the clinically relevant (mutant/broken) allele generates or leads to the accumulation of a damaging product that the body cannot adequately clear. Alternatively, it may occur because it contains a 'gain of function' mutation, which progresses the cell cycle too swiftly, evading check points, and leading to cancer. In a case of simple Mendelian monogenic autosomal dominant inheritance, anyone inheriting the causative allele will show the phenotype (see **Figure 2.16**).

In **Figure 2.16**, notice how alleles of the 'A' gene (note: any letter of the alphabet could be used to denote a gene) are denoted with a capital letter for the dominant allele and lowercase for the recessive allele. The parent on the left is homozygous recessive. They have inherited two recessive alleles and therefore show the recessive (wild-type/healthy/normal) phenotype.

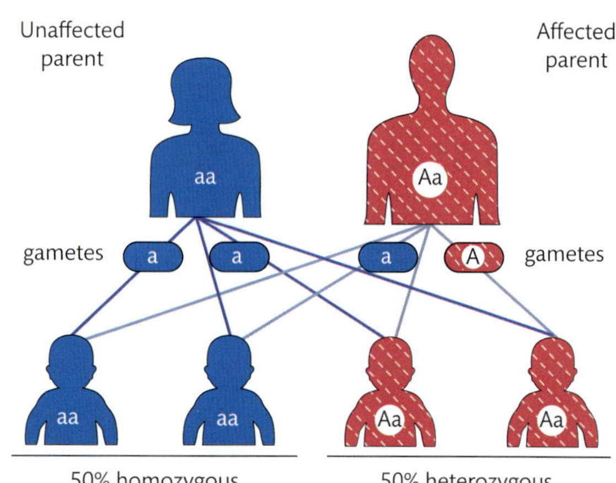

FIGURE 2.16
Autosomal dominant inheritance.

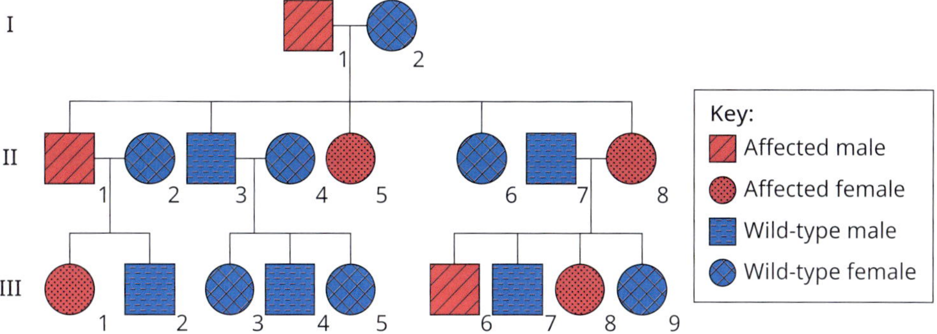

FIGURE 2.17

A simple pedigree diagram for a monogenic autosomal dominant condition.

The parent on the right is heterozygous: they inherited one recessive and one dominant allele from each of their parents. Follow the causative dominant allele into the next (first filial) generation. Each child must inherit one allele from either parent. Therefore, there is a fifty per cent chance that any given offspring will inherit a causative allele from their affected parent. Therefore, there is a fifty per cent chance that any child from this couple will be affected. If this condition were serious and life-limiting, in the UK the couple might be offered *in vitro* fertilization (IVF) and preimplantation genetic diagnosis to screen out affected embryos.

Figure 2.17 shows a simple pedigree diagram for a monogenic autosomal dominant condition. Males are represented by squares and females by circles. (Note: in the UK, special consideration is now given to transgender individuals whereby they are given the symbol of the sex they present as with an annotation below to indicate their sex at birth. Please refer to local guidelines when drafting pedigree diagrams.) The parental generation is shown at the top. Notice how the generations, going down, are labelled in Roman numerals: the first filial generation is II, the second is III, and so on. Individuals across the diagram are labelled in Arabic numerals and every individual, alive or dead, related or not, gets a number. In this family, the grandfather is affected with a monogenic autosomal dominant condition (indicated by full shading). Notice how, of his five children (generation II), three are also affected: his first born (a son), and his third and fifth children (both daughters). Therefore, their genotypes would be denoted Dd. His second and fourth children (a son and a daughter) are unaffected (genotype dd). Notice how monogenic autosomal dominant conditions are inherited by approximately fifty per cent of the offspring, with no sex bias, and they never skip a generation.

Let's look at a monogenic autosomal dominant condition represented using a Punnett square. Punnett squares are named for Reginald Punnett, co-discoverer of genetic linkage in 1904 and co-founder of the *British Journal of Genetics* in 1910. In a Punnett square (**Figure 2.18**), the paternal gametes are placed at the top, and the maternal to the left. The square is then filled across and

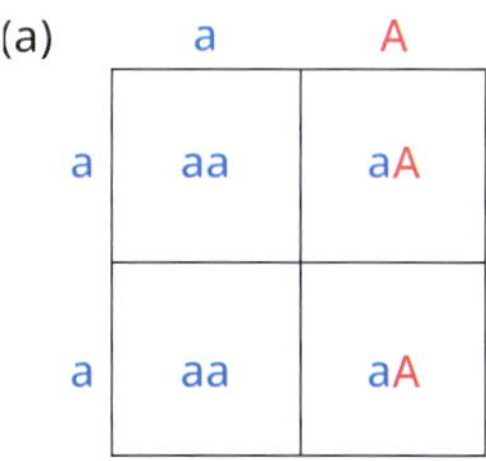

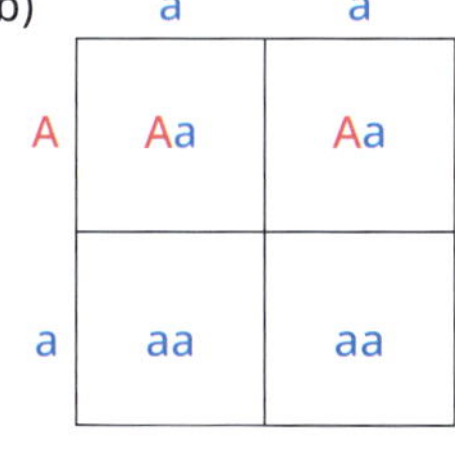

FIGURE 2.18

Monogenic autosomal dominant condition represented using a Punnett square.

then down. From this square (a), you can see that fifty per cent of the offspring have inherited the causative allele and will therefore show the dominant phenotype, as does their father. Square (b) shows the mother as being affected. Notice that the risk ratio to the offspring is the same.

It is important to remember that every new conception is a new 'roll of the dice', so just because a couple have two unaffected children, it does not mean that their next child will be affected. Every conception has a fifty per cent chance of inheriting the causative allele. It is rare for a parent to be homozygous dominant for a clinically relevant allele. However, if that is the case, then they do not have a wild-type allele to offer the next generation and therefore all their offspring will be affected.

SELF-CHECK 2.6

If one parent has a causative allele for an autosomal dominant condition and the couple have six children, how many children are likely to be affected by the condition.

---

epistasis (epistatic)
A situation in which the expression of one gene is affected by the expression of one or more independently inherited genes.

Having become familiar with the concept of dominant inheritance, it is important to add that conditions caused by dominant mutations do not always manifest in the phenotype. Sometimes, due to **epistatic** interactions or for multifactorial reasons a small proportion of people who inherit the causative allele do not go on to show the phenotype. This is called reduced or incomplete penetrance (see **Box 2.2** for definitions of some terms related to complex inheritance).

BOX 2.2 Introducing useful concepts to describe more complex inheritance patterns

Incomplete penetrance Describes the situation when a dominantly inherited genetic condition is not always seen in the phenotype. It can be attributed to the effects of 'modifier genes' (epistatic interactions) or environmental differences. The level of penetrance in conditions that show incomplete penetrance is reported as a percentage or as a proportion of 1.

Variable expression Describes the situation when a genetic condition is not always seen to the same degree in the phenotype. An example is Neurofibromatosis which shows almost complete penetrance (those with causative mutations almost always show the phenotype to some extent), but extreme variability of expression. The phenotypic expression can be anything from just a few faint marks on the skin (referred to as cafe au lait patches) to severe life-threatening complications.

Co-dominance Describes the relationship between two alleles of a gene. If the alleles are different, the dominant allele usually will be expressed, while the effect of the other allele, called recessive, is masked. In co-dominance neither allele is recessive and the phenotypes of both alleles are expressed. An example is the ABO blood groups in humans. A and B are co-dominant alleles, O is recessive to both A and B.

Incomplete dominance Describes the appearance in a heterozygote of a trait that is intermediate between either of the trait's homozygous phenotypes. The situation is sometimes also referred to as semi-dominance.

Complementation Describes a situation where two parents who are homozygous for different recessive causative alleles and therefore show the phenotype for the same recessively inherited condition, albinism for example, will have no affected children because all of the children will be heterozygous at both loci.

Compound heterozygosity Describes a situation in which an individual has two different recessive causative alleles at the same locus, one on each homolog, resulting in the disease phenotype.

2.11.2 Monogenic autosomal recessive inheritance

Let's move on now to consider monogenic autosomal recessive conditions. Again, in these situations, one gene is known to be responsible for the phenotype of interest, and it is located on a non-sex chromosome (monogenic autosomal). However, in recessive conditions, one clinically relevant (mutant) allele is not sufficient to cause the phenotype. This is often because the presence of the non- or partially functional copy can be made up for by the wild-type, functional code. In recessive conditions, we can have 'carriers' who are mostly asymptomatic. One in twenty-five individuals of European descent are carriers for an allele causative of cystic fibrosis. They will never know about it unless they have their DNA sequenced at that locus, or they produce affected offspring.

In **Figure 2.19**, each parent is a carrier; therefore, fifty per cent of their gametes are likely to contain the causative allele. However, there is only a one in four chance that any given conception will result in offspring that has inherited the causative allele from both mum and dad.

Now study the Punnett square in **Figure 2.19**. Notice how both parents have one dominant (A) and one recessive (a) allele. In this case, the causative allele is recessive and is shown in red. There is:

- a twenty-five per cent chance that a child of this couple will be completely unaffected,
- a fifty per cent chance that they will be a carrier,
- and a twenty-five per cent chance that they will be affected.

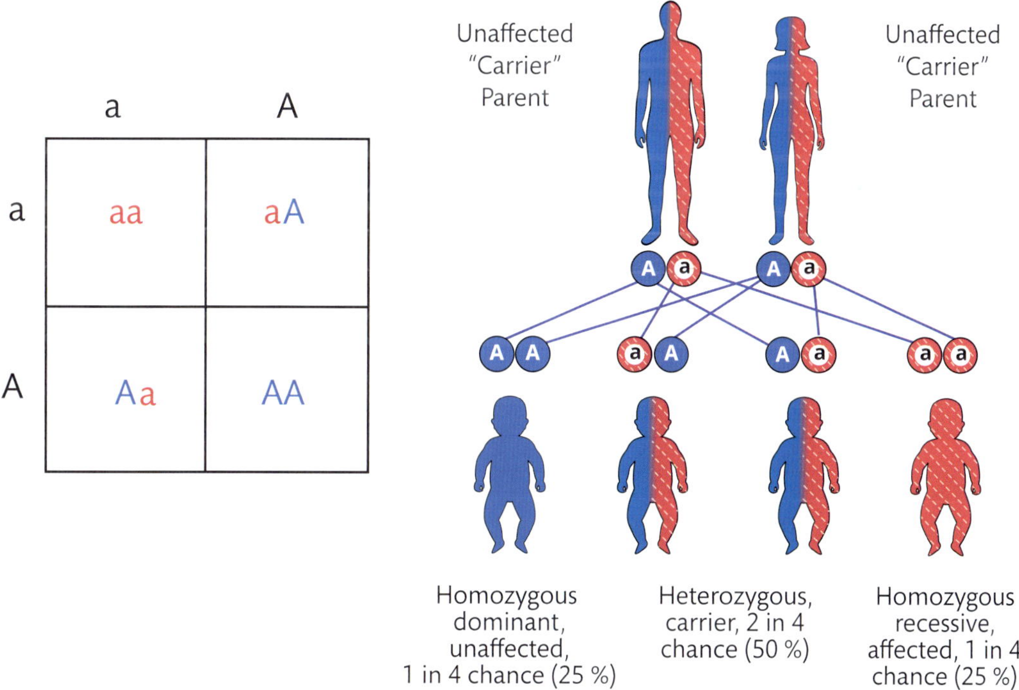

FIGURE 2.19

Monogenic autosomal recessive condition represented using a Punnett square and simple pedigree diagram.

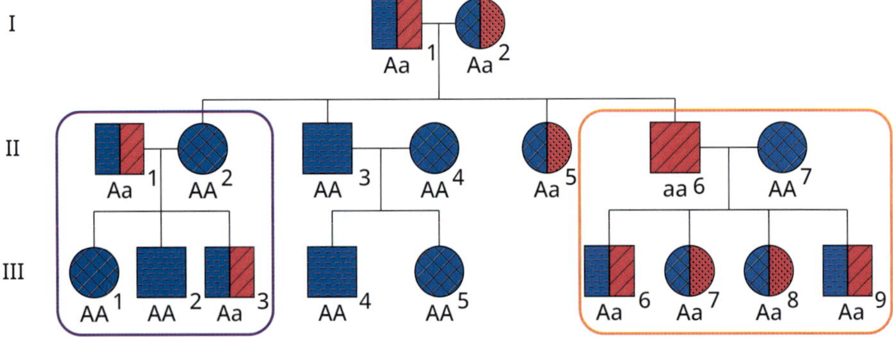

FIGURE 2.20
Monogenic autosomal recessive condition represented using a simple pedigree diagram.

However, remember every conception is a new chance. If a couple already have three unaffected children, there is still only a one in four chance that the next child will be affected.

Figure 2.20 shows a simple pedigree diagram representing autosomal recessive transmission. The parental couple are both carriers (as indicated by half shading). Through chance, their first two children are unaffected, their third child is a carrier, and their fourth is affected. Look back at **Figure 2.18** and notice that in the top-left offspring quarter of the Punnett square, the recessive allele is maternally derived, and in the bottom right, it is paternally derived. When drawing Punnett squares, please fill them across and then down in this way so you can keep track of which allele comes from each parent.

Look again at **Figure 2.20**, notice how in monogenic autosomal recessive conditions affected offspring of carrier parents are seen in approximately twenty-five per cent of the offspring, with no sex bias and they can skip generations. Notice how none of the offspring of individual II.6 will be affected because they will all inherit a wild-type allele from their mother. Fifty per cent of the offspring are likely to be carriers. It is good to remember at this point that real life is not quite so simple and that some recessive conditions can result in symptomatic carriers due to differences in expression levels of wild-type and variant genes and epistatic interactions; however, you will cover these concepts in more detail later in **Chapter 4** of this book.

SELF-CHECK 2.7

Draw a Punnett square for the situation delimited in the right-hand box in **Figure 2.20**.

SELF-CHECK 2.8

Draw a Punnett square for the situation delimited in the left-hand box in **Figure 2.20**.

When considering monogenetic autosomal recessive inheritance, it is important to remember that, although any one individual can only have two germline allelic forms of a given gene, there will be many within a population, several of which are likely to be clinically relevant.

Why does this matter? Because an affected individual (genotype aa) designated homozygous recessive using a simple Punnett square will not always have two copies of the same clinically relevant allele. In fact, they are quite likely to have inherited different causative variants: one from each parent. In which case, they are, in fact, compound heterozygote at the locus in question. For example, there are several known causative variants for cystic fibrosis. The variant common in those of European descent is deltaF508 (which reads as a deletion of phenylalanine at position 508), and common in the Ashkenazi Jewish population is W1282X (which reads as substitution of tryptophan for a stop codon at position 1282). So, if the mother carries one and the father the other, then there is a one in four chance that a child of theirs will be compound heterozygote and therefore affected. In such cases, it is good to be mindful of the fact that the perspective of a genetic counsellor is different to that of a clinician, and different again to that of a molecular geneticist. It matters because some clinically relevant mutations might have partial function and others none. So, knowing the difference between the severity of effects of different alleles and their combinations can enable more accurate prognoses and better treatment planning.

Cross reference

Refer back to **Table 2.2** to see the single-letter codes for amino acids.

CASE STUDY 2.1 A case of compound heterozygosity

Hereditary haemochromatosis (HH) is a condition characterized by iron overload. Most cases of iron overload are caused by variants in the human homeostatic iron regulator protein gene referred to as HFE (for High Fe^{2+}).

Iron overload is a risk factor for other conditions such as arthritis, cirrhosis, and diabetes, but is both preventable and treatable once a diagnosis is given.

Type 1 Hereditary haemochromatosis is typically inherited in an autosomal recessive manner, see **Section 2.11.2**.

In his early 40s, an 80-year-old grandfather was given a diagnosis of primary (Type 1) hereditary haemochromatosis as a result of being homozygous with the C282Y variant of the HFE gene. This means that instead of cysteine at amino acid position 282 both copies of his gene coded for tyrosine (refer to **Table 2.2**). He was unable to make wild-type protein, leading to iron overload; however, his condition was well managed with regular phlebotomy treatments.

His wife was healthy with no previous reported cases of hereditary haemochromatosis in her family.

Their 50-year-old daughter reported to the clinic with symptoms of haemochromatosis. Biochemical tests to ascertain the concentration and ratio of iron and transferrin, and the level of ferritin in her blood, revealed that she did have haemochromatosis.

The diagnosis led to confusion and resulted in several family arguments. The couple had, years ago, received education about the inheritance pattern of type 1 hereditary haemochromatosis and understood that the father's causative alleles were recessive. They had therefore never thought to discuss any risk with their four children.

Genetic tests revealed that the daughter was a C282Y/H63D compound heterozygote. She had inherited the C282Y causative allele from her father and the H63D allele from her heterozygous mother.

This raises the fact that geneticists and families who are aware that they carry causative alleles, even if those alleles are known to be recessive, should be mindful of the possibility of compound heterozygosity.

Q1 What change in the amino acid is denoted by H63D?

Q2 What is the percentage chance of any of the couple's children being compound heterozygous for the causative alleles?

2.11.3 Sex-linked inheritance

gonosomal
Sex chromosomes, also referred to as allosomes, heterotypical chromosomes, gonosomes, heterochromosomes, or idiochromosomes. Gonosomes differ from autosomes in size, structure, and the way that they function within cells.

Having covered the basics of autosomal dominant and recessive inheritance, let's now consider **gonosomal** patterns. This refers to the inheritance of loci on the sex chromosomes. We will begin with monogenic X-linked dominant conditions. These are where one locus within the X chromosome is known to be causative of the phenotype and the presence of only one clinically relevant variant is sufficient to cause the condition.

In **Figure 2.21**, the mother on the left is affected, and therefore the risk to offspring is fifty per cent, irrespective of their sex. However, look at the family to the right where the father is affected. In this second case, all the daughters will be affected. This is because the father passes the causative allele to them when he passes the X chromosome that makes them female.

Figure 2.22 shows two Punnett squares representing the same information. In this case, because the pattern is sex linked, we use the sex chromosomes in the square. The causative allele is denoted by uppercase 'A'; however, we might just as easily use an asterisk to indicate the dominant, causative allele (X*) and just show the wild-type X chromosome as X. Notice how the likelihood of having a male or female child is fifty per cent. Then look back and forth between the two left-hand images of **Figures 2.21** and **2.22** and notice how the causative allele on the mother's X chromosome is passed to fifty per cent of her offspring at a 1:1 male to female ratio. Next, study both right-hand sides and notice how the father has passed the X chromosome with the dominant causative allele on to all his daughters.

Figure 2.23 shows what a simple pedigree diagram of this situation could look like. Notice how the affected father passes the condition to all his daughters, but none of his sons. Notice also how affected female II.3 has passed the condition on to approximately fifty per cent of her offspring, irrespective of their sex.

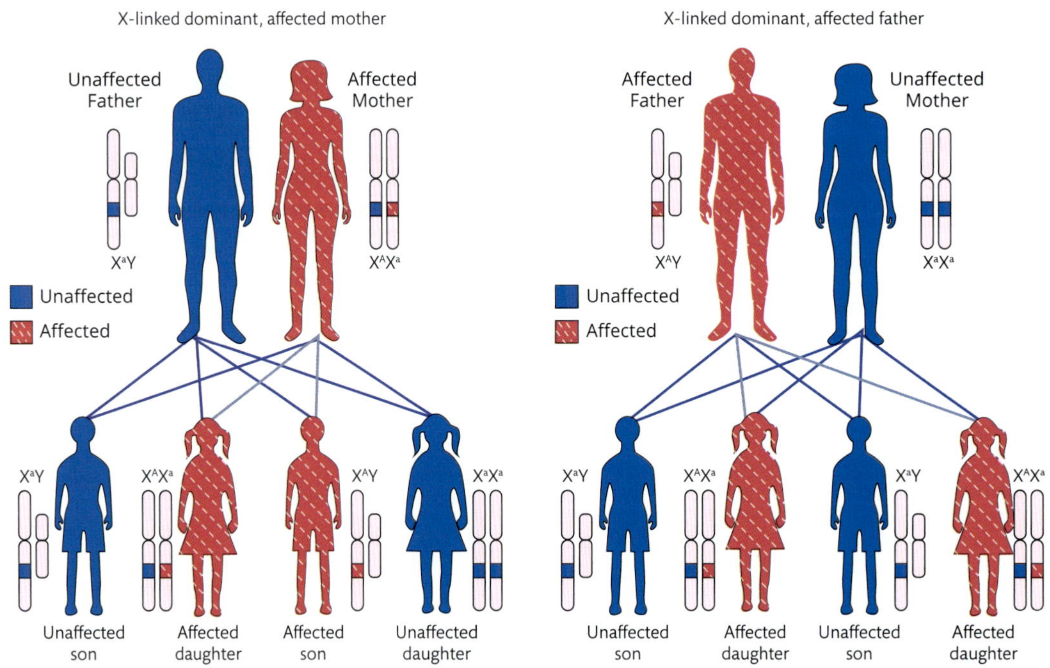

FIGURE 2.21
Illustrating X-linked dominant inheritance.

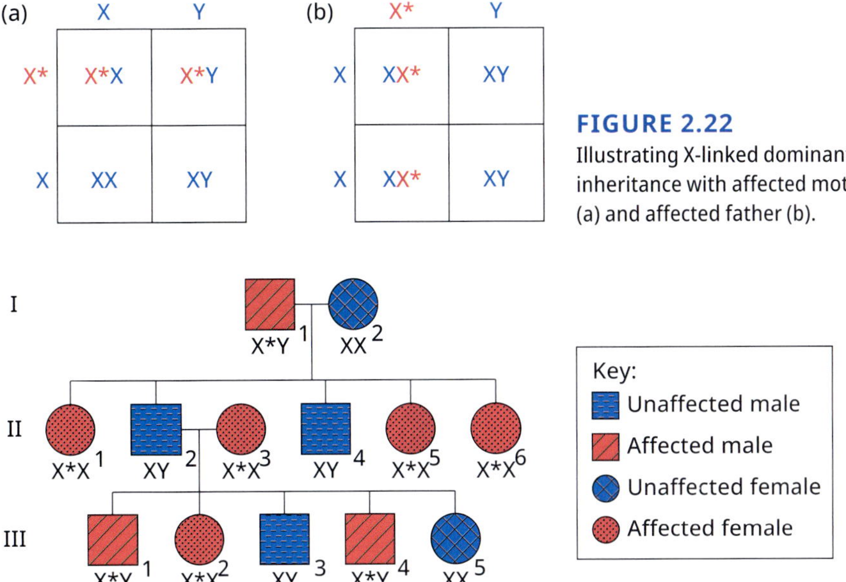

FIGURE 2.22
Illustrating X-linked dominant inheritance with affected mother (a) and affected father (b).

FIGURE 2.23
A simple pedigree diagram to illustrate X-linked dominant inheritance.

Figure 2.24 shows both a Punnett square and a pedigree diagram for a monogenic X-linked recessive condition. Remember that for a recessive phenotype to show, an individual must have two causative alleles. This means that there can be carriers. However, in the case of X-linked recessive situations, there will be no carrier males because they only have one allele on the one copy of their X chromosome, and so that single allele will determine their phenotype. We refer to this as being **hemizygous**: males are hemizygous for all genes on the X chromosome because they only have one copy. However, females can be carriers because they can be **heterozygous** (Aa). Notice in **Figure 2.24** how the carrier mother has passed the causative allele on to fifty per cent of her offspring irrespective of their sex. This means that any male child has a fifty per cent chance of being affected. Daughters are unlikely to exhibit the phenotype because they will all have one wild-type allele from their father. On a pedigree diagram, female carriers are indicated with the addition of a central dot to their symbol.

When working with X-linked cases, we must be aware of the process of **lyonization**, which is also referred to as X chromosome inactivation. Within a few days of conception, a process takes place in the blastocyst where specific genes on the X chromosome are transcribed, and the RNA produced works to trigger compaction and transcriptional shutdown of one of the X chromosomes. The lyonization process occurs at random in each cell, so that either the maternally or paternally derived X chromosome is silenced. From then on, as cells duplicate, all the daughter cells will have the same (M^m or X^p) silenced. In most females, this inactivation process results in approximately fifty per cent of cells keeping either the M^m or X^p active. However, some females show a skewed lyonization pattern. This can be unfortunate in heterozygous cases and lead to symptomatic carriers. This is the case with some types of haemophilia.

To complete our section on sex-linked inheritance patterns, we must consider the Y chromosome. The human Y chromosome has very few genes, and most of them are related to the production of male gametes. Genes passed from father to son via the Y chromosome are referred to as being **holandric**. The presence of the Y chromosome itself does not determine maleness; that

hemizygous
An individual who has only one member of a chromosome pair or chromosome segment. Hemizygosity is used to describe X-linked genes in males and in somatic cell genetics where cancer cell lines are often hemizygous for certain alleles or chromosomal regions.

heterozygous
Having two different alleles of a particular gene or genes.

lyonization
The inactivation (through condensation and epigenetic mechanisms) of one of the X chromosomes in females to achieve dosage compensation of X-linked genes with males.

holandric
Transmitted as part of or being a gene in the Y chromosome.

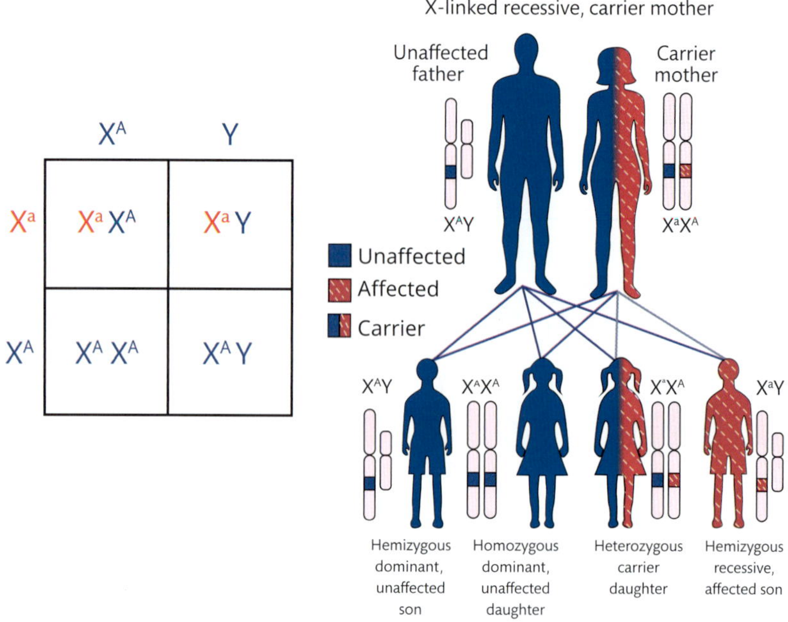

FIGURE 2.24
Illustrating X-linked recessive inheritance with a carrier mother. Left-hand side is Punnett square, right-hand side, simple pedigree diagram.

is down to a single gene on the X chromosome referred to as sex-determining region Y or SRY, for short. *SRY* is only 850 base pairs long and its product acts as a molecular master switch, acting at the right time and place within developmental processes to switch on all the other genes on many of the **autosomes** that confer 'maleness'. There have been cases of XX individuals who are phenotypically male due to the translocation of the *SRY* gene-containing region to another of their chromosomes and XY individuals who are phenotypically female due to an inactive SRY. Most Y-linked clinically relevant conditions in humans are associated with chromosomal loss or with epigenetic silencing of genes on the Y chromosome. Therefore, although holandric inheritance exists, it is not often seen in the context of single gene disorders in the way that X-linked and autosomal conditions are.

autosome
A non-sex chromosome.

2.11.4 Anticipation in inheritance

The final inheritance pattern attributable to variations in autosomal DNA that we will consider in this chapter is **anticipation**. Most genetic conditions caused by repeat expansions in the genome cause neurological conditions. **Table 2.5** lists a selection of those most well understood—see how many you recognize.

anticipation
A situation in which a genetic disorder becomes more severe and/or appears earlier as the causative genetic variant is passed down through the generations. Most often seen in disorders caused by trinucleotide repeat expansions.

In the past, these repeat sections of our DNA were very difficult to accurately, swiftly, and cheaply measure. The process had to be conducted slowly and painstakingly using the gold standard Sanger sequencing methods. The kind of next-generation sequencing techniques used for the 100,000 Genomes Project and many diagnostic purposes are based upon sequencing short sections of DNA swiftly and then using machine learning and bioinformatics to piece the data together. These processes were not well designed to ascertain the length of repeat sections. Think about how difficult it is to put a plain blue sky or green field together in

TABLE 2.5 Repeat expansions causing neurological disease

CAG – at least 10 diseases including Huntington disease, Spinal and Bulbar muscular atrophy, Dentatorubral-pallidoluysian atrophy and seven spinocerebellar ataxias
CGG – Fragile X, Fragile X tremor ataxia syndrome, other fragile sites (GCC, CCG)
CTG – Myotonic dystrophy type 1, Huntington disease-like 2, spinocerebellar ataxia type 8, Fuchs corneal dystrophy
GAA – Friedreich ataxia
GCC – FRAXE mental retardation
GCG – Oculopharyngeal muscular dystrophy
CCTG – Myotonic dystrophy type 1
ATTCT – Spinocerebellar ataxia type 10
TGGAA – Spinocerebellar ataxia type 31
GGCCTG – Spinocerebellar ataxia type 36
GGGGCC – C9ORF72 frontotemporal dementia/amyotrophic lateral sclerosis
CCCCGCCCCGCG – EPM1 (myoclonic epilepsy)

a jigsaw puzzle. However, progress is being made and advances in sequencing and machine-learning technology have enabled researchers to establish new and robust protocols to aid in the diagnosis of conditions caused by these repeat expansions.

The issue with repeat expansions is that when they are inherited, they become enlarged, and so the clinical manifestation of the phenotype worsens with each successive generation until the mutation prevents viability. They are never removed from the gene pool because shorter, survivable versions are always present.

2.11.5 Mitochondrial inheritance

The last inheritance pattern that we will consider in this chapter is mitochondrial (sometimes referred to as maternal). Mitochondrial inheritance follows a unique pattern because, as you will recall from **Section 2.4**, mitochondria are almost entirely maternally derived, via the oocyte. Punnett squares are not appropriate here because there are no alleles to consider, and the DNA is not involved in the same crossing over and independent assortment that chromosomal DNA is. In **Figure 2.25**, notice how an affected female passes the condition on to all her offspring, but an affected male does not transmit the causative mitochondrial sequence.

In the last few sections, you have learned about basic inheritance patterns and begun to explore some complexities presented by repeat expansions and by mitochondrial DNA.

Box 2.2 discusses some terms that define a few of the more complex and nuanced relationships between causative alleles and phenotypic expressions.

There is a lot more to discover and many nuanced ways in which our genetic material is passed down and expressed. Some of these will be discussed in the later chapters of this book, some can be found in the suggested further reading after the end of this chapter, some will require you to go wider still in your search for information, and some have yet to be discovered.

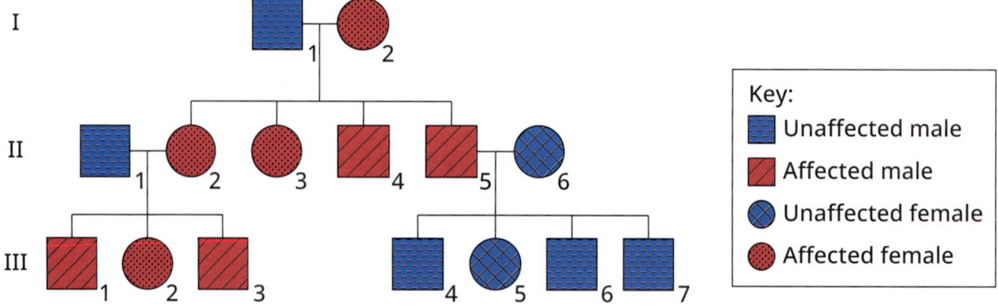

FIGURE 2.25
Illustrating mitochondrial (maternal) inheritance using a simple pedigree diagram.

Chapter summary

- This chapter has introduced you to the structure of DNA.
- You have discovered basics of how DNA is replicated and repaired.
- Explanations of transcription, messenger RNA, and non-coding RNA molecules have been provided.
- You have learned how transfer and ribosomal RNA works, together with messenger RNA, to bring about translation of the DNA code.
- We have covered basic inheritance patterns, and some complexities presented by repeat expansions and by mitochondrial DNA.

Discussion questions

2.1 What particular structural features enable DNA to perform its coding and replicative functions?

2.2 What events cause changes in DNA and how is it that these changes persist into the next generation of cells or individuals?

2.3 Are RNA molecules just as important for correct gene expression as DNA?

2.4 How do the complexities of inheritance patterns and differences in gene expression make it difficult to inform and advise non-scientists of risk to offspring?

Further reading

- Alberts *et al.* (2022) ***Molecular Biology of the Cell.*** 7th edition, Norton & Co., New York. ISBN: 0393884856.

  This excellent book gives a detailed overview of all the key molecular processes within cells. It is a foundational text for anyone wishing to understand the molecular processes of life.

- Yeeles JTP, Janska A, Early A, Diffley JFX (2017) ***How the eukaryotic replisome achieves rapid and efficient DNA replication***. *Molecular Cell*, 65(1), 105–116 ISSN 1097-2765. https://doi.org/10.1016/j.molcel.2016.11.017.

  This paper gives a clear overview of eukaryotic replisomes in much greater detail than could be achieved in the short chapter here.

- Berti M, Cortez D, Lopes M (2020) ***The plasticity of DNA replication forks in response to clinically relevant genotoxic stress***. *Nat Rev Mol Cell Biol*, 21, 633–651. https://doi.org/10.1038/s41580-020-0257-5.

  This paper is an extension to the information supplied in this chapter and in Joseph et al. above. Given that this book is aimed at Biomedical Scientists, interested in pathology, it gives extra depth to the relevance of these topics in clinical practice.

- Falkenberg M (2018) ***Mitochondrial DNA replication in mammalian cells: overview of the pathway***. *Essays Biochem.*, 62(3), 287–296. doi: 10.1042/EBC20170100. PMID: 29880722; PMCID: PMC6056714.

  This paper allows the reader to focus exclusively on replication of the mitochondrial genome. Given that the mitochondrial genome is implicated in many human genetic conditions it is a valuable addition for the Biomedical Scientist specializing in genetics and genomics.

- Bourque G, Burns KH, Gehring M, *et al.* (2018) ***Ten things you should know about transposable elements***. *Genome Biol*, 19, 199. https://doi.org/10.1186/s13059-018-1577-z.

  This article is a beautifully written guide for anyone wishing to know important basics about transposable elements. The chapter on Repeat Expansion Diseases by Professor Henry Paulson in the *Handbook of Clinical Neurology* presents an excellent and detailed description of the many ways in which repeat expansions in the genome lead to genetic disorders (*Handb Clin Neurol.*, 147, 105–123 (2018). doi: 10.1016/B978-0-444-63233-3.00009-9. PMID: 29325606; PMCID: PMC6485936).

- Lakhani N, Kulkarni K, Barwell J, Vasudevan P, Dorkins H (2024) ***Clinical Genetics and Genomics at a Glance—At a Glance Series***. Wiley-Blackwell, Oxford.

  The perfect companion book for the laboratory-based biomedical scientist who wishes to understand clinical genetics and genomics from a medical perspective. Reading it in addition to this book will enable a biomedical scientist practitioner to better understand the clinician's viewpoint.

- Alexeyev M, Shokolenko I, Wilson G, Ledoux S (2013) ***The maintenance of mitochondrial DNA integrity: critical analysis and update***. *Cold Spring Harbor Perspectives in Biology*, 5, 10.1101/cshperspect.a012641.
- Altun A, Garcia-Ratés M, Neese F, Bistoni G (2021) ***Unveiling the complex pattern of intermolecular interactions responsible for the stability of the DNA duplex***. *Chemical Science*, 12. 10.1039/D1SC03868K.
- Bagchi B (2013) ***Water in and around DNA and RNA.*** In: *Water in Biological and Chemical Processes: From Structure and Dynamics to Function (Cambridge Molecular Science*, pp. 151–166). Cambridge University Press, Cambridge. doi:10.1017/CBO9781139583947.013
- Bębenek A, Ziuzia-Graczyk I (2018) ***Fidelity of DNA replication—a matter of proofreading***. *Curr Genet.*, 64(5), 985–996. doi: 10.1007/s00294-018-0820-1. Epub 2018 Mar 2. PMID: 29500597; PMCID: PMC6153641.
- Bennett RL, French KS, Resta RG, Austin J (2022) ***Practice resource-focused revision: Standardized pedigree nomenclature update centred on sex and gender inclusivity: A practice resource of the National Society of Genetic Counsellors***. *Journal of Genetic Counselling*, 31, 1238–48. https://doi.org/10.1002/jgc4.1621
- Cvetic CA, Walter JC (2006) ***Getting a grip on licensing: mechanism of stable Mcm2-7 loading onto replication origins.*** *Mol Cell*, 21(2), 143–144. doi: 10.1016/j.molcel.2006.01.003. PMID: 16427002.
- Dahm R (2008) ***Discovering DNA: Friedrich Miescher and the early years of nucleic acid research.*** *Hum Genet*, 122, 565–81. https://doi.org/10.1007/s00439-007-0433-0
- Forsdyke DR, Mortimer JR (2000) ***Chargaff's legacy.*** *Gene*, 261(1), 127–137. ISSN 0378-1119
- Gammage PA, Frezza C (2019) ***Mitochondrial DNA: the overlooked oncogenome?*** *BMC Biol*, 17, 53. https://doi.org/10.1186/s12915-019-0668-y
- Grasgruber P, Cacek J, Kalina T, Sebera M (2014) ***The role of nutrition and genetics as key determinants of the positive height trend***. *Economics & Human Biology*, 15, 81–100. ISSN 1 570–677. https://doi.org/10.1016/j.ehb.2014.07.002
- Harris J, Haas E, Williams D, Frank D, Brown J (2001) ***New insight into RNase P RNA structure from comparative analysis of the archaeal RNA***. *RNA* (New York, N.Y.), 7, 220–32. 10.1017/S1355838201001777.
- Levene PA (1919) ***The structure of yeast nucleic acid. IV. Ammonia hydrolysis***. *J. Biol. Chem.*, 40(1), 415–24.
- Rochette L, Rigal E, Dogon G, Malka G, Zeller M, Vergely C, Cottin Y (2022) ***Mitochondrial-derived peptides: New markers for cardiometabolic dysfunction***. *Archives of Cardiovascular Diseases*, 115(1), 48–56. ISSN 1875-2136, https://doi.org/10.1016/j.acvd.2021.10.013
- Mishra S, Muthye V, Kandoi G (2020). ***Computational methods for predicting functions at the mRNA isoform level***. *International Journal of Molecular Sciences*, 21, 5686. 10.3390/ijms21165686.
- Pennacchio LA, Bickmore W, Dean A, Nobrega MA, Bejerano G (2013) ***Enhancers: five essential questions***. *Nat Rev Genet*, 14(4), 288–95. doi: 10.1038/nrg3458. PMID: 23503198; PMCID: PMC4445073.
- Shokolenko I, Alexeyev M (2022) ***Mitochondrial DNA: consensuses and controversies***. *DNA*, 2, 131–48. https://doi.org/10.3390/dna2020010

- Stewart JB, Chinnery PF (2021) ***Extreme heterogeneity of human mitochondrial DNA from organelles to populations***. *Nat Rev Genet*, 22, 106–118. https://doi.org/10.1038/s41576-020-00284-x
- Torres AG (2019) ***Enjoy the silence: nearly half of human tRNA genes are silent***. *Bioinform Biol Insights*, 8(13), 1177932219868454. doi: 10.1177/1177932219868454. PMID: 31447549; PMCID: PMC6688141.
- Travers A, Muskhelishvili G (2015) ***DNA structure and function***. *The FEBS Journal*, 282(12), 2279–95. https://doi.org/10.1111/febs.13307
- Tsuge K, Shimamoto A (2022) ***Research on Werner Syndrome: trends from past to present and future prospects***. *Genes*, 13(10), 1802. https://doi.org/10.3390/genes13101802

3

Chromosomes in Health and Disease

Emanuela Volpi

Analysis of chromosomes at different stages of the cell cycle and different resolution levels can provide important clues on cellular genomes and their functioning, often supplying crucial information on the pathogenesis of medical conditions. Indeed, chromosomal abnormalities have long been invaluable biomarkers for diagnosis and prognosis of human diseases. This chapter will cover aspects of chromosome structure and function of relevance for understanding the genetic basis of human disease, along with examples of elements of modern chromosome research that have been translated into advancements in precision medicine.

Learning Objectives

After studying this chapter, you should confidently be able to:

- Explain the principles of chromosomes' structural organization
- Illustrate chromosomes' dynamics during the cell cycle
- Discuss the importance of chromosomes' structural integrity, stability, and correct functionality for cellular homeostasis
- Demonstrate how chromosomal numerical and structural abnormalities lead to disease
- Compare methods for the study of chromosomes at different resolution levels
- Illustrate how chromosomal analysis is applied to the diagnosis and prognosis of human disease.

3.1 Chromosome structure and function: how research has been filling the knowledge gaps

Eukaryotic **chromosomes** can be universally described as microscopically visible, macromolecular aggregates of DNA and proteins residing in the cell nucleus. In different organisms, chromosomes can vary in shape, size, and number. However, across the animal and plant world, they share numerous standard features, the most prominent of which is the mutability of their appearance during the cell life cycle.

In popular scientific imagery, chromosomes tend to be schematically portrayed as X-shaped figures consisting of two rods (the **chromatids**) held together by a central constriction (the **centromere**), an overly simplistic illustration of chromosomes' appearance during cell division. In reality, the structure of eukaryotic chromosomes is far from simple. Complex, dynamic modalities of assembly of long DNA molecules, in the order of hundreds of millions of base pairs per chromosome, together with hundreds of different types of proteins, underpin the very essence of chromosome structure and confer on chromosomes their mutable appearance during the cell cycle. The availability of increasingly sophisticated, high-throughput research and computational techniques has propelled the study of chromosome biology towards new, exciting levels of resolution and provided fascinating insights. However, our understanding of the full intricacies of chromosome structure remains imperfect.

chromosomes
Microscopically visible macromolecular aggregates of DNA and proteins residing in the cell nucleus.

chromatids
The two identical copies of a chromosome that has replicated in preparation for cell division.

centromere
Constricted region of a chromosome at which the two chromatids are joined.

3.1.1 Chromosomes in the nucleus: to see is to believe

Chromosomes were first identified as 'vectors of heredity' by Theodor Boveri in the late nineteenth century. Boveri, considered one of the founders of modern cytology, had the opportunity to observe and describe the regulated function of some enigmatic subcellular structures—which he initially named 'chromatic elements'—while studying cellular events that followed the fertilization of eggs in *Ascaris megalocephala*, a nematode that parasitizes the guts of horses. It is beyond any doubt that Boveri's early insights into chromosome segregation during cell division enabled him to draw the connection between nuclear content and heredity for the first time and laid the foundations of modern chromosome biology (reviewed by Maderspacher 2008). With Walter Sutton—an American geneticist who studied chromosomes in grasshoppers—Boveri is credited for the chromosome theory of heredity, according to which chromosomes' behaviour can explain Mendel's inheritance laws.

Because of the inherent limitations of staining techniques available in those early days of biology for the microscopic visualization of nuclear material (which we now call **chromatin**), chromosomes had been mistakenly deemed transitory structures that would disassemble and completely disperse between cell divisions. However, based on his observations that in the early divisions of *Ascaris,* chromosomal arrangements were preserved in daughter cells, Boveri was the first to formulate the principle of chromosomal continuity and individuality that proposes chromosomes to be independent entities that—albeit with changed appearance—continue to exist throughout the cell life cycle with their identity preserved across cell generations.

Undisputable evidence in support of Boveri's theories was finally provided by advances in DNA cloning and sequence analysis combined with fluorescence microscopy in the late twentieth century. Specifically, it was **chromosome painting**—a variation of the **fluorescence *in situ* hybridization (FISH)** technique—to enable the discovery of chromosome territories or

chromatin
A complex of DNA and proteins found in the nucleus of eukaryotic cells.

chromosome painting
A variant of FISH in which chromosome-specific DNA libraries can be used as probes to map chromosome territories within the nuclear space.

fluorescence *in situ* hybridization (FISH)
A combined molecular and cytological hybridization technique to quantify and evaluate genetic sequences within cellular specimens.

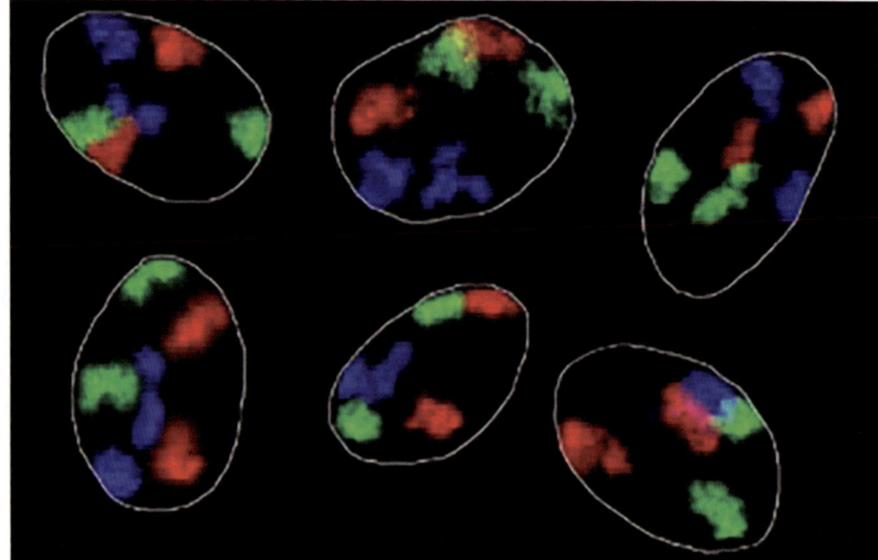

FIGURE 3.1
Chromosome territories by multicolour chromosome painting. Springer Nature.

well-defined, mutually exclusive volumes occupied by specific chromosomes within the cell nucleus for the first time. In chromosome painting, chromosome territories are mapped within the nuclear space by using chromosome-specific DNA libraries as probes for fluorescent hybridization and microscopy analysis. An example of chromosome painting applied to visualizing domains in the cell nucleus is provided in **Figure 3.1**. In the image, differential fluorescent staining shows how chromosomes retain spatial distinctiveness between cell divisions, albeit losing morphological definition by adopting a less compact configuration. The principles and essential procedural aspects of FISH are presented in **Box 3.1**.

Cross reference
Please refer to **Chapter 6** for an overview of different genomic techniques for diagnostic testing and screening.

Chromosome painting has enabled us to gather crucial evidence supporting the persistence of chromosomes as individual entities through the entire cell life cycle (reviewed by Cremer and Cremer 2001). Chromosome painting has also provided early, vital insights into the dynamic organization of the genome in 3D by evidencing changes in chromosomes' appearance and intranuclear positioning according to specific cellular circumstances. A seminal piece of research in this field from both the conceptual and technical perspectives was a study conducted by M-FISH (or 24-colour FISH) and confocal microscopy that led to the first 3D maps of all human chromosomes at once in single nuclei of different types of cells (Bolzer et al. 2005).

Before applying FISH and other immunofluorescence-based assays for nuclear studies, the nucleolus was the only example of a visually recognizable, defined, yet mutable nuclear landmark with a critical functional significance. The **nucleolus** can be defined as the site of ribosome synthesis, which occupies within the nucleus an area created by the physical convergence of ribosomal genes-harbouring chromosomes and their coordinated transcription. It is important to mention that the nucleolar aspect (number of nucleoli per nucleus, as well as size and shape) is an established biomarker in histopathology as nucleolar changes are usually suggestive of aberrant ribosomal synthesis, which is necessary to sustain accelerated proliferation and growth in cancer cells. In other words, the nucleolus provides a paradigm of how specific cytomorphological changes in nuclear and chromatin appearance can provide valuable diagnostic and prognostic insights reflective of altered genomic functionalities brought about by pathological changes in the cell.

nucleolus
The primary site of production and assembly of ribosomal units within the nucleus of a eukaryotic cell.

The growing availability of FISH probes—either single genes or large-scale probes composed of many different DNA sequences—and multiple fluorescent dyes for combined labelling or multiplexing has sparked extensive investigations into the three-dimensional organization of the genome within the nuclear volume. It has opened up a new era for chromosome biology

BOX 3.1 Fluorescence *in situ* Hybridization

Fluorescence *in situ* hybridization (FISH) is a combined molecular and cytological technique that can be applied to evaluate the numerical and structural integrity of genetic sequences on cellular specimens. For its ability to conceptually and experimentally link 'cell and sequence', the introduction of FISH many decades ago prompted the reinvention of **cytogenetics**, the branch of genetics concerned with the study of chromosomes, into **molecular cytogenetics**. As with all molecular hybridization techniques, FISH relies on the accuracy of base complementarity (adenine–thymine or A–T and cytosine–guanine or C–G) for DNA molecules or probes to recognize corresponding target sequences *in situ* (either on chromosomes or nuclei). Before the hybridization stage, probes and targets must undergo denaturation for the molecular pairing of complementary sequences. Denaturation—generally via exposure to high temperature—breaks up the hydrogen bonds that hold the DNA double helix together and readies the single strands or helices for recognition and annealing by complementary sequences. Before denaturation, the DNA probes are labelled with fluorophores, molecules that present with the property of absorbing light of a specific wavelength and emitting light of a different, longer wavelength. The probes are applied to fixed specimens on microscope slides, and the slides are placed to hybridize at 37 °C for a duration that varies according to the type of specimens and probes. The hybridization temperature is crucial to ensure the specificity and sensitivity of the assay. After the hybridization incubation, the excess probe is removed by washing the slides with phosphate buffer solutions in Coplin jars. Slides are then stained with 4′, 6-diamidino-2-phenylindole (or DAPI), a fluorescent stain with DNA affinity. DAPI confers a distinctive fluorescent blue colour to chromosomes and nuclei, providing a defined background for the smaller, sequence-specific FISH signals, which generally emit green and/or red light. See **Figure 3.2** for an overview of the steps in the FISH procedure.

The genetic sequences of interest that have been probed for, either single genes or more substantial targets up to the length of entire chromosomes, can be visualized by fluorescence microscopy. FISH probes can be designed to maximize the efficiency and accuracy of the hybridization. They can vary in sequence length and complexity according to the DNA target(s) under investigation, the biological question to be answered, or the specific disease biomarker to be identified. For instance, we have seen how the probes used in chromosome painting are chromosome-specific

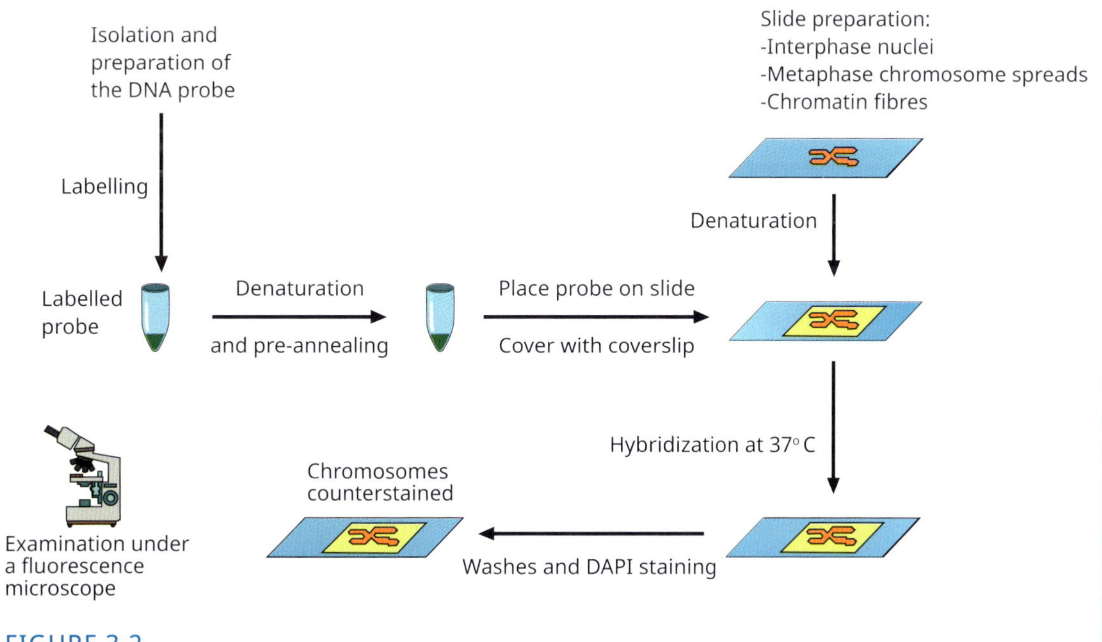

FIGURE 3.2
The critical steps of the fluorescence *in situ* hybridization (FISH) procedure.

genomicelibraries designed to cover the entire chromosomal length *in situ*. According to the size of the target DNA, FISH signals might vary in appearance from small, well-defined spots (e.g. single genes) to variably contoured, more significant fluorescent signals (e.g. entire chromosomes).

The resolving power of conventional FISH analysis, or its ability to resolve two small fluorescent signals as separate or anything smaller in size than the distance between the two distinct points, is constrained optically by the light microscopy Abbe diffraction limit (0.25 μm) and biologically by the intricacy of chromatin packaging. This inherent resolution limit makes FISH unsuitable to detect small targets or genomic changes such as single-nucleotide mutations or small deletions and duplications (<5–10 kbp). For this reason, in diagnostic laboratories, FISH has been slowly phased out in favour of alternative genomic techniques with higher resolving power, such as qPCR, chromosomal analysis by microarrays (CMA), and sequencing. However, for its ability to retain information on a cell-by-cell basis and its relative affordability, FISH remains highly popular worldwide as a 'single cell' genomic approach, instrumental in investigations of clonal evolution, karyotypic instability, and intercellular heterogeneity, often in connection with complex diagnostic and prognostic circumstances. FISH can also be adapted to detect RNA. The main adjustments to the protocol in RNA FISH are around cell fixation (RNA molecules are more prone to degradation than DNA) and omitting the nucleic acid target denaturation step (RNA is single-stranded). RNA FISH is the backbone of modern, single-molecule spatial transcriptomics technologies, which are currently powering the escalation of spatial biology as the methodological approach of choice to study tissues at the molecular level in health and disease (Gerber et al. 2023).

cytogenetics

A branch of genetics concerned with the study of chromosomes and chromosomal abnormalities.

molecular cytogenetics

The study of chromosomes and chromosomal abnormalities at the molecular level.

Cross reference

For more discussion on FISH, see **Chapter 6**, **Section 6.3.1**.

and nuclear architecture studies. Adaptations of the FISH protocol for better preserving and capturing aspects of three-dimensional genome organization within 'intact' cellular contexts include optical sectioning by confocal microscopy (3D FISH) (Solovei and Cremer 2010) and hybridization on super-thin tissue cryosections (CryoFISH) (Branco et al. 2008). Research has shown that chromosomes are not free to move substantially within the nuclear volume as their movements are restrained by interactions with the nuclear membrane and other nuclear structures, such as the nucleolus. However, the radial positioning of chromosomes and genes within the nuclear volume is non-random (Tanabe et al. 2002). For instance, human chromosomes with the highest gene density are preferentially located in the centre of the nucleus. In contrast, gene-poor chromosomes tend to occupy the nuclear periphery, and nuclear positioning can correlate with gene expression levels (Croft et al. 1999; Finlan et al. 2008). In some circumstances, genes have also been shown to modify their position within their respective chromosome territories in connection with transcriptional changes (Volpi et al. 2000).

However, establishing clear-cut associations between changes in nuclear architecture and altered genome functioning has proved challenging overall. Specifically, the potential existence of a causal link between nuclear organization and the regulation of gene expression, in particular the directionality of the possible cause–effect relationship, is still debated, the recurring question being: does structure influence function in the nuclear context or are structural changes merely the effect of functional changes in the nucleus?

The development of high-throughput approaches—such as hiFISH (Finn and Misteli 2021)—which combine multicolour combinatorial hybridization with automated image acquisition and analysis for the localization of hundreds of genomic loci in up to millions of cells, has added further layers of complexity by exposing the extent of intercellular heterogeneity or differences in between cells, even within the same cell population or tissue. There is no doubt that the generation of large datasets of intra-nuclear spatial distances will be crucial in revealing and providing statistical evidence supporting positioning patterns with functional significance.

Chromosome Conformation Capture (3C)

A method based on protein cross-linking that allows us to infer three-dimensional physical proximity between genomic loci within or across chromosomes.

More recently, a suite of molecular techniques collectively referred to as **chromosome conformation capture**, or **3C methods**, has enabled investigations of chromosome structure at the submicroscopic level and revealed details of the organization of the chromatin within the nuclear space at an unprecedented level (reviewed by Sati and Cavalli 2017). The 3C approach allows the quantification of interactions (inferred from three-dimensional physical proximity) between genomic loci or specific sites within one chromosome or across different chromosomes. A crucial initial step in all the chromosome conformation capture method variants is chromatin cross-linking by formaldehyde, the result of which is that chromatin domains that are physically close within the nuclear space in living cells become covalently linked. The cross-linking step is followed by DNA digestion with restriction enzymes and ligation of the ends of the cross-linked fragments. Ligation is followed by de-crosslinking and analysis of the new 'chimeric' DNA sequences, artificially generated by joining a linear sequence of spatially adjacent chromatin structures (**Figure 3.3**). The new chimeric DNA sequences are then amplified

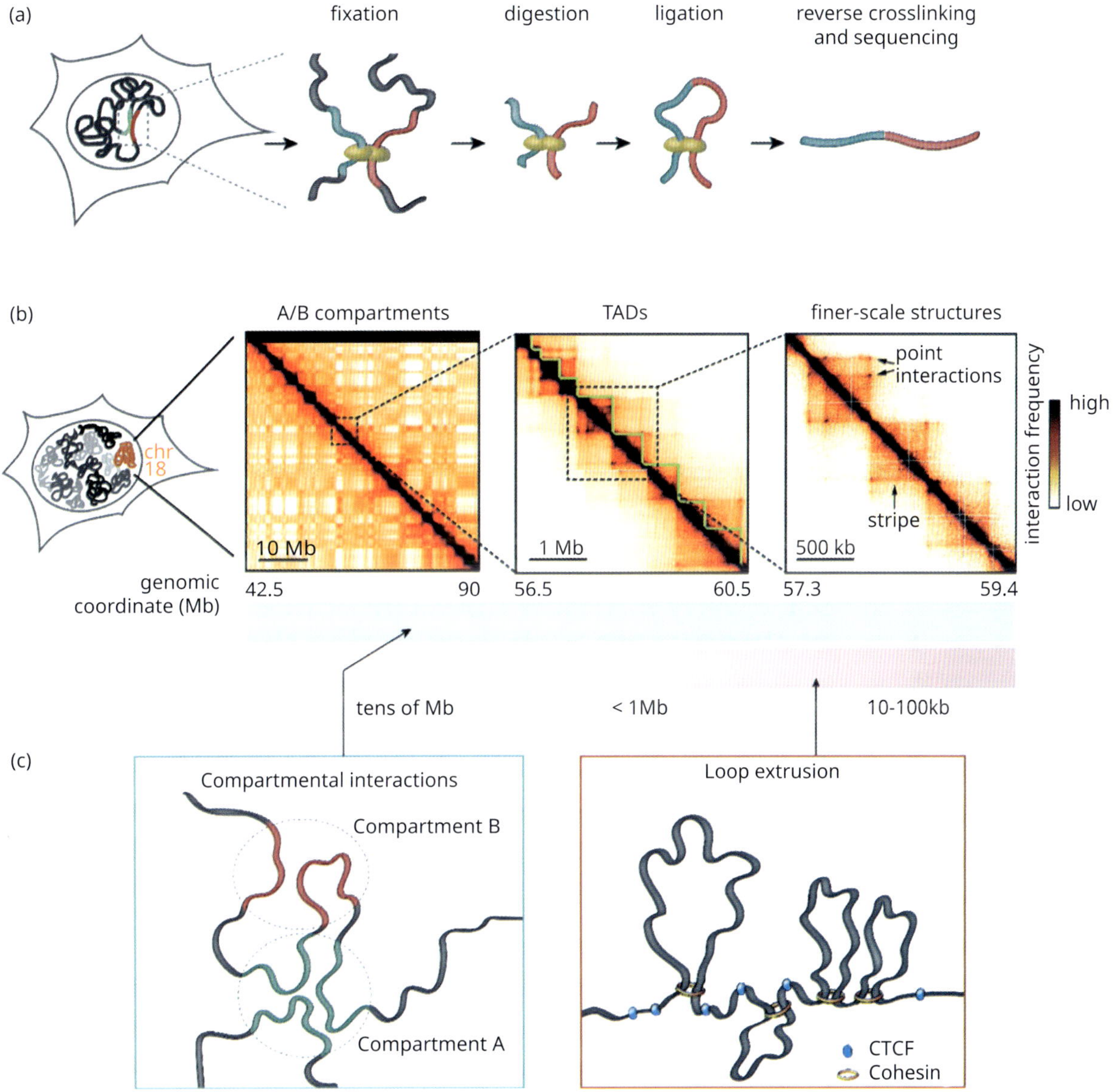

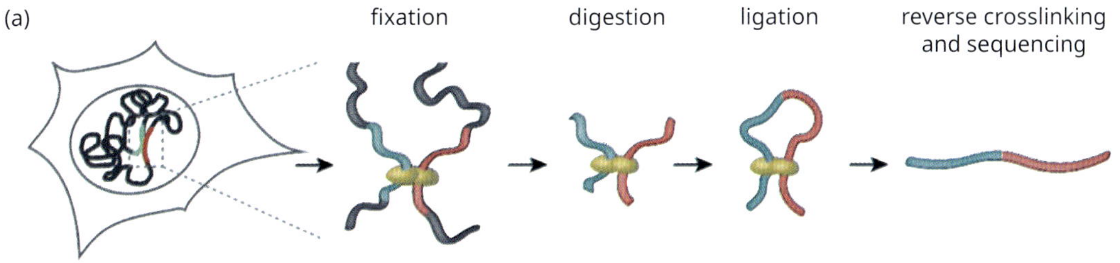

FIGURE 3.3

Chromosome conformation capture: (a) Core steps in 3C methods; (b) Hi-C contact map illustrating the folding of chromosomes into checkerboard-like A/B compartments; (c) Chromosome folding hierarchies and compartmental interactions.

topologically associating domains (TADs)
Chromosomal subdomains with well-defined and evolutionarily conserved boundaries characterized by local chromatin interactions.

Cross reference
Please refer to **Chapters 5** and **6** to learn more extensively about PCR, microarrays analysis, and whole genome sequencing.

by polymerase chain reaction (PCR) and analysed by microarrays or sequencing to identify interactions that are present more frequently than random events. A significant advantage provided by the 3C-based assays—mainly when combined with whole genome sequencing as in Hi-C—is their ability to generate high-throughput, genome-wide chromatin interactions maps based on precise interactions down to the 0.1 megabase scale. Interaction maps display checkboard patterns (**Figure 3.3**) that represent different types of chromatin domains: an active, open A-type of chromatin domain and an inactive, more closed B-type of chromatin domain. At <1 megabase scale, chromosome territories appear to be partitioned into **topologically associating domains** (**TADs**), subdomains with well-defined and evolutionarily conserved boundaries characterized by local chromatin interactions. TADs are often described as self-interactive as DNA sequences within a TAD interact with each other more frequently than with sequences outside the TAD. The domains are stable across cell types, and their boundaries are highly conserved across species, suggesting TADs to be an inherent property of mammalian genomes (Dixon et al. 2016). Although recent research findings suggest possible regulatory functions, understanding their full functional significance is still incomplete (reviewed by da Costa-Nunes and Nordermeer 2023).

Key Points

Chromosomes maintain their individuality during the entire cell cycle by occupying well-defined and mutually exclusive subnuclear volumes called 'chromosome territories'.

SELF-CHECK 3.1

Which modern cytomolecular techniques have progressed our understanding of chromosome organization within the cell nucleus?

3.1.2 Why and how does the appearance of chromosomes change during the cell cycle?

So far, we have learned about chromosome organization in three dimensions (the nuclear space). Now, we shall focus on chromosomes in four dimensions by following chromosomes' behaviour as the cells progress through their life cycle. Over the following few sections, we will learn about chromosomes as self-organizing entities that dynamically morph to support genomic functions necessary for cell cycle progression. The role of chromosomes is to ensure the accurate transmission of genetic information and its correct expression. Transformation in chromosomes' appearance reflects rounds of condensation and de-condensation driven by changes in metabolic requirements as cells prepare for, undergo, and complete cell division.

interphase
The interval phase between rounds of cell division during which the cell grows and replicates its DNA.

The still commonly used definition of **interphase** as the resting phase between cell divisions is somehow misleading; a legacy from when the complex interplay of form and function of macromolecular aggregates in the cell nucleus and their dynamic behaviour during the cell cycle had only been partially understood. Indeed, we now know that the interphase, or the constant interval between subsequent rounds of cell division, is by no means a period of cellular inoperativeness, but rather a phase of conspicuous molecular activity required ahead of cell division or M phase. Cell division occupies only a tiny part (~1 to 2 hours) of the entire cell cycle duration (~24 hours in cultured mammalian cells, but highly variable from one cell type to another *in vivo*). It will become apparent in the following sections how changes in chromosomes' appearance during the cell cycle—with chromosomes reaching maximum condensation during cell division while partly unravelling during interphase—are reflective of molecular and conformational changes happening at the chromatin level, which are required for accurate transmission and expression of genetic information.

Mitosis, or somatic cell division, is a type of cell division that ensures that a single cell gives rise to two daughter cells genetically identical to the parent cell (and, as expected, identical to each other). Mitosis is essential for tissue formation and maintenance, and it occurs in all embryonic tissues (growth/development) and at a lower rate in most adult tissues (replacement). DNA transcription, replication, and repair are critical metabolic activities that must be completed in interphase before mitosis.

mitosis
A process of equational division by which a single cell gives rise to two daughter cells genetically identical to the parent cell.

Interphase can be subdivided into three phases: G1, S, and G2. Transcription and translation of genes encoding for enzymes and structural proteins required by the cell to complete cell division happen in the G1 phase (the gap between mitosis and S phase) and the G2 phase (the gap between S phase and mitosis). DNA replication (the precise duplication of the entire nuclear DNA content ahead of cell division) and much of the DNA repair happens in the S phase. Most proteins synthesized in G1 are enzymes required for DNA replication and repair, while proteins synthesized in G2 are necessary for chromosome condensation ahead of the M phase. Cells spend most of their life in G1 and will enter S only if committed to mitosis (generally in response to growth stimuli), while non-dividing cells will remain in a modified G1 phase called G0.

DNA damage is a constant problem that cells must deal with to maintain viability. DNA damage can come from endogenous sources (e.g., reactive oxygen species and errors during replication) or exogenous forces (ionizing and UV radiations). Maintenance of a genome requires DNA repair. Mitotic accuracy is ensured by checkpoints in the cell cycle, which rely on complex molecular interactions mediated by cyclins and cyclin-dependent kinases (CDK). Checkpoints ensure orderly progression through the cell cycle, whereby one process must be completed before the following process is initiated. In case of unrepaired DNA damage, the G1/S and G2/M DNA damage checkpoints prevent cell-cycle progression into S-phase and M-phase, respectively. **Figure 3.4** provides a graphical representation of the cell cycle.

Key Point

The cell cycle is a highly regulated process comprising four phases: G1, S, G2 (collectively known as interphase), and M.

SELF-CHECK 3.2

What are the critical cellular activities that happen during each of the different phases of the cell cycle?

To understand the biological relevance of changes in chromosome appearance during the cell cycle, we must start by considering the logistical difficulties that eukaryotic cells encounter in having to package billions of base pairs of genomic DNA within the small volume of the nucleus, a membrane-bound organelle with a diameter of only a few micrometres. As an answer to that problem, evolution has conferred on DNA the property of folding and coiling its long molecules within the nuclear space through association with proteins to form chromatin fibres of increasing complexity and thickness.

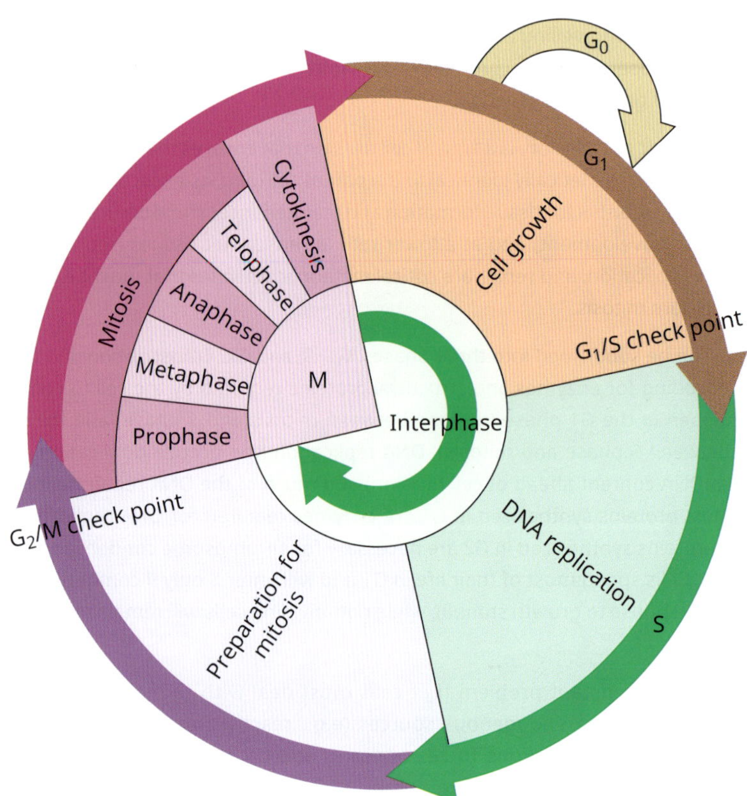

FIGURE 3.4
A diagrammatic representation of the different phases of the cell cycle.

nucleosome
Basic unit of DNA packaging in eukaryotic chromosomes comprising a core of eight histone proteins around which a DNA segment of 147 bp is wrapped.

histone proteins
Positively charged proteins, rich in arginine and lysine, with a critical role in chromatin assembly and compaction.

The fundamental unit of DNA packaging is the **nucleosome**, consisting of ~150 base pairs of double-stranded DNA wrapped around an octameric core of four highly conserved, positively charged **histone proteins** (H2A, H2B, H3, and H4). Adjacent nucleosomes are connected by short stretches of linker DNA, the length of which is usually 20–70 bp but which varies between different genome regions. In addition to the four core histones, a fifth histone (H1) binds to the linker DNA and has a role in chromatin condensation. Electron microscopy images of nucleosomes spaced along the chromatin filament suggest their likeness to beads on a string (**Figure 3.5**). The use of electrons (rather than photons) as a source of illumination confers electron microscopy a higher resolving power (0.1 nm), which enables the visualization of subcellular features, such as the 11 nm in diameter nucleosomes filament, which would not be resolvable by conventional light microscopy, the resolution limit of which we have covered before in **Box 3.1**.

This first level of DNA packaging is the only one that allows transcriptional activity. During periods of inactivity, the elementary, 11 nm in diameter nucleosomal fibre spirals into a 30 nm thick solenoid, also known as the 30 nm chromatin fibre, which is further coiled and looped into increasingly dense structures supported by a scaffold of non-histone proteins. The chromatin reaches maximum compaction when chromosomes are ready to undergo mitosis (**Figure 3.6**). Post-translational modification of the histones, principally the reversible acetylation of the N-terminal histone tails, but also methylation and phosphorylation, contribute to mitotic chromatin compaction.

Cross reference
Please refer to **Chapter 2, Section 2.1** for more details on epigenetic modifications of the chromatin.

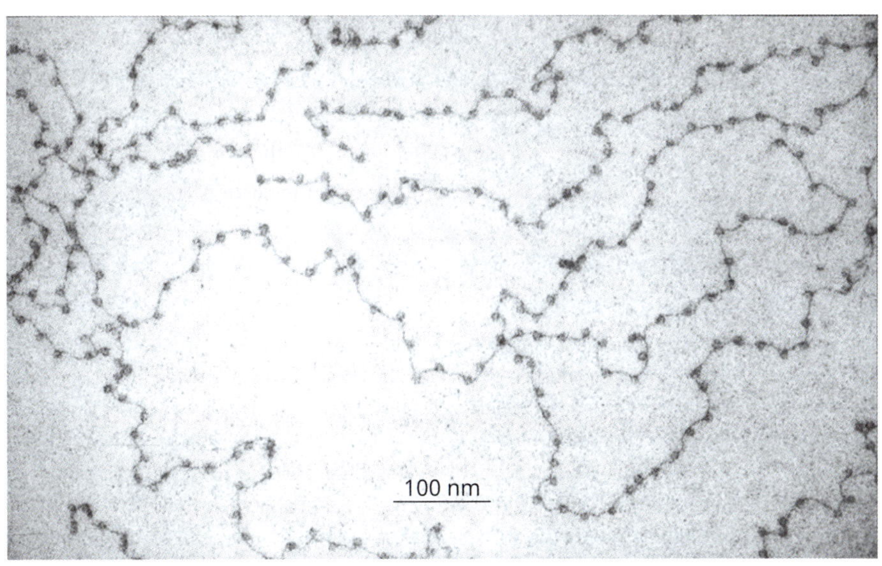

FIGURE 3.5
Nucleosomes as beads on a string. ORNL History.

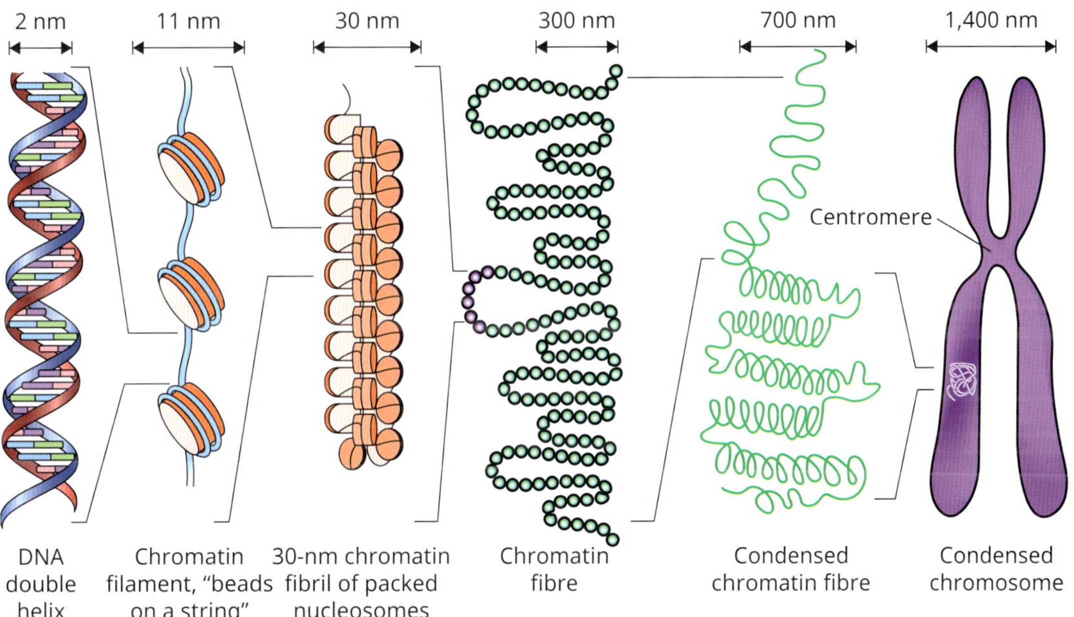

FIGURE 3.6
Progressive levels of DNA packaging in eukaryotic chromosomes.

Key Points

The persistence of chromosomes as individual entities during the cell cycle, with maximum condensation reached at mitosis, ensures efficient storage and faithful segregation of the genetic information.

SELF-CHECK 3.3

How is the orderly packaging of billions of DNA base pairs into a subcellular organelle of a few micrometres in diameter, such as the nucleus, accomplished?

Extreme chromatin compaction at mitosis ensures the effective parting of the parental cell genome into two genetically equivalent daughter cells. The following section describes how effective parting is achieved as the cell progresses through the four phases of mitosis: prophase, metaphase, anaphase, and telophase.

Following replication in the S phase, each chromosome consists of two molecularly identical sister chromatids held together by specialized proteins called **cohesins** (reviewed by Brooker and Berkowitz 2014). Initially, chromatids are held together by cohesins along their entire length. As the cell enters mitosis, the nuclear membrane breaks down. By now, chromatids are only held together at the centromere. As the cell moves from the first mitotic stage or prophase to the next stage of mitosis or metaphase, chromosomes reach maximum compaction and line up in the centre of the cell (the 'equatorial plane'). During the next stage of mitosis or anaphase, cohesion at the centromere is enzymatically cleaved by the anaphase protein complex (APC), and the mitotic spindle fibres, polymers of tubulin or microtubules, pull the sister chromatids apart. Microtubules extend from the **centrioles** located at the opposite poles of the cell and attach to the **kinetochore**, a macromolecular protein complex located at the surface of the centromere of each chromosome.

cohesins
Specialized proteins that hold chromatids together after DNA replication in the S phase.

centrioles
Cellular organelles made of microtubules involved in spindle formation and cell division.

kinetochore
A large protein structure that connects chromosomes to the spindle fibres.

Centromeres present with a variant of histone H3, known in humans as CENP-A, which is essential for attachment to spindle microtubules. A surveillance mechanism known as 'spindle assembly checkpoint' operates to prevent anaphase onset if centromeres are not correctly attached to the spindle microtubules.

Once the chromatids have separated at anaphase, they become individual, fully fledged chromosomes. Following their migration toward the opposite poles of the cell, each of the two daughter cells will end up inheriting an exact copy of the chromosome set of the original cell. During the final stage of mitosis or telophase, the nuclear membrane will reform around the two daughter nuclei, and chromosomes will start to unwind. Division of the cytoplasm or **cytokinesis** will complete cell division.

cytokinesis
The physical process that divides the cytoplasm of a parental cell into two daughter cells.

Key Points

Mitosis or somatic cell division comprises four phases: prophase, metaphase, anaphase, and telophase, followed by cytokinesis.

SELF-CHECK 3.4

What are the key activities during each phase of the cell division that ensure the effective parting of the parental cell genome into the two daughter cells?

The central axis of each sister chromatid in mitosis is enriched in **condensins,** proteins that confer stiffness to the chromosome, allowing resistance to pulling forces exerted by the spindle, resilience to breakage, and mechanical stabilization (reviewed by Skibbens 2019). Condensins are not static but rather move along the chromatid axis, enabling via chromatin cross-linking a fluid process of outwards looping of chromatin fibres, which confers on the mitotic chromosome structure significant lateral compaction and a bottlebrush appearance, observable via electron microscopy (**Figure 3.7**). The overall condensation of the DNA in mitosis

condensins
Proteins that play a critical role in the structural and functional organization of chromosomes.

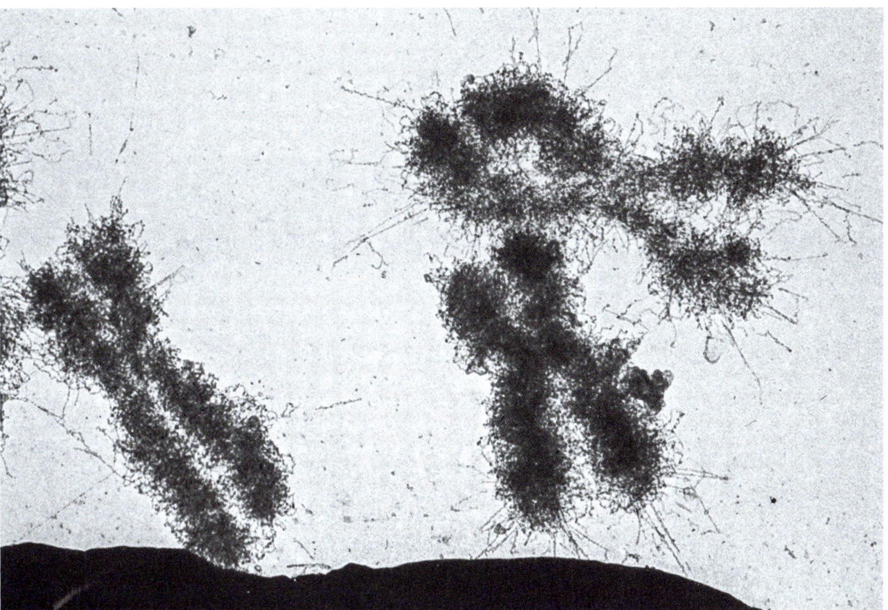

FIGURE 3.7
Transmission electron micrograph of human metaphase chromosomes. Magnification 14,000×.
Richard J. Green/Science Photo Library

reaches 10,000- to 20,000-fold (relative to the *naked* DNA molecule), and gene expression is homogeneously shut down.

Upon mitosis completion, condensins disassociate from chromatin. In interphase, cohesins replace condensins to re-establish looping patterns, also using cross-linking. However, when compared to mitosis, interphase loops created by cohesins are more minor, short-lived, regulated by boundaries imposed by the protein CTCF, and interspersed with gaps, conferring on chromatin in interphase a less constrained and more mobile state (reviewed by Spicer and Gerlich 2023). **Figure 3.8** presents a graphic representation of the difference in loop extrusion in interphase and metaphase and ensuing differences in chromatin stiffness.

Key Points

The persistence of chromosomes during the cell cycle, with maximum condensation reached at mitosis, ensures efficient storage and faithful segregation of the genetic information.

SELF-CHECK 3.5

What is the role of cohesins and condensins in chromosome structure?

During interphase, most of the chromatin is transcriptionally active (**euchromatin**). It exists in an extended state to facilitate access for key molecules (e.g., transcription factors) and to control regulatory interactions of distant DNA elements (e.g., enhancer–promoter interactions). Euchromatin is marked by weak binding of the histone H1 molecules and extensive acetylation of the histones in the nucleosomes. In contrast, transcriptionally silent chromatin (**heterochromatin**) retains a high level of compaction during the entire cell cycle.

euchromatin
Transcriptionally active chromatin.

heterochromatin
Transcriptionally silent chromatin.

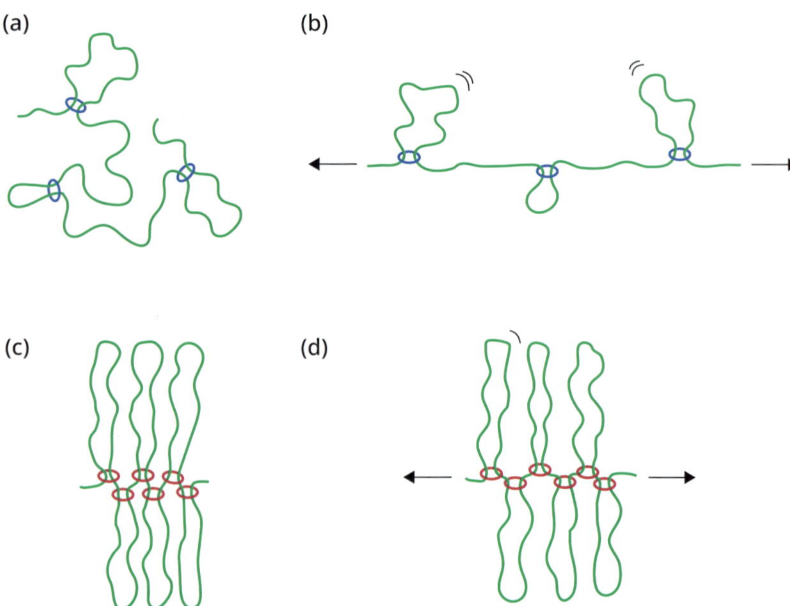

FIGURE 3.8

Chromatin stiffness is modulated by loop extrusion (a) In interphase, cohesin (represented by circles) forms loops that are interspersed with gaps. (b) Owing to gaps, interphase chromatin can be deformed by application of forces. (c) In mitosis, condensin (represented by circles) forms arrays of consecutive, larger loops without large gaps in between them. (d) Mitotic chromatin is stiffer than interphase chromatin in response to pulling along the mitotic chromosome axis. Arrows indicate the direction of pulling forces.

Heterochromatin is characterized by tight histone H1 binding and histone methylation (and its recognition by heterochromatin protein 1).

constitutive heterochromatin
Transcriptionally silent chromatin found at centromeres and telomeres.

telomeres
Structures made of DNA and proteins which protect the end of chromosomes.

alpha-satellite DNA
A type of non-coding DNA formed of multiple repetitions of short sequences and located at the centromere of all chromosomes.

facultative heterochromatin
Transcriptionally active or potentially active chromatin that becomes heterochromatic in a developmentally regulated manner.

Constitutive heterochromatin comprises repetitive DNA and is mainly found at chromosomes' centromeres and **telomeres**. In **Figure 3.9**, centromeric and telomeric sequences located at the primary constrictions and the extreme ends of the chromosome arms are evidenced via dual-colour FISH on metaphase chromosomes. Centromeres contain long, repetitive DNA sequences that can extend over several mega base pairs. The most common DNA component of all human centromeres is the **alpha-satellite DNA**, comprised of tandem repeats of a 171 bp monomer and marked by a recognition site for CenP-B, a highly conserved centromere binding protein with a role in chromatin maintenance. Telomeres are chromosome-protective ends consisting of thousands of tandem repetitions of the hexameric nucleotide sequence TTAGGG, which are necessary to maintain chromosome integrity. Long stretches of repetitive, non-coding DNA at the chromosome ends protect from the inevitable shortening of end sequences caused by DNA replication. Suppose a telomere is lost due to attrition (shortening due to repeated cycles of cell division) or chromosome breakage caused by unrepaired DNA damage. In that case, the chromosome ends fuse with other broken chromosomes, leading to chromosomal instability. We shall look into the causes and effects of chromosomal instability in connection to human disease later in the chapter.

Facultative heterochromatin is a type of chromatin that can appear to be either condensed (transcriptionally inactive) or decondensed (transcriptionally active). For example, in somatic cells of female mammals, a randomly chosen X chromosome—either the paternal X or the maternal X—is permanently inactivated and acquires a condensed, distinguishable appearance,

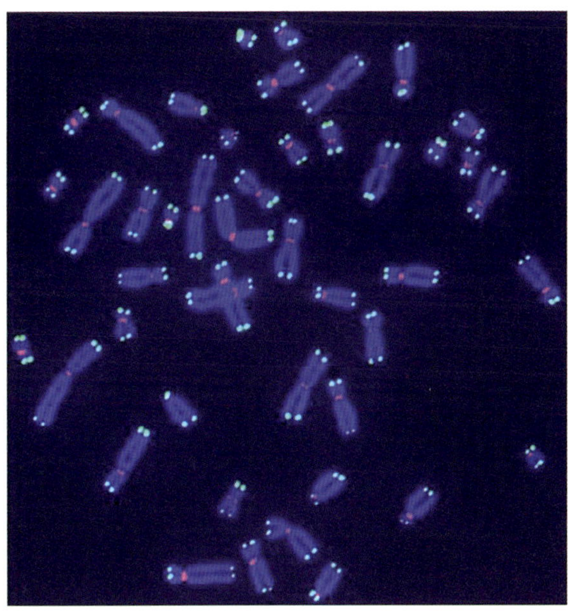

FIGURE 3.9
Centromeric and telomeric DNA sequences evidenced by dual-colour FISH on human metaphase chromosomes. Shay Lab UT Southwestern

Barr body
Condensed, inactivated X chromosome that appears as a densely stained small chromatin mass at the nuclear periphery when visualized by light microscopy in mammalian female somatic cells.

mosaicism
The presence in an individual of cell lines or clones derived from a single zygote that have evolved to contain different chromosomal complements.

identifiable with bright-field microscopy as an intensely stained body of highly compacted chromatin at the nuclear periphery, known as the **Barr body**. To learn more about X inactivation in female mammals, please refer to **Box 3.2.**

X-inactivation is initiated in the inactivation centre (XIC), a 1 Mbp region in the long arm of the X chromosome. Then, it spreads along the chromosome (reviewed by Augui et al. 2011). The X inactivation centre contains the *XIST* gene, transcribed only on the inactive X copy. The *XIST* gene encodes a 17 kbp spliced and poly-adenylated long non-coding RNA (lncRNA)

BOX 3.2 X-inactivation in female mammals

Dr Mary Lyon first formulated the theory of X-inactivation in mammalian early female embryonic development in the 1960s (reviewed by Blewitt 2024). The X and Y chromosomes present with significantly different size and gene content. The Y chromosome is small, primarily heterochromatic, and contains a limited number of genes (the fewest of any chromosome), mainly involved in male sex determination and development. The X chromosome is three times larger than the Y, accounts for almost 5% of the human genome, and contains many important protein-coding and RNA genes. Because of karyotypic differences concerning X chromosome copy number, it would be reasonable to expect males and females to differ in terms of X-linked gene products. Yet, this is different. Lyon proposed X-inactivation via the heterochromatinization of one X chromosome copy as the mechanism of gene dosage compensation that equalizes the amount of X-linked gene products in male and female cells. During early female development, in each somatic cell, one of the two X chromosomes is randomly chosen for inactivation. Once an X chromosome—either the maternal or the paternal copy of it—is inactivated in a cell, it will remain inactive in all descendants of that cell. As a result of X inactivation, a mammalian female body will be a **mosaic** composed of distinct cell clones, all retaining the pattern of X inactivation of the progenitor cell, so some with an active X of paternal origin and others with an active X of maternal origin. Calico cats (**Figure 3.10**) provide the most striking (as in phenotypically evident) example of **mosaicism** caused by X-inactivation. Heterozygosity at an X-linked coat colour locus confers these almost invariably female cats their tortoiseshell appearance. Different colour patches reflect clones in which either one allele (orange) or the other allele (black) is inactivated (white patches result from an unrelated gene).

FIGURE 3.10
The tortoiseshell appearance of the Calico cat's fur coat is a visually detectable example of mosaicism caused by X-inactivation.

that remains in the nucleus and coats the chromosome. The RNA coating recruits polycomb proteins that inhibit transcription, leading to heterochromatinization or condensation of the inactive X chromosome. X-inactivation is an example of an **epigenetic** effect whereby altered gene expression is not caused by a change in the DNA sequence but by changes in the chromatin environment, which impact the DNA properties and its transcription.

epigenetic
A change in gene expression caused by chromatin modification, and not by a change in DNA sequence.

Key Points

Based on transcriptional activity, chromatin can be classified as either euchromatin (active) or heterochromatin (inactive).

SELF-CHECK 3.6

Providing specific examples, explain the difference between the two types of heterochromatin: constitutive and facultative.

3.1.3 Chromosomes and sexual reproduction

meiosis
A process of reductional division by which haploid gametes (sperms and eggs) are formed from precursor diploid cells.

In the previous section, we learned how in mitosis—the process through which two identical daughter cells are formed from a single diploid cell—chromosome behaviour facilitates the accurate preservation and transmission of the genetic material from one cell generation to another. We shall now be looking at chromosome dynamics during **meiosis**, a specialized type of cell division that has great evolutionary significance as it is vital for sexual reproduction and also as a source of genetic variation.

Before learning about meiosis, acquiring some basic terminology around ploidy is crucial. Ploidy is the number of complete sets of chromosomes in a cell. **Haploid** cells contain only one set of chromosomes, which in humans is equivalent to twenty-three chromosomes (n=23). **Diploid** cells have two complete sets of chromosomes, so two copies of each chromosome (2n=46). Chromosome numbers vary across species. However, the same terminology applies. For example, the haploid genome of the mouse (*Mus musculus*) comprises twenty chromosomes (n=20), while its diploid genome comprises forty chromosomes (2n=40). In diploid cells, chromosomes within pairs are called **homologous**. Homologous chromosomes are chromosomes that are identical to each other. A complete set of human chromosomes or a *normal* human **karyotype** consists of twenty-two pairs of **autosomes** or homologous pairs (pairs of identical chromosomes) and one pair of **sex chromosomes**, XX in female and XY in male cells. To learn more about chromosome number, size, and DNA content, please refer to **Box 3.3**.

Meiosis is the process by which haploid gametes (sperms and eggs) are formed from precursor diploid cells (oogonia in females and spermatogonia in males) during sexual life cycles in eukaryotes. The origin of meiosis remains an enigma, as identifying the selective scenario that

haploid
Refers to the presence of a single set of chromosomes.

diploid
Refers to the presence of a double set of chromosomes.

karyotype
A description of the whole set of chromosomes of an individual that provides genome-wide information on chromosomal

autosomes
Any chromosome that is not a sex chromosome.

sex chromosomes
Chromosomes involved in sex determination.

BOX 3.3 Some facts and numbers to prop up chromosome knowledge

It has recently been confirmed that there are, on average, 36 trillion somatic cells in the male human body and 28 trillion in the female body (Hatton et al. 2023).

All cells derive from an initial diploid zygote—ensuing from the fertilization of a haploid egg by a haploid sperm—after multiple rounds of cell divisions (estimated number of mitotic divisions over a lifetime equal to 10^{17}).

The DNA content (C) of each haploid set of human chromosomes (n=23) is equivalent in weight to 3.5 pg (3.5×10^{-12} g). Accordingly, a diploid cell's DNA content (2C) (2n=46) will be 7 pg. After undergoing DNA replication in the S-phase before entering mitosis, the total DNA content of a human somatic diploid cell will double up to 4C before returning to 2C in each of the two genetically identical daughter cells after cell division is completed. The number of chromosomes (2n=46) remains constant during mitosis. However, after replication, each chromosome comprises two identical chromatids joined at the centromere. Once the chromatids part, they become fully fledged chromosomes. In meiosis, as in mitosis, the total content of a diploid cell will double up to 4C ahead of cell division. However, as in meiosis one round of replication is followed by two cycles of cell division, each of the four daughter cells will inherit a haploid set of chromosomes and half the DNA content of the original cell. That is why meiosis is called reductional cell division, while mitosis is called equational cell division.

In terms of DNA length packaged in the nucleus, each diploid cell contains 6×10^9 DNA base pairs, which, given the single base pair size of 0.34 nm in length, translates into an equivalent of 2 metres of DNA per diploid cell. Accordingly, it has been estimated that the human body contains over 60 trillion metres of DNA, enough to go around the Earth's equator over 1.5 million times.

Human chromosomes vary in size, gene content, and shape. An average-sized human chromosome comprises 140 Mbp (140 million base pairs) of DNA. Chromosome 1 is the largest in the human karyotype, both in DNA length (8.5 cm) and content (248 Mbp), while the Y chromosome is the smallest (2 cm, 62 Mbp). The human genome contains around 26,000 genes, but gene loci are not distributed proportionately on chromosomes. For example, chromosomes of similar sizes (such as chromosomes 21 and 22) may contain a significantly different number of genes and be classified as gene-rich, like chromosome 22 with its 448 genes, or gene-poor, like chromosome 21 with its 234 genes. As we shall see in the following sections, chromosomal gene content impacts the severity of chromosomal abnormalities.

endoreduplication
Replication of the nuclear genome without subsequent cell division.

syngamy
The fusion of two haploid cells.

led to its early evolution is complex (Lenormand et al. 2016). It has been postulated that a form of 'proto-meiosis' might have originated in early asexual unicellular eukaryotes, either to correct the accidental occurrence of diploidy as a result of **endoreduplication** (replication of the nuclear genome without subsequent cell division) or in response to **syngamy** or the fusion of two haploid cells.

Meiosis is characterized by one round of DNA replication followed by not one but two successive rounds of cell division (M1 and M2), as shown in **Figure 3.11**. This means that the replicated genome content will be ultimately shared not by two but by four daughter cells with obvious consequences regarding the final DNA content/ploidy in each daughter cell. For this reason, meiosis is referred to as reductional cell division. Similarly to mitosis, each meiotic cycle comprises four phases: prophase, metaphase, anaphase, and telophase. Ploidy reduction (from diploidy or 46 chromosomes to haploidy or 23 chromosomes) is a process that initiates with the pairing of homologous chromosomes during the prophase stage in the first meiotic cycle or M1. This pairing process is called synapsis. The perfect alignment of the chromosome pair due to synapsis is a prerequisite for the efficient parting and migration of each homologue toward the opposite spindle pole during anaphase. Please note that it is whole chromosomes with intact centromere parting during anaphase in meiosis I and not chromatids, as in mitosis. As a result, the two daughter cells ensuing from M1 are haploid.

Meiosis I is immediately followed by a round of equational cell division in meiosis II. In M2, the metaphase and anaphase stages resemble the corresponding phases in mitosis, with chromosomes aligning on the equatorial plane of the cell and, following the splitting of the centromere, chromatids migrating towards the opposite poles. After telophase II, when the chromatids—now fully fledged chromosomes—have reached the opposite poles and start to unwind, the nuclear membranes will begin to reform, and cytokinesis will occur. The result of meiosis in males will be four functional haploid daughter cells, each with the same amount of cytoplasm and half the number of chromosomes of the originating cell. Nearly all of the cytoplasm in females goes into one haploid daughter cell, which will then form the egg. At the same time, the other three cells become polar bodies, small, non-functional cells that eventually degenerate. Like male gametes, female gametes contain half the number of chromosomes as the original cell. Following fertilization of an egg by a sperm, a diploid number of chromosomes will be re-established in the zygote.

Key Points

Meiosis is the process by which haploid gametes are formed from precursor diploid cells during sexual life cycles in eukaryotes.

SELF-CHECK 3.7

How is ploidy reduction achieved in meiosis?

independent assortment
Refers to the random alignment and parting of homologous chromosomes of maternal and parental origin in meiosis I.

recombination
Refers to the physical swapping of genetic material that can happen between chromosomes during meiosis.

From an evolutionary perspective, the importance of meiosis as a genomic process is not limited to the attainment of ploidy reduction necessary for sexual reproduction but crucially around meiosis as a source of genetic variation through **independent assortment** and **recombination**. Independent assortment refers to the random alignment and parting of homologous chromosomes of maternal and paternal origin in meiosis I. Chromosomes from homologous pairs segregate independently so that each of the resultant daughter cells will inherit a haploid set of chromosomes consisting of a random mixture of chromosomes of paternal and maternal

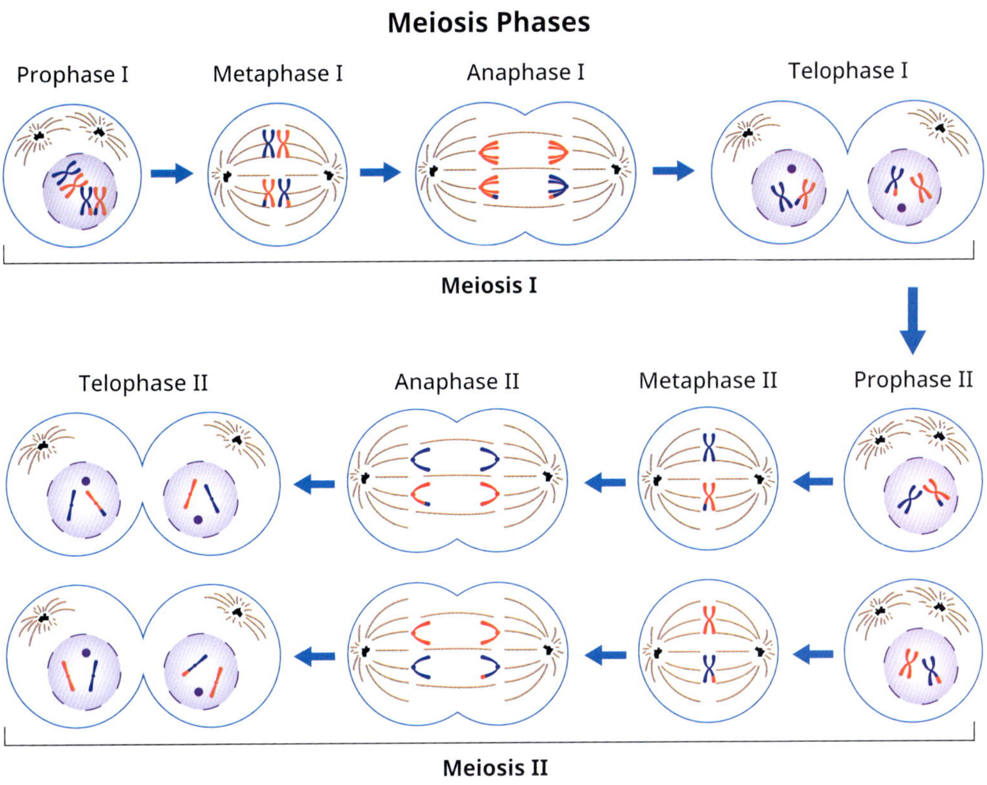

Meiosis Phases

Prophase I → Metaphase I → Anaphase I → Telophase I

Meiosis I

Telophase II ← Anaphase II ← Metaphase II ← Prophase II

Meiosis II

FIGURE 3.11
During meiosis, four daughter nuclei are formed from one parent nucleus after two stages of nuclear division.

origin. Recombination refers to the physical swapping of genetic material that can happen during late prophase I. At this stage, the chromatids of the two homologous chromosomes are entwined, and the chromosome pair acquires a distinctive configuration called either **bivalent** (or **tetrad**). The exchange of genetic material between homologous chromosomes or **crossing over** happens at points marked by cross-shaped structures called **chiasmata**. Each chiasma (singular of chiasmata) is visual proof of the ongoing switching of genetic material at that specific chromosomal locus at that particular point in time. In summary, using a random assortment of chromosomes of maternal and paternal origin and through physical swapping of genetic material between homologous chromosomes, meiosis produces new allelic combinations and, in so doing, it creates genetic variation. Genotypic and phenotypic variation in natural populations provides an evolutionary advantage by conferring the ability to respond and adapt to changing environmental circumstances.

Understanding chromosome dynamics in meiosis and mitosis is relevant not only from an evolutionary perspective but also to appreciate how errors during cell division lead to chromosomal abnormalities and ultimately cause different types of pathologies. Gametes with the incorrect number of chromosomes caused by defects in meiosis or mitotic errors that happen early in the embryo's life can cause significant genetic disorders. In contrast, mitotic errors occurring at any point in one's lifetime can cause cancer. These aspects will be expanded in the following sections.

bivalent
Structure resulting from the pairing of homologous chromosomes during the first division of meiosis.

tetrad
Structure resulting from the pairing of homologous chromosomes during the first division of meiosis.

crossing over
The exchange of genetic material between homologous chromosomes.

chiasmata (singular **chiasma**)
Points of protracted contact between paired homologous chromosomes at which exchange of genetic material can happen during the first metaphase of meiosis.

Key Points

Meiosis generates genetic variation using independent assortment and recombination.

SELF-CHECK 3.8

What happens specifically during the processes of independent assortment and recombination, and what are the outcomes?

3.2 Chromosome analysis in clinical settings

3.2.1 Traditional cytogenetic analysis

Cytogenetic laboratories (or cytogenetic services within larger clinical genetics laboratories) are clinical laboratories that conduct investigations to detect chromosomal abnormalities in diseases or suspected conditions. Cytogenetic laboratories or services typically offer analysis and clinical interpretation of various prenatal, postnatal, haematological, and cancer samples using techniques including karyotyping and FISH.

According to the type of pathology under investigation, chromosomes can be investigated either in peripheral blood lymphocytes stimulated to divide in culture or in rapidly dividing cells that can grow in the laboratory without stimulation, such as cancer cells or bone marrow cells. Peripheral blood is used to examine the chromosomes of individuals with congenital anomalies and developmental impairment. At the same time, cancer and bone marrow cells will be analysed to identify somatic changes of relevance for cancer diagnosis and prognosis. Foetal cells derived from amniotic fluid (amniocytes) or chorionic villus biopsy can also be cultured successfully for chromosomal analysis as part of prenatal care.

Cross reference

Please refer to **Chapter 6** for a detailed description of laboratory procedures for chromosome analysis in clinical settings.

For traditional cytogenetic analysis, chromosomes must be harvested from dividing cells (Howe et al. 2014). A critical step in the chromosome harvesting procedure is the use of a 'spindle poison' such as colchicine or colcemid (a plant extract that inhibits tubulin polymerization) to be added to the dividing cells in culture to prevent spindle formation and halt chromosomes in metaphase when they are maximally condensed and gathered on the equatorial plane of the cell. Following chemical fixation and staining, digital microscopy enables us to image and analyse chromosomes. A digital image of a chromosome *spread* can be automatically converted into a karyotype with individual chromosomes ordered and displayed in a numbered sequence called a **karyogram** in which homologous chromosomes are paired and organized in descending order according to their size (and shape). Most chromosomes present with a short arm (or p arm) and a long arm (or q arm), separated by the primary constriction or centromere, the most distinctive feature in chromosome morphology. Conventionally, chromosomes in karyograms are oriented so that their short arms point upwards, and their long arms point downwards (**Figure 3.12**). Chromosomes can be classified as **metacentric**, **submetacentric**, and **acrocentric**, based on whether the centromere is positioned in the middle of the chromosome (with chromosomal arms of approximately equal length), displaced to one end of the chromosome (with chromosome arms of clearly different sizes) or near one end of the chromosome. It is worth remembering that, as mentioned before, the centromere is not merely a characteristic landmark of the chromosomal shape but a vital kinetochore component. This functional macromolecular structure is indispensable for

karyogram
A photograph of an entire set of chromosomes of a cell, ordered in pairs and in numbered sequence based on size and centromere position.

metacentric
A chromosome with chromosomal arms of approximately equal length.

submetacentric
A chromosome with chromosome arms of clearly different sizes.

acrocentric
A chromosome in which the centromere is located near one of the chromosome ends.

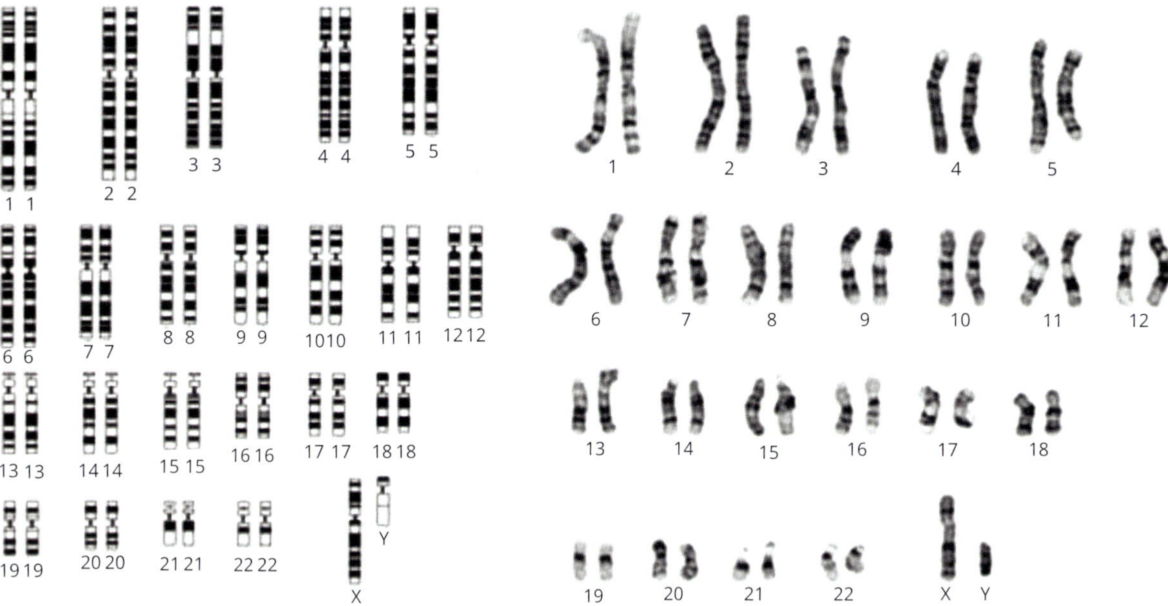

FIGURE 3.12

A human idiogram and a normal karyogram. Springer Nature

accurate chromosome segregation during cell division. Chromosomes provide an excellent example of the close, evolutionarily explicable link between form and function in biological systems.

Key Points

In clinical cytogenetic laboratories, chromosomes for analysis are harvested from dividing cells in culture.

SELF-CHECK 3.9

Based on which principles are chromosomes paired and ordered in a karyogram?

The etymology of the word *chromosomes* (from the Greek words for 'coloured bodies') refers to their property of being able to be stained. The most common type of stain for chromosome analysis is Giemsa, a mixture of methylene blue, eosin, and Azure B with nucleic acid affinity. Giemsa banding or **G-banding** for chromosomal analysis involves a light proteolytic treatment with trypsin before Giemsa staining, the result of which is a unique and reproducible pattern of alternating dark and pale chromatin bands, like a barcode, along the length of each chromosome. G-banding permits the distinction and pair-up of homologous chromosomes in karyograms. Crucially, it also enables the microscopic identification of chromosomal breakages and rearrangements. A standard G-banded karyotype will present with 400 to 550 bands per haploid set, allowing the identification of abnormalities along the length of the chromosomes and by extent anywhere in the genome at a resolution of 5 to 10 Mbp. For ease

G-banding
A technique that produces a longitudinal staining pattern along the length of the entire chromosome, which is unique for each chromosome pair.

idiogram
A diagrammatic representation of a karyotype.

of interpretation and representation, a karyogram can be accompanied by an **idiogram** or a diagrammatic chromosomal map with a clear indication and numbering of chromosomal bands for reference (**Figure 3.12**). In clinical cytogenetics, the attainment of a karyogram or the pairing of chromosomes in an orderly array is a fundamental early step in the diagnostic procedure as it allows a prima facie evaluation of genetic anomalies across the whole genome. It was indeed in this way that in 1959, the connection was made between Down syndrome as a clinical condition and the presence of an extra chromosome 21 for the first time (Antonarakis et al. 2020).

Being able to interpret a karyotype accurately and describe it unequivocally is paramount for both clinical diagnostic work and research. To do so, geneticists worldwide follow a standardized nomenclature (International Standard Cytogenetic Nomenclature or ISCN) regularly updated by a selected global expert committee. For example, the correct way to describe a normal male karyotype is 46, XY, while a normal female karyotype should be 46, XX. A female karyotype with Down syndrome will be 47, XX, +21, while a male karyotype will be 47, XY, +21.

Despite the push in high-income countries to escalate the adoption of next-generation sequencing as a one-stop diagnostic solution in healthcare settings, for its relatively low cost and infrastructure accessibility, karyotype analysis by G-banding remains part of the current portfolio of diagnostic techniques in most cytogenetics laboratories, all over the world. To learn more about the clinical indications for chromosome analysis, please refer to **Box 3.4**.

Key Points

G-banding is a Giemsa-based staining technique that generates a reproducible pattern of alternating dark and pale chromatin bands along the chromosome length.

SELF-CHECK 3.10

What is G-banding used for in cytogenetic analysis?

BOX 3.4 Clinical indications for chromosome analysis

Chromosome abnormalities occur in approximately 1 of every 150 live births and are responsible for a significant fraction of genetic diseases. As well as for routine diagnosis of specific conditions (e.g., chromosomal disorders), chromosome analysis can also be undertaken for explorative purposes in clinical situations such as high-risk pregnancies (e.g., increased maternal age or known or suspected family history of chromosomal abnormality), history of infertility/recurring miscarriages, for cases of stillbirth and neonatal birth, in children presenting with developmental delays and dysmorphic features, in children with intellectual disability, and to gather diagnostic, prognostic, and predictive insights in cancer. Regularly updated guidelines provide a framework of reference that enables consistency of practice and quality for genetic laboratories across Europe (Silva et al. 2019).

3.2.2 Disease diagnosis through unusual karyotypic findings

CASE STUDY 3.1 Swyer syndrome

A 30-year-old patient with a female external phenotype presented to the clinic with complaints of primary amenorrhea. There was no similar family history of infertility, amenorrhea, abnormal external genitalia development, or cryptorchidism. On physical examination, the patient presented with normal female external genitalia and normal breast development. However, the axillary and pubic hair were underdeveloped. After a preliminary, only partially informative ultrasound, the patient was further evaluated with an MRI of the pelvis that confirmed a hypoplastic uterus along with a dysplastic cystic left gonad with no evidence of any ovary or ovary-like structure/testes/testes-like structures in the abdomen. Hormonal assays further confirmed complete gonadal dysgenesis. The patient was referred for cytogenetic testing. The cytogenetic laboratory report showed a 46, XY karyotype, confirming the Swyer syndrome imaging diagnosis (**Figure 3.13**). The clinician advised the patient to undergo prophylactic gonadectomy of the left dysplastic gonad. Early diagnosis is crucial in these patients since prophylactic gonadectomy reduces the risk of developing malignancy. The pathogenesis of this syndrome includes mutations in or deletions of the *SRY* gene or sex-determining region located on the Y chromosome, which plays a crucial role in testicular differentiation.

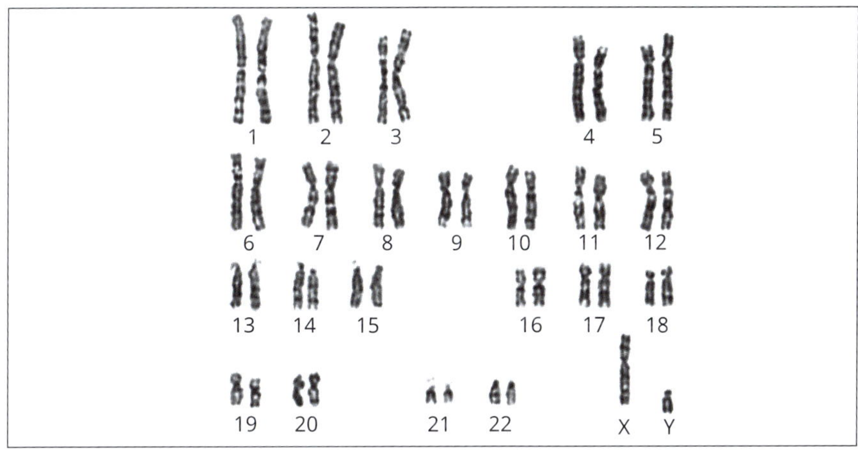

Technique: Standardized lymphocyte culture with GTG banding
Metaphase Analyzed: 25 Karyotyped: 10
Autosomes: 44 Sex Chromosomes: XY
Co-ordinates X/Y: 23.3/50.3 Estimated Band Resolution: 500
Karyotype ISCN 2020: **46,XY female**

Impression: **XY female**

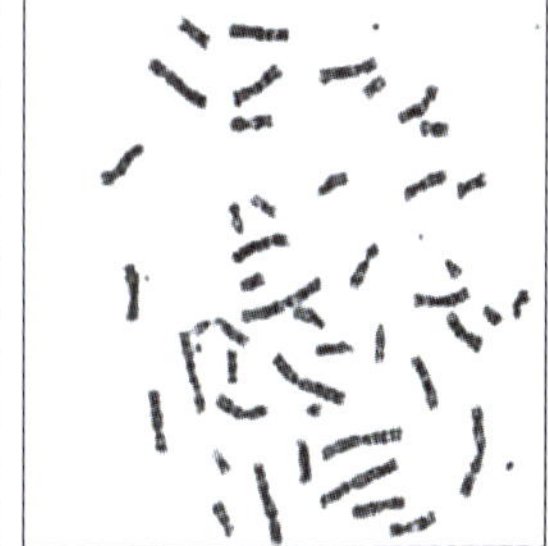

FIGURE 3.13

Cytogenetic analysis showing a 46,XY karyotype. Elsevier.

CASE STUDY 3.2 Immunodeficiency, centromeric region instability, and facial anomalies (ICF) syndrome

The patient was a 42-year-old British man of Indo-Caribbean origin with mild bird-like face abnormality but no other congenital malformations, cognitive impairment or developmental delays. From a very young age, the patient suffered from frequent infections. Investigations showed that the patient had hypogammaglobulinaemia (decreased IgG, IgA, and IgM) but with a normal lymphocyte count and normal proliferation of T cells to phytohaemagglutinin. From age 2, the patient was given 2-weekly intramuscular immunoglobulin injections, long courses of antibiotics, and regular chest physiotherapy. By the age of 7, he had developed significant bronchiectasis. Analysis of sweat electrolytes was carried out to see if the patient had cystic fibrosis, but this was ruled out. When the patient was in his early 30s, he had genetic tests to identify BTK gene mutations for X-linked agammaglobulinaemia, which was negative. Finally, in his late 30s, he was referred for cytogenetic testing. The karyotype analysis revealed chromosomal findings consistent with a diagnosis of ICF syndrome (**Figure 3.14**).

ICF syndrome is diagnosed by standard chromosomal analysis of peripheral blood cells during metaphase. Cells will have instability in the form of, for example, deletions, breaks, and elongation/decondensation of the region adjacent to the centromere in chromosomes 1, 16, or 9. Other findings include multi-branched chromosomes made up of duplicate arms of (usually) chromosomes 1 and 16. Most patients die at a young age, but this patient has survived into his 40s and might be the oldest known survivor of ICF syndrome. The patient is currently under the care of haematologists and respiratory physicians to manage immunodeficiency and bronchiectasis. He receives 6-weekly intravenous infusions of immunoglobulin (IVIG). The patient continues to have chest physiotherapy and frequent antibiotics.

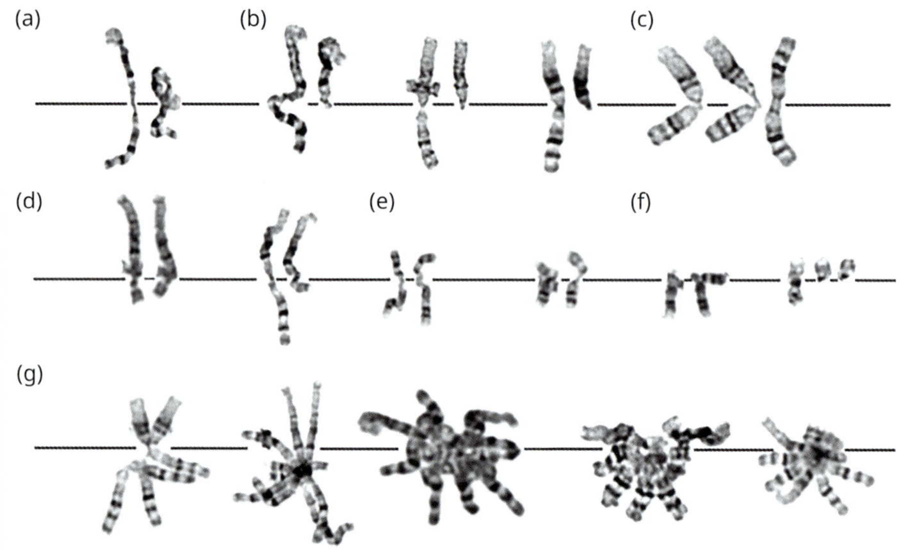

FIGURE 3.14

Chromosomal aberrations. (a) Decondensation of the pericentromeric region of chromosome 1. (b) Whole-arm deletion of the long arm of chromosome 1. (c) Isochromosome consisting of two copies of the long arm of chromosome 1 (third copy of chromosome 1). (d) Translocation between the short arm of chromosome 1 and the short arm of chromosome 16. (e) Decondensation of the pericentromeric region of chromosomes 9 and 16, respectively. (f) Whole-arm deletions of the short arm of chromosome 9 and the long arm of chromosome 16, respectively. (g) Multi-branched radials. Springer Nature

3.3 Chromosomes' integrity, stability, and dynamics are essential aspects of cellular fitness

The causative link between nucleotide-level mutations and other types of physically inconspicuous yet potentially detrimental genomic changes (e.g., deletions or duplications or insertion of a small number of bases) and well-known genetic disorders (e.g., cystic fibrosis and sickle cell disease) is well established. Similarly, the direct impact of large-scale changes to chromosomes' integrity, stability, and dynamics on the cell's or cellular homeostasis's physiological equilibrium and human health has been ascertained. Conventional examples include the link between chromosomal anomalies and infertility, the pervasiveness of karyotypic complexities and instabilities in cancer, and changes in chromosome number or **aneuploidies** as causes of chromosomal disorders such as Down Syndrome. This section will explore the fundamental role of chromosomes in providing physical context for the appropriate expression of genetic information and the consequent link between chromosomal anomalies and disease aetiology.

aneuploidies
Abnormal numbers of chromosomes in a cell that deviate from the haploid set or a multiple of the haploid set.

loss-of-function mutation
A type of genetic mutation that leads to the loss of essential functions.

gain-of-function mutation
A type of genetic mutation that leads to the acquisition of unnecessary functions.

structural chromosomal abnormalities
Microscopically visible changes to the content and/or order of DNA sequences on chromosomes.

deletions
Physical losses of DNA.

duplications
The presence of an extra copy of a specific genetic sequence.

inversions
A 180-degree rotation of the normal orientation of a DNA sequence caused by a double DNA break along the chromosome structure.

heteromorphisms
Mitotically stable, natural variations of chromosomal size and morphology within the general population.

genetic drift
A change in the frequency of a genetic variant due to random chance.

3.3.1 Different types of chromosomal abnormalities

Chromosomal abnormalities can be either of a structural or numerical type. Both categories of chromosomal changes can lead to either loss or gain of genetic material that, similarly to traditional, small-scale variants (e.g., single nucleotide changes), can cause detrimental imbalances in gene expression, ultimately leading to either loss of essential functions (**loss-of-function mutation**) or acquisition of unwarranted functions (**gain-of-function mutation**).

Structural chromosomal abnormalities—such as **deletions**, **duplications**, and **inversions**—are sizeable, microscopically visible changes to the content and/or order of DNA sequences on chromosomes. As for small-scale variants, the functional impact of structural abnormalities is dictated by the temporal and local transcriptional or regulatory significance of the specific chromosomal site(s) and gene loci contained, spanned by the chromosomal change. The larger the variant, the more likely a detrimental impact. However, similarly to other types of genetic polymorphisms (e.g., single nucleotide polymorphisms or SNPs), structural abnormalities involving transcriptionally inactive chromosomal regions—for instance, the constitutive heterochromatin adjacent to the centromeric regions of human chromosomes 1, 9, and 16—can persist as **heteromorphisms** within the population as an effect of **genetic drift**, and contribute to the genetic variation load.

Structural chromosome abnormalities are often the result of the incorrect joining of two broken chromosome ends. Deletions are physical losses of DNA, usually caused by imperfect DNA damage repair mechanisms following a double break on a chromosome and the incorrect joining of double-strand breaks. Duplications denote the presence of an extra copy of a specific genetic sequence and are generally caused by unequal crossing-over events that occur during meiosis between misaligned homologous chromosomes. Gains or losses of genetic material will likely cause severe pathological consequences. The Cri du Chat Syndrome or Cat Cry Syndrome is an example of a rare genetic disorder caused by a sizeable deletion on the short arm of chromosome 5. The deletion is denoted as 5p− or 5p minus. Babies with Cri du Chat have a high-pitched cry. Developmental delays and intellectual disability characterize the disorder. The disease can be diagnosed cytogenetically by karyotyping or FISH testing (**Figure 3.15**).

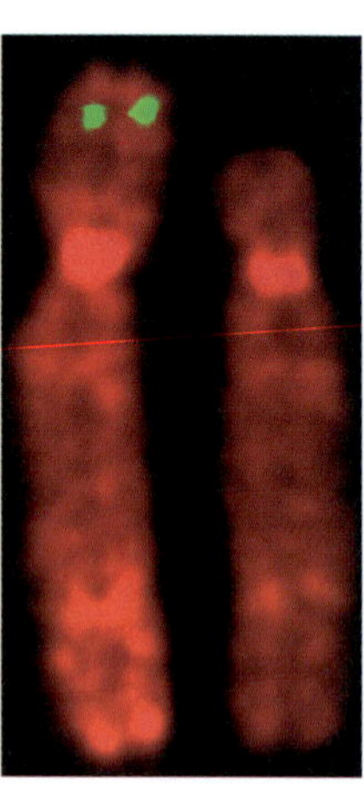

FIGURE 3.15
Cri du Chat chromosomes showing a deletion on 5p (G-banding on the left and FISH on the right). Science Photo Library

homogenously stained regions (hsr)
Sizeable structural anomalies that present as interstitial or terminal chromosomal segments not showing differential banding following Giemsa staining.

ring chromosomes (r)
Defective chromosomes with a circular shape ensuing from the fusion of their broken ends.

double minutes (dmin)
Small, circular fragments of extrachromosomal DNA, lacking centromeres which segregate randomly during cell division.

amplification
A process leading to the gain of multiple copies of a genetic sequence.

Cross reference
For a more extensive cover of cancer genetic biomarkers, please refer to **Chapter 7**.

paracentric
A chromosomal inversion that does not include the centromere.

Homogeneously stained regions (hsr), **ring chromosomes (r)**, and **double minutes (dmin)** are physical, microscopically visible chromosomal biomarkers which harbour multiple gene duplications, a phenomenon known as gene **amplification**, that leads to overexpression of genes providing a selective advantage for the progression of cancer. Homogeneously stained regions are sizeable structural anomalies that present as interstitial or terminal chromosomal segments not showing differential banding following Giemsa staining. Ring chromosomes are defective chromosomes with a circular shape ensuing from the fusion of their broken ends. They usually arise as a result of DNA damage. Double minutes are small, circular fragments of extrachromosomal DNA, lacking centromeres, which segregate randomly during cell division, generating unequal partitioning of the amplified DNA sequences and contributing to the genetic heterogeneity of cancer cells. The unequal distribution and frequency of these three distinctive biomarkers of neoplastic transformation and progression in different malignancies remains an open question (reviewed by Mandahl et al. 2024).

Inversions result from a 180-degree rotation of a sequence's normal orientation caused by a double DNA break along the chromosome structure. In some rare instances, although no genetic material might be lost or gained, the inversion process might still disrupt gene sequences with phenotypic consequences. Inversions can be referred to as **paracentric** when they do not include the centromere and **pericentric** when they involve the centromere.

Unlike deletions, duplications, and inversions, structural abnormalities can involve, at times, more than one chromosome and the exchange of material between chromosomes (**translocations**). Usually, translocations arise as reciprocal exchanges. If genetic material is neither lost nor gained in the translocation process, the translocation is called balanced and will likely be asymptomatic (no phenotypic effect). Indeed, balanced chromosomal rearrangements such as balanced translocations and balanced inversions are compatible with life and can be found as genetic variants in the population. A specific example of balanced rearrangements is **Robertsonian translocations**, which ensue from centromeric fusions of acrocentric chromosomes. A carrier of a Robertsonian translocation will have a numerically abnormal karyotype, with a total number of chromosomes equal to 45, but will be phenotypically normal.

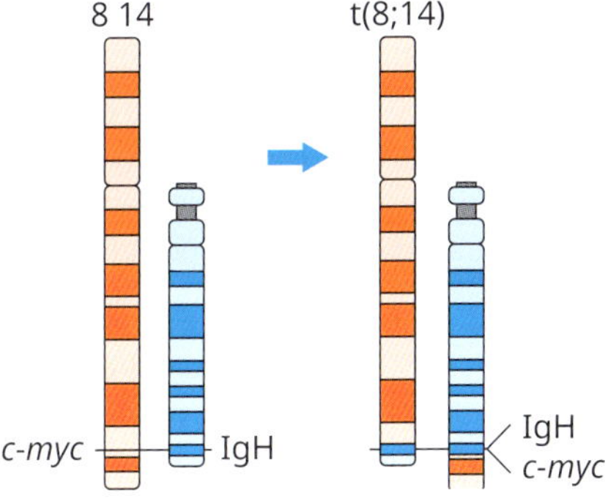

FIGURE 3.16
In Burkitt lymphoma, *Myc*, an oncogene normally found on chromosome 8 is translocated to chromosome 14.

However, as with the inversions, in some instances, even without gain or loss of genetic material, the 'balanced' translocation process can disrupt genetic sequences with phenotypic consequences. Burkitt lymphoma provides an example of how such a chromosomal occurrence—a balanced translocation—can lead to oncogene activation and adverse outcomes, even without DNA gain or loss. Burkitt lymphoma is an aggressive non-Hodgkin B-cell lymphoma associated with chromosomal translocations that cause over-expression of the oncogene *c-myc*. The most common chromosomal **biomarker** in Burkitt lymphoma, which occurs in 80% of cases, consists of a translocation between chromosome 8—the chromosome on which the *c-myc* oncogene is located—and chromosome 14. The written description of the translocation as t(8;14)(q24; q3.2) suggests chromosomal breaks happen in correspondence of band 2.4 on the q (long) arm of chromosome 8 and band 3.2 on the q (long) arm of chromosome 14 (**Figure 3.16**). As a result of the chromosomal swap, *c-myc* is relocated to chromosome 14 and juxtaposed to the promoter sequence of the immunoglobulin IgG heavy chain gene that leads to constitutive activation of the oncogene, disrupting its function in controlling cell growth and proliferation.

Numerical chromosomal abnormalities can be categorized as polyploidies and aneuploidies. The term **polyploidy** describes occurrences wherein rather than the expected number of chromosome sets (one complete set or n=23 in a haploid cell such as a sperm or an egg, and two complete sets or 2n=46 in a diploid somatic cell), a cell will contain one or more extra sets of chromosomes. For example, **triploidy** indicates the presence of an additional set of chromosomes (3n=69). Triploidy commonly arises from defective fertilization and is incompatible with life. **Tetraploidy** suggests the presence of two entire extra sets of chromosomes, 92 chromosomes (4n) in a cell.

Aneuploidy describes instances whereby a cell presents with a non-integral multiple of the haploid number of chromosomes, usually one more copy (**trisomy**) or one less copy (**monosomy**) of a specific chromosome. In addition to trisomy 21, there are only two other autosomal trisomies compatible with live birth, although presenting with distinctive, severe congenital malformations and developmental disabilities: Patau syndrome due to trisomy 13 and Edwards syndrome due to trisomy 18. Indeed, aneuploidies have been identified as recurrent in early spontaneous abortions. Other aneuploidies compatible with life involve sex chromosomes, such as Turner syndrome (45, X0), the only whole-chromosome monosomy compatible with life. Having an extra X or Y chromosome—as in Klinefelter syndrome (47, XXY), Triple X

pericentric
A chromosomal inversion that includes the centromere.

translocations
Structural abnormalities that involve the exchange of genetic material between chromosomes.

Robertsonian translocations
Translocations that ensue from centromeric fusions of acrocentric chromosomes.

biomarker
A measurable indicator of a disease state.

numerical chromosomal abnormalities
Changes to the expected number of chromosomes in a cell.

triploidy
An instance in which a cell presents with an additional set of chromosomes.

tetraploidy
An instance in which a cell has acquired two extra sets of chromosomes.

trisomy
An instance in which a cell presents with one more copy of a specific chromosome.

monosomy
An instance in which a cell presents with one less copy of a specific chromosome.

(47, XXX), and XYY (47, XYY)—is compatible with survival, though specific phenotypes occur as a consequence. The developmental abnormalities associated with monosomies and trisomies result from imbalances in the level of gene products encoded by genes on different chromosomes. Aneuploidies of sex chromosomes are better tolerated as the Y chromosome carries a relatively small number of genes, and the phenomenon of X-inactivation in female mammalian cells enables the control of the level of X-encoded gene transcription independently of the number of X chromosomes in the cell.

Aneuploidies usually arise from non-disjunction at anaphase or failure of paired chromosomes or sister chromatids to disjoin during cell division, either meiosis or mitosis. As a result of non-disjunction, two cells are produced, one with an extra copy of a chromosome (trisomy) and one with a missing copy of that chromosome (monosomy). Mitotic non-disjunction in a somatic cell might result in mosaicism or the presence in an individual of cell lines or clones derived from a single zygote that have evolved to contain different chromosomal complements. About 1% of patients with trisomy 21 or Down syndrome are mosaics with normal and trisomic cell lines. Trisomy 21 mosaicism often presents with a milder clinical expression of the phenotype.

Key Points

Chromosomal anomalies can be classified as numerical or structural.

SELF-CHECK 3.11

Can you provide examples of different chromosomal anomalies which can potentially lead to either loss or gain of function?

3.3.2 Large-scale mutational events in cancer (and other disorders)

According to the somatic mutation theory of cancer, malignancies arise due to the sequential accumulation of mutations in genes that control the cell cycle and cell proliferation (oncogenes and tumour suppressor genes). However, recent discoveries in oncogenomics around **chromoanagenesis**, enabled by high-resolution, high-throughput approaches, challenge the conventional view of cancer as a stepwise process (reviewed in Shah 2024). Chromoanagenesis is an all-encompassing word that describes catastrophic occurrences generating complex chromosomal rearrangements (reviewed in Shorokova et al. 2021). **Chromothripsis** is a chromoanagenetic phenomenon consisting of a single mutational event causing the localized shattering of one or more chromosomes into up to thousands of different fragments resulting from many different chromosomal breaks in clusters. **Figure 3.17** schematically shows how the DNA repair machinery's random, aberrant reconstitution of the fragmented chromosomes can lead to complex structural rearrangements and DNA losses and gains. Chromothripsis was initially described in chronic lymphocytic leukaemia and bone cancer, and it was originally believed to be a rare genetic phenomenon (Stephens et al. 2011). However, we now know that chromothripsis occurs in about 22% of cancers, with its incidence reaching up to 100% in malignant tumours of the peripheral nerve sheath and 70% of germ cell tumours. Chromothripsis has also been found in benign tumours and the genomes of non-cancer patients with severe congenital abnormalities. **Chromoplexy** is

chromoanagenesis
A catastrophic occurrence generating a complex chromosomal rearrangement.

chromothripsis
A single mutational event causing the localized shattering of one or more chromosomes.

chromoplexy
An event in which multiple chromosomes are broken and the DNA strands ligated in new configurations.

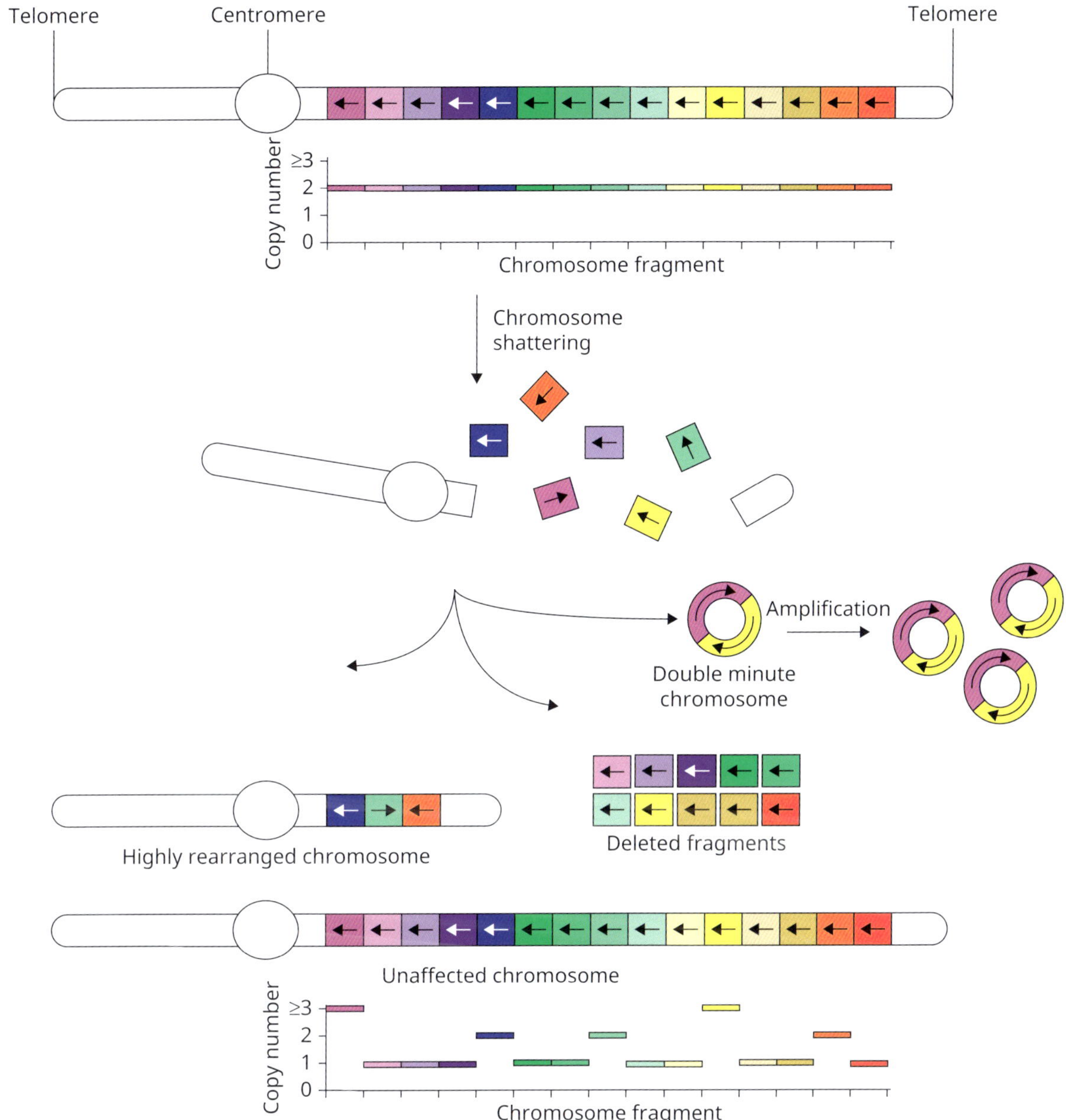

FIGURE 3.17
Schematic overview of chromothriptic chromosomal rearrangements.

a chromoanagenetic event in which multiple chromosomes are broken, and the DNA strands are ligated in new configurations, leading to gene disruptions near the chromosomal breakpoints. **Kataegis** is the most common large-scale mutational event, present in 61% of cancers. It can be described as a focal hypermutation process leading to a high rate of mutations in small chromosomal regions. The discovery of these large-scale, catastrophic genetic events generating alternative mutational routes to malignancy has significantly expanded our understanding of oncogenesis.

kataegis
A focal hypermutation process leading to a high rate of mutations in small chromosomal regions.

Key Points

The discovery of large-scale, catastrophic mutational events has challenged the conventional view of cancer as a stepwise process.

SELF-CHECK 3.12

Explain the mechanistic differences between the three main chromoanagenetic occurrences of chromothripsis, chromoplexy, and kataegis.

3.3.3 Chromosomal instability as a driver of malignant transformation

Chromosomal instability (CIN) is an abnormal cellular phenotype characterized by the acquisition of chromosome gains and losses at an increased rate, which has been observed in virtually every type of cancer. CIN ensues from segregation errors in mitosis and leads to intercellular genetic heterogeneity that can provide survival and adaptation advantages; in the same way as in response to environmental challenges, genetic variation fuels evolution by natural selection in populations of living organisms in nature. Genetic heterogeneity and genomic evolution in chromosomally unstable cancers are considered to be causing the development of resistance to cancer treatments. For all the reasons listed so far, it is not surprising that over the last two decades, CIN as a process (causes and mechanisms), as well as a status (outcomes such as aneuploidies and consequences for cellular homeostasis), has generated growing interest in the fields of cancer pathobiology and clinical oncology, particularly in connection to applications for cancer prognosis and response to treatment, and of course new therapeutic developments (Thompson et al. 2017). As a testimony to the growing scientific interest in this area, genomic instability (CIN is a particular subtype) was proclaimed an emerging 'enabler of cancer' in Hanahan and Weinberg's 2011 edition of the 'Hallmarks of Cancer' paper. This seminal publication in cancer research, based on a periodic, extensive review of published evidence, aims to refine and consolidate our understanding of cancer cells' critical capabilities. Fluorescence *in situ* hybridization (FISH), particularly its 24-colour variant called M-FISH or multiplex FISH, has been instrumental in identifying karyotypic differences at the single cell level within cancer cell populations, which have revealed the full extent of genetic heterogeneity within tumours. **Figure 3.18** provides a schematic representation of numerical and structural abnormalities found in chromosomally unstable cells. Despite the prevalence of CIN in cancer and its association with adverse prognosis, the use of CIN as a biomarker has been restricted so far to pathological classification (Dhital and Rodriguez-Bravo 2023). However, because of its implications for drug resistance, determining CIN levels and characteristics by cytomolecular techniques might become part of routine cancer diagnosis and clinical follow-up.

Key Points

Chromosomal instability is a heightened rate of acquisition of chromosomal abnormalities.

SELF-CHECK 3.13

Can you explain how chromosomal instability in cancer can lead to therapy resistance?

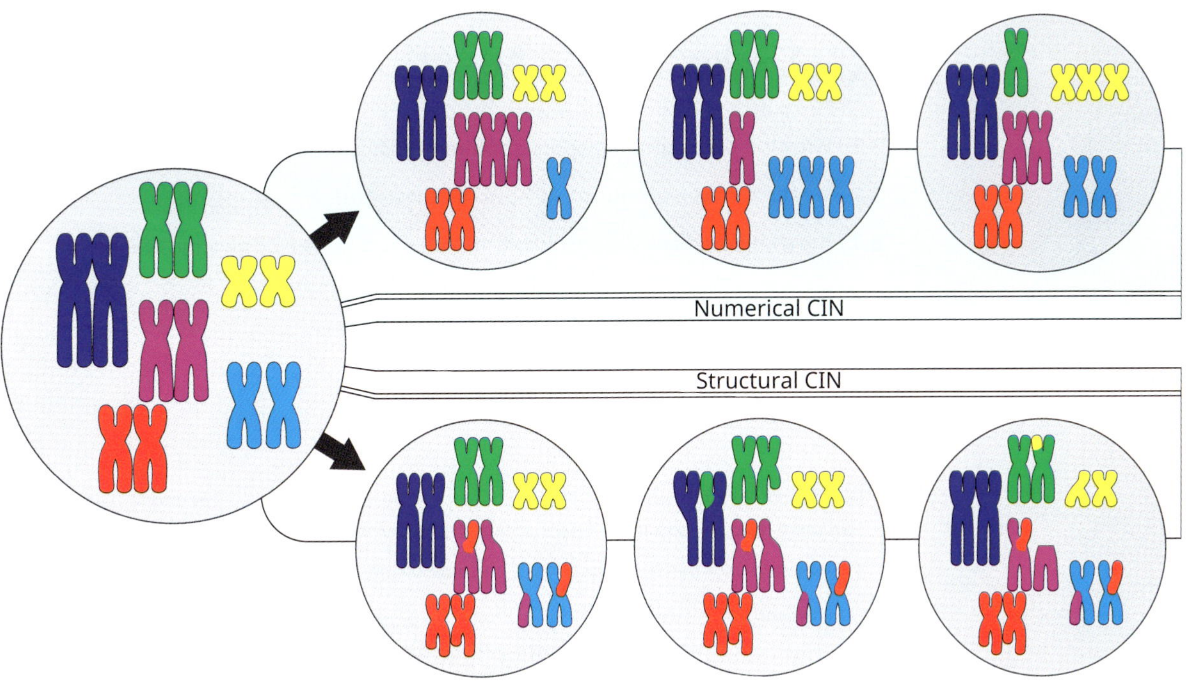

FIGURE 3.18
Numerical and structural chromosomal instability.

Chapter summary

- This chapter outlines the principles of functional organization of eukaryotic chromosomes, including changes in chromosome appearance and behaviour during the cell cycle.
- The complex, dynamic modalities of chromatin packaging within the cell nucleus have been considered in relation to cell division requirements and evolutionary significance.
- Fundamental research discoveries in chromosome biology have been presented in context and chronological order to reflect the progression of thought and conceptual development in logical sequence.
- The causes and consequences of chromosomal anomalies have been explored with reference to human disease and applications for precision diagnostics.
- The advantages and limitations of different methodological approaches for studying chromosomes have been illustrated.
- Examples of using chromosomes as biomarkers in diagnostic and prognostic contexts have been provided.

3.1 What is the biological importance of chromosomes?

3.2 What is the evolutionary relevance of chromosomes?

3.3 Are chromosomal anomalies still valid biomarkers in the post-genomic era?

- Foster HA, Bridger JM (2005) ***The genome and the nucleus: a marriage made by evolution. Genome organisation and nuclear architecture***. *Chromosoma*, 114(4), 212–229.

  This review explores the principles of genome organization within the nuclear context.

- Hochstenbach R, Liehr T, Hastings RJ (2021) ***Chromosomes in the genomic age. Preserving cytogenomic competence of diagnostic genome laboratories***. *Eur J Hum Genet.*, 29(4), 541–552.

  This paper discusses the importance of preserving cytogenomic competencies in diagnostic genomic laboratories.

- Hu Q, Maurais EG, Ly P (2020) ***Cellular and genomic approaches for exploring structural chromosomal rearrangements***. *Chromosome Res.*, 28(1), 19–30.

  This paper presents an overview of experimental methods at different resolution levels and sensitivity scales for studying chromosome structure.

- Liehr T (2021) ***Molecular cytogenetics in the era of chromosomics and cytogenomic approaches***. *Front Genet.*, 12, 720507.

  This paper explores the role of molecular cytogenetics in diagnostics and research in the post-genomic era.

- Meaburn KJ (2016) ***Spatial genome organization and its emerging role as a potential diagnosis tool***. *Front Genet.*, 7, 134.

  This review discusses the potential role of genome organization as a diagnostic tool.

- Vance GH, Khan WA (2022) ***Utility of fluorescence in situ hybridization in clinical and research applications***. *Clin Lab Med.*, 42(4), 573–586.

  This paper presents a review of fluorescence in situ hybridization (fISH) applications in clinical settings.

- Hanahan D, Weinberg RA (2011) ***Hallmarks of cancer: the next generation.*** *Cell*, 144(5) 646–74. doi: 10.1016/j.cell.2011.02.013.

- Hatton IA, et al. (2023) ***The human cell count and size distribution.***

- Lepage CC, Morden CR, Palmer MCL, Nachtigal MW, McManus KJ (2019) ***Detecting chromosome instability in cancer: approaches to resolve cell-to-cell heterogeneity.*** *Cancers (Basel)*, 11(2), 226. doi: 10.3390/cancers11020226.

- Luijten MNH, Lee JXT, Crasta KC (2018) ***Mutational game changer: chromothripsis and its emerging relevance to cancer.*** *Mutation Research/Reviews in Mutation Research*, 777, 29–51. https://doi.org/10.1016/j.mrrev.2018.06.004.
- Pathak S, Raj G, Pratap R, Singh S (2023) ***Late presentation of Swyer syndrome: a case report.*** *Radiology Case Reports*, 18(9), 3295–98. https://doi.org/10.1016/j.radcr.2023.06.061.
- Sathasivam S, Selvakumaran A, Jones QC, et al. (2013) ***Immunodeficiency, centromeric region instability and facial anomalies (ICF) syndrome diagnosed in an adult who is now a long-term survivor.*** *Case Reports*, 2013, bcr2013200170. https://casereports.bmj.com/content/2013/bcr-2013-200170
- Silva M, de Leeuw N, Mann K, Schuring-Blom H, Morgan S, Giardino D, Rack K, Hastings R (2019) ***European guidelines for constitutional cytogenomic analysis.*** *Eur J Hum Genet*, 27(1), 1–16. doi: 10.1038/s41431-018-0244-x.

Useful websites

- The Genome Reference Consortium **https://www.ncbi.nlm.nih.gov/grc/human/data** An online resource providing bioinformatic tools that facilitate the curation of genome assemblies.
- The Human Protein Atlas **https://www.proteinatlas.org/search/chromosomes** A programme that aims to map all the human proteins in cells, tissues, and organs using an integration of various omics technologies, including antibody-based imaging, mass spectrometry-based proteomics, transcriptomics, and systems biology.
- Human Cell Treemap **https://humancelltreemap.mis.mpg.de/** Data on this website are intended to provide a holistic view of cell size, count, and aggregate biomass of all the dominant cell types grouped into the tissues of the human body.
- The Mitelman Database of chromosome aberrations and gene fusions in cancer **https://mitelmandatabase.isb-cgc.org/** A database that relates cytogenetic changes and their genomic consequences, in particular gene fusions, to tumour characteristics, based either on individual cases or associations.

4

Complex Disorders and Polygenic and Multifactorial Inheritance

Nirmal Vadgama, Shivani Shah, Hussein Sheikh Ali Mohamoud, and Jamal Nasir

The inheritance of traits controlled by a single gene with two alleles follow the principles explained by Mendel's observations. The division process of meiosis results in each gamete cell receiving just one gene copy (allele). The **law of segregation** states that the two alleles of a single trait are separated randomly during gamete formation. The **law of independent assortment** states that segregation of different alleles occurs independently of each other.

Mendelian inheritance patterns are well established, and often serve as 'textbook' examples of many single gene diseases. The book *Mendelian Inheritance in Man* by Victor McKusick, Professor of Medicine at the Johns Hopkins Hospital, provides a knowledge base of human genes and phenotypes that are inherited in a Mendelian manner. The inheritance patterns include:

- autosomal dominant,
- autosomal recessive,
- X-linked dominant,
- X-linked recessive,
- Y-linked, and
- mitochondrial.

Today, the online version, called The Online Mendelian Inheritance in Man (OMIM) database, contains over 25,000 entries.

law of segregation
From Mendel's observations, stating that the two alleles of a single trait are separated randomly during gamete formation.

law of independent assortment
From Mendel's observations, states that segregation of different alleles occurs independently of one another.

However, there are many situations where a disorder has a clear familial recurrence and therefore a strong genetic component but does not follow a neat Mendelian inheritance pattern. In such situations, we must consider and work with concepts of multifactorial inheritance and complex trait genetics. An important factor to consider is the extent to which the condition in question has a genetic component. This is done by calculating **heritability** using family, adoption, and twin studies. Another factor to consider is that often conditions are effected and affected by more than one gene and, when this is the case, we must consider the impacts of additive, synergistic, and antagonistic genetic influences, collectively referred to as ***epistatic** interactions*. As we strive to understand more about multifactorial and multi-locus conditions, we are beginning to develop polygenic risk scores for complex traits which may one day have significant clinical utility. In this chapter we will explore what is known about complex inheritance, review the ways in which we are investigating it, and look to the future of what might be achieved.

heritable (heritability)
Capable of being inherited, or passed on from parent(s) to offspring, not necessarily genetic.

epistasis (epistatic)
A situation in which the expression of one gene is affected by the expression of one or more independently inherited genes.

Learning Objectives

After studying this chapter, you will be able to:

- Contrast single gene with polygenic and multifactorial conditions
- Briefly describe the key design points in and explain the power of the following three types of study: family, adoption, and twin studies
- Explain the concept of epistatic interactions
- Explain the concept of reduced penetrance and contrast it with variable expression
- Describe the key methodological considerations for identifying *de novo* variants
- List basic epigenetic mechanisms and describe how they might affect gene function
- Describe the use of 'omics' technologies in searching for downstream consequences of genetic changes
- Describe the impact of big data and machine learning on the improvement of polygenic risk scores and the growing area of personalized medicine.

4.1 Single-gene disorders

To further understand the mechanisms behind multifactorial disorders it is worth reiterating the principles of single-gene disorder inheritance. Many inherited diseases can be traced to variants in a single gene. Cystic fibrosis, haemochromatosis, Tay–Sachs, and sickle cell anaemia are examples of the most common disorders with significant pathology in the population. Although these diseases are primarily caused by a single gene, multiple mutations can result in the same disease but with varying degrees of severity.

For instance, cystic fibrosis is caused by variants in the cystic fibrosis transmembrane conductance regulator (*CFTR*) gene on chromosome 7. The Cystic Fibrosis Mutation Database currently lists >2000 mutations in the *CFTR* gene that cause cystic fibrosis. Different combinations of mutations in the gene have different effects on its function and thus cause a spectrum of phenotypic expression. The most common mutation seen in cystic fibrosis patients, accounting for ~70% of cases, is ΔF508. This corresponds to the deletion of the amino acid phenylalanine at position 508 of the CFTR protein.

CFTR is a membrane protein and chloride channel in vertebrates. It regulates transport of chloride ions and the movement of water in and out of cells that make up the surface of many organs, such as the lungs, bowel, and skin. An aberrant CFTR protein causes a number of problems including the formation of thick mucus in the lungs, which leads to recurrent respiratory infections and limits the ability to breathe over time.

Molecular gene mutations tend to affect one or a few bases of a gene. They can be deletions, insertions, or substitutions of one or more bases. They can be in regulatory regions (such as promoters or enhancers), introns (affecting mRNA splicing), or in exons (affecting the sequence of the coding region).

newborn blood spot screening
UK screening programme enabling early identification, referral, and treatment of babies with an approved list of rare but serious conditions, also referred to as the 'heel prick test'.

Newborn blood spot screening (or the **heel prick test**, as it's sometimes called) is offered to all babies born in the United Kingdom, and screens for nine rare but serious conditions (six of which are inherited metabolic diseases). They include:

- Phenylketonuria (PKU)
- Medium-chain acyl-CoA dehydrogenase deficiency (MCADD)
- Homocystinuria (pyridoxine unresponsive) (HCU)
- Maple syrup urine disease (MSUD)
- Glutaric aciduria type 1 (GA1)
- Isovaleric acidaemia (IVA)
- Sickle cell disease
- Congenital hypothyroidism
- Cystic fibrosis

The aim of this initiative is to reduce incidence of diseases in the population and improve long-term health outcomes in children. With population screening, individuals could know their status before conception (or marriage in some cultures), which could also avoid the need for terminations. There has been considerable success in the Cypriot population for β-thalassaemia and among Ashkenazi Jews for Tay–Sachs disease. These apparent advantages are aside from the social and ethical ramifications, of which there are many. There could be stigmatization of unaffected carriers, a risk of losing children to untreatable genetic disorders, or genetic incompatibility between couples. Further, allocation of resources to screening could divert focus from treating affected individuals.

The newborn blood spot screen is not infallible. Because it screens for biomarkers, it can only indicate that a baby is at high risk of having one or more of the conditions tested for. Molecular genetic testing is the gold standard and a fast-growing diagnostic discipline. With next generation sequencing (NGS) technologies becoming more affordable and feasible, diagnostic labs are increasingly being able to offer a more accurate and comprehensive molecular diagnostic test for genetic disorders.

100,000 Genomes Project
A UK-led project established to sequence 100,000 genomes from around 85,000 National Health Service (NHS) patients affected by a rare disease, cancer, or infectious disease.

whole genome sequencing (WGS)
A method for determining the complete DNA sequence of an organism's genome, used to identify genetic variations associated with diseases.

Our Future Health
A UK initiative funded by a collaboration of the public, charity, and private sectors, to build a resource that reflects, amongst other things, the genetic diversity of the UK population.

The United Kingdom has been at the leading edge of genomics research (see **Figure 4.1**), initiating studies such as the UK Biobank (Sudlow et al. 2015) and Deciphering Developmental Disorders (Firth and Wright 2011). More recently, the **100,000 Genomes Project** was established to sequence 100,000 genomes from around 85,000 NHS patients affected by a rare disease, cancer, or infectious disease. Combining **whole genome sequencing (WGS)** data with detailed medical records has created a ground-breaking research resource. As part of its recruitment target, Genomics England has highlighted the importance of engaging with the United Kingdom's increasingly diverse Black, Asian, and minority ethnic populations, who are often under-represented in clinical studies and research. **Our Future Health** is a recent UK initiative funded by an ambitious collaboration of the public, charity, and private sectors.

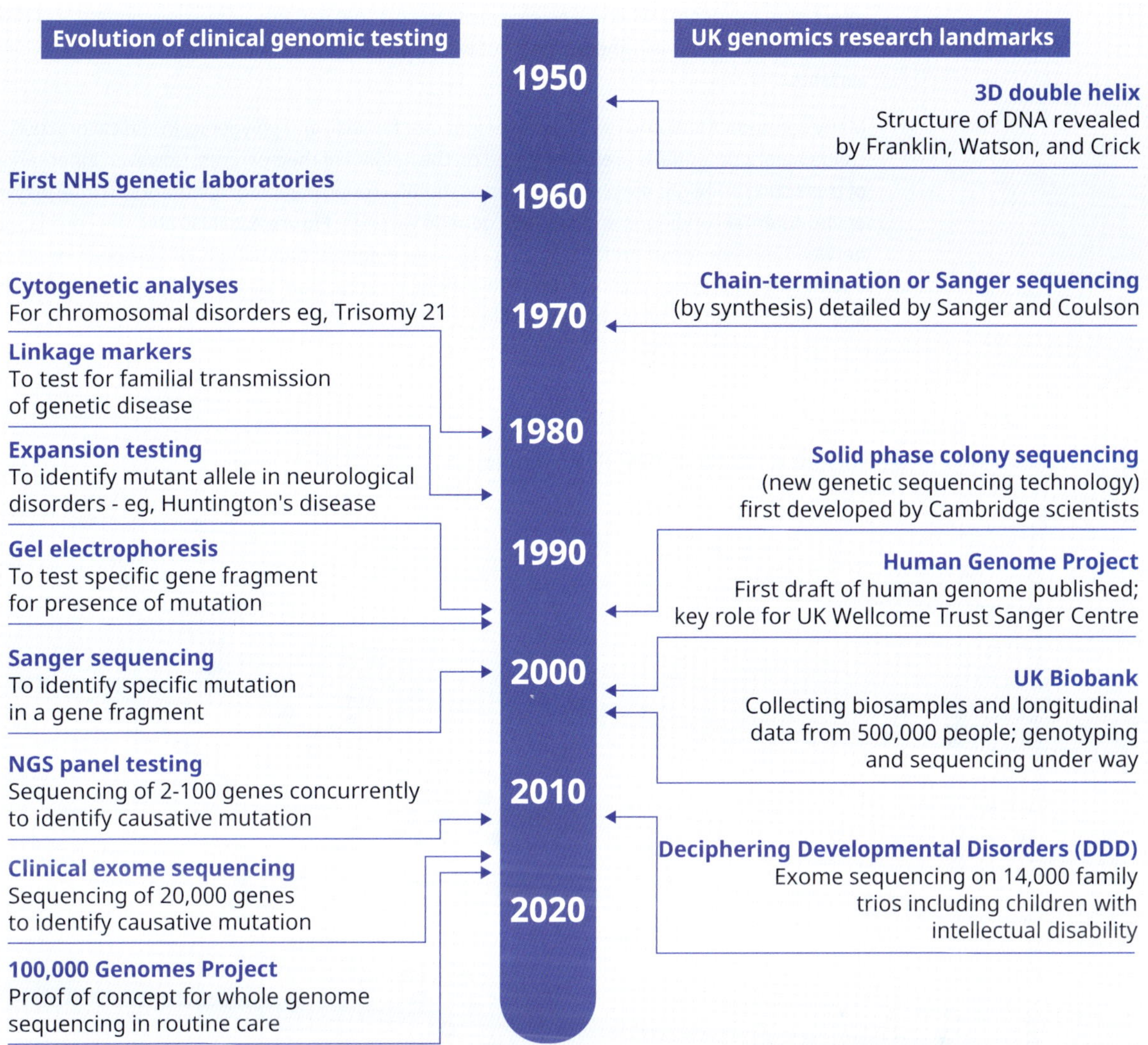

FIGURE 4.1
Genomics in the United Kingdom: Timelines of clinical testing and research achievements.

It is committed to building a resource that truly reflects the UK population, so that we can identify differences in how diseases begin and progress in people from different backgrounds.

Rare hereditary disorders collectively affect ~5% of the population, but most patients do not receive a molecular diagnosis (Turro et al. 2020). These large-scale studies allow the implementation of NGS technologies to help identify novel genetic aetiologies in complex disorders.

4.2 Complex disorders

Pull up any recent study on the genetics of a common human disease, and chances are the first few lines will contain the words *complex*, *multifactorial*, *polygenic*, and/or *heterogeneity*. These words are often used interchangeably, but they are in fact distinct terms. *Polygenic*, although self-explanatory, implies that a particular trait arises in an individual because of the combined effect of many genes. Whereas *heterogeneity* means that variants at multiple genetic loci produce the same or similar phenotypes, but where individual cases are caused by a single or few variants.

Many common traits, such as physical height, are familial, and polygenic. In clinical practice, a child's height potential is predicted based on the heights of their parents. However, inheritance of traits such as height are also described as multifactorial, meaning that there's an interplay of environmental conditions and multiple genetic variants. **Figure 4.2** illustrates the correlation between protein consumption and average male height in populations of European descent. The height differences are in part due to genetic variables, but clearly socio-economic factors and quite probably diet are playing a role.

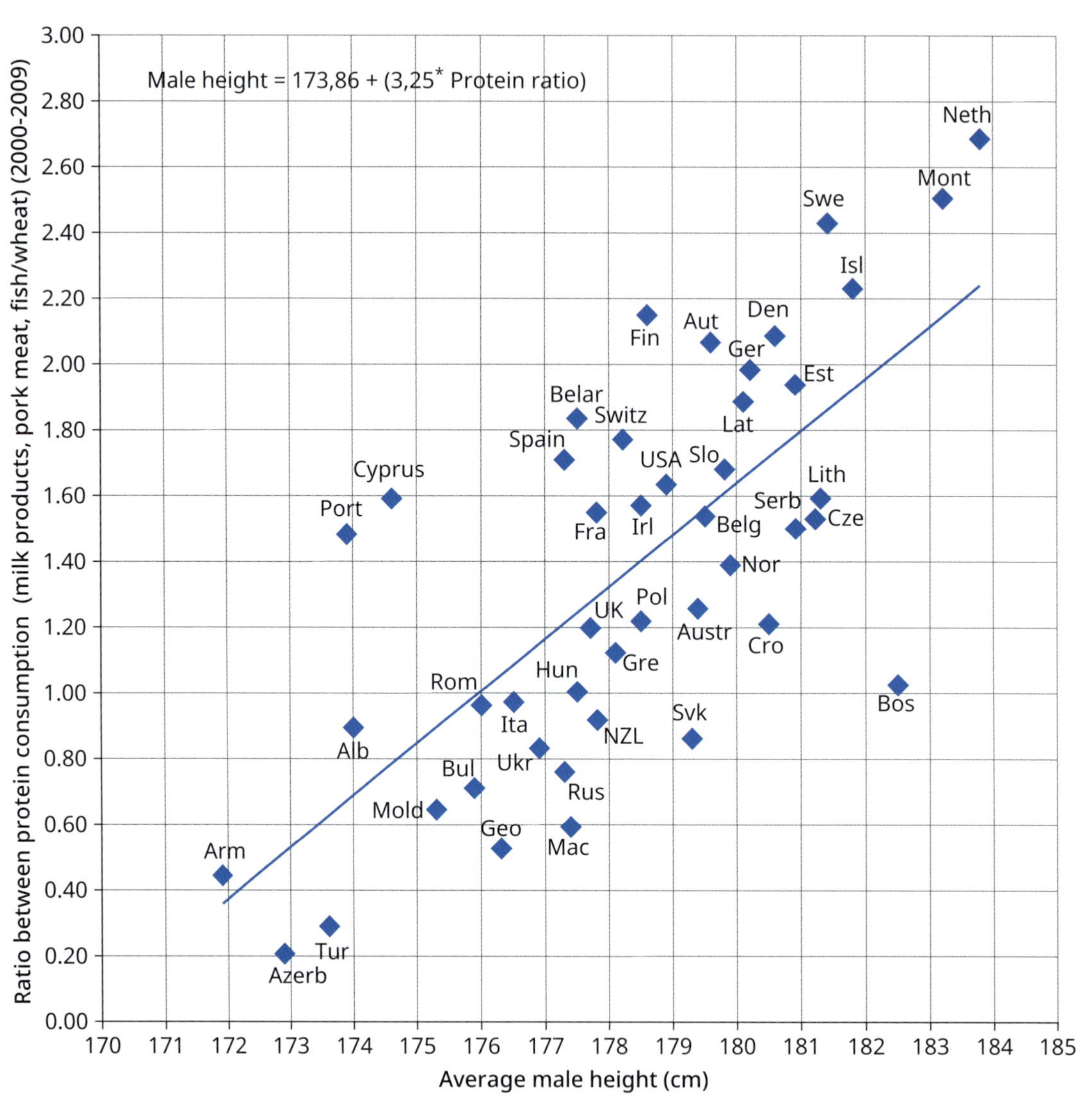

FIGURE 4.2

Relationship between average GDP per capita (current international USD, by purchasing power parity; 2000–2012) and male height ($r = 0.38$; $p = 0.011$). The correlation would reach $r = 0.66$ ($p < 0.001$) in the 'West' and $r = 0.54$ ($p = 0.008$) in the 'East'.

These terms have been used to describe many disorders, including schizophrenia, hypertension, Tourette syndrome, and Crohn's disease. These and potentially hundreds of other disorders are classified as complex because they are influenced by multiple genetic and environmental factors. Stroke is a leading cause of disability and death worldwide, characterized by the sudden loss of blood circulation to an area of the brain. It is an example of a complex disorder, with different aetiologies and prognoses. Researchers have associated common variants at approximately 35 genetic loci to stroke risk. Most of these variants are dispersed across the genome and do not cluster on one specific chromosome (Dichgans, Pulit, and Rosand 2019). Additionally, environmental factors, such as smoking, diet, and certain medications, play a considerable role in stroke risk.

It has been argued that complex disorders actually represent a collection of symptoms caused by rare mutations in an array of genes (Mitchell 2012). If stroke, to use the example again, is really a syndrome comprising several pathophysiological mechanisms, rather than a distinct disease, it would preclude the search for common variants conferring disease risk. One way to determine the extent to which a particular disorder is genetic is by calculating its heritability. It has already been mentioned that heritability can be estimated using the twin study design. In the following sections, some of the established statistical genetics methodologies underlying the search for disease-associated common and rare variants will be discussed.

SELF-CHECK 4.1

What are the differences between single gene, polygenic, and multifactorial conditions?

4.3 Family relationships and heritability

The fact that phenotypic traits run in families was recognized long before we knew anything about genes or DNA, long before we even understood sexual reproduction. However, just because a condition or trait runs in families does not mean that it has a genetic cause. Families tend to live together, sharing the same environment and influencing one another's development. Thus, a condition can be heritable without being entirely genetic. If a condition is known to arise from a combination of both genetic and environmental influences, it is referred to as **multifactorial**. The first pie chart in **Figure 4.3** illustrates the proportion that recent research attributes to environmental and genetic causes for Alzheimer's disease. The second gives more detail as to the specific nature of the genetic contribution.

multifactorial
Describes phenotypic traits that occur because of a combination of both environmental and genetic factors.

Those interested in the genetic component of complex traits have benefited from family, adoption, and twin studies to investigate what proportion of a phenotype can be attributed to genes and what proportion is attributable to environmental factors.

4.3.1 Family studies

A family study begins with the identification of one family (or preferably several), in which one or more individuals show the phenotypic trait of interest. The incidence of the given phenotype is then calculated. Care must be taken to achieve full ascertainment. This can be tricky because some individuals might not disclose mild signs or symptoms and there can be misunderstanding amongst teams of assessors as to precisely what those signs and symptoms are. Next, the results of the count are adjusted according to the degree of relationship of the affected

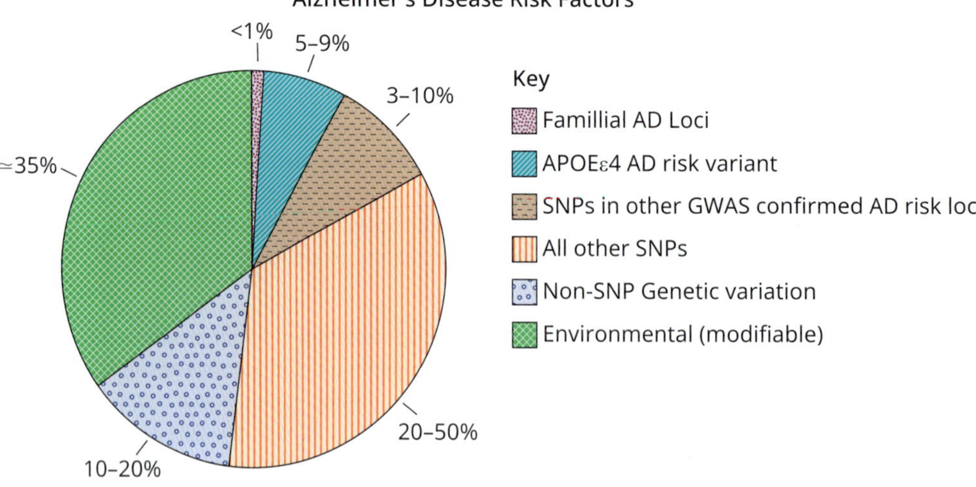

FIGURE 4.3
Alzheimer's disease is a multifactorial, heritable condition. The pie chart illustrates the fact that modifiable or environmental factors account for approximately 35% of the disease risk and about 65% can be attributed to various genetic factors. Familial Alzheimer's Disease (FAD) loci are known to contribute less than 1% of the disease risk. The ε4 allele of Apolipoprotein E (APOE) gene accounts for approximately 5–9%. A further 3–8% is attributed to other Alzheimer's disease associated loci as determined by genome-wide association studies (GWAS). The remaining risk fractions are divided between other assessed single nucleotide polymorphisms and genetic variations that are not single nucleotide polymorphisms.

TABLE 4.1 Illustrating degrees of consanguinity and approximate shared amount of DNA

Relationship examples	Degree of relationship	Average proportion of shared genes
Parents (mother and father) Siblings (brother and sister) Progeny (children/offspring)	FIRST	50 % (1/2)
Grandparent (grandmother and grandfather) Grandchildren (grandsons and granddaughters) Uncles and aunts (siblings of parents) Half siblings	SECOND	25 % (1/4)
Great grandparents Great grandchildren Half uncles and half aunts (half brother/sister of parent) First cousins (children of aunt/uncle)	THIRD	12.5 % (1/8)

individuals in any given family to the 'index case'. **Table 4.1** shows the approximate proportion of genes that are likely to be shared by relatives.

Corrections and adjustments must be made for late onset disorders/conditions. Once this is done, a lifetime risk (lifetime expectation/morbidity risk) is reported for every individual in each family pedigree. Lifetime risk is given as a percentage or as a relative risk in comparison to an individual in the general population. Relative risk is denoted as lower-case Lambda r (λr). If an individual's λr is 2, then their risk is double that of an individual in the general population. λs specifically denotes the relative risk for a sibling. It is this kind of information that has been used to set up complex algorithms to assess lifetime risk for individuals developing certain conditions. An example is Cambridge University's CanRisk, a risk prediction model used to calculate future breast and ovarian cancer risks in women.

4.3.2 Adoption studies

Proving that a particular phenotype is clustered in families does not prove multifactorial inheritance. It could be that the phenotype is simply the result of shared environment. Therefore, further analysis in the form of twin or adoption studies is needed to provide sufficient evidence for the genetic component. Adoption studies allow two important comparative measures.

Firstly, the measurement of occurrence rates in children adopted away from affected biological parent(s). If these remain the same as for children who stay with their biological parents, then this evidences genetic causes. If the incidence is lower, that is evidence of causative environmental factors.

Secondly, the measurement of occurrence rates in children adopted away from unaffected biological parent(s) to affected parent(s). If these remain the same as for children who stay with their biological parents, then this evidences genetic causes. Higher incidence is strong evidence that the phenotype is affected by environmental factors.

4.4 Twin studies

Family and adoption studies can only go so far in helping us to determine genetic influences of a given phenotypic trait. The ideal for a geneticist trying to distinguish the 'nature vs nurture' components in a multifactorial condition would be multiple sets of monozygotic twins separated at birth. For ethical reasons, this is impossible with humans. Multiple births have captured the interest of researchers for centuries, and proposals to use them in empirical studies originate from as early as 415 CE by Augustine of Hippo (Kelley 2012). Many ethical considerations surround the use of multiple births in research, and historically many mistakes have been made, not least the famous 1960s case of the triplet brothers Galland, Kellman, and Shafran. Researchers, clinicians, and those working in diagnostics must always be mindful of the ethical implications of their work.

Cross reference

For an introduction to ethics in human genetics and genomics, please refer to **Chapter 9**.

4.4.1 The history of twin research

Hermann Werner Siemens, a German dermatologist, was one of the pioneers of the twin study. He made significant contributions to the systematic analysis of similarity between **monozygotic** ('identical') and **dizygotic** ('fraternal' or 'non-identical') twins. Siemens independently formulated the twin rule of pathology: identical twins are more likely to be concordant for a heritable disease than non-identical twins, and concordance will be even lower in non-siblings (Siemens 1924). Siemens' study of moles (skin blemishes, not the animal) led him to come up with an innovative idea of merging correlation analysis with twin data. He correlated mole counts within twin pairs and contrasted this correlation between monozygotic and dizygotic pairs. Monozygotic twins, who share all—or nearly all—of their genetic material had a mole count correlation of 0.4. Whereas dizygotic twins, who are on average 50% genetically identical, had a correlation of only 0.2. The results demonstrated the importance of genetic factors to the variation in mole count. More generally, the genetic similarity of monozygotic twins is associated with their larger resemblance for the phenotype under study.

monozygotic
Identical (in twins).

dizygotic
Fraternal or non-identical (in twins).

4.4.2 The classical twin study design

The classical twin study compares phenotypic resemblances of monozygotic and dizygotic twins. Comparing the resemblance of monozygotic twins for a trait or disease with that of dizygotic twins offers an initial estimate of the extent to which genetic variation determines phenotypic variation of that trait. Heritability (h^2) can be defined as the proportion of variance of a phenotype that is due to genetic influences between individuals in a population. If a gene has a high degree of heritability, it indicates that environmental factors have little impact on the phenotype expressed, and vice versa. Heritability can be estimated from twice the difference between monozygotic and dizygotic correlations—that is, [2(rMZ—rDZ)], where rMZ and rDZ are the estimated correlation coefficients of monozygotic and dizygotic twins, respectively. In a meta-analysis of twin correlations, Polderman et al. (2015) reported variance components of all published twin studies of complex traits. They report that the heritability estimate for all 17,804 traits investigated is 49%. The largest heritability estimates were for traits classified under the ophthalmological domain (h^2=70%), whereas traits in the reproduction, environment, and social values domains had the lowest heritability estimates. In a population-based study of 1,033 female–female twin pairs, the average monozygotic and dizygotic correlations for depression was reported to be around 0.4 and 0.2, respectively (Kendler et al. 1992). Thus, the estimated heritability is calculated as approximately 40%. Further, in a meta-analysis of published twin studies of autism spectrum disorders (Colvert et al. 2015), the correlation for monozygotic was almost perfect at 0.98, whereas for dizygotic twins it was 0.53. Thus, the meta-analytic heritability estimates are considerable, ranging up to 90% (see **Table 4.2**).

The application of classical twin studies led to substantial changes in the way we think about the determinants of health and disease and the causes of behavioural differences. During the last few decades, a shift has taken place from strict environmental explanations to a more balanced view that recognizes the importance of genes, for example in autism spectrum disorders in children, or in the development of dependence on nicotine, alcohol, and other drugs in adults (Manuck and McCaffery 2014). Recent advancements in both scanning during pregnancy and DNA sequencing techniques has led to significant methodological improvements, allowing the comparison of **monochorionic** with **dichorionic** monozygotic twins and genomes down to base pair level. This is especially true with the co-evolution of next-generation sequencing (NGS) platforms and DNA **microarray** strategies (Schuster 2008). This has ineluctably spurred greater interest in research into ***de novo*** mutations and cases of mosaicism in relation to monozygotic twins.

monochorionic
A pregnancy where multiple babies share a single placenta.

dichorionic
A multiple pregnancy in which each foetus has its own placenta and amniotic sac.

microarray
A technique that uses thousands to millions of short single-stranded nucleic acid fragments bound in specific positions to a solid surface (referred to as a 'chip'), to identify the presence of complementary molecules within samples of interest.

de novo
A previously identified mutation or clinically relevant variant that arises in a family member for the first time.

SELF-CHECK 4.2

With reference to **Table 4.2**, which are the most and least heritable conditions respectively?

SELF-CHECK 4.3

With reference to **Table 4.2**, is major depression more heritable in males or females?

4.4.3 The case-control co-twin design

Monozygotic twins are particularly useful in case-control studies as they are perfectly matched for genotype and family background. An earlier use of this approach was illustrated by Martin et al. (1982) who studied the effect of vitamin C on the common cold by administering it in one twin and a placebo in the co-twin. Contrary to popular belief, it was determined that vitamin C is

TABLE 4.2 Twin concordance for complex disease

	Proband-wise concordance* (%)	
	MZ twins	DZ twins
Type 1 diabetes	42.9	7.4
Type 2 diabetes	34	16
Multiple sclerosis	25.3	5.4
Crohn's disease	38	2
Ulcerative colitis	15	8
Alzheimer's disease	32.2	8.7
Parkinson's disease	15.5	11.1
Schizophrenia	40.8	5.3
Major depression	31.1[‡] or 47.6[§]	25.1[‡] or 42.6[§]
Attention-deficit hyperactivity disorder	82.4	37.9
Autism spectrum disorders	93.7	46.7
Colorectal cancer	11	5
Breast cancer	13[§]	9[§]
Prostate cancer	18	3

*Defined as 2C/(2C + D), where C is the number of concordant affected twin pairs, and D is the number of discordant twin pairs. [‡]Concordance in male twin pairs. [§]Concordance in female twin pairs. DZ, dizygotic; MZ, monozygotic (Van Dongen et al. 2012).

no more effective than a placebo. In line with the Barker hypothesis (which states that adverse nutrition in early life, including prenatally, increases susceptibility to metabolic syndrome (Hales and Barker 2001)), differences in birthweight in monozygotic and dizygotic twin pairs and their association with cardiovascular and metabolic parameters have been investigated. If these associations are due to shared genetic factors, difference scores are uncorrelated in monozygotic, but not in dizygotic twins. IJzerman, Stehouwer, and Boomsma (2000) determined that the association between low birth weight and high blood pressure in later life seems to be mediated by common genes.

Discordant monozygotic twins, where only one twin has a disease, are often used to investigate which non-shared environmental influences are related to the disorder (Asbury et al. 2003). Such discordant twin pairs might also form the perfect case-control design for gene-expression studies to distinguish between genes that are related to the causes of disease and genes that are expressed as a consequence of disease. These differences might also indicate causal genes that are differentially activated by epigenetic factors (Petronis et al. 2003).

More recently, the classical co-twin design has received much interest in the field of molecular biology. Identifying genomic differences between twins and appreciating the subtle nuances in choosing the ideal analytical approach, relies on an understanding of the mechanism and aetiology of the twinning process (Vadgama et al. 2015b).

Key Point

The specific nature of twin, family, and adoption studies has been of critical importance as geneticists have worked to ascertain which conditions have a purely genetic basis and which are multifactorial.

4.5 Genomic approaches to deciphering complex disorders

4.5.1 Genetic linkage analyses

Traditionally, the search for disease-associated genes began with linkage analysis, a genetic mapping approach for estimating the locations underlying a trait. The premise of linkage analysis is that alleles located closely together on a chromosome are more likely to be inherited together through meiosis. Genetic linkage was first described in the early 1900s by British geneticists Bateson, Saunders, and Punnett who noted certain phenotypes were inherited together rather than independently. This was quite controversial at the time as it seemed to contradict Mendel's law of independent assortment (Bateson, Saunders, and Punnett 1909).

recombination
The formation of new combinations of DNA sequences, either naturally (by crossing over and independent assortment during meiosis) or, in the laboratory (by the direct manipulation of genetic material).

marker
A specific DNA sequence with a known physical location on a chromosome.

Recombination (crossing over) occurs in meiosis and the pairs of chromosomes exchange DNA segments with each other. Two regions in close proximity on a chromosome are likely to be inherited together. However, the further apart the two regions are, the greater the chance that recombination occurs between them during meiosis, preventing their co-inheritance in the offspring.

Linkage analysis therefore uses family data to determine the degree to which a disease phenotype co-occurs with genetic **markers** that are dispersed throughout the genome. Markers on the chromosomes close to the gene (such as microsatellites) are genotyped for all family members under study and these results are used to look for co-segregation of markers with disease status. Microsatellite markers, also called short tandem repeats, contain repeated nucleotide sequences that often vary in length. These markers should ideally be close to the gene of interest or recombination can pose an issue in the analysis.

If the linked satellites used for disease analysis are close to the gene of interest, the chance of recombination occurring between the marker and the mutation is virtually zero, giving a negligible recombination uncertainty. If the marker genotypes are farther apart, recombination is more likely to occur between them, rendering them unsuitable for prenatal diagnosis in the future (**Figure 4.4**). If there is a statistical association between a marker and the disease allele within a population, the allelic variants are said to be in **linkage disequilibrium (LD)** with each other.

linkage disequilibrium (LD)
The non-random association of alleles at two or more loci in a general population.

As we shall see, this is contrary to genetic association analyses, which test whether individual genetic markers are directly associated with a trait of interest (see **Section 4.5.2**). Linkage methods have been extensively used for gene discovery in monogenic Mendelian diseases, but previous attempts limited by family size lacked statistical power. With a limited number of genetic markers and small sample sizes, there is a higher probability of making a Type II (false negative) error. Despite these limitations, linkage analyses have been successful in identifying single-gene diseases with clear patterns of Mendelian inheritance (Myers et al. 1993) and complex diseases with highly penetrant mutations (Bird 2008). However, it has failed to identify casual genetic loci for most complex disorders, where disease variants have modest effect sizes. This apparent weakness may not be intrinsic to the theoretical foundations of linkage

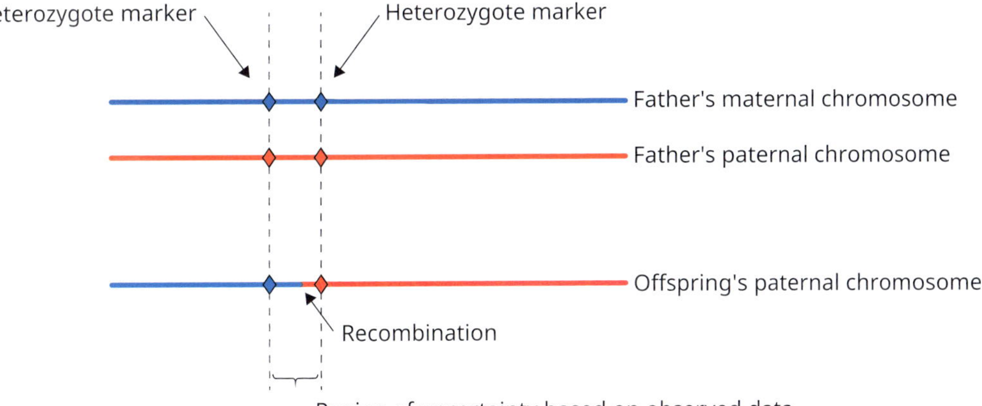

FIGURE 4.4
Recombination occurring between two markers. Linkage analysis can be used to determine which of the two chromosomes the mutation is on by genotyping markers close to the gene of interest. The results are used to work out who else in the family has inherited that chromosome and thus the defective gene. In this example, recombination has occurred between the markers. If our gene of interest is between them, normally co-segregating with them both, it would no longer be clear who has inherited the defective gene. Only the father's chromosome is shown, but the same principle applies to a mother–offspring pair.

methods, but rather due to categorizing clinical diagnoses based on superficial criteria. For example, assuming blindness is a disease rather than a syndrome could pose issues. The aetiology for blindness is broad, ranging from age-related macular degeneration, through cataracts, to multiple sclerosis. For polygenic disorders, data from multiple families is often combined. If the clinical diagnosis is not well characterized, it would be especially unlikely that all families have the same collection of underlying genetic factors. This could cause conflicting linkage results and decreased logarithm of the odds, a statistical parameter that indicates how close two genes are located to one another.

4.5.2 Genome-wide association studies

To overcome the limitations presented by combining family relationship data with linkage analysis, genome-wide association studies were developed. The basic principle of a **genome-wide association study (GWAS)** is a case-control design requiring a comparison of the genomes of usually thousands of unrelated individuals with a particular disease, with genomes of individuals without the disease, in an attempt to identify genetic variations that may confer risk of disease. Many GWAS involve large-scale use of **single nucleotide polymorphism (SNP)** microarrays that contain hundreds of thousands of unique nucleotide probes to evaluate results from different individuals. By determining which SNPs co-occur in individuals with the disease, scientists can calculate the disease risk associated with each variation. The frequency of each individual marker identified by GWAS requires statistical analysis, usually with **Pearson's chi-square test**, to predict the relative potential impact (association or risk) on development of a disease phenotype. However, very few markers, if any, reach genome-wide significant levels. This is usually attributed to stringent **Bonferroni correction** for multiple testing, which set a *P*-value threshold of 5×10^{-8}. The recent development of polygenic risk scores (reflecting the sum of all known risk alleles) overcomes this limitation. Polygenic risk scores currently have very reliable predictive ability (Lewis and Vassos 2020). **Figure 4.5** shows the workflow of a typical GWAS.

genome-wide association study (GWAS)
A technique in which researchers scan genetic markers across the complete genomes of many individuals to find genetic variations that are statistically associated with a particular phenotype of interest.

single nucleotide polymorphism (SNP)
Variation found at a single position in a DNA sequence when individual genomes are compared.

Pearson's chi-square test
A non-parametric mathematical tool designed to analyse group differences. Used when the dependent variable is measured at a nominal level.

Bonferroni correction
A statistical test used to reduce the instance of a false positive when performing a hypothesis test with multiple comparisons.

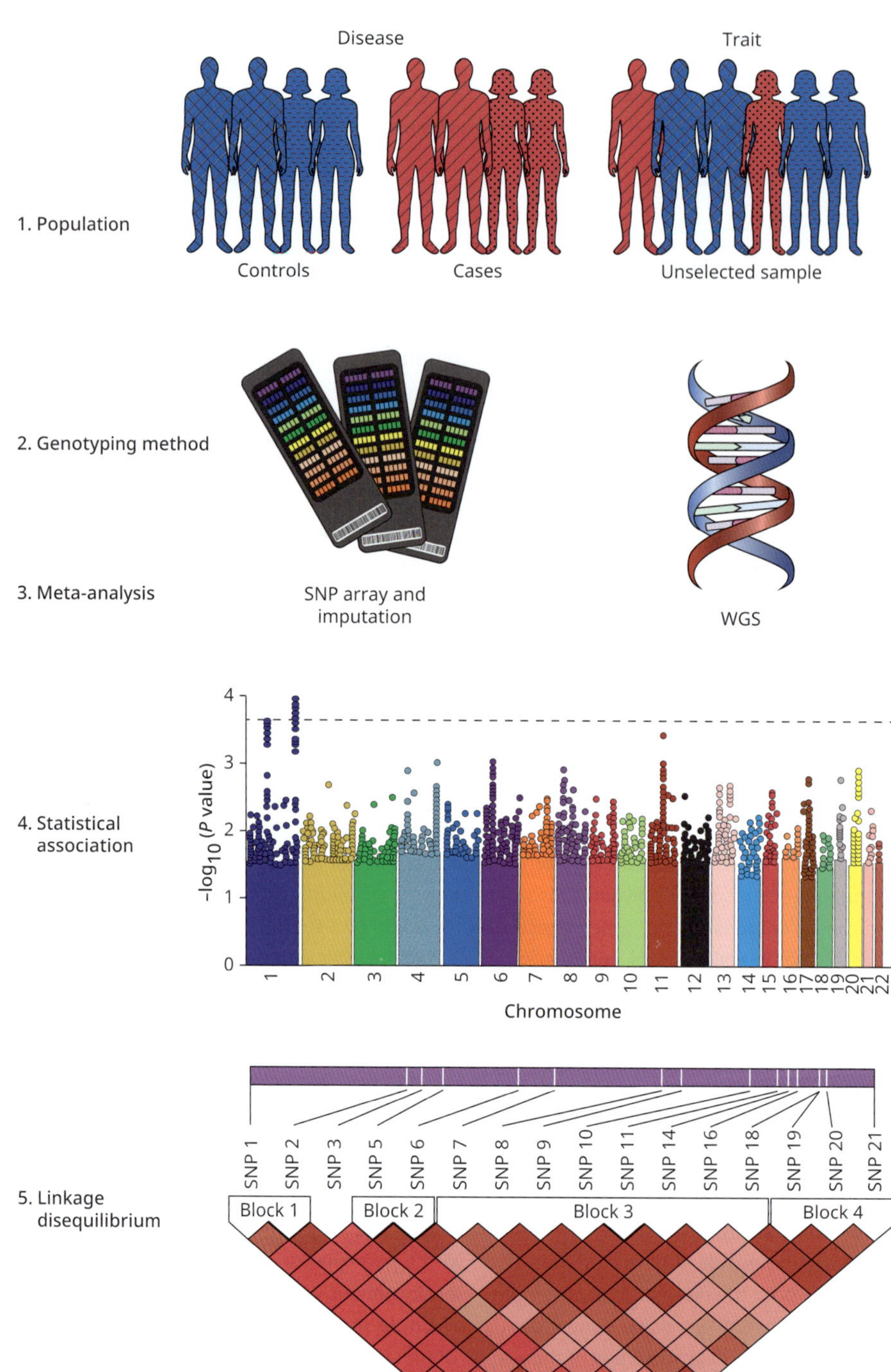

FIGURE 4.5

Genome-wide association study design. The first step involves identifying the disease or trait to be studied and collecting appropriate samples. Genotyping is often performed using single nucleotide polymorphism (SNP) arrays, but whole genome sequencing (WGS) has been recently adopted. Regions of the genome statistically significantly associated with the phenotype of interest are identified. Causal variants are usually not directly genotyped but are found to be in linkage disequilibrium with the **polymorphisms** statistically associated with the phenotype of interest. Findings are often replicated with an independent sample set.

polymorphism
A common variation in the DNA code.

But there are some concerns about the GWAS approach. For instance, what is the significance of having one, three, or 30 risk alleles for a particular condition? Put differently, how many variations have a significant effect on risk? Questions like these have uncertain answers, and that can be an issue when it comes to disease diagnosis and treatment. But the number of risk genes identified by most GWAS shows us that, unlike single-gene disorders, complex genetic disease conditions involve a multitude of genetic factors contributing to the total risk for developing a condition.

Despite large GWAS meta-analyses having been conducted, the genetic contribution of many disorders studied remains unexplained. For example, in a Crohn's disease GWAS consisting of more than 21,300 people, around 70 genetic loci were identified at genome-wide significance, but that could explain only 23% of heritability (Franke et al. 2010). GWAS cannot determine whether the alleles with positive associations contribute to causality, and at best, can only identify loci that have modest effects on disease risk. Recent methodological approaches integrate data from diverse omics platforms, translating GWAS signals into functional understanding. Combined genetic–epigenetic mapping, followed by functional assays, can help delineate regulatory regions that causally underlie an association signal (see **Section 4.8**). These post-GWAS approaches are helping to unravel causal variants underlying disease risk, thus honing population-based risk to individualized risk.

Analysis of cardiovascular diseases and breast cancer offer some of the strongest evidence for the clinical utility of polygenic risk scores. However, there is still some way to go before they become a routine tool for clinicians.

Over 90% of GWAS signals map outside of coding regions (Schaid, Chen, and Larson 2018). The non-coding genome is mostly composed of regulatory elements that control gene expression, but these have remained largely unexplored. Given that the majority of disease-associated GWAS signals are located in non-coding regions of the genome, such as promotors and enhancers, integrating expression data could help interrogate the complex intricacies of biological networks between specific tissues and the pathogenesis of complex disorders.

4.5.3 Rare variants association analyses

The consensus in GWAS interpretation is that SNPs that reach statistical significance are associated with disease risk. Under the polygenic model, many common variants of small effect sizes act in concert to contribute to the phenotype. In the previous section, it was eluded that rare variants of large effect sizes or non-additive effects can play a predominant role in the inheritance of complex disorders.

GWAS are typically performed in genetic variants that have minor allele frequencies (MAFs) of > 0.05 (5%). The genetic architecture of complex disorders remains elusive, leaving several possible explanations for the 'missing heritability' (Manolio et al. 2009). Because GWAS have focused on common genetic variants (MAF > 5%), the majority of which make up ethnicity-specific differences, rare genetic variants could explain a significant portion of this missing heritability.

Many studies have shown that risk alleles were enriched in minor alleles, especially for variants with MAFs < 0.01 (1%), which is considered low. It is very unlikely to have a MAF of more than 1% for Mendelian disease.

According to the joint consensus recommendation of the American College of Medical Genetics and Genomics (ACMG) and the Association for Molecular Pathology (AMP) guidelines for the interpretation of sequence variants, a MAF greater than the disease prevalence but less than

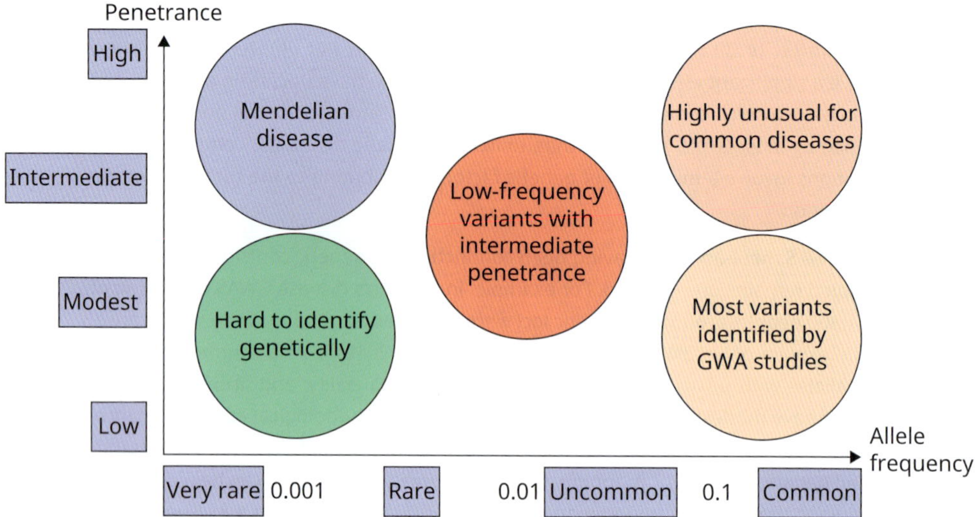

FIGURE 4.6

Allele frequency, penetrance, and disease. GWAS are effective in detecting variants with MAFs >5%. The SNPs identified by this approach often have small effect sizes and much of the genetic component remains to be identified. Rare variants (MAF <1%) may explain a significant proportion of 'missing heritability'. In contrast to complex disorders, Mendelian diseases have a high penetrance and low allele frequency.

5% is strong evidence that the variant is benign. However, a MAF of <0.001 is recommended for a dominant disease variant discovery in Mendelian diseases (**Figure 4.6**).

The rapid advance of NGS has enabled comprehensive assessment of rare genetic variants, and statistical methodology has emerged for the analysis of these data. One of the most difficult challenges with rare variants is that, by definition, they are only found in a small fraction of the overall population, and as a result, extremely large sample sizes are necessary to study these variants. Large population-based cohorts are ideal for these studies.

Prioritization of the identified putative disease-associated rare variants is carried out based on their mutational architecture and biological relevance to disease. The next section will review the currently available computational tools used to predict the functional impact of variants and evaluate pathogenicity.

4.5.4 *In silico* prediction tools

In silico, in reference to biological experiments, means 'performed on the computer'. There are many peer-reviewed computational tools used in bioinformatics analyses that can predict the pathogenicity of a variant or give useful information on a gene's function and essentiality.

Gene essentiality can be estimated based on loss-of-function (LoF) mutation intolerance. To class a gene as 'essential' is highly dependent on the circumstances in which an organism lives. For example, *INS*, the gene for insulin, is only essential if an organism has no other way of controlling blood glucose levels. Gene essentiality correlates with intolerance to variation and has been directly assessed in several species using high-throughput techniques in cellular and animal models. Many definitions of essentiality are derived from these different model approaches, but put simply, it is 'the requirement of a gene for an organism's survival' (Cacheiro et al. 2020).

In silico prediction tools can determine what impact certain variations can have on health. An evolutionarily well-conserved gene is similar across different species. If a gene has remained mostly unchanged throughout evolution, it suggests that changes in the gene are likely to be lethal. The **Genomic Evolutionary Rate Profiling (GERP)** score measures the conservation of each nucleotide in multi-species alignment and predicts the deleteriousness of variants (Davydov et al. 2010).

Genomic Evolutionary Rate Profiling (GERP)
The score measures the conservation of each nucleotide in multi-species alignment and predicts the deleteriousness of variants.

Other computational methods measure whether a gene is intolerant to LoF mutations. This can be determined through the Residual Variation Intolerance Score (RVIS), or the probability of being loss-of-function intolerant (pLI). An RVIS below 0.0 means that a given gene has less common functional variation than expected and is referred to as 'intolerant'; whereas an RVIS above 0.0 indicates that a gene has more common functional variation. Genes with pLI scores above 0.9 are extremely LoF intolerant, whereas genes with pLI scores below 0.1 are LoF tolerant.

Let us consider the gene phospholamban (*PLN*), a substrate for the cAMP-dependent protein kinase in cardiac muscle. Pathogenic mutations in *PLN* cause inherited cardiomyopathies due to its role in calcium homeostasis. *PLN* has an RVIS score of 0.12 and a percentile of 62.38%, meaning that it is among the 62.38% most intolerant of human genes, which is not very intolerant. For *PLN*, the constraint metrics from the Genome Aggregation Database (gnomAD) are below threshold (0.44947), albeit the coding sequence merely has one exon, possibly being too small a target to reliably detect population levels of LoF variants.

However, the **haploinsufficiency (HI)** score of *PLN* is 10.25%. HI is when one copy of a gene is inactivated or deleted, and the remaining functional copy of the gene is not adequate to produce the necessary gene product to maintain the standard phenotype and thus normal functioning. A low HI score (0–10%) indicates that the gene is more likely to exhibit haploinsufficiency (Huang et al. 2010).

Haploinsufficiency (HI)
The condition in which heterozygosity for a genetic variant leads to insufficient product and the consequence of this is observed in the phenotype.

Taken together, measures of genetic intolerance to functional variation, which correlates with gene essentiality, indicates that *PLN* may not be critical for human survival. Moreover, studies show that the absence of *PLN* can be compensated by a complex interplay of non-essential and essential genes. The ablation of an essential gene could require non-essential genes to act as a compensatory mechanism and vice versa. Nevertheless, PLN aberrations lead to the development of cardiomyopathy, a type of progressive heart disease that can significantly reduce the quality of life. Interestingly, some pathogenic mutations in *PLN* exhibit incomplete penetrance, meaning that individuals with the same mutation express different degrees of disease severity, even among individuals of the same family (Landstrom et al. 2011).

Inherited cardiomyopathies, a group of disorders that cause heart failure, are genetically heterogeneous with multiple associated genes and several different mutations within each category. More than 60 genes have been attributed to disease causation, but the number of variants of uncertain significance (VUS) detected by screening these genes remains high, making clinical correlation difficult. As we saw earlier, commonly used computational tools may not always provide a clear-cut picture of a gene's role in disease, suggesting that there is much room for improvement. Powerful genetic techniques, such as animal models and CRISPR-based *in vitro* perturbation, have helped determine disease-causing variants and what genes are required for the life of a human cell.

Key Points

Linkage analysis and genome-wide association studies have helped geneticists to identify causative loci for phenotypes of interest. However, even with powerful genome-sequencing technology and *in silico* prediction tools the genetic basis for many complex disorders still eludes us.

4.5.5 *De novo* mutations and somatic mosaicism

novel
A genetic mutation or variant that has not been discovered before—contrast with *de novo*.

A *de novo* mutation is one that arises in a family member for the first time. These might be **novel** (previously unrecorded variations), or changes already seen in other people/families. It can occur in a gamete (an ovum or sperm) of one of the parents and be passed on to their offspring, or postzygotically (after fertilization). If a *de novo* mutation occurs in a sperm or egg during meiosis, fertilization would result in all body cells of the offspring possessing the mutation. Postzygotic mutations accumulate during the aging process—the older we get, the more we acquire new mutations through environmental damage and stochastic events. Unlike inherited *de novo* mutations, postzygotic mutations only affect a fraction of cells, resulting in somatic mosaicism (**Figure 4.7**). The earlier the postzygotic mutation occurs during development, the higher the fraction of affected cells will be in the organism. Somatic mutations are also known to be an early event in carcinogenesis. A smoker in adulthood can acquire new mutations due to damaged cells in the lungs. In fact, tobacco smoke contains thousands of chemicals, over 70 of which are known to cause cancer. Smoking can cause hundreds of mutations, leading to cumulative damage to the genome. The greater exposure to risk factors, the greater the chance that mutations will occur in genes responsible for cancer.

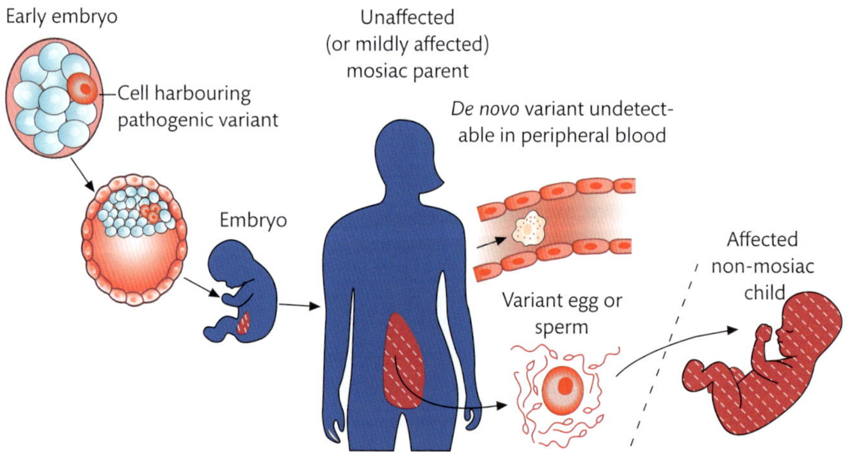

FIGURE 4.7
A postzygotic *de novo* mutation results in somatic mosaicism in the parent. This mosaicism is undetectable in peripheral blood, thus highlighting the importance of sequencing the affected tissue from the individual under study. A postzygotic *de novo* mutation can be inherited by the offspring as a germline variant and become incorporated into the DNA of every cell in the body. Figure from Morrow (2020).

A central dogma in biology is that all cells within an organism contain identical genetic material. However, mosaicism, a term used to describe the occurrence of two or more populations of cells with different genotypes in a given individual, is a violation of this dogma. Single nucleotide variants (SNVs), **insertion-deletion mutations (indels)**, copy number variants (CNVs), and other structural variants continually accumulate as cells divide during growth and development.

insertion-deletion mutations (indels)
Insertion into and/or deletion of short sequences (< 1kb) of nucleotides in genomic DNA.

With the sheer number of mitoses taking place, this could result in many cells with their own unique personal genome. It is therefore almost certain that all humans are mosaic. Most of the time mosaic mutations go unnoticed. But they can also account for a wide range of disorders, including Klinefelter syndrome, mosaic Down syndrome, or Proteus syndrome (popularized by the movie *The Elephant Man*). By definition, all cancer patients are mosaic.

In the following sections, we will review the mechanisms by which *de novo* mutations arise, how they result in somatic mosaicism, and the methods for detecting them.

SELF-CHECK 4.4

What is the difference between a *de novo* and a novel mutation?

4.5.6 Chromosomal aneuploidy and structural abnormalities

Chromosomal abnormalities can be organized in two groups, numerical or structural. The most common type of chromosomal abnormality is known as aneuploidy, defined as an abnormal number of chromosomes in a cell. Structural chromosomal abnormalities result from breakage and incorrect re-joining of chromosome segments.

Effectively, all disease-causing aneuploidies arise as *de novo* events. Only a select few aneuploidies are compatible with human life in the constitutional state. In humans, the most common aneuploidies are trisomies. A recognized example is trisomy 21, first identified in 1959 as the cause of Down syndrome (Lejeune, Gauthier, and Turpin 1959). Nondisjunction during meiosis will result in all cells being affected in the offspring, with meiotic risk increasing with maternal age; while postzygotic nondisjunction (during mitosis) will result in mosaicism.

A wider range of aneuploidies has been observed in the mosaic state, with human sexual mosaics being good examples. These include familiar conditions such as Turner syndrome (45,X) and Klinefelter syndrome (47,XXY). Depending on when the chromosomal segregation defect occurred in the early zygote, these conditions can have varying percentages of cells affected, prevalence, and clinical features.

Ring chromosomes and isochromosomes are two other large structural abnormalities that are observed in the mosaic state. Ring chromosomes occur when two terminal breaks in both chromosome arms, resulting in 'sticky ends', fuse together forming a ring (**Figure 4.8**).

Isochromosomes are caused by centromere mis-division and occur frequently enough that recognizable syndromes can be established, the most common being i(Xq), where no Xp material is present (Wolff et al. 1996). Another example includes i(12p), which causes Pallister–Killian syndrome. Pallister–Killian syndrome is an extremely rare genetic disorder caused by the presence of four copies of the short arm of chromosome 12 (12p). The two additional copies of 12p appear as a single chromosome, instead of one long (12q) arm and one short (12p) arm. It is only seen in the mosaic state, limited to specific cell types, because such an aberration is lethal constitutionally (Campbell et al. 2015), but tolerated if absent from tissues that are vital to life (Raffel, Mohandas, and Rimoin 1986).

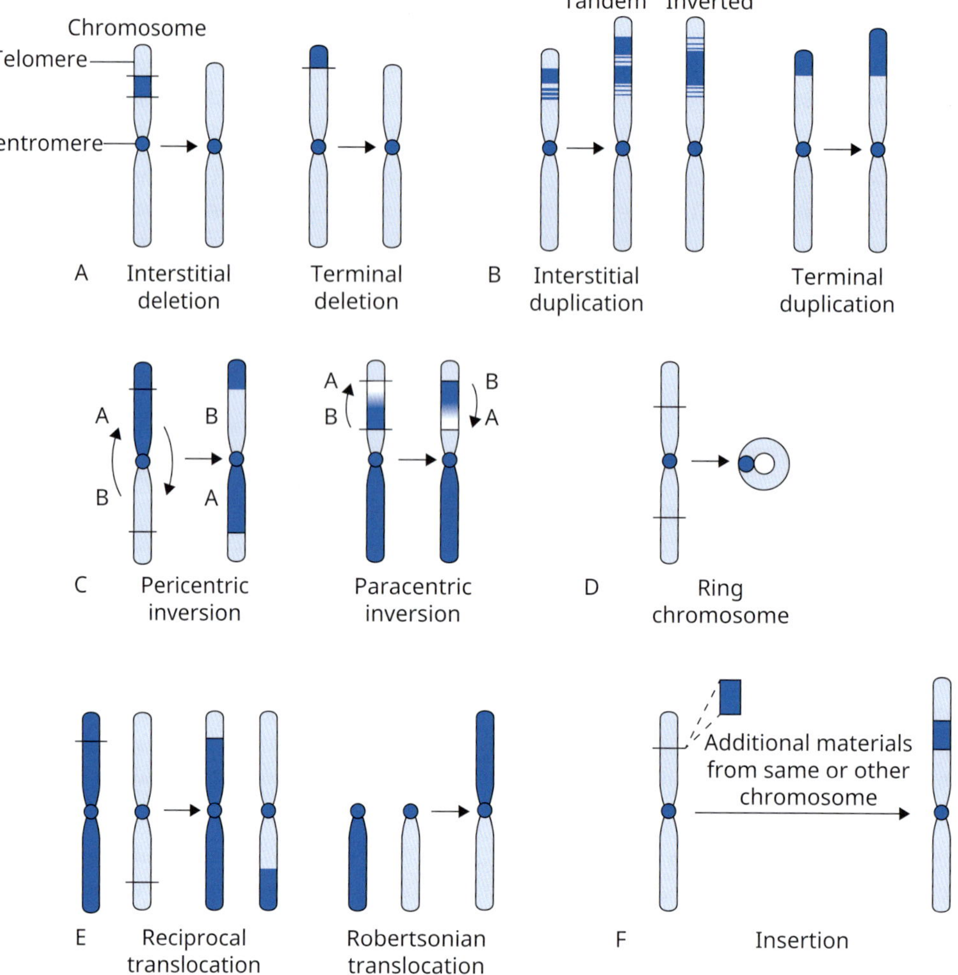

FIGURE 4.8
Structural chromosomal abnormalities. Chromosomal instability can be severe, resulting from large structural alterations that affect the expression of many genes. In nature, examples include deletions, duplications, inversions, ring chromosomes, translocations, and insertions.

4.6 Finding DNA 'needles' in a genetic 'haystack'

4.6.1 Single nucleotide variations

Studies based on observation of *de novo* single nucleotide variants (SNVs) in parent–offspring trios have identified a low rate estimated to range between 0.82–1.70 × 10^{-8} mutations per base per generation (Campbell et al. 2012; Kong et al. 2012; Dal et al. 2014). Based on the mutation frequency of these studies and given that there are approximately 3.0 × 10^9 base pairs in the human genome (Abdellah et al. 2004), it can be inferred that the average individual acquires 37 to 51 *de novo* single nucleotide mutations during development.

Comparisons of monozygotic twins were used to calculate the proportion of early postzygotic mutations to the overall human *de novo* SNV rate. It was estimated to be 0.34×10^{-8} and 0.04×10^{-8} in each twin (Dal et al. 2014), which translates into 1 to 10 single nucleotide mutations in the genome. This entails that although postzygotic SNVs are a relatively rare occurrence, they comprise a considerable proportion of the rate of *de novo* mutations in humans. Acuna-Hidalgo et al. (2015) have even shown that an important fraction of *de novo* mutations presumed to be germline in fact occurred either postzygotically in the offspring or were inherited because of low-level mosaicism in one of the parents. These mitotic events during embryogenesis resulting in somatic mosaicism may thus have an important and significant contribution in disease discordance observed in monozygotic twins (Vadgama et al. 2019b). It is possible, however, that individual mutation rates vary considerably. This necessitates much larger studies if the true extent of variation in mutation rates is to be determined.

4.6.2 Copy number variations

Germline and somatic *de novo* genetic alterations have been implicated in human disease for decades. Over the past several years, genomic microarray technology has provided insight into the role of *de novo* copy number variations (CNVs) in disease (Veltman and Brunner 2012). A CNV is a section of DNA that is ≥ 1 kb and present at variable copy number when compared to a reference genome (Redon et al. 2006). They are structural variations that comprise deletions, insertions, duplications, and complex multi-site variants, which alter the diploid status of DNA. They have been found in all human populations and are widespread throughout the genome (Freeman et al. 2006).

Large CNVs typically occur at a low rate, arising at a frequency of only 0.01 to 0.02 events per generation (Veltman and Brunner 2012; Campbell and Eichler 2013; Kloosterman et al. 2015), but their disruptive effect on many genes contributes considerably to severe congenital malformations and neurodevelopmental disorders (Weischenfeldt et al. 2013).

Previous studies have demonstrated that postzygotic CNVs can cause discordant phenotypes in monozygotic twin pairs that otherwise are genetically identical (Bruder et al. 2008). There is now overwhelming evidence that CNVs play a significant role in normal population variation and evolution (McCarroll et al. 2008; Ionita-Laza et al. 2009), and significant CNV associations have been found to underlie disease. Although postzygotic *de novo* CNVs have been detected in discordant monozygotic twins, many studies testing this hypothesis have reported no or limited evidence (Laplana et al. 2014; Abdellaoui et al. 2015). Additionally, there are reports of copy number differences in monozygotic twin pairs without discordant phenotypes (Abdellaoui et al. 2015; Magnusson et al. 2016).

Different types of mutations are summarized in **Table 4.3**. The abundance of missense and nonsense variants suggests that Mendelian disorders are primarily associated with alterations in the coding region. However, as our analysis protocols improve, we are likely to see an increase in the regulatory variants, the functional roles of which are just beginning to be explored.

4.6.3 Mitochondrial DNA

When we talk of the human genome, we often overlook the fact that humans possess at least two distinct genomes: nuclear DNA, which has been the centre of discussion thus far, and mitochondrial DNA (mtDNA), a circular structure of double-stranded DNA found inside mitochondria. Mitochondria are vital to cellular energy production, and thus crucial to healthy human

TABLE 4.3 Relative frequency of gene variants responsible for human disorders

Change	Frequency	% of total
Missense/nonsense	15,9705	57.9
Splicing	23,868	8.7
Regulatory	4,575	1.7
Small deletions	39,822	14.4
Small insertions	16,881	6.1
Small indels	3,652	1.3
Gross deletions	19,491	7.1
Gross insertions	4,945	1.8
Complex rearrangements	2,231	0.8
Repeat variations	546	0.2
Total	**27,5716**	**100.0**

heteroplasmy
A condition in which a cell or individual has more than one type of organellar genome. Typically used to describe the situation in which two or more mitochondrial DNA (mtDNA) variants exist within the same cell.

homoplasmy
The condition in which all copies of the mitochondrial genome are identical. This might be either wildtype or mutated sequences.

physiology. mtDNA consists of merely 37 genes, which code for tRNAs, rRNAs, and proteins involved in cellular respiration.

Hence, disorders that are linked to tissues with high energy demands, such as brain or skeletal muscle, have been associated with defects in the mitochondria (Mattson, Gleichmann, and Cheng 2008; Rygiel et al. 2015). The pathogenesis of these disorders may involve mutations in either mtDNA or nuclear-encoded mitochondrial genes (Rollins et al. 2009).

In contrast to the nuclear genome, double-stranded circular mtDNA is present in multiple copies within the same cell, sometimes as many as thousands, and is exclusively maternally inherited. Among the hundred to several thousand copies of mtDNA in a single cell, segregation of the mitochondrial populations during cell division can lead to mtDNA variation as they divide. The coexistence of different mtDNA sequences is known as **heteroplasmy** (Stewart and Chinnery 2015; **Figure 4.9**); whereas **homoplasmy** describes a cell that has a uniform collection of mtDNA.

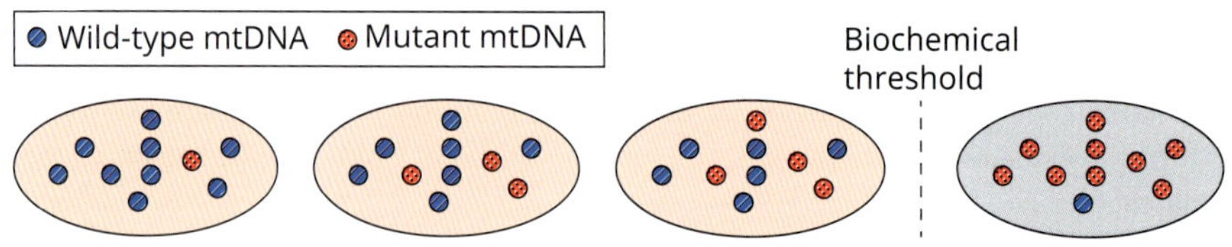

FIGURE 4.9
mtDNA heteroplasmy and the threshold effect. The same cell can contain different populations of mtDNA. If a mutation is damaging, the cell may be able to tolerate a certain level until a threshold is exceeded leading to disease. Heteroplasmy is notoriously difficult to quantify accurately, so it is not clear what threshold is considered pathogenic. Generally, however, a low mutation load of < 10% is considered benign, and a high mutation load of > 80% often leads to disease, suggesting that mtDNA mutations are haploinsufficient or recessive.

Incomplete penetrance of several mitochondrial diseases can be attributed to heteroplasmy of pathogenic mtDNA mutations (Stewart and Chinnery 2015). Disease onset is (partly) dependent on whether the allele frequency of a pathogenic mtDNA variant exceeds a certain threshold (DiMauro and Schon 2003).

mtDNA is known to have a higher mutation rate owing to apparent decreased replication fidelity (Song, Wheeler, and Mathews 2003). mtDNA is known for its lack of protective histones and limited repair capacity, which can render it susceptible to insults that can overcome its replicative machinery or stimulate mitochondrial fusion causing mtDNA levels to differ. Various forms of cancer, toxin exposures, aging, and oxidative stress can either increase or decrease mtDNA copy number, depending upon the nature of the damaging process (Wrede et al. 2015).

Deep sequencing is increasingly being used to detect genomic variation. However, quantifying mtDNA heteroplasmy can be challenging (Yao, Kajigaya, and Young 2015). Mitochondrial variation can be misinterpreted given the presence of nuclear mitochondrial sequences (NumtS), which are copies of mtDNA found in the nuclear DNA. Studies have reported difficulties in estimating variation due to co-amplification of NumtS with the mtDNA (Bouhlal et al. 2013). Recent studies, however, have demonstrated significant improvements in increasing the sensitivity of heteroplasmy detection, and NGS has been recognized as an effective method for the identification of low levels of mtDNA heteroplasmy (Kloss-Brandstätter et al. 2015).

4.7 Methodological considerations for identifying *de novo* variants

4.7.1 Why tissue type matters

Reliably detecting *de novo* somatic mutations is more complex than calling *de novo* germline mutations, because somatic mutations will vary between tissue types and appear in percentages that are akin to false-positive sequencing rates. High-level mosaicism can result from early postzygotic mutations in an organism, leading to a wide distribution of the variant in many different tissues. In contrast, mutations that occur late during embryogenesis, or postnatally, can be present in only a select few somatic cells (Acuna-Hidalgo, Veltman, and Hoischen 2016). The difficulty in obtaining a wide variety of tissues in humans, due to ethical and practical reasons, makes a comprehensive study of somatic mosaicism challenging. Still, inter-tissue variation of somatic mutation frequencies has been determined by analysing routinely sampled tissues, such as blood, buccal epithelium, skin fibroblasts, hair follicles, and urine. Lindhurst et al. (2011) led one of the first studies to detect somatic mosaicism in monozygotic twins using exome sequencing. The authors identified the cause of Proteus syndrome, in which a *de novo* mutation was found in *AKT1* in multiple tissues, but not in DNA from peripheral blood.

4.7.2 Methods in single nucleotide variant and indel analysis

Although NGS has proven to be a powerful approach, there remain numerous technical challenges in obtaining an accurate and complete record of sequence variation from the copious data generated, and in converting raw sequence reads into biologically meaningful information (Pirooznia et al. 2014) (**Figure 4.10**). Granted accurately mapped and calibrated reads, identifying SNVs and indels, not to mention more complex variation such as insertions, deletions,

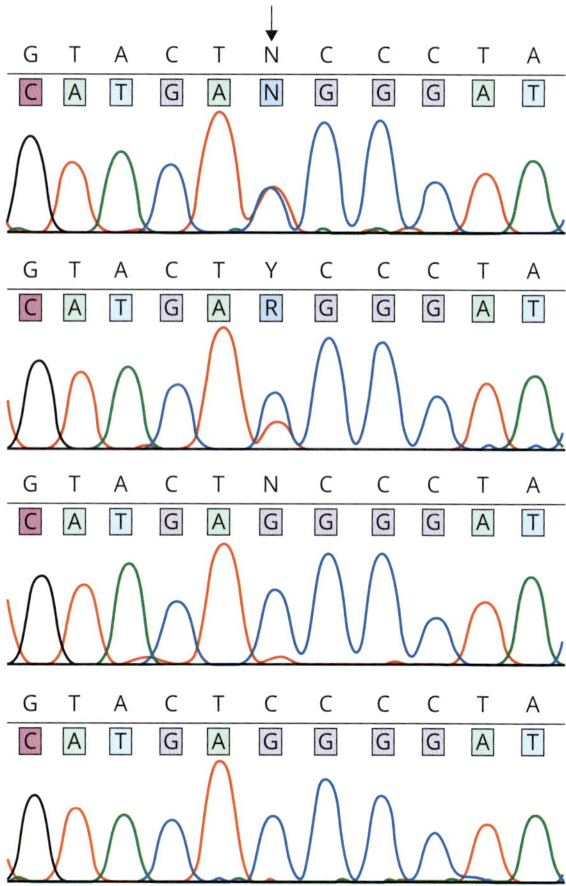

FIGURE 4.10
Sanger sequencing detection of mosaicism for point mutations. A Sanger sequence trace can indicate heterozygosity by approximately equal peak heights of the two nucleotides at a single position (C and T in this example). However, with decreasing levels of allele frequency, a lower level of mosaicism is difficult to quantify, hence necessitating sequencing technologies with higher resolution.

inversions, CNVs, and multiple base pair substitutions, requires complex statistical models and sophisticated bioinformatics tools (Pirooznia et al. 2014).

Several methods have recently been developed to enhance variant calling accuracy, including GATK, SAMtools, and MuTect2 (**Figure 4.11**). Unfortunately, however, there can be considerable disagreement among the algorithms (O'Rawe et al. 2013), thus highlighting the different error prediction models or assumptions underlying each method. Taking the union of two or more variant callers will help reduce the probability of calling false positive SNVs and indels and avoid biases from one particular caller. However, choosing the ideal tools for variant calling will depend on a specific set of data type and experimental conditions (Xu et al. 2014).

SELF-CHECK 4.5

Why is it so difficult to detect and quantify mosaicism?

4.7.3 Methods in copy number variant analysis

Genotyping arrays (or SNP arrays) have outpaced the use of aCGH in CNV detection. SNP array genotyping usually has a higher probe density, allowing better resolution of CNV breakpoints (Legault et al. 2015). In fact, there has been a significant upsurge over the past decade in the development of algorithms and technical resolution that are applied across platforms and

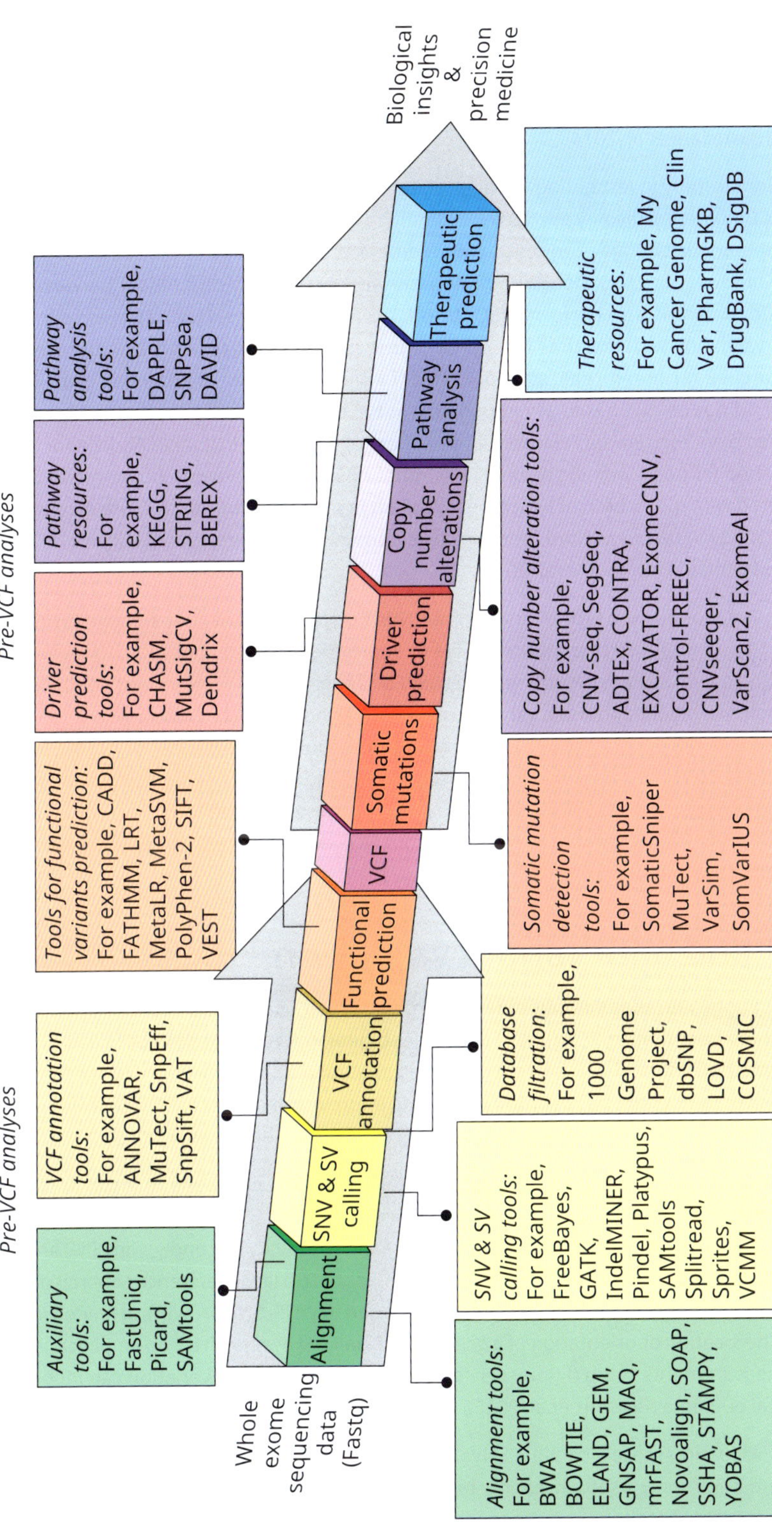

FIGURE 4.11
Whole exome sequencing data analysis steps. Many computational pipelines and tools have been developed to analyse the full spectrum of exome sequencing data, translating raw **FASTQ** files to biological insights and precision medicine.

FASTQ
A text-based data file format for storing both a biological sequence (usually nucleotide sequence) and its corresponding quality scores.

programs. Various software packages are available, including Birdsuite, dChip, cnvPartition, Genotyping Console, and PennCNV. In a study assessing putative CNV calls made using multiple software, Castellani et al. (2014) suggest using a combination of three programs to optimally identify true CNV calls, giving a trade-off between sensitivity and specificity. The authors also suggest the use of PennCNV over all other methods when the use of only one tool is preferable (Castellani et al. 2014).

In a longitudinal study investigating the age-related acquisition of somatic structural variants, four mosaic uniparental disomy events were identified in a cohort of elderly Danish twins, suggesting that somatic changes can result within different body tissues of the same individual (Magaard Koldby et al. 2016). Detection of non-mosaic somatic CNV differences between the monozygotic twins was also performed. Of these, however, only one CNV was eligible for experimental validation with real-time PCR (qPCR), but it could not be validated. Numerous studies have searched for non-mosaic somatic CNV differences between monozygotic twins, with the overall findings being unfruitful. Although this is typically attributed to a small sample size, some studies have recruited a relatively large cohort of twins. For instance, Abdellaoui et al. (2015) searched for post-twinning *de novo* CNV mutations in 1,097 co-twins, but only two CNVs were validated with qPCR, both of which were present in the same individual. More recently, in a cohort of 100 twin pairs enriched for neurodevelopmental disorders, no postzygotic *de novo* CNVs were identified (Stamouli et al. 2018).

Another explanation for the lack of discordant CNVs between twins is that somatic CNVs are more likely to be mosaic, necessitating the use of appropriate copy number calling tools. This is in line with several studies (Bruder et al. 2008; Forsberg et al. 2012; Magaard Koldby et al. 2016). For microarray data, mosaic variants can be detected using the Mosaic Alteration tool (González et al. 2011), and estimation of the mosaic proportion of cells can be performed based on the study by Rodríguez-Santiago et al. (2010). Exome and genome sequencing can also be harnessed to detect mosaic structural abnormalities. A method called MrMosaic, which uses deviations in allele fraction and read coverage from NGS data to detect structural mosaic abnormalities, has proven to be successful (King et al. 2017).

high-resolution melting analysis
A post-polymerase chain reaction analysis method in which characteristic melting temperatures are observed and used to identify genetic variation in amplicons.

allele-specific PCR
A method for detecting any known mutations involving single base changes or small deletions, which is based on the use of sequence-specific PCR primers that allow amplification of test DNA only when the target allele is contained within the sample. Also referred to as amplification refractory mutation system (ARMS).

immunohistochemistry
A method for localizing specific antigens within fixed tissue sections based on antigen-antibody recognition.

pyrosequencing
A non-gel-based DNA sequencing technique that detects inorganic pyrophosphate (light) released during DNA synthesis.

SNaPshot
Single Nucleotide Polymorphism (SNP) genotyping technology developed by Applied Biosystems (ABI). The assay involves using fluorescently tagged dideoxynucleotides (ddNTPs) to terminate the extension of DNA primers after the addition of one nucleotide and enables the precise determination of SNPs based on emitted fluorescence.

4.7.4 Candidate variant validation

It has been suggested that without independent validation using bench confirmation techniques, such as quantitative PCR (qPCR) or droplet digital PCR (ddPCR), CNV calls using computational methods should be at best considered tentative (Castellani et al. 2014). Further, incorporation of family data has been shown to help improve the quality of CNV calls alongside the use of multiple CNV calling methods (Castellani et al. 2014; Legault et al. 2015).

There are several reasons why microarray CNV calling methods are prone to error, and thus why experimental validation is essential. Firstly, CNV calling using SNP arrays is a relatively new technology, and each platform has its own sensitivity and specificity. Secondly, with SNP arrays, the sample in question is compared to reference samples or to a large reference cohort, so only the relative copy number is determined. Third, compared to WGS, SNP arrays give limited information on the location or orientation of a given CNV. Finally, SNP arrays rely on amplification of DNA and measurements using (fluorescence) intensity, thus technical variation can influence experimental outcome (Brosens et al. 2016).

In terms of validating mosaic somatic SNVs and indels, other methods alongside Sanger sequencing are commonly used, including **high-resolution melting analysis**, **allele-specific PCR**, **immunohistochemistry**, **pyrosequencing**, and **SNaPshot**. These techniques are especially useful in candidate variant screening and result verification in the NGS era (Gajecka 2016). **Figure 4.10** shows that low levels of mosaicism are difficult to quantify with Sanger sequencing.

CASE STUDY 4.1 Discovery of a mosaic basis for auto-inflammation

Identifying the causes of systemic inflammatory diseases is challenging. Somatic mutations have long been suspected to play a role but are as yet poorly understood.

A baby boy presented with dysmorphia and neonatal onset immunodeficiency with auto-inflammation. The child has healthy parents and two healthy older brothers. A peripheral blood sample was taken from the patient for DNA extraction and exome sequencing.

The sequence data revealed an in-frame four amino acid deletion in the interleukin 6 cytokine family signal transducer (*IL6ST*) gene, which, further experiments confirmed, led to constitutive GP130 cytokine receptor signalling. This meant the deletion caused IL6ST to always be in an active state even in the absence of an agonist.

The deletion was confirmed by amplicon deep sequencing and was found to be present in white blood cells, buccal cells, hair follicles, and urine sediment. Most samples showed a variant allele frequency (VAF) of between 30 and 40%. A lower VAF of 15% was seen in the white blood cells, which suggested that the variant might be a disadvantage to peripheral blood immune cells.

To search for the other variants potentially contributing to the patient's condition the sequence data was analysed for all known pathogenic variants, rare variants in known disease-associated genes, and also rare variants in potentially novel disease-causing genes.

To simplify the interpretation of any variants of unknown significance (VUS), whole exome sequencing was also carried out on DNA samples from the patient's parents. Apart from the reported mosaic IL6ST c.560_571del (Tyr186_Tyr190del), no known or potentially novel variants which could plausibly contribute to the patient's phenotype were found.

Loss-of-function variants in the *IL6ST* gene had been previously reported in patients with elevated IgE levels. However, this mosaic IL6ST syndrome was distinguishable from its germline equivalent because this patient had normal IgE levels, persistent elevation of inflammatory markers, and constitutive activation of IL6 signalling.

The IL6-independent activation of GP130 in this patient led to hyperphosphorylation of Tyr705 in STAT3, and the clinical team were able to advise the use of ruxolitinib or tofacitinib to combat this hyperphosphorylation in the hope of easing the symptoms.

Studying the variant allele frequencies in different body tissues may help to ascertain the impact of the variant on different cell types and thereby reveal more about the molecular biology of the condition of interest.

Variants of unknown significance can be compared in close family members of affected individuals to ascertain those that are likely benign.

Revealing the downstream molecular consequences of a variant can potentially lead to treatment for a condition.

Mosaicism is implicated in many conditions and with recent advances in sequencing technology we can begin to understand its impact on a case-by-case basis.

Q1 Do you think the affected child would have been born had the mutation been in all their cells?

Key Points

In cases of mosaicism, a number of different tissue types should be tested to ascertain the presence of the variant allele.

4.8 Searching for downstream consequences

We have discussed how genetic alterations—whether SNVs, indels, or CNVs—can lead to disease, and how to detect these alterations. Patients that are interrogated at multiple omics levels or investigated under a variety of environmental conditions are offering remarkable insights (**Figure 4.12**). The following section will outline how the epigenome, transcriptome, proteome, and metabolome of patients can be studied in an integrative manner to provide a more systematic exploration of gene–environmental interaction, and what the current research in this area has to offer.

4.8.1 Epigenetics

Epigenesis, a term from which *epigenetics* derives, is an early embryological theory postulating the development of an embryo from the successive differentiation and elaboration of an originally undifferentiated fertilized egg (as opposed to preformation, an outmoded developmental model that suggested complex organisms are already completely formed in the germ cell and develop merely by enlargement). The term *epigenetics* was initially coined by developmental biologists who sought to explain the mechanism by which gene–gene and gene–environment interactions fashioned the phenotype of an individual during development (Youngson and Whitelaw 2008). However, the term has found new meaning in modern language. Epigenetics can be generically described as alterations in genomic function that are controlled by potentially reversible heritable factors, without altering the DNA sequence.

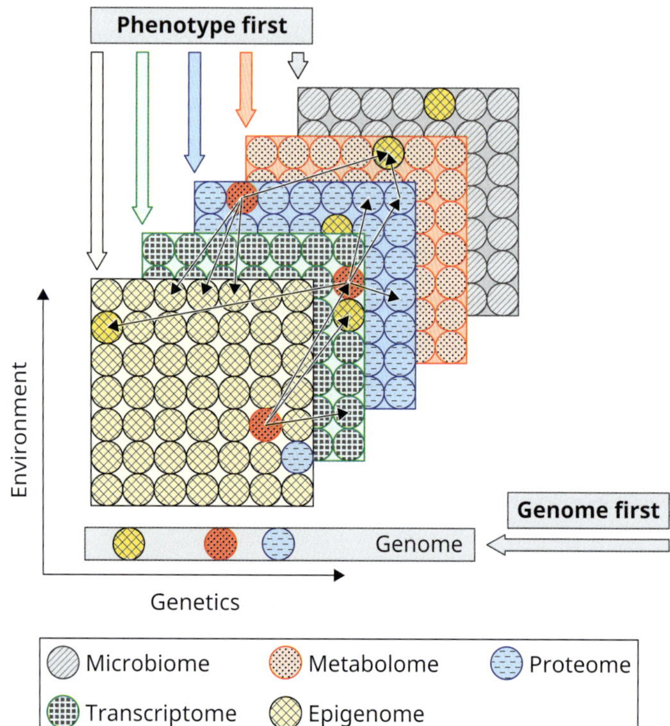

FIGURE 4.12
Multi-omics approach to disease. Layers depict different types of 'omics' data and circles represent molecules from which samples are collected. Layers above the genome reflect downstream effects of genetic regulation and environmental influences. The thin arrows represent interactions between molecules in different layers. The thick arrows at the top indicate different omics approaches to investigate disease.

In other words, epigenetic regulation of gene expression uses reversible modifications of DNA and chromatin structure to mediate the interaction of the genome with a variety of environmental factors and to generate changes in the patterns of gene expression in response to these factors. Such mechanisms include DNA methylation, histone modifications, ATP-based chromatin remodelling, transcription factor-binding mechanisms, and non-coding RNA-mediated gene silencing (Bell and Spector 2011; Ketelaar, Hofstra, and Hayden 2012).

Data from the ENCyclopedia of DNA Elements (ENCODE) project suggest that biochemical functionality could be assigned to 80% of the human genome, affecting regulatory and tissue-specific expression patterns (Dunham et al. 2012). This undermines old genetics textbooks, which claim that the genome is mainly composed of 'junk' DNA. The ENCODE project used a range of different experimental assays to analyse what they referred to as 'sites of biochemical activity'. The ENCODE research shows that non-coding variants can modulate cis-regulatory machineries via several mechanisms, such as DNA methylation, disrupting or creating transcription factor binding sites, altering DNA–DNA interactions and chromatin looping, or miRNA recruitment. The advent of large-scale and high-throughput sequencing technologies has helped to unravel chromatin biology and genome function.

There are many experimental approaches that allow the dissection of the regulatory landscape. For example, transcription factor binding sites and chromatin states (histone modifications) can be assessed using ChIP-seq; chromatin interactions can be identified using chromosome conformation capture techniques (such as 3C, 4C, 5C, Hi-C, and ChIA-PET); and open chromatin, indicative of active regulatory regions, can be analysed using DHS-seq, FAIRE-seq, or ATAC-seq (**Figure 4.13**). As discussed earlier, current computational tools are based on evolutionary conservation and have been calibrated on variants associated with disease. However, novel approaches are emerging that integrate multiple layers of genomic information to assess and predict the impact of disease-risk variants.

Epigenetic changes can be observed to variable degrees in patients afflicted with disease, estimates for which have been shown to be affected by sample size, tissue type, age, and CpG island selection (Czyz et al. 2012). Indeed, a major challenge in epigenetic research is that these differences are often tissue- and cell-specific, and sometimes it may be impractical to obtain the ideal tissue source.

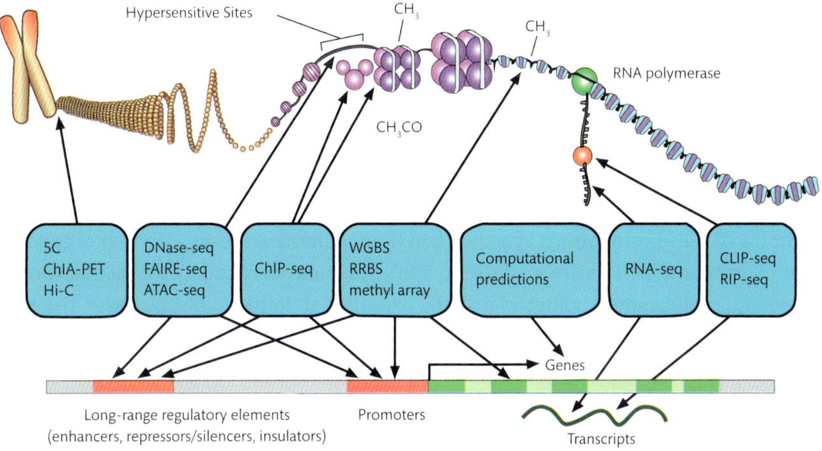

FIGURE 4.13
Methods to study the epigenome.

Interestingly, epigenetic and transcriptomic modifications have been reported in twins discordant for a range of complex traits, including cancer (Heyn et al. 2013), Alzheimer's disease (Poulsen et al. 2007), autism spectrum disorders (Loke, Hannan, and Craig 2015), multiple sclerosis (Handunnetthi, Handel, and Ramagopalan 2010), pain sensitivity (Bell et al. 2014), psoriasis (Gervin et al. 2012), and type 1 diabetes (Stefan et al. 2014). Quite often, these differences would be found despite no differences detected at the genetic sequence level.

4.8.2 Proteomics

Where whole genome and exome sequencing has been less beneficial, attention is increasingly switching to RNA sequencing and epigenetics to explore potentially missed disease-causing mechanisms. Proteomics, however, has the potential to capture all the variations arising from genomic, transcriptomic, and epigenetic changes (**Figure 4.10**).

Alternative splicing and post-translational protein modifications entail that the number of proteins can be two orders of magnitudes higher than the number of genes (Zierer et al. 2015). Proteomic techniques in current practice, such as immunoassays, protein arrays, or mass spectrometry, are limited in that they can measure only a small fraction of the proteome. The most comprehensive analysis of the human proteome to date consists of over 18,000 proteins composed from 10,000 mass spectrometry experiments, across various tissues (Wilhelm et al. 2014).

Analysing the proteomes of human biological fluids (such as serum, urine, saliva, synovial, and cerebral spinal fluids) among individuals with a shared genetic background, but possibly different environmental and/or epigenetic influences, may cast welcome new light on the identification of putative disease-associated biomarkers. Although comprehensive proteomics studies on discordant monozygotic twins are still lacking, the few studies that have implemented the co-twin design have provided some important insights. Disorders investigated include systemic autoimmune diseases (O'Hanlon et al. 2011), bipolar disorder (Kazuno et al. 2013), and chronic fatigue syndrome (Ciregia et al. 2013). In a recent study, a proteomic profiling of serum samples from monozygotic twins discordant for ischaemic stroke was analysed through a label-free pipeline. The finding of a distinct proteomic profile associated with ischaemic stroke raises the possibility of patient-centred diagnostic, prognostic, and therapeutic strategies in future (Vadgama et al. 2015a, 2019a).

4.8.3 Metabolomics

The metabolome refers to the complete set of low molecular weight compounds in a sample. These compounds are the substrates and by-products of enzymatic reactions and have a direct effect on the phenotype of the cell. Thus, metabolomics aims at determining a sample's profile of these compounds at a specified time under specific environmental conditions. Similar to proteomics, there are, to date, no analytical methods at our disposal that can determine and quantify all metabolites in a single experiment. Remarkably, however, the Human Metabolome Database (Wishart et al. 2012) contains more than 40,000 distinct metabolites from different tissues. These low molecular weight compounds are the closest link to phenotype. Although genomics and proteomics have provided extensive information regarding the genotype, it may be difficult to elicit direct information about the clinical phenotype being investigated.

Several new omics technologies have emerged recently, which deserve to be mentioned but cannot be described in depth due to their branching themes lying beyond the scope of this chapter. These include glycomics (post-translational modifications), microbiomics, and phenomics (Zierer et al. 2015). The currently employed approach to data analysis in the omics era is to take the reductionist stance by focusing on individual factors. However, these emerging technologies have the capacity to broaden investigations into complex traits by an unprecedented amount. The 'holistic' approach, or systems biology, integrates data from different experiments to gain an understanding of the system as a whole (Zierer et al. 2015).

Large-scale multi-omics investigations into monozygotic twins are currently being carried out. For instance, the MuTher study, consisting of several hundred female twins, has been evaluated globally at the genome, transcriptome, metabolome, and microbiome levels (Fizelova et al. 2016). The data has given valuable insights into the genetic control of molecular traits, biological pathways involved in metabolic syndrome, and the heritability of gut microbiota (Heinig et al. 2010). Another human reference population study, dubbed Metabolic Syndrome in Man (METSIM), consists of a cohort of about 10,000 Finnish men. Like the MuTher population, METSIM participants have been characterized clinically for a variety of metabolic and cardiovascular traits at the genomic, transcriptomic, and metabolomic levels (Fizelova et al. 2016; Civelek et al. 2017; Laakso et al. 2017).

4.9 Big data and translational medicine: from the bench to the bedside

To date, thousands of GWAS have identified numerous genetic loci for hundreds of traits. Despite the explosive sample sizes in recent years, these GWAS risk loci can encompass multiple genes. Some of these genes may be related to the disorder studied, while others may not. The increasing availability of multi-dimensional omics data, such as chromatin accessibility, histone modifications, proteomics, and transcriptomics now makes it possible to connect genetic signals more precisely to specific genes. Interrogation of functional information from large repositories, such as ENCODE, Roadmap, and the Genotype–Tissue Expression (GTEx) project has helped identify potentially causal variants and genes.

Polygenic risk scores are not yet routinely used in the clinic because much improvement remains in how they are generated and exactly how they should be utilized. Nevertheless, improvements in data collection and analysis have allowed increased predictive accuracy and are already being offered by direct-to-consumer genetic testing companies for some complex disorders. An acknowledged limitation, however, is their dependency on linear regression and inability to account for complex data interactions (Ho et al. 2019).

Machine learning, the objective of which is to develop computer algorithms that improve with experience, can aid in the analysis of large, complex datasets. In contrast to polygenetic risk scores, machine learning data modelling in complex disease prediction is superior given its ability to model multi-dimensional data and its proven capacity to classify disease risk with high precision (Ho et al. 2019). These promising new developments could pave the way for improved healthcare outcomes, and ultimately bridge the translational gap from the lab bench to the patients' bedside (**Figure 4.14**).

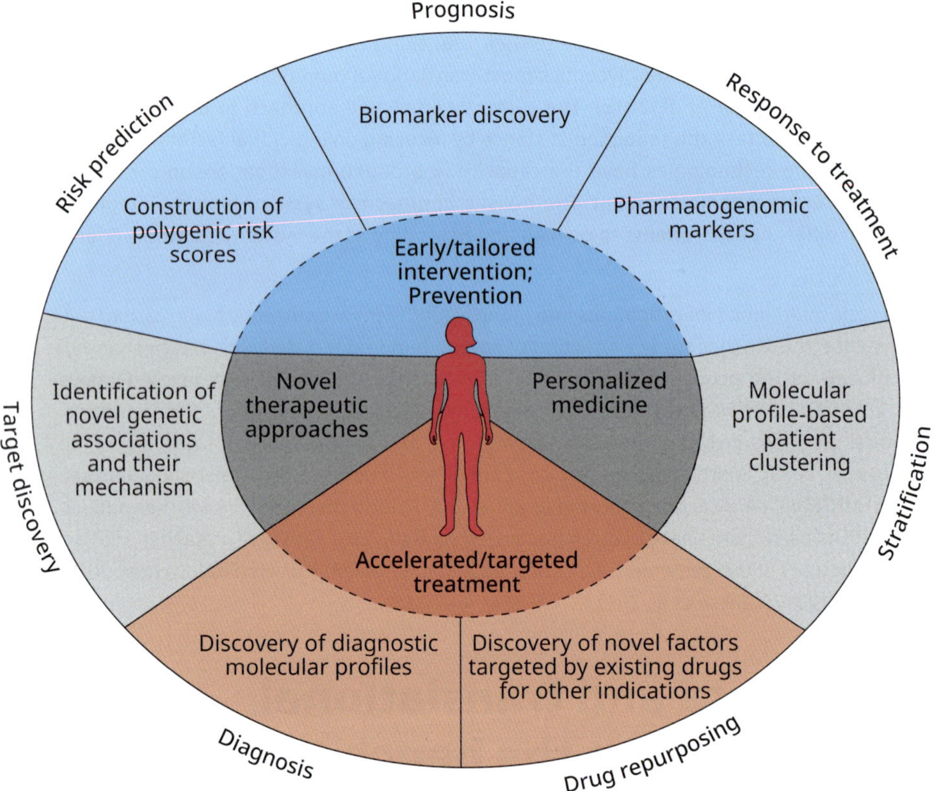

FIGURE 4.14

Complex disease and the potential for translational genomics. We are entering a new era of individualized medicine that incorporates the development of novel, gene-based diagnostics and preventive therapies. This could lead to significant improvements in human health and well-being.

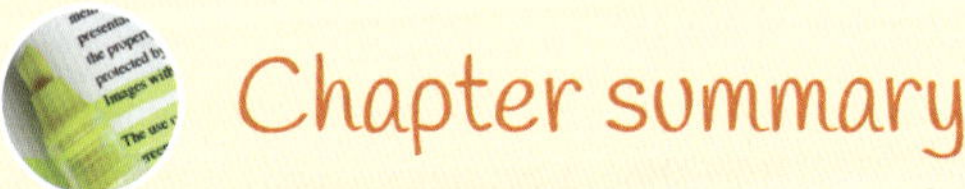

Chapter summary

- Single gene disorders can be contrasted with polygenic and multifactorial conditions.
- Twin studies were one of the first attempts to estimate the degree to which a disease is genetic, and linkage analyses and large-scale association studies quickly followed.
- Although GWAS have successfully uncovered several susceptibility loci for complex disorders, these loci account for only a small proportion of estimated heritability. You have learned about the limitations of GWAS, and the realization that rare variants also play an important role in the genetic architecture of complex disorders and how this knowledge resulted in the development of rare-variant analysis methods.
- Recently, single-cell epigenetic and transcriptomic profiling across multiple tissues and cell types in large cohorts has given us a better understanding of gene–environment interactions and the aetiology of complex traits.

- Other 'omics' technologies, such as proteomics and metabolomics, have provided insights into downstream consequences of disease, and are increasingly being incorporated into personalized medicine.
- The future of genetic studies will likely be based on 'multi-omic' and machine learning approaches and might radically improve prediction of complex disorders and how this will require a conceptual shift in the research and diagnostics paradigm.

Discussion questions

4.1 Talk through the key design points in family, adoption, and twin studies and relate these points to the type of information that can be gathered from each.

4.2 How is family information combined with linkage data used to elucidate loci associated with particular phenotypes?

4.3 Why has this approach proved successful for single gene disorders and complex conditions with only a few major susceptibility loci but not for conditions influenced largely by many genes of small effect?

Further reading

- Mills MC, Barban N, Tropf FC (2022) ***An Introduction to Statistical Genetic Data Analysis***. The MIT Press, Cambridge, MA. ISBN: 9780262538381.

  This book gives a comprehensive introduction to modern applied statistical genetic data analysis and is accessible even to those without a strong background in molecular biology or genetics.

- Jorde LB, Carey JC, Bamshad MJ (2019). ***Medical Genetics.*** 6th Edition, Elsevier, Amsterdam. ISBN: 9780323597371.

  Giving solid background information and written in an accessible way, this book covers recent advances in the genetics of common diseases, as well as current progress in gene therapy. *Medical Genetics* integrates key concepts with clinical practice and will help the Biomedical Scientist to appreciate how their work fits into the wider context of patient-facing practice.

- Strachan T, Lucassen A (2022). ***Genetics and Genomics in Medicine.*** 2nd Edition, CRC Press, Boca Raton, FL. ISBN 9780367490812.

  This excellent book places emphasis on the modern techniques that are revolutionizing the use of genetic information in medicine. Amongst other topics the book explores the complex role of genetics in common diseases, epigenetics, and the roles of non-coding RNA molecules. It also gives up-to-date information about cancer detection and genomics and approaches to treatment and prevention, including pharmacogenomics, genetic testing, and personalized medicine.

- Biesecker LG, Spinner NB (2013) ***A genomic view of mosaicism and human disease***. *Nat Rev Genet*, 14, 307–20.
- Gajecka M (2016) ***Unrevealed mosaicism in the next-generation sequencing era***. *Mol Genet Genomics*, 291, 513–30.
- Hasin Y, Seldin M, Lusis A (2017) ***Multi-omics approaches to disease***. *Genome Biol*, 18, 1–15.
- Hay W, Levin M, Deterding R, et al. (2009) ***Current Diagnosis and Treatment Pediatrics***. 19th Edition. McGraw-Hill Professional, New York.
- Hintzsche JD, Robinson WA, Tan AC (2016) ***A survey of computational tools to analyze and interpret whole exome sequencing data***. *Int J Genomics*, 2016. doi: 10.1155/2016/7983236.
- Kong A, Thorleifsson G, Gudbjartsson DF, et al. (2010) ***Fine-scale recombination rate differences between sexes, populations and individuals***. *Nature*, 467, 1099–103.
- McCarthy MI, Abecasis GR, Cardon LR, et al. (2008) ***Genome-wide association studies for complex traits: consensus, uncertainty and challenges***. *Nat Rev Genet*, 9, 356–69.
- Tam V, Patel N, Turcotte M, et al. (2019) ***Benefits and limitations of genome-wide association studies.*** *Nat Rev Genet*, 20, 467–84.
- Turnbull C, Scott RH, Thomas E, et al. (2018) ***The 100 000 Genomes Project: bringing whole genome sequencing to the NHS***. *BMJ*, 361. doi: 10.1136/bmj.k1687
- Zeggini E, Gloyn AL, Barton AC, et al. (2019) ***Translational genomics and precision medicine: Moving from the lab to the clinic***. *Science*, 365, 1409–13.

Useful websites

- Mendelian Inheritance in Man (OMIM®). **https://www.omim.org/** This continually updated online resource was developed from Victor McKusick's seminal catalogue of Mendelian phenotypes and their associated genes entitled, 'Mendelian Inheritance in Man'. OMIM now describes over 16,400 genes and priority is given to those whose function is understood and/or are known to be associated with particular defined phenotypes. Moving this resource online has allowed for much expansion of information and enabled the addition of many search, cross-referencing, and link-out features that cannot be achieved with hard copy.

5

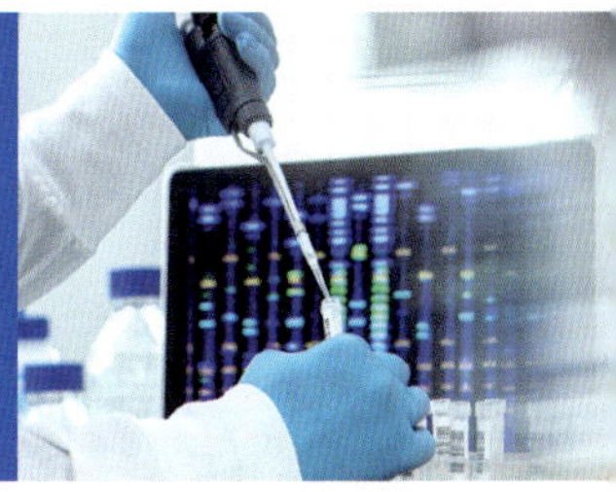

Genome Sequencing and Variant Interpretation

Applications in diagnostics and healthcare

Dianne Newbury

When we talk about sequencing, we are referring to the characterization of the order of the four bases (A, G, C, and T) along the DNA strand. The DNA sequence is simply a string of these four bases, which encodes crucial biological information. In order to understand this code, we need to be able to decipher it. This can be a very difficult task, but through the study of the human genome, we have started to understand the significance of some of the sequences. For instance, by comparing sequences between individuals and across the genome, we can see patterns that are repeated. For example, the starts and ends of genes are marked by characteristic sequence motifs. These patterns tell us more about the way in which the genome works and how it varies between individuals, and help us to develop better methods to identify variations in the sequence that contribute to disease. In this chapter, we will talk about different methods of sequencing and applications of sequence interpretation in diagnostics and healthcare.

Learning Objectives

By the end of the chapter, you should confidently be able to:

- Describe the differences between Sanger sequencing and next-generation sequencing (NGS) methods
- Describe how the first human genome was sequenced
- Explain what we mean by a reference genome
- Be aware of different resources that we can use to look at the Human Genome sequences
- Understand how Mendelian diseases are diagnosed in a clinic
- Be able to say what NGS misses and how third-generation or long-read sequencing may fill in this gap.

5.1 Sequence motifs

sequence motifs
Characteristic patterns in the sequence that act as signals for specific biological functions.

The DNA sequence alone can provide a lot of information about the genome structure and function through the identification of **sequence motifs**. These are recognizable patterns in the sequence that act as signals for specific biological functions. For example, the start or end of a gene or the binding sites for regulatory transcription factors or ribosome binding.

In eukaryotes, there is not one universal sequence motif that marks the start or end of a gene. However, regulatory motifs that mark the binding sites for basal transcription factors are often associated with the start of the gene. These include the core promoter sequences or the TATA box, the initiation sequence, the B-Recognition Element (BRE), and the Downstream Promoter Element (DPE), as shown in **Figure 5.1**.

Sequence motifs can be uncovered by bioinformatics algorithms that search sequence data for repeating patterns and are represented by graphical representations known as sequence logos (as shown in **Figures 5.2** and **5.3**).

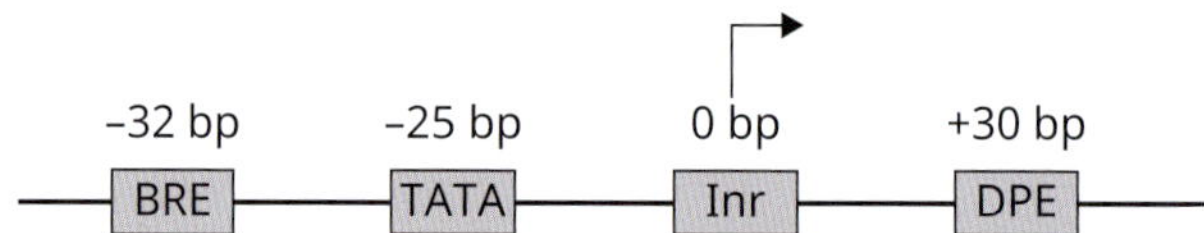

FIGURE 5.1
Positions of some consensus sequences in the promoter region. The basal transcription machinery binds to these elements to initiate transcription. Transcription begins within the initiator sequence at 0bp. All other positions are given in relation to this.

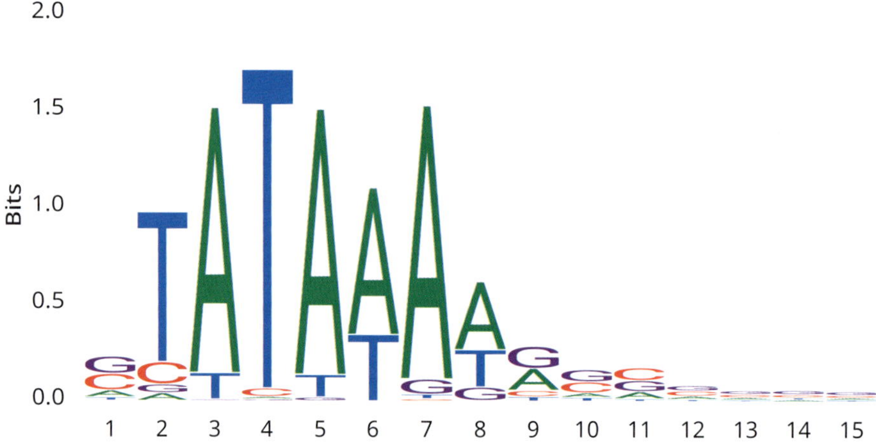

FIGURE 5.2
The TATA Box falls approximately 25bp upstream of gene coding sequences and has the consensus sequence TATAWAW (where W denotes a weak base pair (A or T)). By aligning all TATA boxes across the genome, we are able to derive a consensus sequence that shows how variable each base is between binding sites. This representation is called a sequence logo. In the logo, the height of each base represents its conservation across the genome. For example, the T base at position 4 is usually a T in most TATA boxes (as this is the highest character) but can sometimes be a C. In comparison, the bases at positions 9 to 15 of the TATA box are more variable across the genome. (JASPAR 2024)

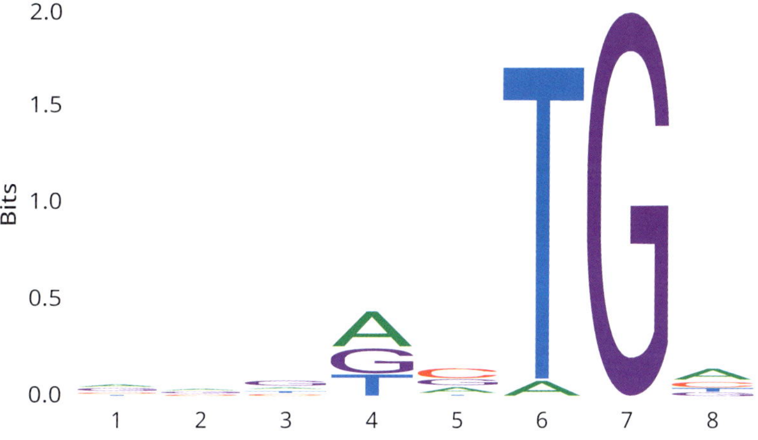

FIGURE 5.3
The initiator sequence marks the transcription start in the gene and has the consensus ATG. In this motif, the seventh base is always a G (as shown by the height of this base in the logo) while the other bases vary across the genome. Just as the start of the gene is marked with consensus sequences, so are the start and ends of exons and the end of the gene. Exons end in AG, and introns start with GT (donor site). At the other end, the intron ends with CAG and the intron starts with G (acceptor site). These sequence motifs are recognized by the spliceosome machinery, which removes the introns from the mRNA (as shown in **Figure 5.4**). (JASPAR 2024)

Exon 1
A
AG GT A AGT
C G
INTRON
CAG G
Exon 2
G
T

FIGURE 5.4
Splice site consensus sequences mark recognition sites for the spliceosome machinery, which removes the intronic sequences from the RNA.

Similarly, the end of the gene is usually marked by a polyadenylation signal, which has the consensus AAUAAA, a Poly A cleavage signal. These sequences are recognized by the polyA polymerase which cleaves the growing mRNA strand and adds the polyA tail onto the cleaved mRNA.

Cross reference
Look back at **Chapter 2** to read more about mRNAs and polyA tails.

By searching for the various consensus sequences mentioned above, it is possible to identify the position of genes from sequence data alone. Once the first draft of the Human Genome was complete, computational methods were used to scan the sequence for these motifs and position genes and functional elements. These have told us a lot about the coding regions of the genome, but we still have a lot to learn about motifs in non-coding regions.

There are similar consensus sequences that mark out many of the functional regions of the non-coding portions of the genome. New motifs are being mapped all the time, and even after the Human Reference sequence is complete, they play an important role in understanding non-coding regions of the genome through projects like the Encyclopaedia of DNA Elements (ENCODE).

5.2 Sanger sequencing

Even though it is now relatively cheap and easy to sequence whole genomes, this is sometimes overkill; often we are particularly interested in a single gene or a single region of DNA. In these cases, Sanger sequencing still represents the most efficient method of sequencing. Sanger sequencing is named after Fred Sanger who first developed this method in 1975. In a Sanger sequence reaction, we first amplify our region of interest in a **polymerase chain reaction (PCR)** (**Box 5.1**).

polymerase chain reaction (PCR)
A method of DNA amplification.

BOX 5.1 The polymerase chain reaction (PCR)

The polymerase chain reaction uses DNA polymerase to copy a DNA template. The double-stranded template is denatured to form single-stranded DNA. The region of interest is bound by small complementary single-stranded DNA fragments (~20bp in length) known as primers. The DNA polymerase then extends the free end of DNA and copies the DNA template to form two double-stranded copies of the template. This process uses enzymes and processes that occur naturally in cells to mend breaks in the DNA. The cycle is repeated thirty to forty times. In each cycle, the amount of starting template DNA doubles, fuelling an exponential increase in the amount of DNA present (**Figure 5.5**).

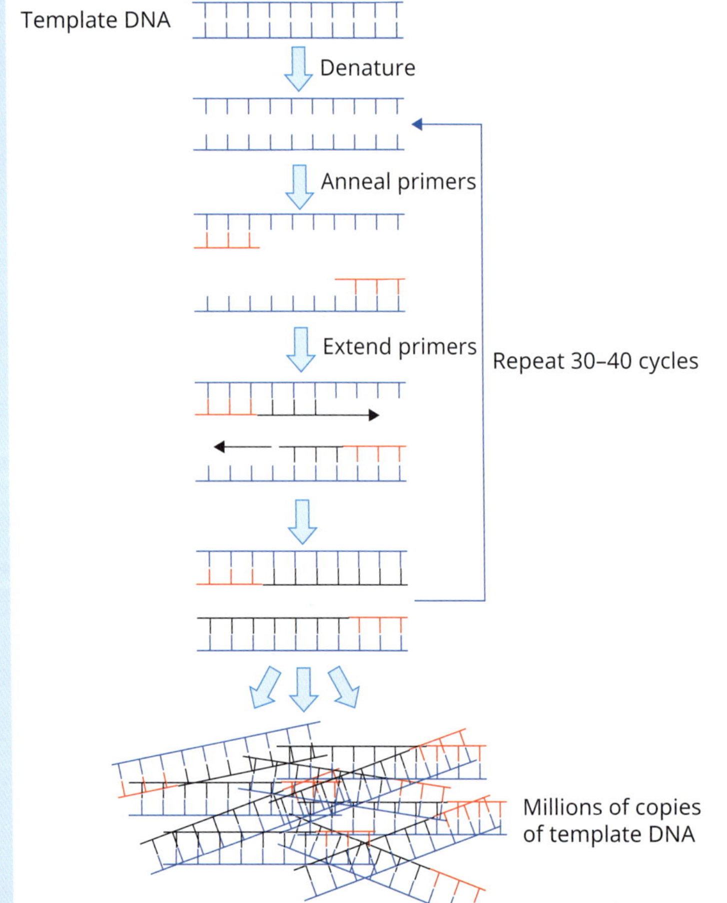

FIGURE 5.5
The polymerase chain reaction (PCR) consists of repeated cycles of denaturation, primer annealing, and extension, the outcome of which is the generation of multiple copies of the template DNA.

The PCR products provide the substrate for the sequencing reaction itself. The whole genome is still present in the tube, but the region of interest is over-represented compared to the rest of the genome because of the millions of copies produced during the PCR reaction. The sequencing reaction is not dissimilar to PCR—it requires primers, Taq polymerase, dNTPs, $MgCl_2$, and buffer. Unlike a PCR, in a sequencing reaction, we only use a single primer (called the Forward or Reverse primer). The PCR products are denatured into single-stranded fragments. This primer sticks to one end of this fragment and is extended by the polymerase using the free dNTPs in solution. The PCR products act as a template for this synthesis. However, in a sequencing reaction, one extra special component is **dideoxynucleotides (ddNTPs)**. These are free nucleotide bases (A, G, C, and T) that lack a hydroxyl group at position 3 of the sugar moiety. This modification means that when they are incorporated into the growing DNA strand, no further bases can be added, and the synthesis reaction stops at this base.

dideoxynucleotides (ddNTPs)
Free nucleotide bases that lack a hydroxyl group at position 3 of the sugar moiety and so cannot be extended. These play an important role in Sanger sequencing.

Each ddNTP is tagged with a fluorescent dye, and all are at a lower concentration in the reaction mix than their corresponding dNTPs. Think of the reaction mix like a bag of coloured balls. Let's imagine there are four different colours in the mix: green balls for C, red balls for T, black balls for G, and blue balls for A. Each colour comes in two different shades, with light balls representing the dNTP and dark balls representing the ddNTP. For every dark ball, there are 1000 light balls. The polymerase pulls a ball out of the bag and checks to see if it complements the template. If it does, the ball becomes incorporated. If it does not, the polymerase pulls another ball out of the bag. This continues until a dark-coloured ball is incorporated, at which point the reaction ends. The chances of the polymerase pulling a dark coloured ball out of the bag are 1 in 1,000 so, on average 1,000 bases will be incorporated before the reaction is terminated. Sometimes, by chance, a dark ball will be pulled out in the very first trial. In others, it will happen much later. Therefore, over subsequent rounds of sequencing, we end up with a mixture of different length products, each terminating in a ddNTP that is tagged according to the identity of the final base incorporated. In the tube, all of these different fragments are mixed up, but we can separate them out according to size using a process known as **gel electrophoresis**. Read **Box 5.2** if you need a reminder of how gel electrophoresis works.

gel electrophoresis
A lab method used to separate DNA fragments according to size.

polyacrylamide
A low toxicity chemical derived from acrylamide that allows us to separate DNA fragments at the base pair level.

Key Point

Sanger sequencing is also known as the chain termination method because it uses ddNTPs to create tagged sequence fragments that are terminated at random bases.

BOX 5.2 Gel electrophoresis

In the lab the process of gel electrophoresis can be used to separate DNA fragments of different sizes. This involves a slab of a jelly-like substance made of agarose. The gel slab is placed inside a buffer solution, and the PCR products are injected into a well in the slab. Once the products are loaded, an electric current is placed across the gel. This charge pulls the negatively charged DNA through the gel towards the positive electrode. The speed at which the fragment moves through the gel is relative to its size—larger fragments get caught up in the matrix-like gel while smaller fragments wiggle through the gel easier. This means that if we apply a constant voltage for a certain amount of time, the fragments will separate across the gel with the smaller fragments towards the bottom of the gel and the larger fragments at the top. The DNA fragments themselves cannot be seen by eye but we can use fluorescent dyes which bind to the DNA and a UV box to visualize the fragments in the gel. We can change the gel substance according to the size of the fragments we want to visualize. Often in the lab, we use agarose gels which have a resolution of ~50bp. In a sequencing gel, we need to be able to separate fragments to the base pair level. This is done using a gel made from a substance called **polyacrylamide**.

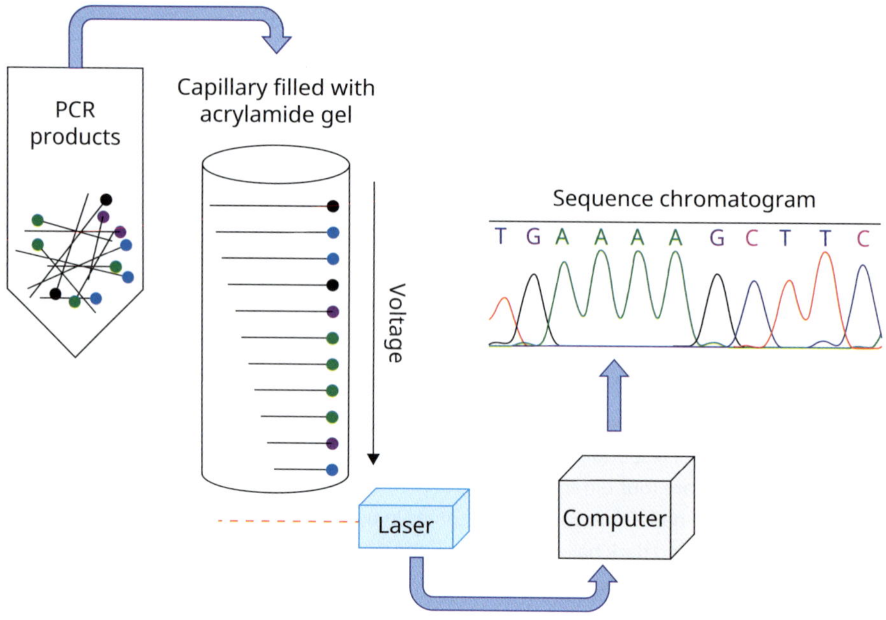

FIGURE 5.6

Fragments to chromatogram. Following the PCR and sequencing reactions, the different-sized fragments, each terminating in a fluorescently tagged ddNTP, are all mixed up in the tube. A capillary tube filled with polyacrylamide is used to separate the fragments according to size, and the fluorescent information is collected by a laser at the end of the capillary. A computer is used to turn this information into a trace called a chromatogram from which the sequence can be read.

chromatogram
A trace that represents the DNA sequence from Sanger sequencing.

In a sequencing gel, we use polyacrylamide. This allows us to separate fragments that differ in size by only a single base pair. We run the DNA fragments so that they migrate right through the gel and fall off the end. A laser is positioned at the end of the gel allowing us to capture the colour of the tag on the terminator base for each fragment. A computer then converts this information into a sequence trace known as a **chromatogram** as seen in **Figure 5.6**.

Human Genome Project (HGP)
A collaborative scientific effort to sequence the entire human genome.

The Sanger sequencing method allows us to generate a trace for a small fragment of DNA (usually less than 1000bp or 1 kbp). Recall that the starting point of the sequencing reaction is a single primer designed to sit next to the region of interest. We, therefore, know where this fragment is positioned in the genome, what gene or motif it represents and, often, what change we are looking for in the sequence. This is obviously a really useful tool in a clinical setting but what if we want to sequence an entire chromosome or if we do not already know the sequence of the region we are targeting? This is the problem that faced scientists sequencing the first Human Genome in the **Human Genome Project (HGP)**.

Key Point

Although Sanger sequencing is automated, the process involves many steps before you can see a sequence:

1. **PCR amplification of the region of interest**
2. **Sequencing of the PCR products**
3. **Separation of sequence products on a gel**
4. **Collection of fluorescence information from the gel**
5. **Generation of the chromatogram.**

5.3 The Human Genome Project

The Human Genome Project was an ambitious project that started in the 1990s to sequence for the first time every single nucleotide in the human genome, from the first base pair of chromosome 1 to the last base pair of the sex chromosomes (Lander et al. 2001). To be clear, the sequence generated by the Human Genome Project did not come from one single person's DNA but represented a mix of DNA from several people. Amazingly, this project was completed using Sanger sequencing technologies as described earlier. The entire genome, which is over 3 billion bases in length, had to be split into small fragments of approximately 1,000 bases each. That is over 3 million sequencing reactions! It perhaps comes as no surprise, therefore, that this project took over ten years to complete, involved thousands of researchers, and cost billions of pounds. On top of the sheer scale of the project, add the fact that because the sequence of the genome had not yet been compiled, there was no way to design targeted primers for the sequencing reactions. So, how was the sequence generated? Researchers generally adopted one of two approaches: shotgun sequencing and clone-by-clone sequencing, as discussed in **Box 5.3**.

A draft of the first Human Genome sequence was made available in 2000, allowing scientists to see the base-by-base sequence across chromosomes for the first time. However, because of the way it was built, this first draft had some gaps, and the full sequence was not published until April 2003. The finished sequence consisted of 27,332,000 base pairs of DNA and covered 99% of gene-containing regions but, until recently, there were gaps within this sequence. These usually occurred in repetitive regions at the telomeres and centromeres where small sequence fragments can be hard to position and stitch back together. One big shock to come from the Human Genome Project was the fact that the Human Genome includes only 20–25,000 genes, much fewer than previously thought. Everyone carries the same 20–25,000 genes and, notwithstanding rearrangements, these are in the same positions along the chromosomes from one person to the next.

cloned contig method
A method of sequencing that first involves the creation of clone libraries from small regions rather than sequencing everything all in one go.

BOX 5.3 Shotgun and cloned contig sequencing

Imagine you are planning a trip across London and a friend suggests that you take the Central Line as this will allow you to see many of the sights. You know the names of all the stations on the Central Line, but cannot remember what order they are in, going from East to West. If you chopped the Central line into short segments, you would see that Bond Street and Oxford Circus are often found on the same segment, as are Oxford Circus and Tottenham Court Road, but Bond Street and Tottenham Court Road seldom, if ever, occur together. This information would allow you to fix the order of these three stations without viewing the whole line: Bond Street—Oxford Circus—Tottenham Court Road. You would be able to give directions from one end of the line to the other, even if you had not done this exact journey yourself. This approach represents the **cloned contig method** (**Figure 5.7**). At the beginning of the Human Genome Project, although researchers did not have a sequence for the entire genome, many genes had been positioned within the genome and it was possible to use these known features as landmarks. The genome was fragmented and cloned into large vectors known as BACs. Each BAC contained landmarks that were physically close to each other. Each BAC was then fragmented further until a region-by-region itinerary was built with each region being small enough to sequence. Once all the small regions were sequenced information about the relative positions of the landmarks helped to stitch all the regions together into a complete map.

Now pretend that you cannot get two weeks off work and decide to visit landmarks in London over a few weekend trips. You do not have a map and do not have time to plan an itinerary, but instead, you can wander around the city and visit landmarks as you go. This approach represents a **shotgun sequencing** method (**Figure 5.8**). Here, the genome was not split into regions, but instead, the entire thing was

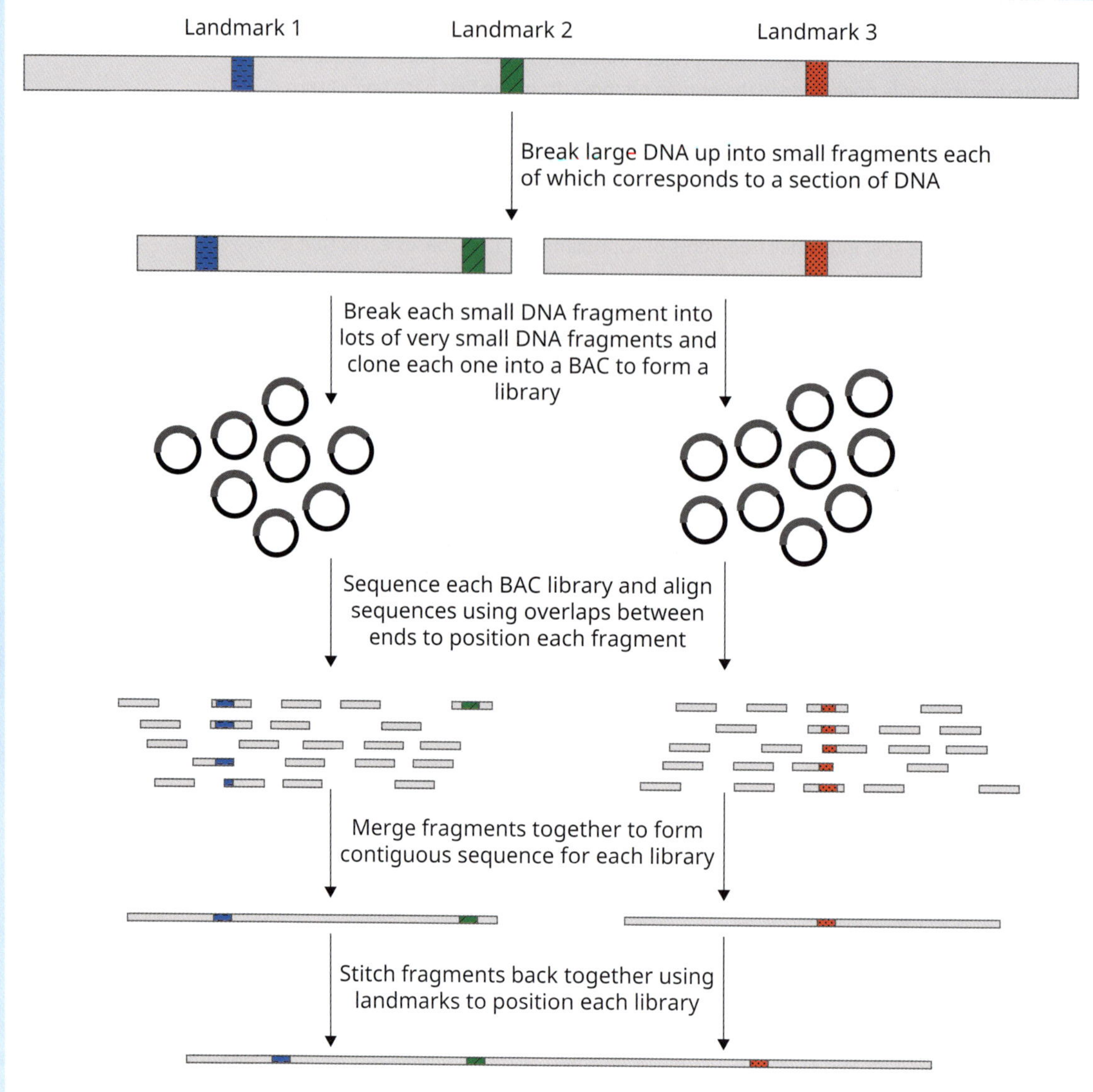

FIGURE 5.7
Cloned contig method. Large DNA fragments are broken up into small libraries. Each library is fragmented and individually sequenced. Fragments are aligned using overlaps between sequences to form a contiguous sequence for each library. Contiguous sequences are then organized relative to each other using the landmarks. Genetic landmarks include known sequence motifs, genes, and features such as microsatellites.

sheared into short fragments, all of which were sequenced in a random way. Overlaps between the fragments were then used to recreate the sequence. The first couple of times you visit London, everything would seem unrelated, and it would be hard to build a picture, but as you explore more and more, you start to recognize the landmarks, and you will be able to build a map of where these are relative to one another. This second approach seems like it might take longer, but actually, because of the work involved in planning the itinerary, the shotgun method is a more efficient means of building the genome. This was the way in which the first Human Genome was sequenced and assembled.

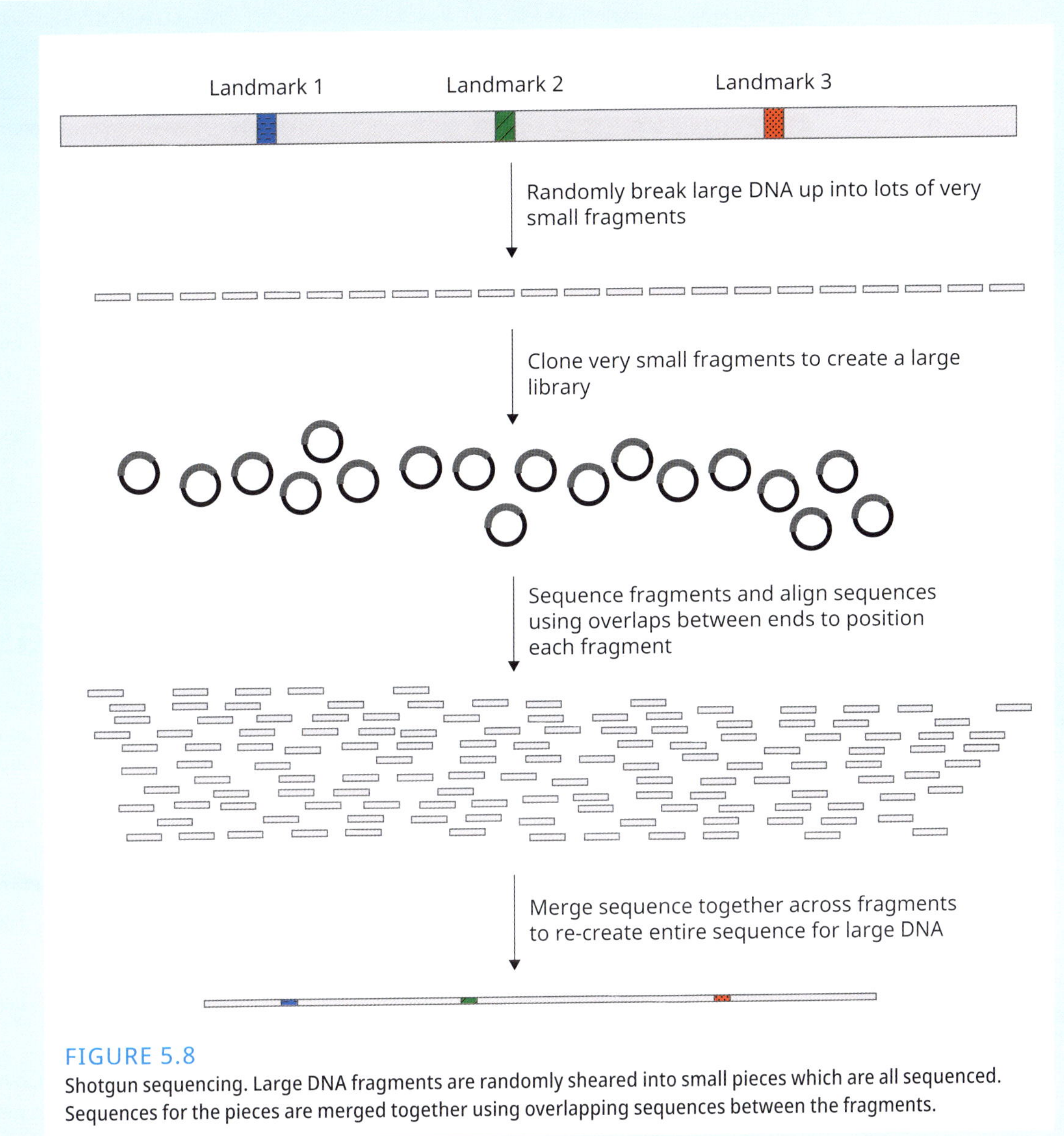

FIGURE 5.8
Shotgun sequencing. Large DNA fragments are randomly sheared into small pieces which are all sequenced. Sequences for the pieces are merged together using overlapping sequences between the fragments.

Key Points

The Human Genome Project was the first time that the sequence for a complete human genome was compiled. Before this, we knew the sequences of some chromosome regions and genes but did not know how all the sequences went together.

Everyone has the same number of genes in their genome (20–25,000). This means that it makes no sense to refer to the gene for (e.g.) sickle cell anaemia. Everyone has the beta-globin gene. People affected by sickle cell anaemia have variations that prevent this gene from functioning properly.

shotgun sequencing
A method of sequencing where the entire genome is sequenced randomly. In this method, the sequence fragments need to be reassembled after sequencing. This is done using a reference genome or overlapping sequences.

SELF-CHECK 5.1

What was the aim of the Human Genome Project?

SELF-CHECK 5.2

Did the Human Genome Project generate a complete sequence of the human genome?

Key Point

Shotgun sequencing requires much more sequencing than cloned contig sequencing. This is because genetic landmarks are not used to position fragments in shotgun sequencing and the only clue to identity is the overlaps between fragments.

SELF-CHECK 5.3

Why do we need to fragment the genome before sequencing it?

5.4 Next generation sequencing (NGS)

Most modern sequencing technologies, known as **next generation sequencing (NGS)**, build upon the lessons learnt from the Human Genome Project. They apply a shotgun method in which the whole genome is sheared into millions of small fragments (usually 100–150bp in length) that are sequenced side-by-side in a **massively parallel** approach. Unlike Sanger sequencing, NGS methods do not usually require specific primers or gels to separate fragments by size. Instead, all fragments from the genome are ligated to adapter sequences which allow the fragments to be stuck onto a solid surface and sequenced using a single universal primer. Each primer is extended using the DNA fragment as a template, and the information about which base is incorporated is captured by automated techniques. Because the fragments are only a few hundred base pairs in length, the sequencing itself can be completed very quickly. The method box that follows discusses the many platforms for NGS, each of which differs in the chemistry used and data collection method.

next generation sequencing (NGS)
Sequencing methods that allow the sequencing of a lot of DNA all in one experiment.

massively parallel sequencing
High-throughput sequencing methods where lots of small DNA fragments are sequenced all in one experiment.

FASTQ file
A file format that stores the sequences of the small DNA fragments in a next generation sequencing experiment.

reference genome
A genome sequence against which all sequence data are positioned. The use of a single reference means that all sequence data can be compared between different experiments.

sequence alignment map (SAM)
A file format that stores the sequences of the small DNA fragments once they have been aligned against a reference genome in a next generation sequencing experiment.

binary alignment map (BAM)
A binary equivalent of the SAM file used to save storage space.

The endpoint of all of these methods is a **FASTQ file**. The FASTQ file contains millions and millions of short sequences known as reads. Just like the Human Genome Project, these reads have to be aligned and combined to create a single genome sequence. But, we now have a map to guide us as the first genome can act as a **reference genome** upon which the reads can be positioned (**Figure 5.12**). The alignment information is contained within a **sequence alignment map (SAM)** or its binary equivalent, the **binary alignment map** (**BAM**) (**Box 5.4**).

Massively parallel sequencing essentially scales up and automates existing sequencing technologies to allow the sequencing of an entire genome within one experiment. The hard part of this method is often the alignment and data storage. This requires a powerful computer that is capable of scanning the entire genome for near matches to each of the millions of short sequence reads produced. This would be impossible to do manually, but it is what computers were built for—an iterative process, repeating a simple action many, many times (**Figure 5.12**).

Although it is possible to sequence the whole genome, most applications are primarily interested in genetic changes that directly affect genes: coding variations. Since the genes only constitute 1–2% of the genome sequence, it is possible to reduce the amount of data storage

METHOD NGS methods

The majority of current next generation sequencing methods are massively parallel. This means that, just like the shotgun methods described earlier, many small sequence fragments are sequenced all in one go. The resultant sequences are then aligned against a reference to create a continuous sequence known as the contiguous sequence or **contig**. The methods of sequence generation differ depending on the sequencing machine, but this principle remains the same. The methods can be largely split into sequencing by synthesis or sequencing by ligation methods, as described next.

Sequencing by synthesis

The most commonly used sequencing method is known as **sequencing by synthesis**. As the name implies, these methods capture the DNA sequence by copying DNA fragments (a bit like Sanger sequencing). The exact ways in which the DNA fragments are isolated and the sequences are captured differ between companies and two variations are shown in **Figures 5.9** and **5.10**.

All sequencing by synthesis methods begin by fragmenting the genome into small DNA fragments of ~100 bp in length. These fragments are denatured to make them single-stranded, and the strands are stuck (ligated) onto a solid surface, like a slide (Illumina, **Figure 5.9**) or a bead within a droplet of oil (known as **emulsion PCR**) used by Ion Torrent and Roche (**Figure 5.10**). Each fragment is clonally amplified, resulting in thousands of copies of each fragment, all in a cluster on the slide or around the bead. The fragments are then copied using DNA polymerase and dNTPs. The copying process is stalled after adding each dNTP so that the sequence can be captured using a reporter molecule. In Illumina machines, each dNTP is fluorescently labelled and the sequence is captured by the colour of fluorescence seen at each fragment cluster (**Figure 5.9**). In Roche machines, each possible dNTP

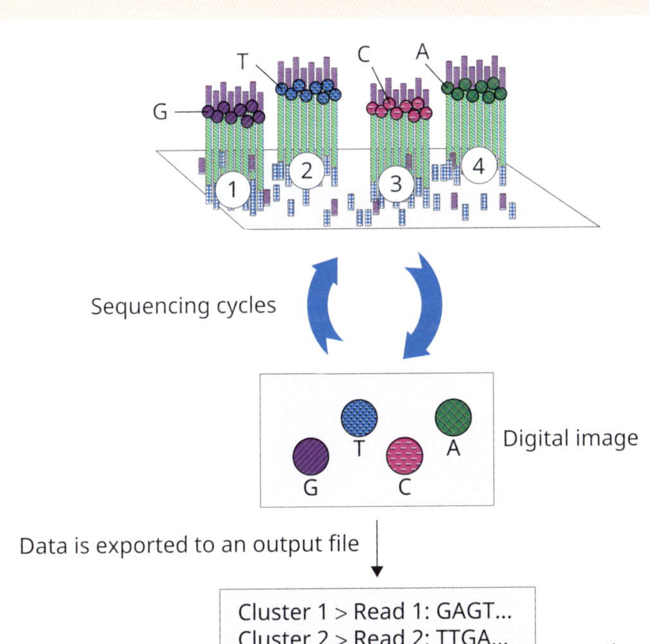

FIGURE 5.9
Sequencing by synthesis on Illumina sequencing machines. DNA fragments are stuck onto single-stranded probes adhered to a slide. Each fragment is clonally amplified and then extended one base pair at a time using fluorescently tagged dNTPs. A computer takes an image of the slide after each base pair extension allowing the reading of the sequence.

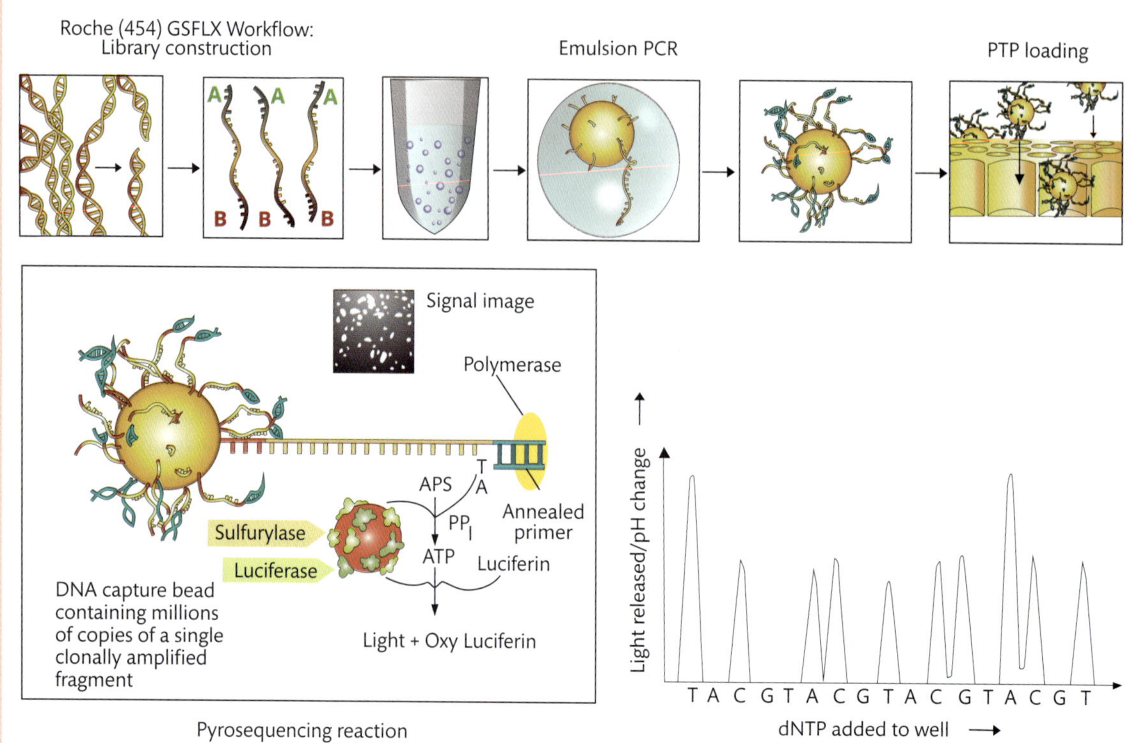

FIGURE 5.10

Pyrosequencing by Roche sequencing machines. In this method DNA fragments are amplified within an emulsion droplet. Each droplet is then separated into a well containing 'clonal' copies of DNA. When dNTPs are incorporated into the DNA fragment, light is emitted and captured on a peak graph. There is one graph for each fragment. In the graph, peaks of double height represent two of the same base in a row.

(dTTP, dATP, dCTP, and dGTP) is added one at a time with a special light-emitting enzyme which is activated when the dNTP is incorporated into the sequence fragment (**pyrosequencing**, **Figure 5.10**). In Ion Torrent sequencing dNTP addition is measured by change in pH instead of light emission. Bases are added one at a time until each 100bp fragment is completed. The computer turns the reporter data into short sequence reads which are then exported into a FASTQ file.

Sequencing by ligation

Sequencing by ligation (**Figure 5.11**) begins with an emulsion PCR (capture of single-stranded DNA fragments on a bead), but the sequencing steps do not use DNA polymerase and dNTPs. Instead, they use short fragments of single-stranded DNA known as oligonucleotides. Each oligonucleotide is tagged at the end with a fluorescent dye which corresponds to its first two bases. A primer, which is complementary to the bead adaptor sequence, is added, along with the oligonucleotides. If the oligonucleotide hybridizes to its complementary DNA fragment and this is adjacent to the primer, then the two are ligated by an enzyme, releasing the dye. A camera captures the fluorescence colour. More oligonucleotides are added extending the ligation fragment. After eight rounds of ligations, the primers and oligonucleotides are then washed off, and the process is repeated using a slightly longer primer. In this way, the sequence is built up by walking across the DNA in 2 base-pair steps.

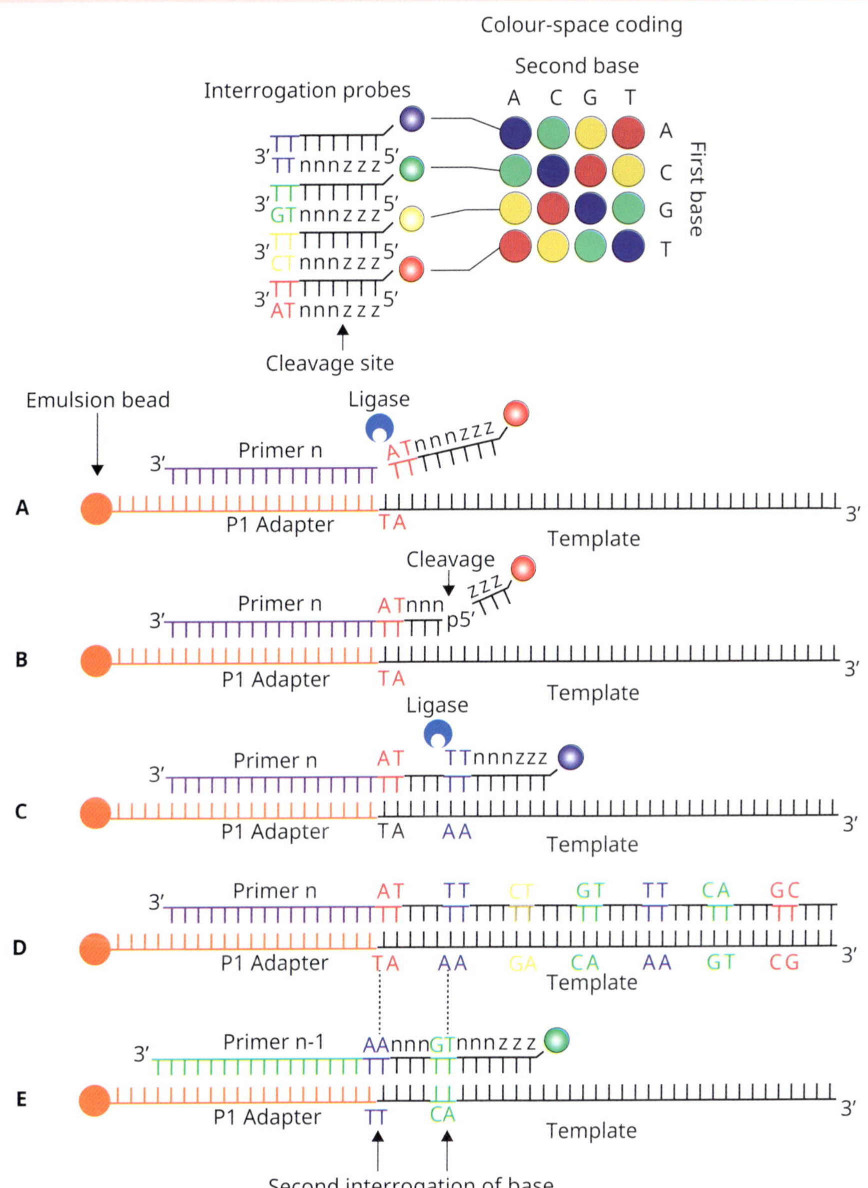

FIGURE 5.11
Sequencing by ligation. Each interrogation probe is an octamer, which consists of 2 probe-specific bases followed by 6 degenerate bases (nnnzzz) with one of 4 fluorescent labels linked to the 5′ end. The 2 probe-specific bases consist of one of 16 possible 2-base combinations. (A) The P1 adapter and template with annealed primer (n) is interrogated by probes representing the 16 possible 2-base combinations. In this example, the 2 specific bases complementary to the template are AT; (B) after annealing and ligation of the probe, fluorescence is recorded before cleavage of the last 3 degenerate probe bases. The 5′ end of the cleaved probe is phosphorylated before the second sequencing step; (C) annealing and ligation of the next probe; (D) complete extension of primer (n) through the first round consisting of 7 cycles of ligation; (E) the product extended from primer (n) is denatured from the adapter/template, and the second round of sequencing is performed with primer (n −1). With the use of progressively offset primers, adapter bases are sequenced, and this known sequence is used in conjunction with the colour-space coding for determining the template. In this technology, template bases are interrogated twice.

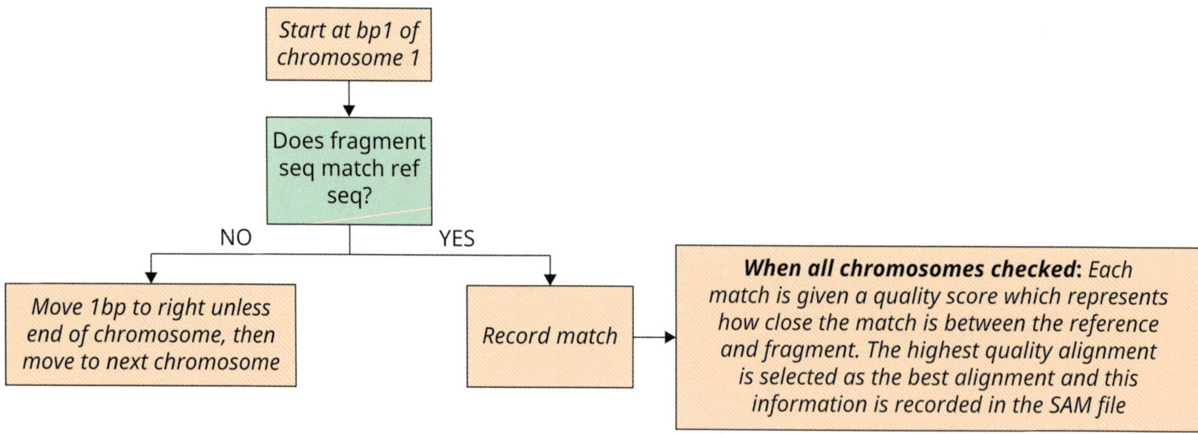

FIGURE 5.12
Alignment of next generation sequence fragments against the reference. Because not all individuals will have the exact same sequence as the reference, the algorithm has to allow for near-exact matches and gaps within the alignments.

and computational power needed by just sequencing known genes. This approach is known as **exome sequencing**. Although the burden of downstream processing is reduced in exome sequencing, the upstream laboratory work is increased. Remember that in whole genome sequencing, we sequence all the DNA within the sample. In exome sequencing, we first need to enrich the coding regions using a **library** preparation. This is a bit like using targeted primers to amplify the region of interest in a Sanger sequencing reaction, but we use multiplex methods to amplify the entire exome. The DNA corresponding to the exome is then fragmented and sequenced in the same way as a genome sequencing workflow.

contig
The contiguous sequence generated once DNA fragments are aligned in a massively parallel sequencing experiment.

sequencing by synthesis
A method of next-generation sequencing.

emulsion PCR
A PCR reaction carried out in a single drop of oil.

pyrosequencing
A method of next generation sequencing.

sequencing by ligation
A method of next-generation sequencing.

exome sequencing
A high-throughput sequencing approach in which only the coding parts of the genome are sequenced.

library
collection of DNA fragments from a particular part of the genome, for example the exome.

Key Point

In next generation sequencing, the genome is fragmented and the fragments are sequenced randomly. After the sequences are generated, all the fragments have to be put back together like a big jigsaw. The reference genome acts like the picture on the jigsaw box to guide the assembly steps.

SELF-CHECK 5.4

What is the difference between Sanger sequencing and next generation sequencing?

Key Points

There are many different methods of NGS, each patented by a different company. The start point of all NGS methods is DNA fragments, and the end point is a FASTQ file. It is just the steps in between these two points that vary.

Even after the sequencing has been completed, the data needs a lot of quality control and reformatting to put it in a form that can be easily read and understood.

BOX 5.4 Sequence file formats

The **raw DNA sequence** (that is, as it comes off the sequencing machine) is represented within a FASTQ file, which includes an identifier, the sequence for a given short fragment and quality information (**Figure 5.13**). Quality is measured as a **Phred** score and each base in the sequence has its own Phred score. Phred scores are based on a logarithmic scale and represent the probability that the base has been incorrectly called. The higher the Phred, the higher the confidence that the base has been correctly 'called'. Phred scores are represented by letters and symbols known as ASCII characters. Each character gives the quality of the call for a given base.

The fragments in the FASTQ file are then aligned against the reference genome to position them. This information is stored in a sequence alignment map (SAM) file. SAM files have a header line which start with @. These provide information about the sequence experiment and alignment algorithm. The header lines are followed by alignment information for the sequenced fragment. The alignment information is split across different columns which give the fragment name, its sequence, and alignment quality in addition to other information.

Once the sequence is aligned against the reference, we can compare the observed sequence with that expected from the reference and identify the variant bases. All the bases that are the same as the reference are discarded and information about the variant bases is recorded in a **variant call file (VCF)**. The VCF file starts with a number of header lines followed by a line for each observed variant. You can see in **Figure 5.14** that each line in the data shows the chromosome and position of the variant, the **reference allele (REF)**, the **alternative allele (ALT)**, and the genotype of the individual sequenced. The genotype includes two alleles and is recorded as 0|0 if the individual is homozygous for the reference allele, 0|1 or 1|0 if the individual is heterozygous

```
@SRR56757.980 SAMPLE1:4:1:2299:1109 length=50              ← identifier for fragment
AAGCTTCATTTACATATTGTTTTGTTCGCTCTAATTCACGTGTATCTCCA         ← sequence of fragment
+                                                          ← denotes blank line
hhhhhhhhhghhhhhhhhhhhffffeeee' eeeebdebVdhhhhhhhc!dd       ← quality scores
```

FIGURE 5.13
Example of a FASTQ formatted file. The identifier line starts with the @ symbol and includes the length of the sequenced fragment. The sequence of the fragment comes in the next line. Then after a line which carries only the + symbol, a quality score is given for each base. The quality scores are represented by letters and symbols. In this example, letters represent higher quality scores and symbols lower quality scores.

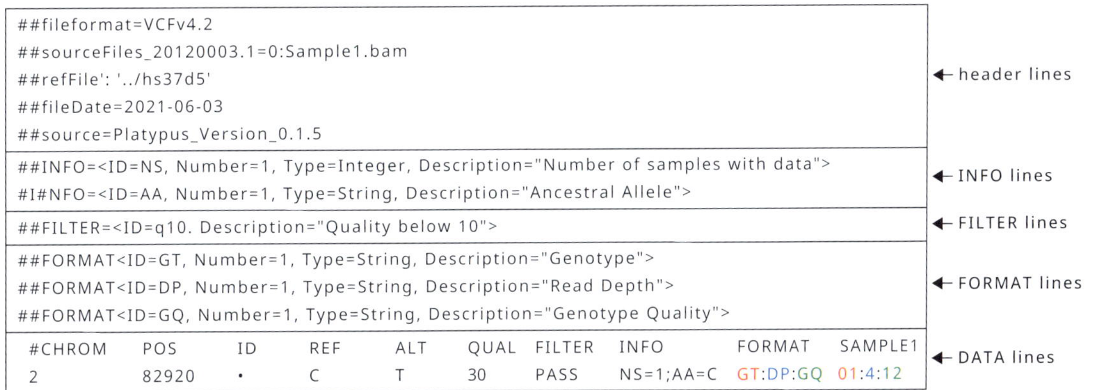

```
##fileformat=VCFv4.2
##sourceFiles_20120003.1=0:Sample1.bam
##refFile': '../hs37d5'
##fileDate=2021-06-03
##source=Platypus_Version_0.1.5
##INFO=<ID=NS, Number=1, Type=Integer, Description="Number of samples with data">
#I#NFO=<ID=AA, Number=1, Type=String, Description="Ancestral Allele">
##FILTER=<ID=q10. Description="Quality below 10">
##FORMAT<ID=GT, Number=1, Type=String, Description="Genotype">
##FORMAT<ID=DP, Number=1, Type=String, Description="Read Depth">
##FORMAT<ID=GQ, Number=1, Type=String, Description="Genotype Quality">
#CHROM  POS    ID  REF  ALT  QUAL  FILTER  INFO       FORMAT    SAMPLE1
2       82920  .   C    T    30    PASS    NS=1;AA=C  GT:DP:GQ  01:4:12
```


FIGURE 5.14
Example of a VCF format file.

(has one reference and one alternative allele) or 1|1 if the individual is homozygous for the alternative allele.

We can see that the in the VCF, lines are marked with ## and can be split into four sections; the header lines (##), info lines (##INFO), filter lines (##FILTER), and format lines (##FORMAT). Header lines include information about what format the VCF is in (VCF version 4.2), the BAM file that the VCF was created from (sample1.bam), the reference genome used (hs37d5), the date of creation (3rd June 2021), and the program used to create the VCF file (Platypus version 0.1.5). The INFO, FILTER, and FORMAT lines provide a key for reading the data lines, shown at the bottom of the file. For example, you can see that the code AA represents the ancestral allele. Looking down at the INFO column in the data, you can see that AA=C; i.e. the ancestral allele of the first variant listed is a C.

The FILTER lines show any filters that have already been applied to the data. Here we can see that the data have been filtered so that any variant with a quality below 10 will be flagged. This flag will appear in the FILTER column of the data. In this example, the FILTER column says PASS so we know the variant has a quality >10. Looking down at the FORMAT column, we can see that the data in the sample lines include three variables given in the order GT (highlighted in red) followed by DP (highlighted in blue) followed by GQ (highlighted in green) and each piece of information is separated by a colon. This information tells us how to read the data presented in the SAMPLE columns that follow (these are highlighted in the same colours here to help you see what each piece of information represents).

Reading across the data line shown at the bottom of the VCF, we can see that there is a variant on chromosome 2 (CHROM) at base pair (POS) 82920. The reference genome (REF) carries a C at this position, but in our sequence data we observed a T base (ALT). The Quality (QUAL) of this variant is 30 and it has passed the quality filters applied (QUAL, i.e. a quality score>10). We have data for one sample (INFO, NS) and the ancestral allele (INFO, AA) at this site is a C. In sample 1, the genotype (SAMPLE 1, GT) is a heterozygote (0|1), the read depth (SAMPLE1, DP; see **Box 5.5**) is 4 and the genotype quality score (SAMPLE 1, GQ) is 12.

Note that the VCF file only includes information about the position and alleles of the observed variant sites. Further information, for example any genes affected by the variant, need to be added by the **annotation** process.

raw DNA sequence
The DNA sequence that is output from a sequencing experiment (before any analysis).

phred
A logarithmic score used to indicate the quality of DNA sequencing.

5.5 Reference genomes

Although we sequence the whole genome or exome, the majority of these bases will be identical between individuals. It is estimated that, if you compare two random genomes from unrelated individuals then, on average, you will see 1–2 million differences between them or one variation every 1500 base pairs. This means that 99.99% of the genome sequence is the same between these two individuals. Look at **Box 5.5** to see what variations in the DNA sequence look like in Sanger sequence traces and NGS data. You can see that they look very different.

If we are looking for variants that contribute to disease, then these invariant bases tell us nothing, and so can be removed from the sequence file. This is achieved by a variant calling

BOX 5.5 Variant identification

Variants can be identified in both Sanger and next generation sequencing by bases that differ from the expected (or reference) sequence. Recall that we are amplifying all the DNA from each individual and so, in samples from heterozygous individuals, two bases will be seen. Samples that are homozygous at a particular locus will show only one base.

In **Case study 5.1**, a mother and father who are both carriers for an autosomal recessive condition have each passed the causative allele to their child. The child (proband, patient) in this case is therefore homozygous for the causative variant.

In **Figure 5.15B**, the Sanger sequencing chromatogram, each base is represented by a peak. The peak colour indicates

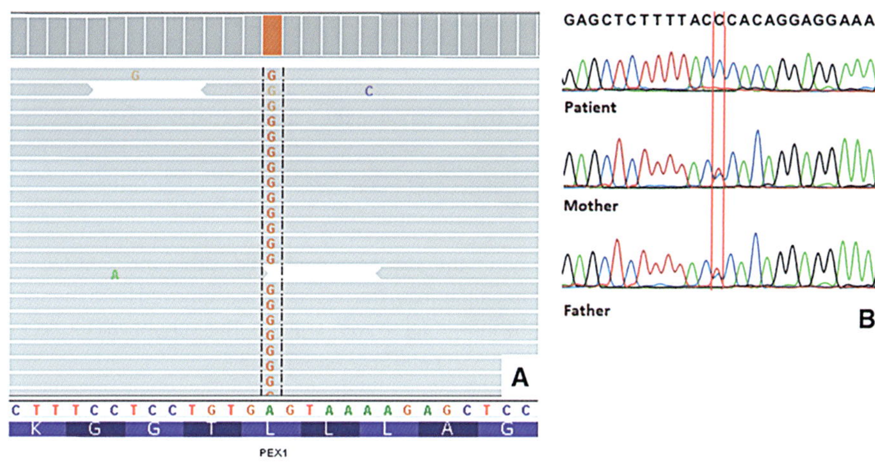

FIGURE 5.15

Variant identification. Results of DNA sequencing. (A) Excerpt of next generation sequencing data from the proband (patient) visualized with Integrative Genomics Viewer (IGV). (B) Sanger sequencing chromatogram showing 790T > C, p.Leu597Pro in the PEX1 gene of the proband (patient) and heterozygous carrier father and mother (indicated by red frame). Havali, C, Dorum, S, Akbas, Y, Görükmez, O, and Hirfanoglu, T (2020). Two different missense mutations of PEX genes in two similar patients with severe Zellweger syndrome: an argument on the genotype-phenotype correlation. Journal of Pediatric Endocrinology and Metabolism. 33. 10.1515/jpem-2019-0194.

which fluorescent ddNTP was excited by the laser and read by the light detector at that particular position. The variant base in this case is indicated by the red frame. The heterozygote samples (from mother and father) appear as two overlapping peaks of reduced height. This shows that half of the fragments at this position terminated in a ddCTP and half terminated in a ddTTP. This is interpreted as a heterozygous A/G variant in the coding strand. Note that this Sanger sequencing shows the sequence of the complementary, non-coding strand and so a T/C is seen. The height of the peaks indicates the strength of the fluorescence. Notice how the peaks in the heterozygous samples (mother and father) are lower, but there are two (red and blue). So, although there are two colours (two different bases represented) the cumulative intensity is a close match to all the other peaks in the sequence. Manufacturers of sequencing reagents and equipment try to keep the intensity of the four fluorophores equal. However, depending on the chemistry of the assay kit used, some fluorophores might have a higher or lower intensity than others; this is not important as long as the readout can be interpreted. For example, notice how in image (B) the last blue ddCTP peak in the mother and father's chromatograms (indicated by the red arow) is taller than most other peaks. This is of no consequence—no matter the height, it still represents a C at that position.

In the NGS dataset (**Figure 5.15A**), each DNA fragment is represented by a grey bar. The reference sequence is shown at the bottom of all of the grey bars, and beneath it the amino acid sequence coded for (if relevant).

The site of variation is marked by coloured letters that correspond to the base seen at that specific position in each fragment. Count down the grey bars and note that, in this experiment, twenty-one fragments were mapped to this position, all of which have a guanine base at the position of interest. Note that what could have been the fourteenth fragment from the top has a blank area where no data was available. The proportion of base calls is indicated at the top, by the bar directly above the position of interest—in this case the bar is 100% orange (G). If the data for one of the heterozygous parents had been shown, then it is expected that approximately ten of the fragments would have shown an adenine at the position of interest and approximately ten a guanine. In that case the bar at the top of the read would be half orange and half green.

The number of fragments at any one base position is referred to as the **coverage** or **read depth**. The next generation sequencing data (image A) has a read depth of twenty-one fragments. A greater read depth increases confidence that the variant we are seeing is true. Imagine if there were only three fragments covering this position: two carrying the G base and one carrying the A base. In that case, it would be very hard to say whether the A was a true variant or just an error in the sequence data.

variant call file (VCF)
A file format that stores the positions at which an individual differs from the reference genome.

reference allele (REF)
The allele found in the reference genome at a given base.

alternative allele (ALT)
The allele of a given variant that differs from that in the reference genome.

annotation
The process of adding genetic information to sequence data.

Cross reference
Please refer to **Table 2.2** for the single-letter amino acid codes.

software which compares the sequence contained within the SAM file to the reference genome. Bases that do not differ from the reference are removed from the sequence file leaving a list of sites at which the individual shows variation. This variant call file (VCF) is greatly reduced in size compared to the FASTQ and SAM files (**Box 5.4**).

The variant calling process, in itself, brings up an interesting question. How do we know if an allele represents the base typically seen at this position or if it is a variant base? There are many ways to approach this problem. All variant calling software compares the sequence generated against the reference genome. Bases that match the reference genome are labelled as reference (REF) and discarded, while those that differ from the reference are labelled as alternative (ALT) and kept for analysis. This is an important step because it ensures that everyone is working to the same standard and provides a static reference across research studies and databases. However, this approach assumes that the reference sequence represents a single 'perfect genome'. Recall from earlier that this is not true—the first reference genome arose from the Human Genome Project which involved DNA from several healthy volunteers. As we sequence more individuals, this reference genome is updated. Different versions are labelled with the prefix hg (e.g. hg38) or GRCh (e.g. GRCh38) and a number which represents the version of the genome. However, this reference does not represent a flawless genome, nor is it the most common form of the genome or the ancestral version or ethnically matched to a given population (**Box 5.6**). Recently, researchers have begun to petition for the development

BOX 5.6 Allele classification

The term *variation* refers to a base in the DNA that can take different forms. The different forms of a given variant are known as alleles.

We have spoken about the reference and alternative alleles in the text, but there are many other ways that we can refer to and classify different variations across the genome. These often depend on the focus of interest of a given researcher.

Ancestral and derived alleles: The **ancestral allele** is the allele that was originally present in the genome of a shared common ancestor before the variation arose. This is usually inferred by comparing sequences across species. This terminology assumes that the observed variant must have arisen through a mutation event in the common ancestor generating a novel **derived allele**.

Major and minor alleles: The **major allele** is the form of the variant that is most common within a given population and the **minor allele** is the form of the variant that is least common within a given population. For a single nucleotide variant, there are usually two possible alleles with a combined frequency of 1.0. The minor allele has a frequency of <0.5 and the major allele has a frequency of >0.5. Note that this classification depends upon your reference population.

Reference and alternative alleles: As mentioned earlier in this chapter, the reference allele (REF) refers to the allele found in the reference genome for the species of interest. Any allele that differs from the reference is called the **alternative allele (ALT)**. Note that this classification depends upon your reference genome. In many species (including humans) there is one reference genome to which all research refers.

of a **pan-genome** that represents all known possible variations at each of 3 billion bases or, alternatively, a **consensus reference** that carries the most commonly found alleles for a given population. These updated references would allow us to compare our sequences to a detailed catalogue rather than a singular reference.

> Key Point
>
> **The reference genome is not a perfect genome. It includes rare variants and disease-causing variants, just like any other genome.**

SELF-CHECK 5.5

What is the difference between the information contained within a FASTQ and a VCF file?

> Key Point
>
> **Positions where an individual genome sequence differs from the reference sequence are called variations.**

5.6 Genome variation

In the same way that not all bases in the genome are variable, not all variations cause disease. Population sequencing studies like the 1000 Genomes Project, which sequenced over 2000 individuals from 26 different populations, help us to understand inter-genome variation. By comparing genetic variation between individuals and across populations, they showed that the majority of genetic variants are widespread across populations, reflecting our shared ancestry. You can see from **Table 5.1** that the average genome includes 3–4 million variations, most of which take the form of **single nucleotide variants (SNVs)**. But, perhaps surprisingly, quite a high number of variations involve larger-scale changes of the sequence through the insertion and deletion (**indels**) or reorganization of DNA. Most of the variations found in the human genome lie outside of the gene sequences (intergenic) or in the non-coding portions of genes (intronic or untranslated regions (UTR)), but variations within genes are still commonplace.

Each genome can be expected to carry about 20,000 variants that directly affect genes and about half of these will introduce a change to the protein sequence (nonsynonymous variants). Such variants can have a big effect on cell function. In fact, it is estimated that about 150 of the coding changes completely knock out the function of the gene they occur in—so called **loss-of-function (LoF) variants**. This class includes particularly disruptive nonsynonymous variants such as stop-loss, start-loss, stop-gain, splice site, and frameshift variants.

These findings indicate that the average person has 150 genes in their genome that do not work at all! How is this possible? Well, some proteins are more essential to cell function than others. For example, the human genome includes over 300 olfactory receptor (OR) genes. The proteins that these genes code for allow us to perceive different odours. These are useful, but unless the odour is the breath of a big lion about to pounce on you, you can probably live without them. Compare this to the *TP53* gene which encodes a protein which regulates the cell cycle. You can see how loss of function of this gene would have dire consequences for the cell. It is estimated that one quarter of the genes in the genome are essential for cell function.

coverage
The number of DNA fragments that capture a given base pair in a next generation sequencing experiment. Also known as read depth.

read depth
The number of DNA fragments that capture a given base pair in a next generation sequencing experiment. Also known as coverage.

ancestral allele
The allele that was originally present in the genome of a shared ancestor.

derived allele
The allele generated by a mutation event in the genome of a shared ancestor.

major allele
The allele that is most commonly found within a given population.

minor allele
The allele that is rarest within a given population.

pan-genome
A reference genome that contains information about every possible base at every possible position.

consensus reference
A reference genome that carries the most commonly found allele at every base position.

single nucleotide variant (SNV)
A variation that involves a single base pair of DNA.

indel
The insertion or deletion of a few base pairs in a DNA sequence.

loss-of-function (LoF) variant
A variant that completely prevents the function of a gene.

TABLE 5.1 Average number of variant classes across a human genome

	African	American	East Asian	European	South Asian	Average
Single nucleotide variants (SNVs)	4.31M	3.62M	3.55M	3.53M	3.6M	3.72M
Small insertions & deletions (indels)	625K	557K	546K	546K	556K	566K
Large deletions	1100	949	940	939	947	975
Copy number variants (CNVs)	170	153	158	157	165	160.6
Inversions	12	9	10	9	11	10.2
Nonsynonymous changes	12.2K	10.4K	10.2K	10.2K	10.3K	10.6K
Synonymous changes	13.8K	11.4K	11.2K	11.2K	11.4K	11.8K
Intronic changes	1.06M	1.72M	1.68M	1.68M	1.72M	1.38M
Changes in promoter regions	102K	84K	81K	82K	84K	86K
Changes in untranslated regions (UTRs)	37.2K	30.8K	30K	30K	30.7K	31.7K
Loss-of-function (LoF)	182	152	153	149	151	157.4
Variants previously reported to cause disease (Clinvar)	28	30	24	29	27	27.6

M=million, K=thousand. Data taken from Table 1 in The 1000 Genomes Project Consortium (2015) A global reference for human genetic variation *Nature* 526:68

Cross reference

Look back at **Chapter 2** to read more about gene structure.

Cross reference

Look back at **Chapters 2** and **3** to read more about different types of genetic variations.

dosage-sensitive gene

A gene that encodes a protein that must be present at a very particular level within a cell.

Critical genes tend to be expressed across multiple tissues and interact with many other proteins forming regulatory hubs of cell function (Karczewski et al. 2020).

Another thing to bear in mind is that the genome includes two copies of every gene. Any single loss-of-function variant will only affect one gene copy meaning that you can still produce the respective protein. The effect of gene loss upon a cell is measured by dosage sensitivity metrics. For some genes, the level of expression must be tightly regulated, and these are known as **dosage-sensitive genes**. Variants that knock out dosage-sensitive genes lead to dominant forms of disease.

Key Point

Everyone has millions of variations across their genome. This is part of what makes you, you!

5.7 Genome annotation

Information about genetic variations found in sequencing experiments can be gathered through the process of annotation. Recall from **Box 5.4** that the VCF file only contains information about the position of observed variations and the reference and alternative alleles. If we want more information about the variant, we need to collect further information from genome databases. As you can see from **Box 5.7**, there are many such databases that can be used. By searching these databases, we can see whether the variant falls within a known gene and, using information about the gene coding sequences, we can tell whether the variant will

BOX 5.7 Genome databases

During the Human Genome Project, it was agreed that all genome information should be made as widely available as possible. This idea is preserved in research today. Sharing information helps researchers to exchange ideas and to generate collaborative research. There are many, many different genetic databases available online. These are more than a big book of As, Gs, Cs, and Ts. The databases link the DNA sequence to user-friendly interfaces that allow us to browse the Human Genome like a catalogue. They include information about the function and expression patterns of genes and proteins, variation across populations, and variations that lead to disease. A few databases are mentioned below, but this only scrapes the surface. Use a search engine to find one that may have the information you are after.

The **Ensembl** and **UCSC** databases provide access to information about the genome sequence and how this relates to genes, proteins, and phenotypes not just in humans but across different species. You can download the entire human reference sequence from these databases, but an easier way to use these portals is to search for your gene of interest using the search bar on the home pages.

Information about how the genome varies across individuals and populations can be found in the **International Genome Sample Resource (IGSR)** and **Genome Aggregation Database (gnomAD)**. These databases compare genetic sequences between healthy volunteers from all around the world. Search bars on these sites allow you to see how your gene of interest varies between populations and whether your variant of interest occurs in the general population. Information given about minor allele frequencies is often based on figures from these databases.

Databases like **Clinvar**, **OMIM**, and **DECIPHER** provide clinical information about the genome. These databases allow you to see whether your gene of interest has been linked to a particular disease or phenotype before and whether there are studies that support a loss-of-function effect for your variant of interest.

These databases mean that you do not have to have an enormous research budget to find out more about gene function and genomics. All of this information is freely available to everyone. Because these databases hold so much information, they can seem daunting at first but there are lots of online tutorials that can help you to find the information you need.

lead to a change in the protein sequence. This is straightforward if we have only one variant to investigate, but imagine the work involved if we had to find this information manually for all of the millions of variants across a given genome. Annotation software cross-references the variant position against a multitude of databases to automatically gather information for any list of variations. This information goes beyond the gene and protein effects. For example, we can annotate whether any of the variants have previously been reported to cause disease, or previously been shown to be associated in genome-wide association studies (GWAS), how common the variant is across different populations (1000 Genomes, gnomAD), whether the variant affects known transcription factor binding sites, whether the variant affects methylation motifs, or how conserved the base is across species. All of these pieces of information help us to build a better idea of individual variation across the given genome and to identify variants that may have clinical relevance.

Key Point

The genome sequence is just a series of letters. It tells us nothing about the genes contained within that sequence or the effects of variations. To infer this information, we need to pull in data from genomic databases. This is done through the annotation process.

Ensembl
A genome database.

UCSC
A genome database.

International Genome Sample Resource (IGSR)
A genome database that includes information about variation across populations.

Genome Aggregation Database (gnomAD)
A genome database that includes information about variation across populations.

Clinvar
A genome database that includes information about clinically relevant genetic variation.

5.8 Variant interpretation

Suppose we are aiming to identify genetic variants causative of clinical disease. In that case, one easy way to filter through the millions of variants within any genome is to focus on coding variants and, in particular, aim to identify genetic variants causative of clinical disease. This approach assumes that that one single variant explains the majority of disease risk and has a high penetrance (a Mendelian model).

However, recalling from earlier, we would still expect the average genome to include 10,000 nonsynonymous and 150 loss-of-function variants. This is a greatly reduced variant set compared to the whole genome, but how do we prioritize among this shortlist if we want to identify variants that might cause disease? By looking at common features of disease-causing variants, the **American College of Medical Genetics** (**ACMG**) has developed guidelines to help us through this process (Richards et al. 2015). These guidelines classify variations as **pathogenic** (a high chance that the variant is disease-causing), **likely pathogenic** (a chance that the variant is disease-causing), **benign** (not likely to be disease causing), **likely benign** (unlikely to be disease-causing) and **variants of uncertain significance** (**VUS**) (everything in between).

These classifications rely upon a points-based system which collates evidence drawn from population, computational, functional, and clinical data which are added to the VCF file during annotation. Variants that cause the loss of function of a dosage-sensitive gene and are rare in populations score highly on the pathogenicity scale. Variants that fall in important protein domains or change highly conserved bases attract additional points. Further marks will be awarded if the variant falls in a gene previously shown to have a relevant function or has previously been related to the disease in question. Other points come from consideration of the disease mechanism; if the variant co-segregates with disease within the family, or if the variant is *de novo* when the parents are unaffected, or if the variant occurs in *trans* (i.e. on different allelic copies of the gene) with another pathogenic variant in the case of recessive disorders. This system provides a structured way to evaluate variations observed through the genome clinically. Variant classifications are stored in a big database known as Clinvar, allowing future researchers to see previous evidence ratings.

Key Point

One of the most difficult parts of clinical genomics is the identification of variants that lead to disease. We need to narrow down a list of over 1 million genetic variants found in a given individual to the one variant that is causing the disease to be treated.

SELF-CHECK 5.6

Which of the following variants is most likely to be pathogenic?

Variant 1 A stop-gain variant in an olfactory receptor

Variant 2 A frameshift in a transcription factor

Variant 3 A lysine to tyrosine variant in a keratin gene

Variant 4 A synonymous variant in a non-coding RNA

OMIM
A genome database that includes information about clinically relevant genetic variation.

DECIPHER
A genome database that includes information about clinically relevant genetic variation.

Cross reference

Look back at **Chapter 2** to read more about methylation.

Look back at **Chapter 4** to read about GWAS.

Look back at **Chapter 2** to read more about Mendelian mechanisms of disease.

American College of Medical Genetics (ACMG)
An organization that represents the interests of medical genetics professionals including clinical geneticists, clinical laboratory geneticists, and genetic counsellors.

pathogenic variant
A genetic variant that is highly likely to directly cause a disease.

likely pathogenic variant
A genetic variant that is likely to directly cause a disease.

benign variant
A genetic variant that is highly unlikely to directly cause a disease.

likely benign variant
A genetic variant that is unlikely to directly cause a disease.

variant of uncertain significance (VUS)
A genetic variant for which there is not enough evidence to classify it as pathogenic or benign.

5.9 Sequencing in the clinic

As the amount of sequence data increases and databases become better linked, we have an ever-growing appreciation of typical and clinically relevant variation through the genome, allowing genomic technologies to become integrated into clinical settings. In 2012, the British Government announced the intention to sequence the genomes of 100,000 NHS patients through a programme known as **Genomics England** (www.genomicsengland.co.uk). This project focused on patients with cancer or with rare diseases that could not be diagnosed through conventional methods. The occurrence and progression of both of these disease classes are strongly influenced by genetics; cancer develops from changes in the genetic sequence acquired over the lifetime (somatic variation) and rare developmental diseases are often inherited in a Mendelian fashion. Next generation sequencing technologies, just like the ones discussed earlier, were used to catalogue all genomic variations in both germline and tumour samples from cancer patients and in patients and their family members for individuals affected by rare disease. The analyses on these samples and datasets are still ongoing, but this project showed that genomics could be helpful in a clinical setting. It identified new diseases, providing diagnoses and genetic counselling for affected families (**Case study 5.1**), and helped to develop new ways to identify tumours that might be sensitive to particular cancer treatments.

Genomics England
A programme to integrate genomic medicine into clinical care in the UK.

CASE STUDY 5.1 Genetic testing by sequencing can provide diagnostic clarity in uncertain medical cases

A 7-year-old boy, AJ, was referred to the genetics clinic due to his moderate developmental delay and unusual facial features. He was the second child of a non-consanguineous Caucasian couple. The older daughter was healthy and developing normally. The parents were separated and the mother had a history of alcohol addiction. The mother's parents, sisters, and nieces were healthy. There was no other relevant family history of note.

AJ was born at 37 weeks, following a normal pregnancy and delivery. The first concerns about him started at around 6 months when it was noticed that his neck control was poor. His motor milestones continued to be delayed, and he only started walking at about 3.5 years old. His speech was also severely delayed, and he had no words in his speech at the age of 7 years. His understanding of speech was limited to a few single words. He was also not toilet-trained at the age of 7.

On examination, he had a relatively large head with a tall forehead, coarse face, deep-set eyes, right-sided ptosis (droopiness of eyelid), and prominent chin. He was drooling constantly. His hands and feet were small, and he had tiny fingers and toes. He walked with a broad-based gait and had tremors in his hands.

Q1 What initial genetic investigations would be appropriate?

Basic investigations for suspected neurodevelopmental disorders include microarray CGH (comparative genomic hybridization) to rule out chromosomal abnormalities and Fragile X testing (especially in males). In this case, testing for Angelman syndrome is also appropriate due to severe neurodevelopmental delay, broad-based gait, and some facial features (deep-set eyes and prominent chin). The test came back negative.

Next, duo exome sequencing in AJ and his father (as mother's sample was unavailable) identified a splice-site c.1497+1_1497+18 variant of unknown significance in *CUL4B* in AJ but not in his father.

Mutations in this gene cause X-linked intellectual disability syndrome, known as Cabezas syndrome. It is characterized by severe intellectual disability, relatively large head size, coarse facial features with prominent chin, broad-based gait, and tremors.

As AJ's mother's sample was unavailable, after much discussion, the maternal grandmother's sample was tested, and she was found to carry the same variant. Fortunately, the mother also came forward for genetic testing at that time, and the same variant was confirmed in her.

In order to move the variant from a VUS to a likely pathogenic, RNA sequencing can be done to check the effect on splicing. RNA analysis showed a deletion of exon 12. Skipping of exon 12 is predicted to lead to a frameshift and premature termination of translation, p.(Leu461Ilefs*8). The variant was now classified as **'likely pathogenic'**.

As this is an X-linked recessive disorder, AJ's sister, maternal aunt, and maternal cousins are at risk of being carriers of the *CUL4B* variant and are at a risk of having affected sons. Genetic testing should be offered to these individuals. His mother is also at a risk of having an affected son as she is a carrier.

Cross reference

Look back at **Chapter 2** to read more about Mendelian inheritance.

In **Chapter 6**, we will talk more about sequencing applied in diagnostic settings.

In **Chapter 7** we will talk more about somatic variations and cancer.

virtual gene panels

A way of analysing only the parts of the genome sequence that are believed to be relevant to the clinical presentation.

In 2018, the NHS launched the Genomic Medicine Service, which incorporates genome sequencing into routine medical care in the UK. Other countries have similar programmes: China launched its own 100,000 Genomes Project in 2018. The Initiative on Rare and Undiagnosed Diseases (IRUD) in Japan uses NGS to investigate complex diseases that cannot be diagnosed in traditional ways. The Australian Government pledged 20 million dollars to fund a pre-pregnancy screening programme in families affected by genetic disorders. The National Institutes of Health (NIH) in the United States is building two genome centres that aim to sequence 100,000 participants each in the first year they are open. Projects like these help to generate a global picture of genetic variation and make it easier to distinguish disease-causing variation from typical genetic variation. Although whole genome sequencing is becoming a reality in the clinic, we still rely upon **virtual gene panels**. These use the symptoms seen in the patient to prioritize the analysis of subsets of genes that have been shown to be relevant in patients with similar signs. In this way, it is possible to perform a staged analysis studying the most likely candidate genes before annotating the whole genome.

Key Point

Next generation methods are being integrated into clinical care across the world.

5.10 So why can't we diagnose everything?

heterogeneity

Variation between individuals carrying the same genetic variant.

background genome

The genetic sequence that an individual carries aside from a variant of interest.

If we have so much genomic data and large databases of known pathogenic variations, why can't we diagnose every disease ever seen in the clinic? The hard part comes in linking the variation we catalogue during sequencing to the symptoms experienced by the patient. The variants that we are identifying through the clinical studies mentioned earlier lead to highly penetrant, Mendelian forms of disease. But even for these diseases, we see that two people carrying exactly the same causal variant can show very different symptoms. Presumably, this phenotypic variation, or **heterogeneity**, represents the action of the **background genome**, which will interact with the causal variant, leading to a worsening or relief of the disease presentation. Heterogeneity can make it hard to form a consistent link between genotype and phenotype, even for the most straightforward of genetic diseases.

Cystic fibrosis is a classic recessive Mendelian disease caused by the loss of function of the *CFTR* gene. This loss of function can be caused in many different ways—there are over 360 different variations in the gene that are linked to disease (https://cftr2.org). Broadly, these can be classified by their effect on protein function; class I variants prevent the production of a functional enzyme and are usually nonsense, frameshift, or splicing variants. These variants carry the highest risk of disease. Class II variants cause problems with protein production and processing. This class includes the classic Phe508del variant. Class III leads to altered protein regulation, and class IV variants alter protein stability. All of these variants lead to breathing difficulties, increased sweat chloride levels, and male sterility when present on both alleles. Additional symptoms, like pancreatic deficiency, intestinal obstruction, diabetes, and liver disease are associated with dual class I-III variants but not in everyone. The age of onset of chronic infection and lung function decline is highly variable even between two individuals carrying exactly the same variant (O'Neal and Knowles 2018). These multi-system differences illustrate an important fact—different tissues, even within the same individuals, can respond to the presence of the same genetic variation in different ways. In some organs, like the sweat glands, the *CFTR* variants are highly penetrant while in others, like the intestines, only a fraction of patients show symptoms. Family studies show us that about 50% of the variation in CF lung disease can be explained by variations in modifier genes such as *TGFB1* and the *HLA* genes, and the other 50% by environmental factors such as climate and smoke exposure (O'Neal and Knowles, 2018). Understanding such modifiers can help us to predict those individuals at the highest risk of severe symptoms and to target personalized medicine.

You can see from this example that even Mendelian diseases have complexities that make it difficult to predict individual outcomes. Remember from **Chapter 4** that many genetic diseases are complex and include multiple genetic factors. The progression of these disorders can be very hard to predict even with a whole genome sequence. You can see from **Box 5.8** that there are many ways that we can apply sequencing technologies, but the clinical application of this knowledge is still limited.

Cross reference

Look forward to **Chapter 6** to read more about the genetic mechanisms underlying cystic fibrosis and its diagnosis.

proteomics
The large-scale study of proteins.

transcriptomics
The large-scale study of gene transcripts.

epigenomics
The genome-wide study of epigenetic marks.

metabolomics
The large-scale study of substrates and products of metabolism.

single-cell RNA sequencing (sc-RNAseq)
The sequencing of all the RNA from a single cell.

BOX 5.8 Applications of sequencing technologies

In this chapter, we have focused upon sequencing as a method of finding clinical variation within the genome. However, high-throughput sequencing technologies can be applied in many ways to investigate cell function. When such technologies are applied at a comprehensive level, they are known as -omics. For example, **proteomics** is the large-scale study of proteins, **transcriptomics** considers the complete set of RNA transcripts in a cell, **epigenomics** describes the genome-wide study of epigenetic marks, and **metabolomics** refers to substrates and products of cell metabolism.

RNA sequencing (RNA-seq) uses high-throughput methods to sequence gene transcripts. Most mRNAs within a cell have a polyA tail. By using probes specific to these tails, it is possible to make libraries of these RNAs which can then be used as a substrate for sequencing. This provides information about variation within the coding sequences of genes but also provides information about gene expression levels; by counting the read depth for a given transcript, it is possible to estimate gene expression levels for that gene. RNA-seq technologies are now being applied at the single-cell level (**sc-RNAseq**), allowing us to investigate differences between gene expression across cells and tissues.

In **Chapter 2**, you read about epigenetic marks and the importance of methylation. It is possible to use a method known as methylation sequencing to capture this information at a genome-wide level. Recall that this epigenetic mark involves the addition of a methyl group to cytosine bases. We can use a chemical called bisulfite to identify methylated C bases. Treatment with bisulfite converts unmethylated cytosines to uracil but methylated cytosines are protected from this change. By sequencing treated and untreated samples, we can identify C bases that are resistant to conversion and are, therefore, methylated.

Another valuable method for genome characterization is the combination of antibody selection (immunoprecipitation) with NGS. For example, when a protein binds to DNA, it creates links. It is possible to fix these links and then use an antibody specific to your protein of interest to select the protein–DNA complexes out of the solution. The DNA is then removed from this solution to create a sequencing library that allows the identification of all protein-bound regions of DNA in the genome. This method is known as **chromatin immunoprecipitation sequencing (ChIP-seq)**. ChIP-seq can be useful for identifying regions of DNA targeted by transcription factors or regions of DNA bound by histone proteins. A similar method known as **methylated DNA immunoprecipitation sequencing (MeDIP-Seq)** can be used to identify methylated regions in the genome.

These are just a few examples of how the sequencing technologies that came out of the Human Genome Project are pushing science forward.

chromatin immunoprecipitation sequencing (ChIP-seq)
A method that sequences all DNA segments bound to a specific protein to catalogue the targets of that protein.

Cross reference
Look back at **Chapter 4** to read more about complex genetic inheritance.

methylated DNA immunoprecipitation sequencing (MeDIP-Seq)
a method that sequences all methylated DNA segments in a genome.

Key Points

The genome includes 3 billion base pairs that all work together to make the cell function. It does not make sense to single out one variation among those 3 billion bases. Most genetic disorders involve a complex interaction between many variants across many genes as well as the environment.

Sequencing methods are still evolving allowing us to generate more data, quicker and cheaper than ever before. The hard part is learning how to interpret and understand this data.

5.11 What does NGS miss?

Finally, we should consider what NGS misses. Whole genome sequencing can give us a complete picture of base-pair variation across the genome, but recall from **Section 5.2** that it can be hard to align genetic sequence data in regions of repetitive sequence. This means that genetic variations in these regions are often missed. Also recall from **Box 5.4** that most current sequencing methods require us first to break the DNA up into small fragments. This means that the genome's large deletions, duplications, and rearrangements can easily be missed because they will not be contained within one small fragment. The alignment of genome sequence data against a reference sequence assumes that the broad organization of the genome is the same in every individual. If a large chunk of the genome is displaced, for example, in a translocation, NGS methods would not pick this up. For this reason, emerging sequencing technologies, known as **third-generation sequencing** or long-read sequencing, focus on sequencing large DNA fragments. Look at **Box 5.9** to see how technologies such as PacBio and Nanopore allow the sequencing of a chromosome from telomere to telomere providing relative positional information that is not afforded by massively parallel methods. The downside of these applications is that they have a higher error rate at the base-pair level. Even third-generation methods do not capture factors that affect genome function from outside of the base-pair sequence such as epigenetic effects, infectious agents, or the microbiome. The capture of these effects is a whole other story.

third-generation sequencing
Sequencing methods that are capable of whole genome sequencing but do not rely upon massively parallel methods.

BOX 5.9 Third-generation sequencing or long-read sequencing

All of the methods we have spoken about in this chapter rely upon the fragmentation of DNA and the massively parallel sequencing of the products. However, these methods come with limitations. By sequencing small fragments of DNA, it can be hard to see the big picture. Suppose that someone had a rearrangement of DNA which meant that the ends of chromosome 9 and 22 were swapped over. This translocation is actually quite commonly seen within the genome and is known as a Philadelphia chromosome. This rearrangement does not lead to the loss of any genetic material but creates a new transcript by the fusion of two genes known as *BCR* and *ABL*. The *BCR-ABL* fusion protein activates signalling pathways within the cell and can lead to chronic myeloid leukaemia (CML), a type of cancer that affects the white blood cells. Because this translocation involves the displacement of a relatively large section of DNA, it can be hard to spot using massively parallel methods. The displaced DNA will be fragmented and sequenced in small chunks. Mapping algorithms will be used to place the fragment sequences against the reference genome and these will assume that they map to their rightful chromosomes. The only clue that they may have moved is if there are fragments that directly cover the breakpoint. In this case, the reads will be split with half of the fragment mapping to chromosome 9 and the other half to chromosome 22. New methods of sequencing, known as third-generation or long-read sequencing, can help to solve this problem as they are capable of sequencing large fragments of DNA.

The Nanopore is a small sequencing device consisting of protein pores as shown in **Figure 5.16**. DNA fragments are passed through the pores. As the DNA moves through, tiny

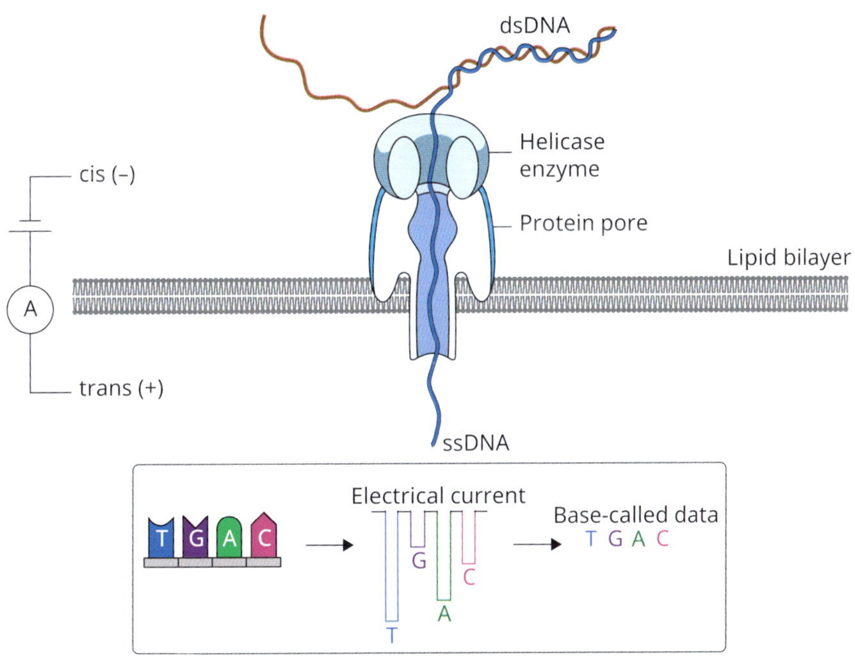

FIGURE 5.16

Nanopore sequencing. The Nanopore sequencer relies upon a very different method from the NGS approaches discussed in this chapter. These machines are made from protein pores through which the DNA is passed (blue region on the figure). At the bottom of the figure, you can see how the current changes as the DNA strand passes through the pore. This signal can be directly converted into a DNA sequence in real time—it does not require any intermediate mapping steps like NGS.

changes in electrical current occur. The change in charge is relative to the DNA base that is passing through and so by monitoring changes in current, it is possible to generate sequence data. These machines are now capable of sequencing whole genomes and, provided the DNA string can be kept intact, they are able to sequence very long fragments of DNA. In 2020, these machines were used to sequence the entire X chromosome from one telomere to another. That is 3.1 million base pairs of DNA! You can see how using this technology could simplify sequencing as little mapping is required and it is possible to sequence very repetitive regions. The Nanopore sequencer is also very small and can plug into a laptop by USB. This makes them very useful in situations where large labs are not possible. For example, the Nanopore sequencer was used to map strains of Ebola in the latest outbreaks. Sequencers could be taken directly to the field and provided DNA sequences to allow strain identification and outbreak mapping. However, these machines still have a high error rate compared to NGS and so are not currently used in clinical settings.

Another third-generation system that is capable of long-sequence reads is the PacBio sequencer. As can be seen in **Figure 5.17**, this method feels a bit more familiar as

A

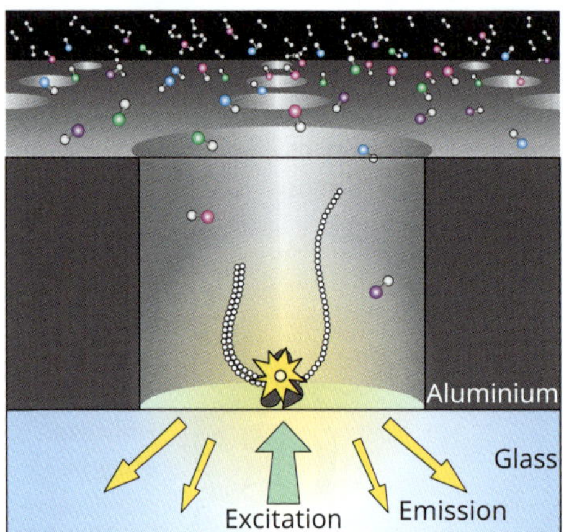

B

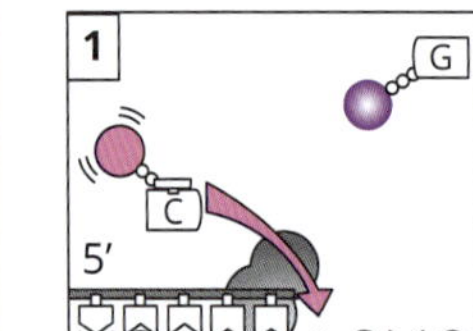

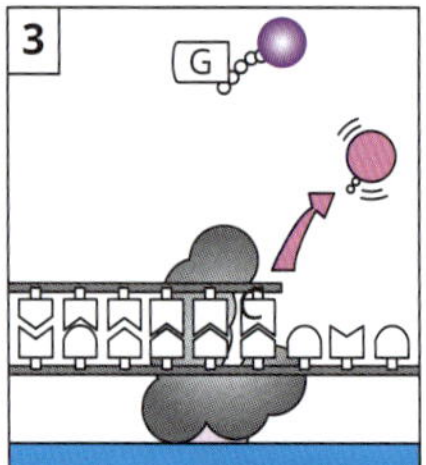

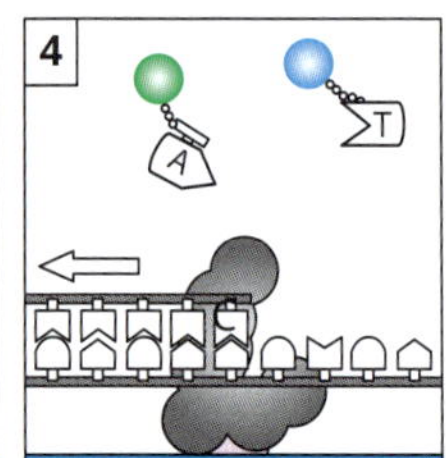

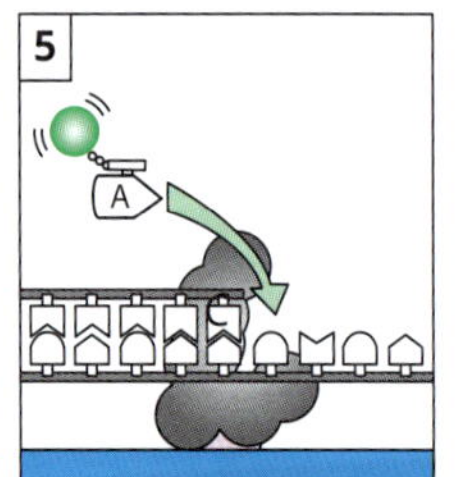

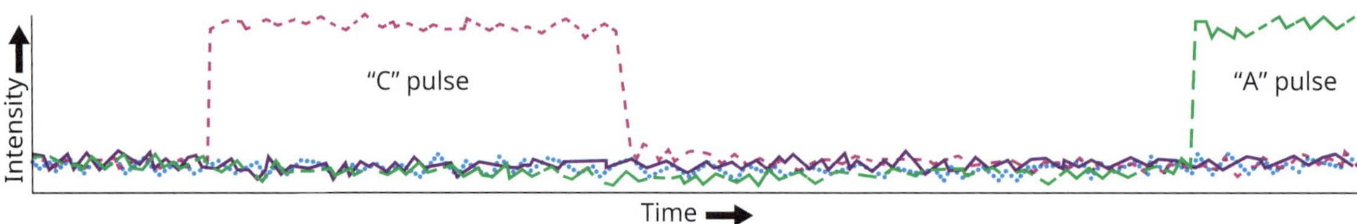

FIGURE 5.17

PacBio sequencing. Like the Nanopore, this method allows the sequencing of long DNA fragments in real time. Single-stranded DNA is stuck in a well (A) and copied by a polymerase (B). As bases are added by the polymerase, light is released (B, top). Light release is captured in a plot over time which can be interpreted as a sequence (B, bottom). Also, like the Nanopore, it has a high error rate compared to NGS.

it involves DNA polymerase copying a DNA strand using fluorescently tagged dNTPs. Hairpin adaptors are ligated to the ends of DNA fragments so that they form continuous loops. Each loop of DNA is injected into a tiny well where the DNA replication occurs. As fluorescent dNTPs are incorporated into the DNA strand, light is emitted. This information is captured in real time by the PacBio sequencer.

Chapter summary

- The term sequencing refers to the characterization of the nucleotide bases along the DNA strand.
- The sequence itself is just a string of A, G, C, and T but we can use patterns in this sequence to infer a lot about the function of the genome and its relevance to disease.
- The Sanger method of sequencing is a high-quality and reliable method of sequencing short DNA fragments.
- The first complete Human Genome was sequenced in the Human Genome Project. The sequence from this project is still being worked on. It forms the reference genome for all sequencing studies today.
- If we want to sequence whole genomes, we need to use a more automated method of sequencing such as next generation sequencing.
- The sequencing itself is just the first step in a next generation sequence experiment. Once the sequences are generated, they need to be aligned and stitched back together into a genome sequence. The sequence then needs to be annotated to identify variants of clinical significance.
- Clinically relevant variations usually (but not always) have a big effect upon protein function and are relatively rare in populations.
- Genomic information is stored in open-access databases. There are many databases, each tailored for different types of information or analyses.
- Sequencing and genomics are being integrated into clinical services in the UK but these are currently limited to very specific forms of genetic disease.
- Sequencing methods are always evolving to make DNA sequencing cheaper, quicker, and more effective. These methods are also being adapted into various -omics technologies.

Cross reference

Look back at **Chapter 3** to read more about diseases that involve rearrangement of DNA.

Cross reference

We will talk more about the genetic mechanisms of cancer in **Chapter 7**.

Cross reference

We will talk more about infectious agents and the microbiome in **Chapter 8**.

Discussion questions

5.1 If everyone had their genome sequenced at birth, how would this direct their healthcare?

5.2 What are the remaining difficulties in interpreting genome information? How can we overcome these?

5.3 Who should be responsible for genome information and who should be allowed to access it?

Further reading

- International Human Genome Sequencing Consortium (2001) ***Initial sequencing and analysis of the human genome***. *Nature*, 409, 860–921.

  The first release of the Human Genome.

- The 1000 Genomes Project Consortium (2015) ***A global reference for human genetic variation.*** *Nature*, 526, 68.

  This paper covers variation across genomes.

- Adams J (2008) ***The proteome: discovering the structure and function of proteins***. *Nature Education*, 1(3), 6.

- Adams J (2008) ***Transcriptome: connecting the genome to gene function***. *Nature Education*, 1(1), 195.

  These two papers are a good starting point for learning more about omics.

- Eid J, Fehr A, Gray J, Luong K, and 50 others (2009) ***Real-time DNA sequencing from single polymerase molecules***. *Science*, 323(5910), 133–138.

- Mardis (2008) ***The impact of next-generation sequencing technology on genetics***. *Trends in Genetics*, 24, 133.

Useful websites

- The Human Genome Project. **https://www.genome.gov/human-genome-project**
- The Genomics England website has more information about The Human Genome Project, along with lots of case studies. **https://www.genomicsengland.co.uk/**

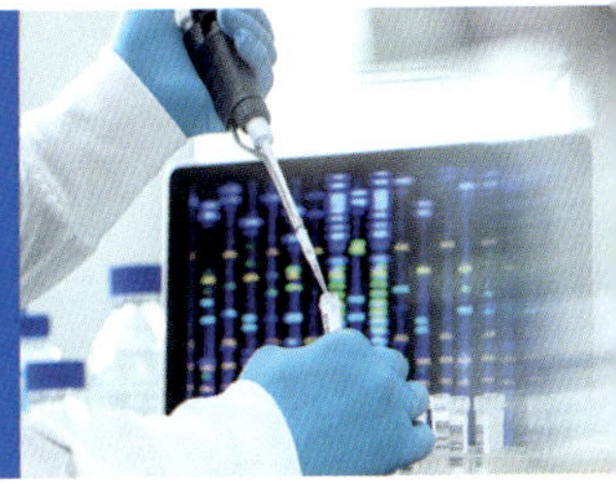

6

The Genetic Diagnostic Laboratory

Drew Ellershaw

This chapter will cover the main aspects of genetic testing services and screening in the healthcare sector. While this chapter focuses on genetic services provision in the UK, content related to workflow and methodological considerations apply globally. Genetic disease is caused by alterations in the normal functioning of DNA and its products, which in turn impacts on bodily systems, resulting in disease such as cystic fibrosis. These can be *de novo*—for example, new **pathogenic** variants which have arisen from conception—or inherited from either an affected parent or from a **phenotypically** normal carrier parent (although some may have minor symptoms). Pathogenic variants can also be acquired later in life through **mutation**, which can result in various cancers such as leukaemia and lung cancer. It is important that the underlying genetic cause is identified not only for **diagnosis**, but also for **prognosis** (predicting the course of the condition), treatment, **recurrence risks** in families, and also for support.

Genetic diseases vary widely in nature both by presentation and underlying genetic mechanism, and testing is focused to provide the highest diagnostic yield by using one (or more) different testing methods. The results of these tests must be carefully interpreted to classify any **variants** found as pathogenic (disease-causing) or benign (non-disease-causing). Sometimes, due to lack of (or conflicting) evidence, a variant may be of uncertain clinical significance. For variants of uncertain significance more investigation may be required such as re-examination of the patient, or family studies to look at segregation of the variant. These variants can also be the target for further research, including functional studies, but this is generally outside the scope of a diagnostic testing laboratory.

pathogenic
Disease-causing.

diagnosis
The physical manifestation or visible characteristics resulting from gene expression.

mutation
A change in the DNA sequence (this should not be used to imply a pathogenic change).

prognosis
The likely course and outcome for a medical condition.

recurrence risk
The chance that an inherited genetic disorder will occur again in a family.

variant
A change in the DNA sequence, more often used in genetic reporting as this is a more neutral term than mutation (see **Box 6.1**).

Learning Objectives

- Outline the organization of a genetics service
- Describe the workflow, from primary sample receipt through to report
- List the main types of genetic testing methods
- Explain why different tests are used for different genetic disorders
- Discuss the complexities of genetic counselling.

6.1 Service provision in the UK

Genetic testing in the UK is predominantly provided by the National Health Service (NHS). Regional genetics laboratories all offer a core list of genetic tests (for common genetic disorders). Whereas specialized tests, for rare conditions, are only offered in a limited number of centres. A network of Genomic Laboratory Hubs (GLHs) delivers the provision of genetic testing in the UK, each hub responsible for coordinating services for their part of the country.

Private laboratories and commercial testing also exist and may be chosen over NHS laboratories if they can offer better pricing or turn-around times, but this is, at present, uncommon in the UK.

Users of genetic (see **Box 6.2** for terminology) services such as GPs, fertility centres, various hospital specialities (including specialist clinical geneticists) and foetal medicine units (FMUs) assess their patients and family history for indicators of a genetic condition. The relevant samples are then sent to the regional laboratory. The laboratory will perform the requests, test, analyse and interpret the results, and send a clinical report directly back to the referring clinician/team who will then discuss the findings with the patient. Where a result identifies a pathogenic (disease causing) variant referral to clinical genetics for specialist counselling is normally recommended.

6.1.1 UK National Genomic Test Directory

As the field of genetics and genomics is continually advancing at a rapid pace, new methods of testing are being developed, offering either increased resolution, faster workflow, or higher throughput. This can lead to changes in the provision of service, and laboratories' uptake of new technology can be variable and dependent upon the size of the laboratory and available

BOX 6.1 Terminology: Mutation versus Variant

Although mutation is defined simply as a change in DNA sequence, without any measure of its effect, the word *mutation* is most often used to mean a change in the DNA sequence which is associated with a negative effect. In genetic testing, the word *variant* is now preferred, to avoid any implicitly negative associations. In the context of genetic disease, the word variant is normally prefixed or suffixed with the following: pathogenic, likely pathogenic, uncertain significance, likely benign, benign.

BOX 6.2 Terminology: Genetics and Genomics

Traditionally, the word *genetics* has been used in reference to the inheritance and function of single genes. In modern genetics, genes are no longer seen as single operating units, but as interconnected elements within a functional continuum (the genome). In this new era, *genomics* testing can now be performed at the whole genome level, including not only all genes, but also non-coding DNA providing a complete set of genomic information.

support and funding. This in turn can lead to different levels of service between laboratories and different costs for tests, which can affect where users of the service send samples to. It is worth reminding ourselves that the goal of the NHS is to offer free healthcare to all, but disparities between labs can mean that some tests offer more information (i.e. larger **gene panels**) than others.

gene panel
Genetic sequence analysis for a given set of genes by NGS.

In view of this, NHS England set forth change to the landscape of genetic testing in the UK, culminating in the National Genomic test directory (both for rare and inherited disease, and for cancer). The test directory sought experts (clinicians, clinical scientists, economists, as well as patient and public bodies) to develop this test directory suitable for the new era of genetic testing in the NHS. This new test directory aims to bring the most optimal testing to get the best value from NHS resources. Testing is now more based upon clinical condition, rather than a specific gene. The directory also lists which specialities should be requesting the test to ensure tests are requested by those with the relevant background and expertise.

The rare and inherited disease directory covers genetic testing for inheritable conditions (including inherited cancers) which may or may not have a family history. In the UK, rare disease is defined as one that affects 1 in 2000 or less of the population, but if all the rare diseases are grouped together then a rare disease will affect 1 in 17 people. Approximately 80% of rare diseases are known to have a genetic cause, thus highlighting the importance of genetic testing. Genetic conditions can be *de novo* in that they arise newly in an individual at conception without any family history. Genetic conditions can also be inherited within families in a dominant pattern of inheritance, where only one of a chromosome pair carries the pathogenic variant for a particular gene or locus, or in a recessive manner where both chromosomes carry a pathogenic variant for a given gene or locus. Genetic disorders can also be sex-linked, whereby only males or only females are expected to be affected by a genetic disorder (e.g. Duchenne muscular dystrophy in males, and Rett syndrome in females).

The cancer test directory (including solid, haematological, and neurological tumours) covers acquired genetic variants, where the patient is not born with a pathogenic variant but acquires it during their lifetime. These can be caused by DNA damage (including UV exposure and chemicals) which is not repaired correctly by the body's own repair mechanisms and gives rise to abnormal populations of cells (a form of **mosaicism**), which can become cancerous.

mosaicism
The presence of two or more cell lines derived from a single zygote.

6.2 Laboratory structure and workflow

Genetics laboratories, where possible, are designed to aid the flow of samples across the laboratory, both to maximize efficiency and to reduce the risk of contamination. Samples are received into the laboratory with a request for testing via a specimen reception area where they will be logged into the laboratory's information management system (LIMS). It is important that all patient demographics are entered correctly as well as any relevant clinical details provided to ensure that the correct tests are performed in a timely manner. Samples (see **Box 6.3**) may be rejected for testing if they do not meet the minimal acceptance criteria according to the laboratory policy, or testing may be delayed if further information needs to be obtained from the referring clinician. Similarly, samples may be rejected if they are too old upon receipt or if sent in the wrong type of container, which can result in cultures failing to grow or PCR reactions failing due to contamination with unwanted additives.

To facilitate the test request and booking in process, laboratories will provide template referral forms for the clinicians to complete with prompts for the relevant patient demographics, clinician details (including contact information), clinical information, and a range of tests that can be requested. Due to the large repertoire of tests a laboratory performs, different types of referral

BOX 6.3 Sample types

Blood samples make up the majority of samples received for genetic testing, with ethylenediaminetetraacetic acid (EDTA) tubes used for DNA extraction and lithium heparin (LH) used for culture. Cord blood may also be taken from babies *in utero* or at birth.

Buccal swabs can be a useful alternative if blood sampling is not possible. Skin biopsies are often sent for investigation of mosaic genetic conditions in which the genetic abnormality may only be present in the skin.

Muscle biopsies may be indicated for functional RNA studies in dystrophinopathies.

Bone marrow aspirates are routinely used in the investigation of haematological malignancies.

For prenatal testing, chorionic villi from the placenta can be biopsied, or amniotic fluid aspirated for culture or DNA extraction, and placental or foetal tissue from miscarriages and stillbirths can be used for genetic testing.

Cell-free foetal DNA (cffDNA) can also be used for non-invasive prenatal diagnosis.

For cancer patients tumour samples as well as routine blood can both be used to test for both inherited and sporadic genetic alterations.

forms will be available limiting the number of tests based on the referral type (e.g. prenatal and postnatal, or cytogenetic and molecular genetic tests, targeted familial testing and genome-wide testing). It is not uncommon for inexperienced clinicians to fill out forms incorrectly (and sometimes will tick to request every test on the referral form when clearly not appropriate!). The opposite may also happen whereby a test may be clinically indicated but has not been requested; in these circumstances a clinical scientist may be required to contact the clinician to discuss the case to make sure vital testing (which may be required urgently) is not missed.

After booking in, samples are usually cultured or undergo RNA/DNA extraction directly. In some instances, the sample may be processed for testing directly. Generally genetic technologists will perform most of the wet lab work with some basic analysis and report writing, whilst clinical scientists will perform the majority of analysis, interpretation, report writing, and authorization (**Figure 6.1**; see **Section 6.2.2** for more details regarding the types of staff in a genetics laboratory).

6.2.1 Laboratory accreditation

All laboratories undertaking clinical testing in the UK are expected to perform to a high level of quality to ensure that testing is performed safely, giving users confidence in the final report a laboratory produces. This is demonstrated through laboratory accreditation for conformance with the ISO 15189 standard for Medical Laboratory accreditation. Accreditation is achieved through an independent external agency, the United Kingdom Accreditation Service (UKAS). Assessment is against two broad laboratory areas: the quality management system (QMS) and technical processes, which are both further subdivided into more criteria. A good quality management system should encourage continual improvement and includes, but is not limited to, staff training, competences, and continuing professional development, routine audit for compliance to standard operating procedures (SOPs) and best practice guidelines (BPGs), document management, reporting of incidents including root cause analysis to ensure the incident does not reoccur, and participation in external quality assurance (EQA). **Figure 6.2** demonstrates the relationship between testing and the quality management system whereby a good quality management system following ISO 15189 will result in the best quality results giving improved satisfaction from the referrer and patient.

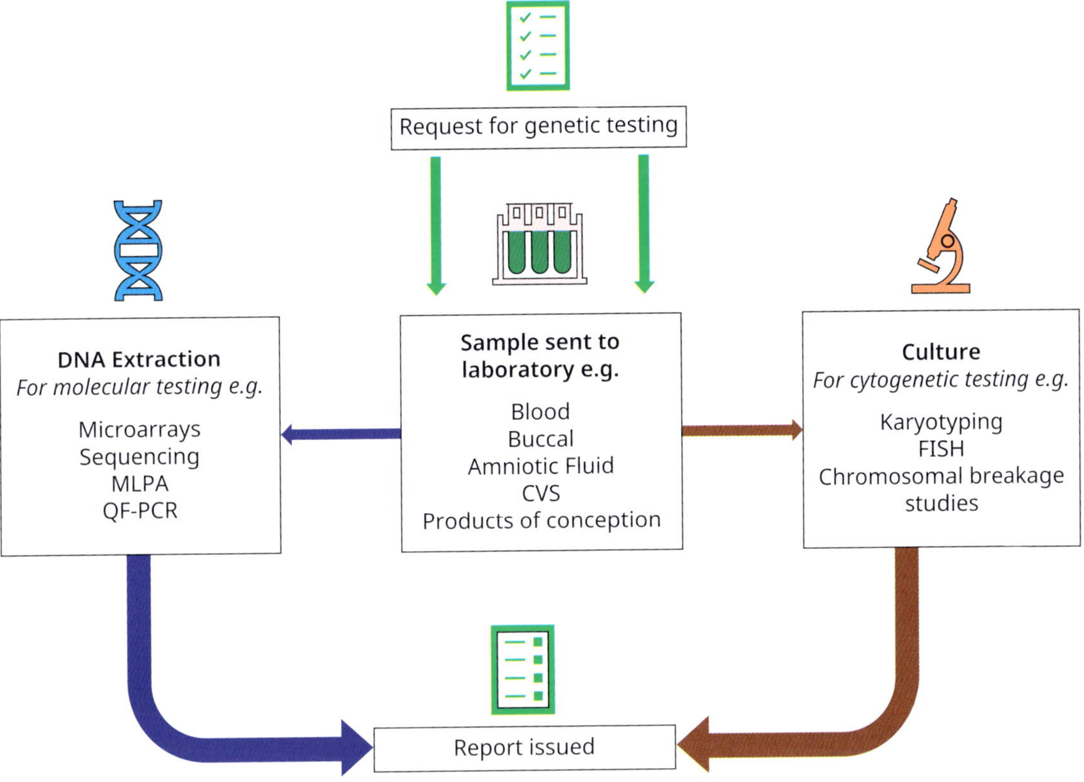

FIGURE 6.1
A simplified sample flow from test request to the issue of a report back to the referring consultant.

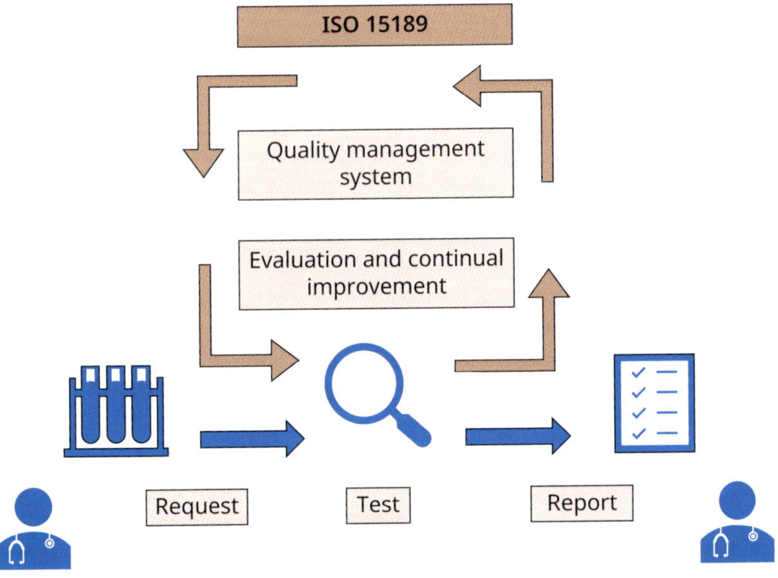

FIGURE 6.2
Sample testing flow and cycle of continual improvement through a QMS.

BOX 6.4 Important professional bodies associated with genetics

BSGM (British Society for Genomic Medicine)

ACGS (Association for Clinical Genomic Science)

HCPC (Health Care Professions Council)

HTA (Human Tissue Act)

RCPath (Royal College of Pathologists)

NSCS (National School for Healthcare Science)

GenQA (Genomics Quality Assessment)

EQA schemes are available for most tests and can assess wet laboratory processes through to data analysis and a full clinically interpreted report (not just a technical report of the data). Therefore, a good QMS should lead to high-quality technical results and a culture of continual improvement. Technical processes not only include the testing procedure itself (the examination process) but also relate to items such as referral forms, information to service users (pre-examination processes), as well as reporting, and storage of clinical material (post-examination procedures). Laboratories must ensure that any new equipment, software, tests, reagents etc. are all fully tested and validated for use—UKAS does not only accredit the lab as a whole but each type of test needs to be individually accredited. Laboratories who are accredited are then permitted to use the UKAS accreditation logo on reports (with their accreditation number) as a sign of accreditation. For any tests performed outside of accreditation, the laboratory must clearly state this in the final report and should not include the UKAS logo. For laboratories validating new tests (which is often the case in genetics due to changes in technology) they must apply for a UKAS extension to scope (ETS) for individual techniques not previously covered by assessment. Other important professional bodies are listed in **Box 6.4**.

6.2.2 Laboratory staff

The laboratory can roughly be divided up into three main staff groups: administrative staff, genetic technologists, and clinical scientists.

Administrative staff will generally be tasked with booking in samples and data entry of patient demographics, sending out reports, taking minutes of meetings, and dealing with basic laboratory enquiries, amongst other duties. This work is essential for the laboratory and enables other staff to focus as much as possible on laboratory and clinical work.

Genetic technologists are the backbone of the laboratory and perform at a range of different grades. At lower levels, assistant technologists are involved in areas such as stock control, routine cleaning of equipment, filing of clinical material, and basic laboratory work. Genetic technologists will be involved with the preparation of samples such as culture set-up and maintenance, DNA extraction, harvesting of cultures, a range of PCR-based testing methods, DNA sequencing, as well as basic analysis, interpretation, and report writing. For more information about the range of testing performed see **Section 6.3** (Genetic Testing and Screening Methods). Senior technologists manage teams of genetic technologists on a day-to-day basis, ensuring a smooth running of the service, as well as performing a range of tests and offering troubleshooting where required. For larger laboratories a lead technologist may be required to oversee and manage the technical team as a whole, across both cytogenetics and molecular genetics.

Clinical scientists in genetics are required to undertake more complex scientific and clinical roles and the varied roles can be grouped into three main areas: clinical, leadership and management, and research and development. On the clinical side clinical scientists are involved in analysing and interpreting a wide range of genetic tests and writing the clinical reports. Testing can be complex and involve multiple tests on a single sample, which then need to be interpreted together considering the clinical information provided. This can also involve discussions at **multidisciplinary team meetings (MDTs)** whereby a range of clinicians and scientists from different disciplines will be present to discuss findings from the most complex cases to help diagnose the patient. As the clinical scientist gains more experience, they may take on a more senior role as a team leader, head of a whole section, become qualified as a consultant, and even become head of a laboratory. Within these senior roles the scientist may also take on more specialist roles such as quality manager, audit officer, training lead, incident manager etc. to further support the laboratory with respect to quality and accreditation. Finally, scientists may have a role within research and development, ensuring the laboratory keeps up to date with the latest developments and testing methods. This is mainly focused on translational research which is aimed at setting up newly published methods as part of a routine clinical diagnostic service involving a full validation of the method including clinical validity.

multidisciplinary team meetings (MDTs)
Formal periodic meetings between healthcare professionals who have different, relevant expertise. In an MDT the group will discuss patients' diagnosis and condition and agree and organize their treatment plan according to the most appropriate evidence-based protocols.

Key Points

Clinical scientists as a profession are involved in 80% of patient diagnosis and cover a large range of fields such as life sciences (including genomics and genomic counselling), physical sciences (including biomedical engineering), physiological science, and clinical bioinformatics.

6.3 Genetic testing and screening methods

Genetic testing has been historically performed by two branches of genetics—cytogenetics (chromosomal level) and molecular genetics (base pair or gene level). Genetic testing is aimed at offering diagnostic testing to find the cause of an underlying genetic condition which can be either syndromic (associated with multiple features in similarly affected individuals, e.g. Down syndrome) or non-syndromic (a single phenotype, e.g. deafness). Screening tests are aimed at testing individuals or populations who are at an increased risk for a genetic condition. This can be the result of other tests (such as prenatal maternal serum biochemistry for increased risk of Down syndrome) or testing can be offered to specific ethnic groups (e.g. Ashkenazi Jewish) who are more likely to carry a pathogenic variant due to **founder mutations**, which although rare in the general population, appears with increased frequency in their population. Screening tests are normally used for pre-symptomatic testing of disease to allow for early monitoring or intervention (such as in inherited breast cancer, or neurodegenerative disorders). For an overview of referral groups and why different genetic test methods are suited to specific types of referral or disorder refer to **Section 6.4**.

founder mutation
A genetic variant found at a higher frequency in a particular population normally due to isolation.

6.3.1 Cytogenetic methods

Cytogenetics traditionally involves the examination and analysis of chromosomes down a microscope, referred to as **karyotyping**, to look for either numerical changes of whole chromosomes or for chromosomal rearrangements and microscopic deletions and duplications. Over

karyotyping
The pairing and analysis of chromosomes.

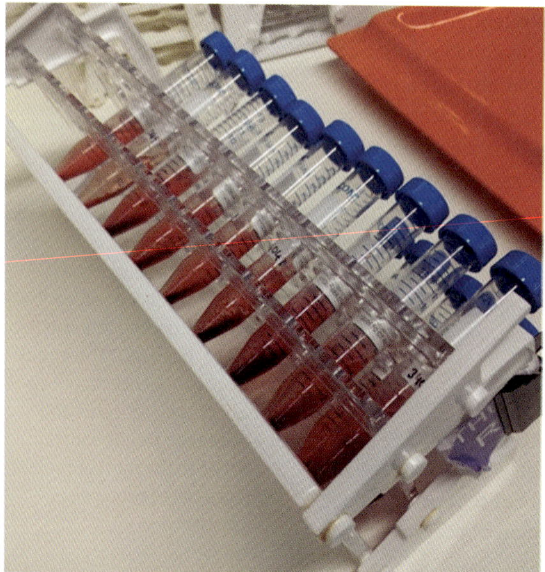

FIGURE 6.3
An example of blood cultures. The racks of tubes are angled to maximize the surface area for growth, as although these are liquid cultures the cells gradually settle to the bottom of the tubes. Arif Biswas/Shutterstock

time molecular cytogenetic methods for targeted testing such as **fluorescence *in situ* hybridization (FISH)** and **quantitative fluorescent polymerase chain reaction (QF-PCR)** were developed, and now the majority of testing is performed by DNA-based high-resolution **chromosomal microarrays (CMA)**. As such, cytogenetics can be thought of as non-targeted genome-wide testing at a relatively low-resolution chromosome level (compared with sequencing), but with advances in technology the line between cytogenetic and molecular genetic testing is increasingly blurred.

Cell culture

For karyotyping (and metaphase FISH), live cells must be grown to yield a sufficient number of cells in **metaphase** for analysis, therefore samples need to be fresh. See **Figure 6.3** for an example of blood cultures. For cytogenetics, blood cultures are routinely set up from lithium-heparin collection tubes; this additive prevents the blood from clotting and does not interfere with the cell culturing process (whereas EDTA restricts growth). Samples should be received within 3 days of being drawn, and samples over 5 days old may be rejected as they are likely to fail to grow in culture. Usually up to 6 drops of blood is added to blood culture media (containing a range of growth supplements and amino acids, including L-glutamine). In order to stimulate the growth of T-cells or B-cells a **mitogen** is added/included with the media—such as phytohaemagglutinin (PHA) or pokeweed mitogen (PWM)—as these cells are normally **quiescent** (dormant). Cultures may be grown from between 24 and 96 hours depending on the required testing, usually 72 hours for constitutional analysis, and variable for testing in **leukaemia** to optimally detect abnormal clones which can quickly die and be lost if cultured for too long. Cultures can also be established from other aspirates such as bone marrow and amniotic fluid, or from tissue samples such as skin, chorionic villus, muscle etc. Cultures from solid tissues and prenatal sample types usually take up to two weeks to grow to a sufficient level before proceeding with downstream processing such as harvesting for karyotyping or FISH, or extraction of DNA.

fluorescence *in situ* hybridization (FISH)
A procedure that uses specific nucleic acid probes and fluorophores to enable researchers to locate and visualize the positions of specific DNA sequences in chromosomes.

quantitative fluorescent polymerase chain reaction (QF-PCR)
A rapid method for the detection of chromosome copy number by amplification of DNA sequences at chromosome-specific loci.

chromosomal microarrays (CMA)
A high-resolution whole-genome microarray in which the chip is designed to detect small genetic alterations, including submicroscopic abnormalities below the size resolved by conventional karyotyping or FISH (fluorescence *in situ* hybridization) analysis.

Culture harvesting

Harvesting of cultures is performed to enable karyotyping and/or FISH analysis and the final product is a clear fixed cell suspension which is enriched for metaphase cells so that chromosomes are available for analysis. The cell cycle is manipulated in order to maximize the **mitotic index** (the proportion of metaphase cells to nuclei) whilst maintaining high resolution of chromosomes. Cells are arrested at metaphase by the addition of a mitotic spindle inhibitor (such as colcemid); thereby, as cells proceed through the cell cycle, they are halted at the stage chromosomes are visible and accumulate. The longer the cells are incubated with the spindle inhibitor the more metaphases a harvest will yield. However, chromosomes will continue to condense and shrink resulting in a lower quality for analysis. DNA intercalating agents such as bromodeoxyuridine (BrdU) can be added to help preserve the length of chromosomes to allow for longer incubation times with the spindle inhibitor.

Alternatively, blood cultures can be synchronized, whereby cell division is blocked overnight so that cells proceed to and are arrested in S phase. The cells are then released before adding the spindle inhibitor; this controls the cell cycle ensuring more cells are at the same stage, reducing the required incubation time with the spindle inhibitor. The result of this is increased metaphase yield whilst maintaining chromosome length for high-resolution karyotype analysis. The cultures can then be harvested to obtain clean and well-spread metaphase cells. First, cultures are centrifuged to pellet the cells and the supernatant is removed. A hypotonic solution is added (e.g. water or potassium chloride) so that water can enter the cells and help them swell up so that the condensed chromosomes in metaphase have more room to spread out, which results in fewer chromosomes crossing over and obscuring each other. The hypotonic solution also helps to lyse red blood cells which assists in cleaning the sample. Next the sample is fixed multiple times which further cleans the sample of any impurities and fixes it so that slides can be made, and the cell suspension stored for use at a later date. The fixation process involves centrifuging the sample again and removing the supernatant before adding fix to the sample (3:1 methanol: acetic acid); this is normally repeated 2–3 times to ensure all impurities are removed and the sample is fully fixed. This method of harvesting is the most basic and variations of the above are commonly found.

metaphase
The stage of the cell cycle where DNA is condensed into visible chromosomes and attach to the mitotic spindle.

mitogen
A protein that induces cell division.

quiescent
A resting or dormant phase of the cell cycle.

leukaemia
Cancer of the blood cells.

mitotic index
The proportion of metaphase cells.

phase contrast microscopy
Used to visualize a specimen based on the light shift when it passes through a transparent object.

Slide making

Slides can then be made from the cell suspension prior to testing (karyotype, or FISH); this involves adding a drop of the cell suspension to a glass slide, usually followed by a drop of fix as the cell suspension begins to dry (indicated by a contracting rainbow sheen known as Newton's rings). The slides are then viewed by **phase contrast microscopy** to assess the quality before making further slides as required. How well the metaphase chromosomes spread is dependent upon several factors which must be controlled to optimize spreading. For example, alterations to drying time (influenced by temperature and humidity), the height from which the cell suspension is dropped, and use of wet or dry slides. A standard environment for slide making in terms of temperature and humidity can, to a limited extent, be controlled by air conditioning. However, special slide-making chambers can also be used where the required air temperature and humidity is precisely set; the slides are made (or dried) within the machine's chambers.

Ideally the chromosomes should be well spread with as few crossovers as possible—if cells appear broken with missing chromosomes this is termed overspread (in extreme cases the cells are broken so badly chromosomes are scattered across the slide with few whole metaphases visible, and can be described as chromosome soup). Conversely, a metaphase is termed under-spread if the chromosomes are packed together with chromosomes lying on top of one another. This means that chromosomes cannot be fully analysed and makes karyotyping more difficult and time consuming. The technologist needs to carefully examine the cell suspension from the case being made and adjust as required to make slides of a suitable quality for testing.

Slides are normally aged prior to any banding, or FISH testing, which helps to improve quality. Ageing methods can include drying in an oven or on a hotplate, placing into hydrogen peroxide, or exposure to UV light.

Banding and karyotyping

For karyotyping, the chromosomes are *banded,* which gives each **homologous chromosome** pair a unique distinctive banding pattern (almost like a barcode). An example of this can be seen in **Figure 6.4**.

homologous chromosomes
Chromosomes exist as homologous pairs, one inherited from each parent.

G-banding
The technique used to stain metaphase chromosomes to produce a unique banding pattern for each chromosome.

The most common banding method for karyotyping is called **G-banding** and the stain traditionally used is Giemsa (alternative Romanowsky stains such as Leishman stain can also be used). The aged slides are first placed into a trypsin solution which differentially removes a range of DNA scaffolding proteins so that when stained it produces the bands unique to each chromosome. After being placed in trypsin for a set time the slides are then rinsed in water, or

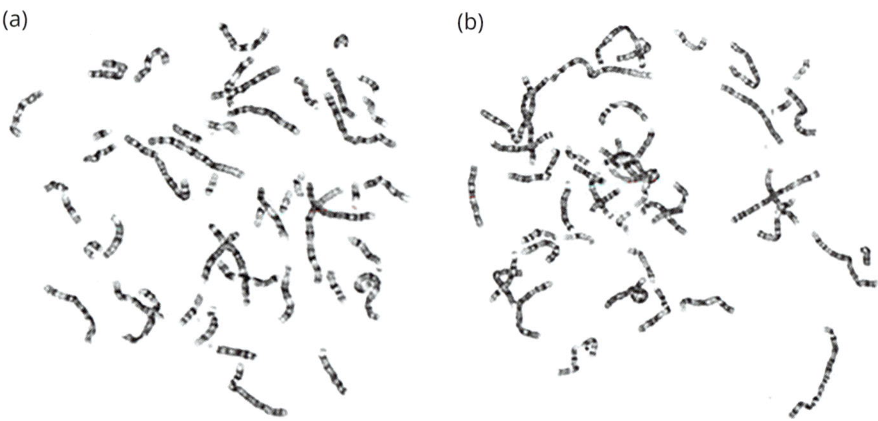

FIGURE 6.4

Banded metaphase chromosomes. The metaphase on the right (b) has longer chromosomes and more clear bands than the metaphase on the left (a). However, with the increased length there is also more crossing over of chromosomes which makes analysis more time consuming.

translocation
Where chromosomes break and transfer to a different chromosome.

aneuploidy
An abnormal number of chromosomes (but not a whole set).

polyploidy
Having more than 2 sets of chromosomes.

triploidy
Having 3 sets of chromosomes (e.g. 69, XXX).

dosage-sensitive
Dosage-sensitive genes cause a phenotypic effect when deleted or duplicated.

heterochromatic variation
Variation in regions of condensed inactive DNA.

centromeres
Specialized chromosomal/DNA structure involved in linking sister chromatids, and the attachment point for spindle fibres during mitosis.

inversion
A segment of DNA, or section of a chromosome, which is reversed.

saline which additionally helps to inactivate any residual enzyme. The slides are then placed into the stain for several minutes before drying. Practically, a test slide is used first to optimize trypsin and stain times. Incorrect trypsin times can lead to the appearance of fuzzy chromosomes or splitting of sister chromatids (appearing as train tracks) which means chromosomes may not be analysable. Similarly, the chromosomes can be too pale for analysis if the stain time is too short, and if the stain time is too long chromosomes can be too dark without distinct bands and it also increases staining debris.

Karyotyping for chromosomal aberrations is performed on G-banded slides. This involves a 'band-by-band' comparison of each chromosome pair (homologues) to look for structural changes such as deletions, duplications, rearrangements (inversions, insertions, **translocations**), **aneuploidy** (gains and losses of a whole chromosome) and **polyploidy** (altered number of chromosome sets e.g. **triploidy** which is 69 chromosomes, 3 of each homologue). See **Figures 6.5** and **6.6** for some examples of chromosomal abnormalities. Deletions and duplications may lead to an abnormal phenotype if the region includes **dosage-sensitive** genes, balanced rearrangements are not normally associated with disease but confer a risk of unbalanced conceptions which can lead to miscarriage or abnormal live birth. Balanced rearrangements are often found in sporadic cancer (such as leukaemia) whereby gene-fusions result in aberrant expression leading to uncontrolled cell division and lack of normal function.

The description of chromosomal aberrations is standardized using the International System for Human Cytogenetic Nomenclature (ISCN) which is available as a published handbook. As the length of the chromosomes affects the resolution of banding, the limit of detection for small aberrations is approximately 5 megabases (5Mb = 5 million base pairs). Any changes below this size may not be detectable as this is below the resolution of the banding and microscopy. Normal variation can occur in the form of chromosomal polymorphisms, most commonly of **heterochromatic** regions of DNA around the **centromeres** of some chromosomes (most commonly the 9q centromeric region). These can be seen as apparent duplications (e.g. 9qh+), deletions and **inversions** which are known to be inherited in patients with a normal phenotype. Karyotyping was for a long time considered the gold standard method of cytogenetic testing for a range of referrals, but for copy number variant (CNV) detection (gains and losses) it has been superseded by FISH and chromosomal microarrays which offer copy number detection at a higher level of resolution.

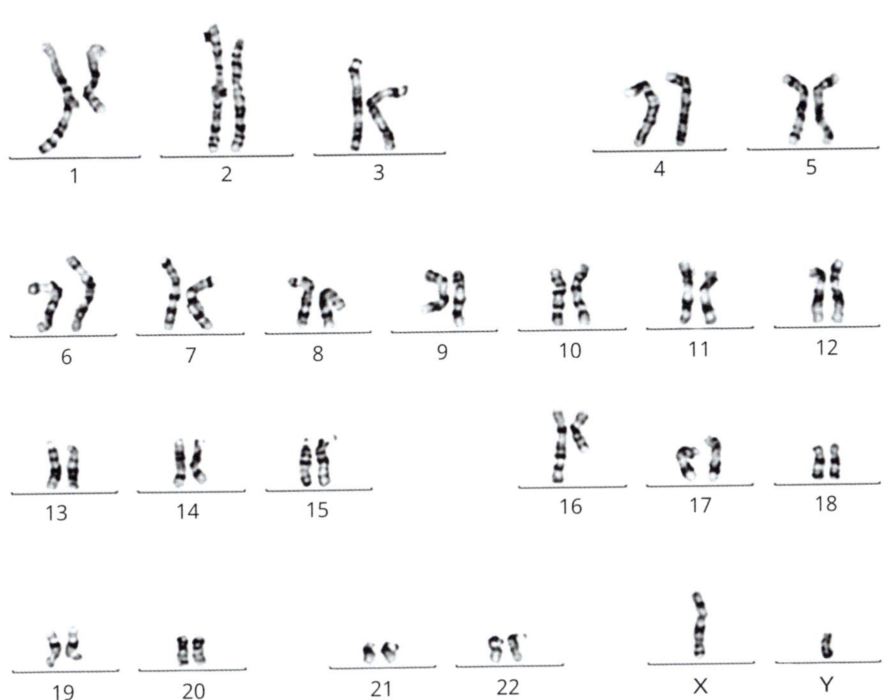

FIGURE 6.5
An example karyogram showing all 22 chromosome pairs and the sex chromosomes. This is an example of a balanced reciprocal translocation between the long arms of chromosomes 1 and 16 in a male (XY).

FISH

FISH involves the targeted testing of specific regions of the chromosomes by fluorescently labelled probes which can be applied to both **interphase** and metaphase cells and are then visualized by **fluorescence microscopy**. In constitutional cytogenetics (for patients born with a genetic abnormality) FISH was previously mainly used for targeted testing of sub-microscopic deletions, such as when there is a suspicion of a very specific genetic syndrome (e.g. DiGeorge syndrome which is commonly caused by a 2.5Mb deletion at 22q11.21). Due to the small size, this deletion is not usually detectable by karyotyping, therefore FISH can be used to test this region. Both a test probe for the region of interest and a control probe on the same chromosome are used together. See **Figures 6.7** and **6.8** for examples of FISH. The use of a control probe is critical for diagnostic purposes to ensure the technique has worked and the probes hybridize to the right location. An absence of a control probe could indicate quality issues with the FISH processing and the test would need to be repeated, otherwise there would be a risk of a false result. In a patient without a deletion of 22q11.21 you would expect to see two signals present for both the control and test probes, whilst in a patient with a deletion only a single signal would be present for the test probe.

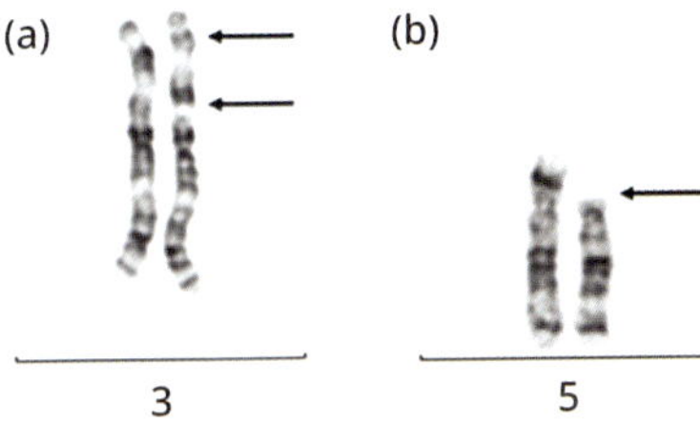

FIGURE 6.6
(a) A paracentric chromosome inversion within the short arm of chromosome 3. (b) A terminal deletion for the short arm of chromosome 5. Arrows indicate the breakpoints.

FISH is also useful for elucidating the origin of **marker chromosomes**, which are small derivative chromosomes where the banding pattern is not enough to confirm the chromosomal origin. M-FISH could be undertaken which applies whole chromosome paints to all chromosomes, each with a unique colour which could then be used to identify the unusual chromosome in question by its colour to determine its identity. Also, telomeric FISH can be used to identify/confirm cryptic, submicroscopic, chromosome rearrangements which can be seen in parents who have a child, or family history, of a submicroscopic abnormality.

interphase
The stage of the cell cycle preparing for cell division.

fluorescence microscopy
Specialized microscopy using fluorescently labelled DNA probes to visualize regions of interest.

marker chromosome
An abnormal small chromosome which cannot be identified by G-banding alone.

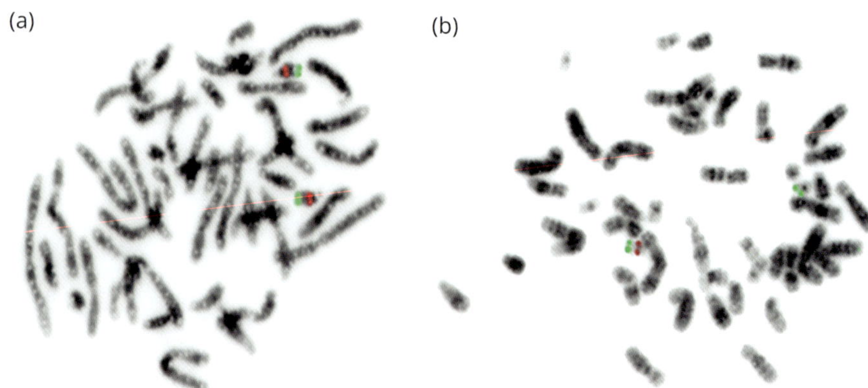

constitutional genetics
Deals with inherited genetic conditions.

Cross reference
Please refer to **Chapter 3, Section 3.1** for an overview of the FISH technique.

FIGURE 6.7
Metaphase FISH image using probes for the DiGeorge critical region at chromosomal location 22q11.21 in red, with a green control probe for the terminal region of chromosome 22. Image (a) shows a normal signal pattern with both 2 red and 2 green signals. Image B age shows a deletion on one of the chromosome 22 homologues with only a single green signal.

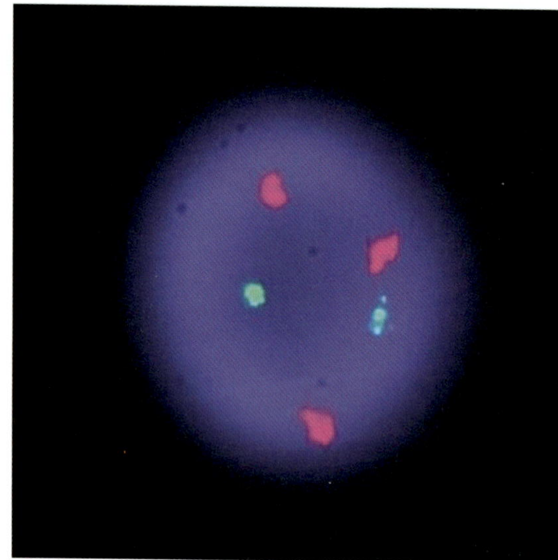

FIGURE 6.8
Interphase FISH for a patient with suspected Down syndrome. Probes for chromosome 21 are in red, and green for the control probe on chromosome 13. There are 3 red signals present consistent with the presence of trisomy 21, and 2 green signals as expected for the control.

In **constitutional** cytogenetics FISH has largely been replaced by chromosomal microarrays as the first-line test (specifically for patients with developmental delay with or without congenital abnormalities and/or dysmorphism), and FISH is now used more often as a secondary test to confirm any deletions found by CMA.

FISH is commonly used in haematological malignancies (acquired cytogenetics) characterized by specific chromosomal rearrangements with prognostic value and requiring urgent treatment (e.g. *PML-RARA* gene fusion in acute promyelocytic leukaemia). A wide range of FISH probes can be deployed to target the specific breakpoint region of the translocation (rather than the distal translocated segments) such as *break-apart* probes and *dual-fusion* probes. Break-apart probes are used when one gene can have multiple fusion partners to see if there is a translocation or not, and if a translocation is seen then further FISH can be performed to identify the partner. Dual-fusion probes are similarly used to target for a very specific gene fusion/translocation product and hybridization of two different coloured probes to the targeted region indicates the rearrangement (by fusion), while individually hybridized probes do not localize and are visible as individual signals.

CMA

CMA is a method for high resolution (~200kb) copy number detection but cannot detect balanced rearrangements and mosaicism cannot be accurately detected. A microarray chip consists of a glass slide with thousands of probes each of which maps to a specific region of the chromosome and are designed so that the probes are tiled across all the chromosomes. The probes can be derived from bacterial artificial chromosomes (BACs) which are large and give less variation in results, but are of lower resolution, or they can be smaller artificially synthesized oligonucleotides in single nucleotide polymorphism (SNP) arrays allowing for higher probe density albeit with a slightly higher individual probe variability. Fluorescently labelled patient DNA is then hybridized to the array. If a region is deleted from a single homologous chromosome (**heterozygous** deletion) you would expect half the amount of DNA from that region (covering multiple probes) to bind resulting in a lower fluorescent signal. If the region was missing from both homologues (**homozygous** deletion) there would be no DNA to bind resulting in only a background level signal from the patient DNA. Similarly, for a duplication a higher fluorescent signal would be seen.

Array Comparative Genomic Hybridization (aCGH) and **SNP arrays** are two types of array technology with a different technical approach and should therefore not be used interchangeably. With aCGH, differentially labelled patient DNA and control DNA are hybridized to the array at equal concentration, and the relative signals (LogR) are compared to determine copy number status. If the signal intensities of the two labels are similar, then there is no comparative difference and hence equal copy number (normal). If the patient signal intensity is significantly lower or higher than the control then this would indicate a copy number loss or gain, respectively. For SNP arrays only labelled patient DNA is hybridized to the array and the absolute fluorescence is recorded for each probe. An ***in silico*** reference file (rather than control DNA) is then used to determine reference ranges for copy number status—see **Figure 6.9** for an example image of a SNP microarray result. As well as providing copy number status the SNP **allele** also provides additional allelic information which can further confirm a copy number loss or gain but may also detect regions with an **absence of heterozygosity (AOH)**. This can be indicative of uniparental disomy (UPD) or can be regions of shared descent (e.g. **consanguinity**) with an increased risk of autosomal recessive genetic disorders. Unusual SNP patterns may also be indicative of mosaicism.

heterozygous
Having two different alleles for a specific locus on homologous chromosomes.

homozygous
Having identical alleles for a given locus on each homologous chromosome.

SNP array
A type of microarray using SNPs for detection of copy number change.

In silico
Computer simulated.

allele
A variant form of a gene.

absence of heterozygosity (AOH)
A region of DNA where only a single allele is present.

consanguinity
Descended from the same ancestor.

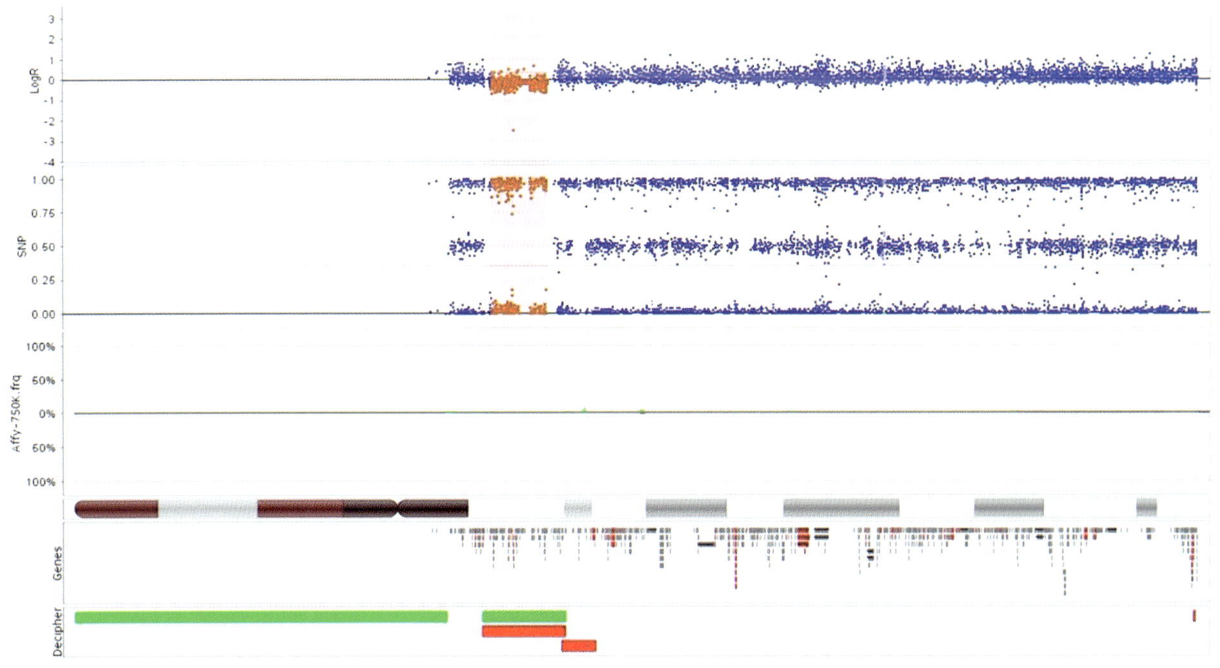

FIGURE 6.9

An example of a SNP microarray showing chromosome 22 only. This shows a 22q11.2 deletion which is associated with DiGeorge syndrome. The deleted region has been highlighted in red with a visible drop in the LogR ratio, and an absence of the middle SNP line.

Other molecular cytogenetic methods

Multiplex ligation-dependent probe amplification (MLPA) is a common molecular PCR-based method for copy number detection. Multiple commercial kits are available which consist of up to ~50 probes and the kits are mostly focused around specific genes (e.g. *DMD*) to rapidly determine the presence of exonic level deletions and duplications. In cytogenetics subtelomeric probe kits (covering all chromosome distal p and q arms) were of interest before CMA was available for screening of terminal subtelomeric deletions as well as unbalanced chromosome rearrangements and aneuploidy. Example MLPA results can be seen in **Figure 6.10**. Each peak represents a specific probe of the kit and is of a specific fragment size (bin size), in this case exons of the *DMD* gene. Blue peaks represent the patient sample and red peaks the synthetic control (an average of multiple samples from within the same run). The patient peaks are compared to the control peaks to produce a probe ratio (as seen in the right-hand side table) which represents the relative copy number. A ratio of ~1 indicates normal copy number, whilst the deleted probes show a ratio of ~0.5. The software automatically highlights probes outside of the normal copy number range in the table, and we can see that probes for exons 46–50 are approximately 0.5, consistent with a heterozygous deletion.

QF-PCR is a method commonly used *prenatally* to rapidly determine aneuploidy status for the most common viable trisomies (chromosomes 13, 18, and 21 associated with Patau, Edwards, and Down syndrome respectively). It involves the amplification of polymorphic **microsatellite DNA** which are small repeat sequences of DNA which vary from person to person, such that you would expect the maternal and paternal alleles to differ in the number of repeats and therefore size. Therefore, when run through a **genetic analyser** you would obtain two peaks for the PCR products of that marker representing the different inherited maternal and paternal alleles. As QF-PCR is a multiplex test, multiple markers are tested for each of the chromosomes of interest, as there is a chance that both alleles are of the same size producing a single uninformative peak. This reduces the risk of a completely uninformative result for a given chromosome, and multiple informative markers are required for analysis and interpretation. As we normally carry two copies of each chromosome a normal result would be indicated by the markers tested showing two peaks with an equal area/height (some markers may be uninformative). If the patient had Down syndrome, they would have three copies of chromosome 21 and therefore you would expect an extra peak (allele). If all three copies of chromosome 21 differ in repeat size, then three peaks will be present for that marker in an equal 1:1:1 ratio. It is also possible by chance that two of the three copies have the same repeat size, whilst the third copy is different. As this is a quantitative method then we expect the shared allele to be twice as large as the non-shared allele, which would show as two peaks with a 2:1 ratio. An example of this can been seen in **Figure 6.11**. Also, by chance all three copies may share the same repeat size and only a single peak will be present (uninformative). Therefore, for trisomy 21 you would expect all informative chromosome 21 markers to show either a 1:1:1 or 2:1 ratio. For any chromosome shown to have a mix of normal and abnormal ratios further work should be undertaken to investigate the possibility of an unbalanced chromosomal rearrangement.

microsatellite DNA
A type of repeat DNA sequence which can be highly variable from person to person.

genetic analyser
Normally refers to capillary electrophoresis machines used to visualize Sanger sequencing and other PCR-based techniques.

Microsatellite analysis also has further applications in genetic testing. In prenatal diagnosis (PND) it is used to rule out maternal cell contamination (MCC) of the prenatal sample. MCC is the presence of maternal cells (or DNA) within a prenatal tissue sample and can lead to incorrect results if present. For MCC testing, both a maternal blood sample and the prenatal sample are tested in tandem and the alleles are compared. The foetus should only show 1 allele in common with the mother and one from the father (which may be the same repeat size as the maternal allele). If there is MCC it will be visible by the presence of both maternal alleles in the foetal sample (however not all markers will be informative). Significant MCC can render the foetal sample unsuitable (as we only want to test the foetus) and a repeat invasive procedure may

MLPA Analysis Report ·	
Software:	**Analysis Type:** MLPA
Project: Untitled	**Compare Type:** MLPA Ratio
Technician:	**Normalization By:** Population Normalization
Report Time:	**Quantification By:** Peak Height
Panel: P034_B2_DMD1	**Classification:** Loss < 0.80 <= Equivalent <= 1.20 < Gain
Control: Synthetic Control Sample	**Report Value Type:** Peak Ratio
Synthetic Used:	

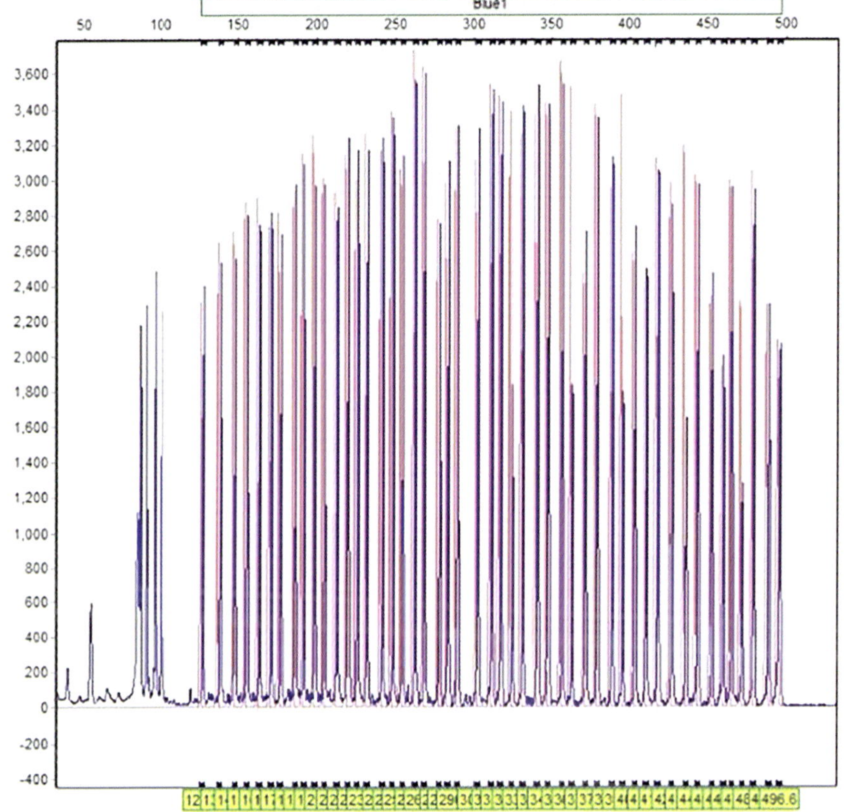

	Probe Na	Bin Size	
1	Ex.1	139.0	0.956
2	Ex.10	466.1	0.987
3	Ex.2	171.7	1.028
4	Ex.21	156.4	0.973
5	Ex.22	192.3	0.980
6	Ex.23	226.2	1.057
7	Ex.24	263.1	0.954
8	Ex.25	303.2	1.057
9	Ex.26	332.0	1.048
10	Ex.27	372.2	1.097
11	Ex.28	403.8	1.058
12	Ex.29	444.5	0.983
13	Ex.3	213.2	0.973
14	Ex.30	480.0	0.966
15	Ex.4	248.7	0.988
16	Ex.41	148.8	0.942
17	Ex.42	187.1	1.012
18	Ex.43	220.2	1.030
19	Ex.44	255.1	1.024
20	Ex.45	290.2	1.002
21	Ex.46	325.1	**0.540**
22	Ex.47	363.6	**0.521**
23	Ex.48	396.0	**0.517**
24	Ex.49	437.0	**0.513**
25	Ex.5	284.1	1.041
26	Ex.50	472.7	**0.551**
27	Ex.6	318.0	0.989
28	Ex.61	164.3	0.947
29	Ex.62	199.4	0.911
30	Ex.63	242.1	1.023
31	Ex.64	269.1	0.988
32	Ex.65	312.2	0.993
33	Ex.66	341.3	1.044
34	Ex.67	379.7	0.976
35	Ex.68	411.7	1.031
36	Ex.69	453.7	1.056
37	Ex.7	357.3	0.964
38	Ex.70	489.6	1.000
39	Ex.8	389.3	1.061
40	Ex.9	428.1	0.956
41	Ref Xp11	127.8	1.036
42	Ref Xp22	205.4	0.988
43	Ref Xq21	348.1	0.999
44	Ref Xq23	496.6	0.990
45	Ref Xq26	419.2	0.978
46	Ref Xq28	460.7	1.030
47	Xp11	177.8	0.956
48	Xp22	232.4	0.968
49	Xq22	278.5	0.992

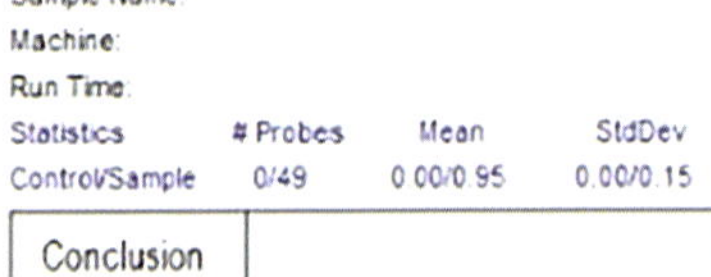

Conclusion		
	Date	Initial
Authorization 1		
Authorization 2		

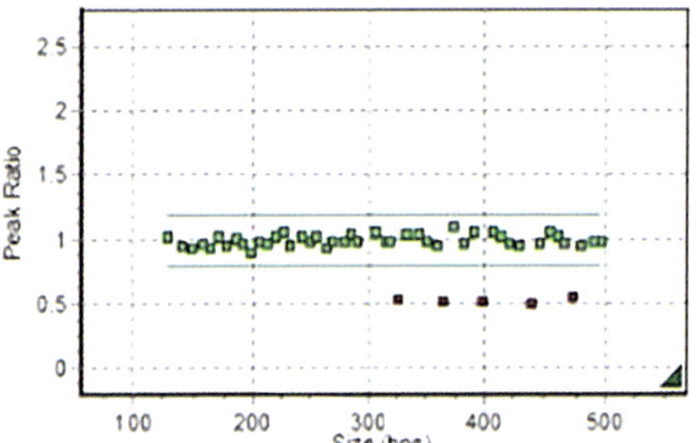

FIGURE 6.10

MLPA results for the MRC-Holland *DMD* probe kit showing a deletion of exons 46–50 in a female carrier.

Marker	Alleles	Allele Length	Peak Area	Peak Ratio	Check
13A-D13S742	2	266:270	34437:31775	1.08	
13B-D13S634	1	409	32610		
13C-D13S628	2	456:460	40816:35275	1.16	
13D-D13S305	1	459	57350		
13K-D13S149	2	119:139	49000:43875	1.12	
18B-D18S978	1	215	62776		
18C-D18S535	2	320.324	23915.20476	1.17	
18D-2-D18S3	2	395:403	34800:29902	1.16	
18J-D18S976	1	462	86053		
18M-GATA17	2	368:380	33976:33096	1.03	
21A-D21S143	3 (1:1:1)	170:186:190	22711:21933:16703		
21B-D21S11	3 (1:1:1)	243:247:251	24341:24294:19984		
21C-D21S141	3 (1:1:1)	296:304:320	37721:34146:24947		
21D-D21S144	3 (2:1)	468:475	31609:16736	1.89	
21H-D21S144	3 (1:2)	372:392	35564:67852	0.52	
21I-D21S1437	3 (1:1:1)	120:124:131	40526:42980:38814		
AMELXY	1	AMELX	41037		
SRY	0				
T1	2	Chr. 7:Chr. X	104837:101700	1.03	
T3	2	Chr. 3:Chr. X	43647:38395	1.14	
X1-DXS1187	1	146	62942		
X3-XHPRT	1	281	76261		
X9-DXS2390	2	321:337	33835:29964	1.13	
XY2-DXYS26	2	191:199	38163:33075	1.15	
XY3-DXY218	1	242	102621		
ZFYX	1	163	105985		

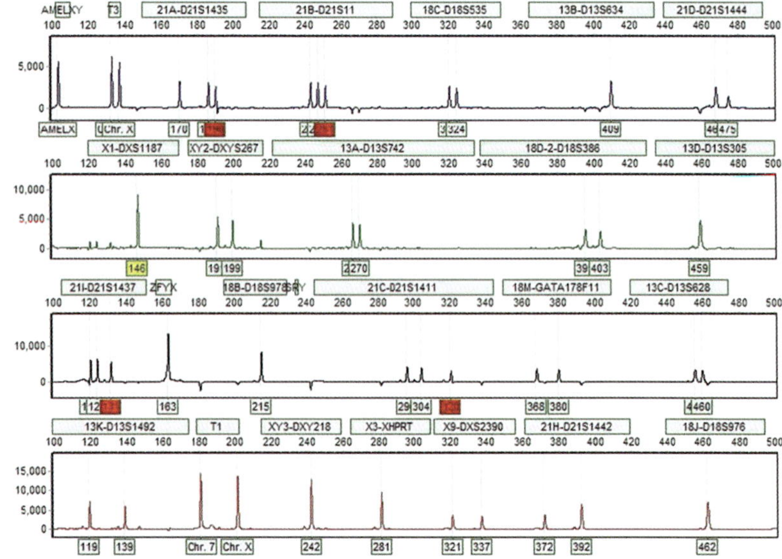

FIGURE 6.11

QF-PCR analysis indicating trisomy 21. The peaks for each marker and fluorophore are shown on the right, with a summary table of the alleles and ratios on the left. For all the chromosome 21 markers (shaded in the table) an abnormal signal pattern is seen showing either 3 peaks in a 1:1:1 ratio or 2 peaks in a 2:1 ratio, consistent with an additional chromosome 21.

chimerism
An individual derived from two or more zygotes.

be required to perform the required testing. Microsatellite analysis is also used in **chimerism** testing for engraftment analysis of leukaemia patients who have received a haematopoietic stem cell (HSC) transplant to determine the percentage of donor cells.

Real-time quantitative PCR (qPCR, not to be confused with reverse transcriptase PCR, rt-PCR) is used as a highly sensitive technique for detection and quantification of a small specific DNA sequence. Its applications can include confirmation of small deletions and duplications (detected by other methods such as microarray or NGS), as well as detection of very low-level signals (such as monitoring residual disease (MRD) in cancer, or detection of low-level foetal DNA in non-invasive prenatal diagnosis (NIPD)). For copy number confirmation, a relative detection analysis is performed where patient DNA is run against normal patient controls and stable housekeeping genes to determine if there is a relative gain or loss. For MRD purposes the patient DNA is run alongside serial dilutions or controls to determine the absolute amount of abnormal DNA present.

6.3.2 Molecular methods

nucleic acid
Complex organic molecule that contains the genetic code.

functional analysis
Analysis of functional effects, e.g. the effect of a genetic variant on gene expression.

For molecular genetic testing the main analyte is **nucleic acid**, usually DNA, but RNA is also required (e.g. from muscle biopsies in dystrophinopathies) for **functional analysis** investigations. High-quality DNA can be obtained from a wide range of samples, the most common type of sample received being EDTA blood due to ease of sampling. The EDTA prevents the blood from clotting and does not interfere with downstream PCRs (whilst LH samples are generally unsuitable as the LH can interfere with PCR). DNA is also routinely extracted from prenatal tissues (such as AF and CV), buccal swabs (for needle-averse patients), skin (usually for investigation of mosaic abnormalities), cultured cells, muscle biopsies, and tumour tissues. DNA can also be extracted from fixed tissue types including formalin fixed paraffin embedded (FFPE) tissues; however, DNA quality is normally compromised due to the fixation process.

DNA extraction and quantification

The basic process of DNA extraction generally involves cell lysis, isolation of the DNA, washing away proteins (and other contaminants), and elution. These steps can vary widely depending on the requirements of the sample and further testing. High-quality DNA is required for tests such as CMA and next generation sequencing (NGS) whilst simpler testing methods such as prenatal QF-PCR may only require minimal low-quality DNA for urgent testing. High-throughput DNA extraction from EDTA blood samples is normally performed through automated robotics using magnetic bead technology. In summary, after the blood cells are lysed, releasing all the cellular contents into solution, a solution of special magnetic beads is added to which the DNA binds. This is then attracted to a magnetic rod so that the waste lysis solution can be washed away; a second solution is then added which causes the DNA to be released from the magnetic beads back into solution. Again, the beads are attracted to a metal rod so that the DNA solution can be aliquoted into a clean tube for quantification and storage.

DNA quantification is normally performed shortly after extraction is complete. This is to ensure there is enough DNA available not only for testing but also for possible further testing in the future. If the quantity of DNA obtained is not sufficient further samples may be requested from the laboratory in order to complete testing. After DNA is quantified, it is normally diluted to a stock working concentration (e.g. 250ng/µl) and when testing is performed an aliquot is taken and further diluted if required.

Spectrophotometry is the most common method for quantifying DNA in laboratories as it is a cheap, quick, and simple method meaning that many samples can be quantified in a short space of time, suitable for laboratories extracting hundreds of samples in a day. Approximately 2 µl of DNA is placed upon a tiny pedestal and a beam of light is passed through the sample to measure the absorption. DNA absorbs at a wavelength of 260 nm (or A_{260}). The A_{260} value is then used to calculate the concentration of DNA with a value of 1 being equal to 50 µg/ml (or 50 ng/µl). However, RNA also absorbs light at 260 nm, phenol (used in some extraction methods) at 270 nm, and protein at 280 nm. Therefore, spectrophotometry readings often overestimate the DNA concentration due to the presence of RNA, or if the DNA extraction is not pure enough protein and phenol carry over. Whilst for some tests the accuracy is not that critical, for complex tests (such as CMA or NGS) the concentration is critical for standardized test methodology, and therefore other methods of quantification are needed. **Fluorometric quantification** overcomes this barrier but is more time consuming and costly due to additional reagents and preparation. Fluorometers detect fluorescent dyes which are only detected when bound to DNA, or other target molecules such as RNA. They give more accurate results because contaminants are not measured, and they are able to detect DNA at much lower concentrations than can spectrophotometers. To quantify the sample, an aliquot of DNA is taken for dilution and the fluorescent dye is added. A reference standard and blank are then used to calibrate the fluorometer, before the actual samples are quantified by measuring the fluorescence of the DNA-bound dye.

spectrophometry
Measuring the quantity of a chemical substance, e.g. DNA, by light absorbance for a given wavelength.

fluorometric quantification
Measuring the quantity of a chemical substance, e.g. DNA, by dyes that bind to the region of interest.

Sanger sequencing

Sanger sequencing has been the traditional molecular genetic test and sequences the individual bases of DNA to look for small variation such as single nucleotide variants (SNVs) as well as small deletions, duplications, and indels (insertion–deletions). See **Figure 6.12** for an example result. Sanger sequencing is useful for targeted testing of a limited number of amplicons for specific genes associated with a very specific genetic condition (e.g. m.1555A>G in aminoglycoside induced deafness). Normally sequencing is performed for targeted exons, or pathogenic variant hotspots, or for single gene screening (sequencing all exons in a gene). Sequencing reactions are generally performed in 96 well plates to increase throughput and can be used

Sanger sequencing
Traditional sequencing of DNA, also known as first-generation sequencing.

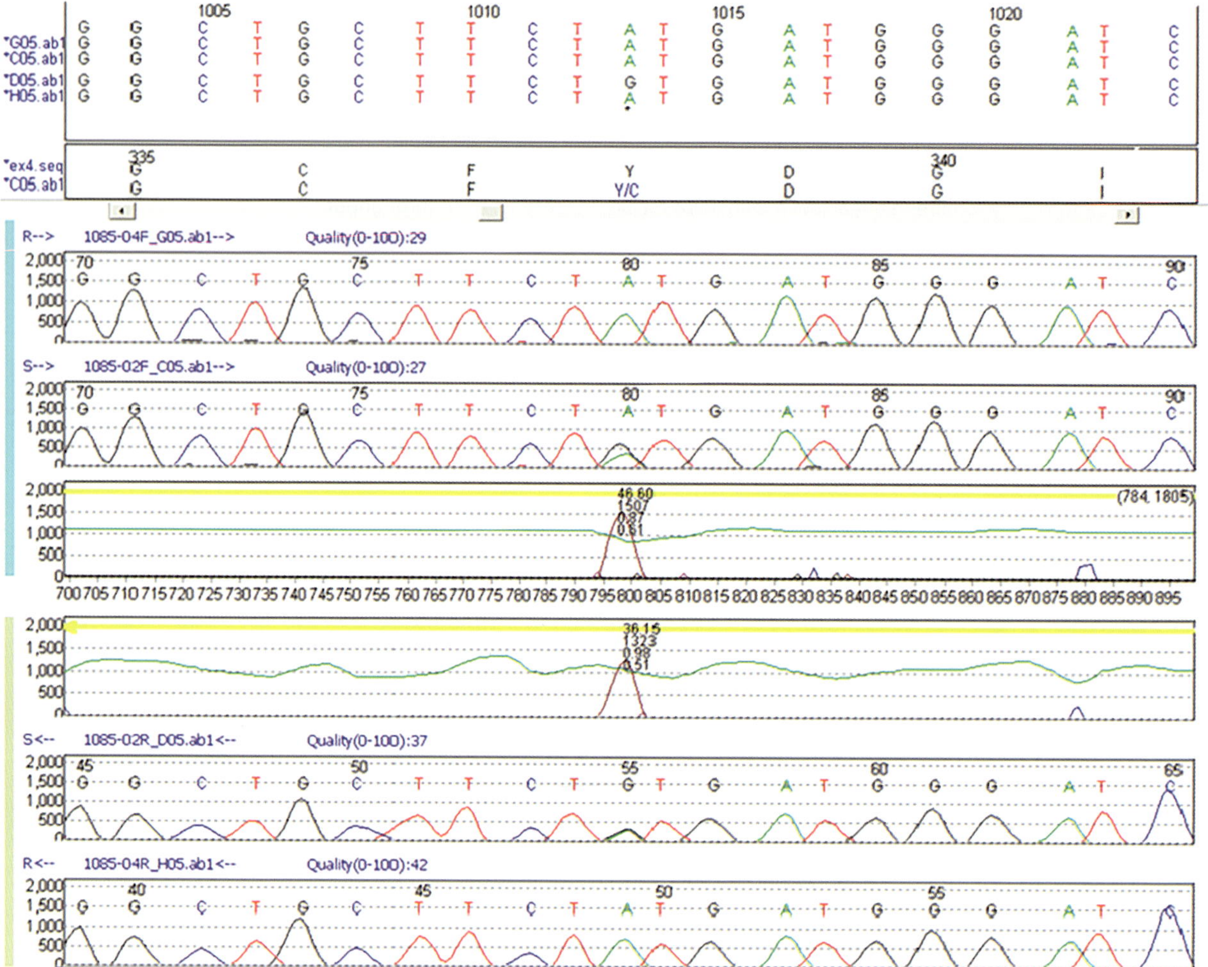

FIGURE 6.12

An example of a Sanger sequencing trace. From top to bottom: the top sequence represents the reference sequence DNA code with G05 and H05 representing a reference control sample, and C05 and D05 the test sample each sequenced in forward and reverse directions respectively. The amino acid code is below.

to sequence single (or few) exons for a large number of samples or sequence a gene (multiple exons) for a small number of patients.

In **Figure 6.12**, each peak in the electropherogram represents a DNA base, and the colour is specific for each base, so we read each coloured peak as the specific DNA base sequence. For the sample we can see that at the 80th base (*c.*1013 according to the reference) both a green and black peak are present representing a heterozygous single nucleotide variant (an A>G transition). In between the sequence peaks a different plot is shown to make it easier to identify any change away from the reference control.

genetically heterogeneous conditions
Genetic conditions which can be caused by a number of genes but presenting with a similar phenotype.

Even with automation, Sanger sequencing is not well suited for testing of patients with **genetically heterogeneous conditions**. These are genetic conditions which could be caused by different genes (rather than a single gene, single condition), all leading to the same or similar phenotype (such as epilepsy, or deafness). In the past, testing was focused on the most likely candidate gene (or exon/s) in a stepwise fashion. If the first suspected gene tested was negative, then the next candidate gene would be tested and so on, often involving multiple clinical reviews to check for differential diagnoses and alternative genes. Patients would often undergo what can be termed a diagnostic odyssey, undergoing multiple expensive genetic tests, and often taking months or even years to find the causative pathogenic variant, with some patients still remaining undiagnosed.

NGS

NGS (next generation sequencing, also known as second-generation sequencing, or massively parallel sequencing) is a relatively new technology which overcomes the issues of sample throughput and sequencing capacity which limits traditional Sanger sequencing.

The vast amount of sequencing data from NGS can be used for different applications. For genetically heterogeneous conditions NGS enables the testing of multiple genes at once through targeted gene panel testing, for example a set panel of 100 genes for epilepsy testing which improves the diagnostic yield compared to Sanger sequencing of single genes which cuts down on the time to diagnosis. One important point to consider, however, is how to decide which loci to include in a panel, and this will often involve the input of clinical scientists, clinical geneticists, and consultants from the relevant medical speciality, and a full review of the literature. This is to ensure that only clinically useful loci are included, their selection being based on clear evidence that the particular locus is causative for the disorder in question. Including genes with uncertain clinical significance makes interpretation of results impossible, and reporting non-clinically relevant variants could lead to incorrect diagnoses, or increased patient anxiety. Gene panel testing is normally used for more prevalent conditions which allows for batch testing of patients with the same panel on a routine basis. NGS can not only be used for detection of SNVs similar to Sanger sequencing but can also be used for copy number detection.

NGS can also be targeted to single genes or exons at an extremely high depth for the detection of low-level DNA changes. This can be used for the investigation of mosaic conditions (e.g. in dermatology for mosaic skin conditions) or for NIPD/NIPT for detecting low-level foetal DNA in the background of a majority maternal cell-free DNA (cfDNA) from plasma. By sequencing at such a high depth NGS can be used as a very sensitive technique. An in-depth coverage of sequencing techniques can be found in Chapter 5.

One of the biggest challenges with NGS is the large amount of sequence data generated. More data means more time to analyse and interpret genetic variants prior to reporting, and the more data analysed the more variants of uncertain significance are found and the higher the chance of an **incidental finding**. Incidental findings are findings secondary to the reason for referral but may still have important clinical significance to the patient (such as finding a pathogenic variant which confers a risk of cancer in someone being tested for epilepsy). Therefore, pre-test counselling and informed consent are critical so that the patient is fully aware of what the test may reveal. Another issue is the safe, long-term storage of NGS data. With individual sequencing runs generating gigabytes of data on a daily basis, storage solutions must be in place to match the pace of data generation and ensure data integrity and security.

incidental finding
Findings of a test unrelated to the initial indication for testing but still of medical value.

Other molecular methods

Additional to sequencing, other specialized methods of detecting genetic abnormalities are still required, either due to the limitations of sequencing for the detection of certain types of variants such as triplet repeat disorders (e.g. fragile X syndrome) or imprinting disorders (e.g. Prader–Willi/Angelman syndrome) or due to efficiencies of targeted mutation testing for very specific genetic syndromes (e.g. CF).

triplet repeat disorders
Genetic conditions caused by the expansion of triplet repeats.

anticipation
For each generation a genetic disorder is passed on the symptoms become more severe, or appear at an earlier age.

Southern blot
Used for detection of specific DNA sequences by running through an electrophoretic gel and transfer to a membrane for visualization.

restriction enzymes
Enzymes which cleave the DNA at specific motifs.

imprinting
The differential expression of genes depending on maternal or paternal origin of the chromosome.

Triplet repeat disorders are caused by an expansion of repetitive sequences (consisting of 3 base pairs) of DNA. These repeats are normally present at a low number and do not cause disease, but these can increase in size upon transmission and when large enough can result in a genetic disorder (e.g. fragile X syndrome and Huntington's disease). As these expansions increase through successive generations patients may show symptoms earlier, or with increased severity. This phenomenon is termed **anticipation**. As Sanger sequencing is not suitable for highly repetitive regions of DNA other sizing-based methods are required to determine the length of these repeats. For many years the main method was **Southern blots**, which requires high amounts of DNA which are then digested by **restriction enzymes** to produce fragments of DNA containing the region of interest, which are then run on a gel to size the region. Not only is this method time consuming, but cannot be used for large numbers of samples, and poses additional health and safety risks due to the chemicals used. Fortunately, newer PCR methods referred to as triplet repeat (or repeat primed) PCR have been developed which allow for accurate sizing of these repeat alleles allowing for much higher throughput and cheaper testing.

Imprinting disorders are caused by abnormal expression of genes based upon the parent of origin for differentially methylated regions (DMRs) of DNA. The most common example of this is for Prader–Willi and Angelman syndromes caused by defects on chromosome 15 (15q11q13). Lack of expression of the paternal allele (e.g. deletions, maternal UPD15) results in Prader–Willi syndrome, whilst lack of expression of the maternal allele results in the more severe Angelman syndrome. **Figure 6.12** shows the imprinted region on chromosome 15. Whilst the deletions are normally detectable by CMA this does not provide any information as to whether it is the paternal or maternal chromosome that has been deleted; therefore, methylation studies are normally used to obtain information regarding the specific mechanism. Methylation-specific PCR (msPCR) involves bisulfite modification of the DNA which converts non-methylated cytosine (C) bases to uracil (U), which during PCR complements with adenine (A). These specific changes at sites of methylation can then be exploited by using PCR primers which target these specific differences. In practice msPCR is used to amplify a specific size maternal PCR product, and a specific size paternal PCR product. The size of the paternal and maternal fragments will differ because of differences in methylation. If only a maternal band is visible when the products are run on a gel, then this would be consistent with Prader–Willi (e.g. paternal deletion or UPDmat) and vice versa. However, this method does not distinguish between a paternal deletion or maternal UPD which have different risk of recurrence for future pregnancies, therefore it is important to make this distinction of the mechanism. msMLPA is a variant of MLPA in which methylation-specific enzymes are used to digest the DNA to determine the parent of origin, and also provides copy number information at the same time. In clinical use msPCR is normally used as the first-line test (as it is cheaper) and any abnormal results are then confirmed by msMLPA to determine the underlying genetic cause. An example of msMLPA can be seen in **Figures 6.13** and **6.14**.

Allele-specific PCR is used to detect specific variants and is mainly used to sequence common variants in disease known to be caused by a small number of high-frequency variants. PCR primers are designed to amplify both the normal sequence and the variant sequence when present, and each produce a product of a different size so they can be distinguished as the wild-type or variant band when run on a gel or sequencer. Primers can also be multiplexed so that multiple variants can be tested in a single reaction, for example in cystic fibrosis where over 1500 variants have been found but a smaller number account for the majority of cases. Commercial kits are available for laboratories to purchase, validate, and use for diagnostic use, reducing the burden of laboratories developing and optimizing their own tests. See **Case study 6.2** on cystic fibrosis for example test results.

Imprinted domain
NIPA1
NIPA2
CYFIP1
TUBGCP5
MKRN3
MAGEL2
NDN
PWRN2
PWRN1
C15orf2
SNURF-SNRPN
HBII-436/13/437
HBII-438A
HBII-85
HBII-52
HBII438B
UBE3A
ATP10C
GABRB3
GABRA5
GABRG3
P (OCA2)
APBA2
Cen
Tel
D15S541
D15S543
IC
BP1
BP2
BP3
Class I deletion
Class II deletion

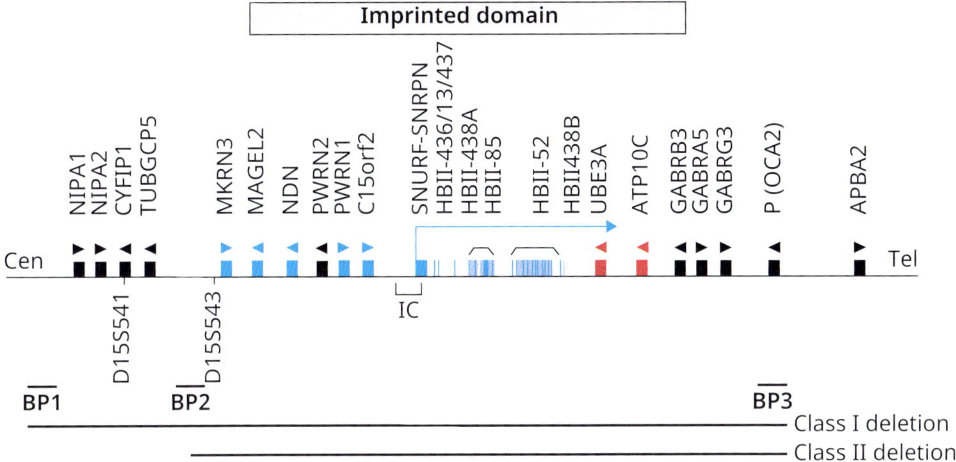

FIGURE 6.13

Genes within the PWS/AS region. Blue boxes represent paternally expressed genes; blue vertical lines, snoRNAs; red boxes, maternally expressed genes; black boxes, biallelically expressed genes; and arrow heads represent the orientation of transcription. IC, imprinting centre; BP, common breakpoint cluster region.

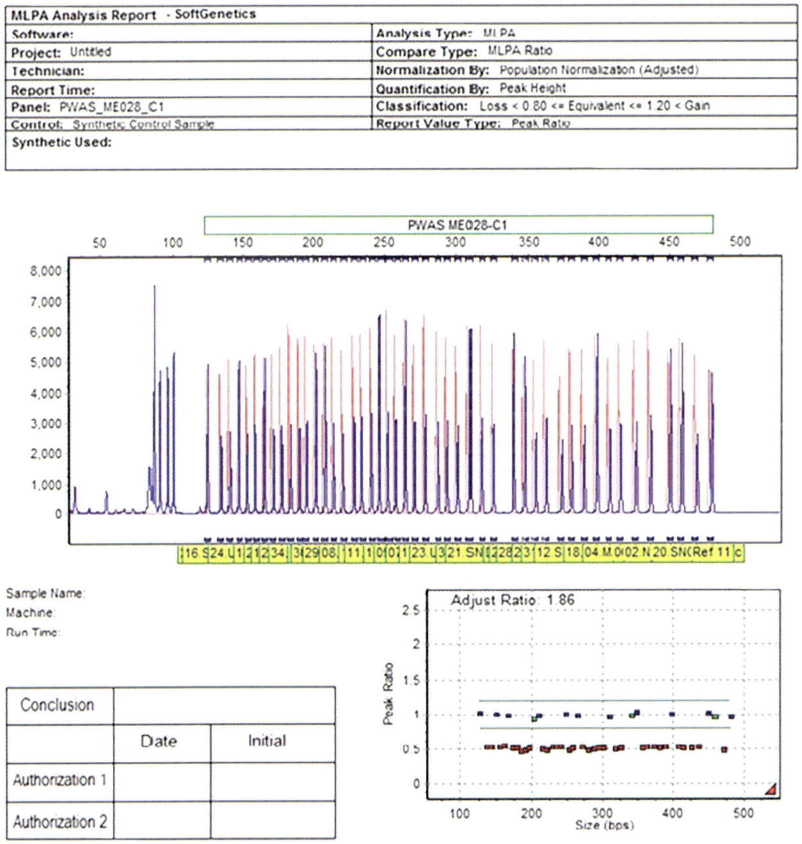

MLPA Analysis Report - SoftGenetics
Software:
Project: Untitled
Technician:
Report Time:
Panel: PWAS_ME028_C1
Control: Synthetic Control Sample
Synthetic Used:
Analysis Type: MLPA
Compare Type: MLPA Ratio
Normalization By: Population Normalization (Adjusted)
Quantification By: Peak Height
Classification: Loss < 0.80 <= Equivalent <= 1.20 < Gain
Report Value Type: Peak Ratio
PWAS ME028-C1
Sample Name:
Machine:
Run Time:
Adjust Ratio: 1.86
Peak Ratio
Size (bps)
Conclusion
Date
Initial
Authorization 1
Authorization 2

	Probe Name	Bin Size	
1	01.TUBGCP5	154.0	0.542
2	02.NIPA1	436.3	0.536
3	03.MKRN3	172.0	0.531
4	04.MAGEL2b	415.3	0.521
5	05.MAGEL2a	234.6	0.536
6	06.NDN	426.0	0.531
7	07.SNRPN ExU1B	287.5	0.506
8	08.SNRPN ExU1B	241.2	0.537
9	09.SNRPN intU2b	278.9	0.497
10	10.SNRPN intU2a	271.4	0.538
11	11.SNRPN ExU5a	258.2	0.529
12	12.SNRPN ExU5b	389.5	0.537
13	13.SNRPNd CpG	252.6	0.494
14	14.SNRPNb CpG	178.0	0.526
15	15.SNRPNc CpG	190.7	0.495
16	16.SNRPNa CpG	141.2	0.534
17	17.SNRPN Ex03	294.0	0.524
18	18.SNRPN Ex07	407.7	0.537
19	19.SNORD116a	214.4	0.515
20	20.SNORD116c	469.3	0.499
21	21.SNORD116b	325.7	0.525
22	22.UBE3A Ex09	355.5	0.529
23	23.UBE3A Ex04	301.0	0.520
24	24.UBE3A Ex03	160.0	0.559
25	25.UBE3A Ex02	195.2	0.524
26	26.UBE3A Ex01b	373.6	0.540
27	27.UBE3A Ex01	184.0	0.474
28	28.ATP10A Ex15	363.0	0.544
29	29.ATP10A Ex01	228.8	0.541
30	30.GABRB3 Ex09	221.2	0.492
31	31.GABRB3 Ex07	380.8	0.526
32	32.OCA2 Ex23	135.0	0.542
33	33.OCA2 Ex03	317.4	0.510
34	34.APBA2 Ex14	202.0	0.948
35	Dig Ctrl 01	340.4	0.990
36	Dig Ctrl 02	458.5	0.975
37	**Ref 01**	125.7	1.033
38	**Ref 02**	148.0	1.009
39	**Ref 03**	166.0	0.996
40	**Ref 04**	208.5	0.991
41	**Ref 05**	246.8	1.015
42	**Ref 06**	264.9	0.993
43	**Ref 07**	309.0	0.981
44	**Ref 08**	347.8	1.047
45	**Ref 09**	398.7	1.006
46	**Ref 10**	450.5	1.031
47	**Ref 11**	479.7	0.979

FIGURE 6.14

Dosage analysis results of msMLPA showing the common deleted region for the 15q11q13 imprinted region of chromosome 15. Probes numbered 01–33 all show a probe ratio of ~0.5 consistent with a heterozygous deletion. However, this does not show the underlying mechanism of the disease.

SELF-CHECK 6.1

1. Why are there numerous methods for genetic testing?
2. What are the advantages and disadvantages of these methods?

6.4 Referral groups

6.4.1 Preimplantation Genetic Diagnosis (PGD)

PGD is a form of fertility treatment aimed to avoid the conception and birth of a child with a genetic condition where the parents are predisposed to having an affected child. They may both be carriers of a recessive pathogenic variant in the same gene, a carrier of a balanced chromosomal rearrangement, one parent may even be affected with a dominant genetic condition themselves and does not wish to have an affected child. The process involves IVF and a few cells are taken from the very early developing foetus (blastomere/trophoblast stage) to test for the specific disorder. PGD is only offered for serious inherited conditions, and currently has been approved for over 600 conditions. The test methods must be highly sensitive to work with limited amounts of starting material and to limit contamination from other samples. Tests can be done by targeted sequencing, but also by CMA.

6.4.2 Prenatal Diagnosis (PND)

chorionic villus
Foetal derived tissue within the placenta.

amniocentesis
The process of sampling amniotic fluid.

PND involves the genetic testing of foetal tissue (or DNA) during pregnancy. Traditionally an invasive surgical procedure is used to sample either foetal tissue (**chorionic villus** biopsy, **amniocentesis**, or less commonly umbilical cord blood sampling) which carry an increased ~0.5–1% risk of miscarriage. Testing methods are dependent upon the referral information. Targeted gene testing (e.g. Sanger sequencing) can be performed for families who are at risk of an affected pregnancy due to previous family history (e.g. CF). Rapid trisomy testing for common trisomies (13, 18, 21) can be performed from pregnancies identified as high risk as part of the UK foetal anomaly screening programme (UKFASP), from maternal serum biochemistry testing (see **Case study 6.1** for a Down syndrome case study), or from NIPT.

Additionally, abnormal ultrasound scan findings (such as heart defects, skeletal abnormalities, and other congenital abnormalities) may be indicative of an underlying chromosomal disorder and therefore microarray testing is indicated to look for submicroscopic copy number changes which offers a 3–5 % increased diagnostic yield over karyotyping. Prenatal exome sequencing is soon to be offered as part of the NHS service and can further identify a diagnostic or clinically relevant variant in 12.5% of foetuses with abnormal scan findings without a pathogenic variant identified by microarray.

Where there is a family history of a genetic condition, genetic testing of the foetus may also be offered to determine whether the pregnancy will be affected or not. This will be done by targeted testing, the method of which will be appropriate to the genetic variant (e.g. targeted Sanger sequencing for SNVs, or targeted microarray testing for a parent who is a carrier of a balanced translocation).

The genetic testing of pregnancy losses (such as recurrent miscarriage, second or third trimester losses, or pregnancies with foetal abnormalities detected) is also routinely performed in genetic laboratories. Approximately 15% of first trimester pregnancies result in miscarriage, up to 50% of which have a cytogenetic abnormality, with the risk of pregnancy loss decreasing with increased gestational age and with a decreasing genetic abnormality rate. Testing is performed in order to determine if the loss of the pregnancy is due to a genetic cause which can inform if parental genetic testing is required to determine the recurrence risk of further affected pregnancies. A parental rearrangement could both predispose to further pregnancy loss or result in an abnormal live birth. Trisomies account for a large proportion of pregnancy losses, followed by monosomy X (Turner syndrome), triploidy, unbalanced rearrangements, and copy number change. Therefore, SNP arrays are suitable as a first-line test which is able to detect all of these abnormalities; however, QF-PCR testing may be used as a more efficient screen for the more common trisomies associated with pregnancy loss.

6.4.3 Postnatal

For patients with developmental delay, intellectual disability, or autism with or without dysmorphism/congenital abnormalities, microarray is normally the first-line test for the detection of copy number variants. Where a specific genetic condition is suspected, targeted molecular testing may be more appropriate, or even NGS panel testing (e.g. for epilepsy). Core molecular testing for the more common genetic disorders may include: fragile X syndrome, cystic fibrosis, Prader–Willi and Angelman syndromes, Duchenne muscular dystrophy, inherited breast and ovarian cancer, and familial hypercholesterolaemia.

Newborn screening (see **Box 6.5**) is also offered for nine genetic conditions where early diagnosis can lead to early intervention and improved patient outcomes. Metabolic or biochemical testing is performed first using blood spots on a Guthrie card from a heel-prick blood sample of the baby. If a positive result is found, then treatment can be started whilst genetic testing is performed to identify the genetic variant causing the disorder.

For newborns with ambiguous genitalia (a disorder of sexual differentiation, DSD) it is important to determine the karyotypic sex and also the presence or absence of *SRY* (or the sex determining region on chromosome Y); therefore, urgent karyotyping is performed as well as FISH for the X and Y chromosomes. For newborns with a suspicion of one of the common trisomies (13, 18, or 21) FISH is performed urgently (within 3 days) for a rapid diagnosis. If a trisomy is detected, then this will be confirmed by karyotype analysis to confirm if this is due to a numerical or structural abnormality. If the result is normal, then a microarray is normally performed due to the associated congenital abnormalities. Molecular testing may also be undertaken for genes associated with sexual differentiation including: *CAH*, *WT1*, and *AR* genes amongst others.

The investigation of infertility can involve both cytogenetic and molecular genetic testing. Sex chromosome aneuploidies (e.g. monosomy X (Turner syndrome), XXY (Klinefelter syndrome)) are often a cause of infertility and may be mosaic; therefore, karyotyping is required. Also balanced translocations involving the sex chromosomes, or more rarely autosomes, are also associated with infertility and currently require karyotyping for detection. Molecular testing can include testing for CF, fragile X syndrome, and deletion of the *AZF* gene on the Y chromosome.

BOX 6.5 Newborn screening

Newborn screening can be performed on babies from 5 days old for sickle cell disease, cystic fibrosis, congenital hypothyroidism, and the following inherited metabolic diseases: phenylketonuria (PKU), medium-chain acyl-CoA dehydrogenase deficiency (MCADD), maple syrup urine disease (MSUD), isovaleric acidaemia (IVA), glutaric aciduria type 1 (GA1), and homocystinuria (pyridoxine unresponsive) (HCU).

6.4.4 Cancer

Testing for inherited cancer syndromes is normally performed by molecular genetic testing on peripheral blood and includes the following: colorectal cancer (HNPCC, FAP, MUTYH), breast and ovarian cancers (*BRCAs*), endocrine (*MEN1*, *MEN2*), and Li-Fraumeni (*TP53*). Testing is normally offered to patients with a known family history, or screening is offered to populations at an increased risk (e.g. Ashkenazi Jewish). If a pathogenic variant is found, the patient is at an increased lifetime risk of developing cancer (but there is still a chance they might not develop cancer), and increased monitoring can be arranged, or prophylactic surgery undertaken to prevent the development of the cancer of a particular organ.

For sporadic cancers, tumour samples are directly tested as the acquired pathogenetic variants will be limited to the cancerous tissue, and not present in the germline tissue. Tumours are heterogeneous in nature with different tumour cells showing different genetic abnormalities, whilst some cells may be normal. Therefore, care must be taken to ensure that biopsies are from the correct region of the tumour, ideally with histological confirmation that cancerous cells are present. Immunohistochemistry can also provide further information to guide genetic testing to the correct gene by identifying the lack of specific protein complexes in the tumour (e.g. loss of *MLH1* in Lynch syndrome [colorectal cancer]).

For haematological malignancies (e.g. leukaemia) bone marrow samples as well as peripheral blood are used for both cytogenetic detection of chromosomal abnormalities as well as for molecular genetic testing. NGS is becoming increasing used over traditional cytogenetic methods to target newly discovered genes with high prognostic significance. The aim of identifying genetic abnormalities helps to stratify patients in prognostic risk groups and help to guide treatment strategy and enrolment into clinical trials of either new drugs or treatment regimes.

combined test
A screening test during pregnancy for Down syndrome, Edwards syndrome, and Patau syndrome performed between weeks 10 and 14.

quadruple test
A screening test during pregnancy for Down syndrome, Edwards syndrome, and Patau syndrome performed between weeks 14 and 20.

Robertsonian translocation
The fusion of the two long arms of acrocentric chromosomes.

SELF-CHECK 6.2

What genetic testing would be suitable for a patient with developmental delay and epilepsy, and why?

CASE STUDY 6.1 Diagnosis of Down syndrome

Down syndrome is caused by having an additional chromosome 21 and is not only clinically recognizable at birth by specific congenital abnormalities and dysmorphism but can also be detected during the antenatal period.

The clinical features of Down syndrome prenatally can include raised nuchal translucency, heart defects (commonly AVSD), shorter than average long bones, or brachycephaly; however, some of the finer features may not be visible by ultrasound scanning and will only be visible at birth. These can include single palm crease, wide sandal gap, Brushfield spots, and macroglossia. Patients with Down syndrome have developmental delay and reduced IQ, but they are normally of a happy demeanour. Later in life they can develop problems with their immune system and acquire leukaemia. The lifespan of a person with Down syndrome has improved from ~25 years in 1983 to 60 years old. **Figure 6.15** shows that heart and lung diseases are the leading cause of death in Down syndrome. Increased maternal age (>35) is a risk factor to having a baby with Down syndrome.

Down screening can occur from as early as 7 weeks gestation through what is known as non-invasive prenatal testing (NIPT), by testing the cffDNA in the mother's blood. Invasive testing, chorionic villus sampling and/or amniocentesis, both

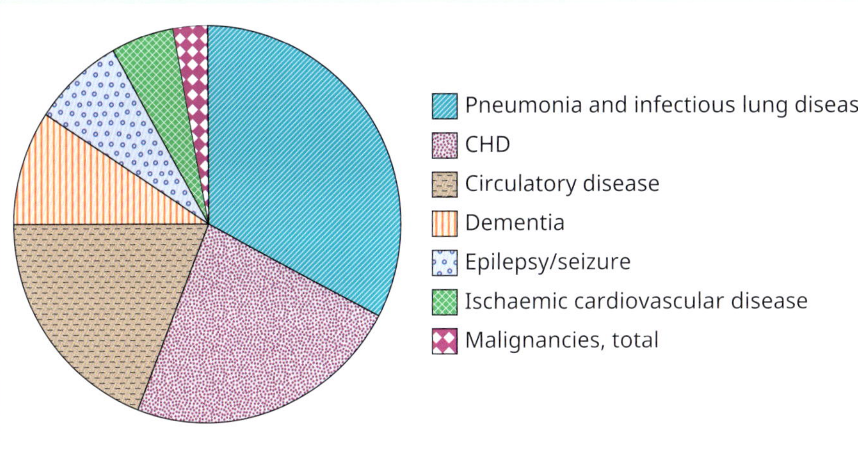

FIGURE 6.15
Causes of death in Down syndrome.

of which carry a risk of miscarriage, are offered to those with a cffDNA positive result. cffDNA analysis for Down syndrome is highly accurate; however, it still carries a small chance for false-negative and false-positive results, and therefore it is not a diagnostic test. The sample is termed non-invasive as a sample from the foetus is not taken directly—instead a maternal blood sample is taken to perform this genetic test. Foetal DNA is present in the plasma of the circulating maternal blood; this originates from foetal cells from the placental trophoblast which naturally apoptose leading to the release of the cellular DNA (cell-free DNA, cfDNA) into the maternal bloodstream. Maternal cells are also releasing DNA into the bloodstream at a much higher quantity, therefore the maternal cfDNA represents a greater proportion of the total cfDNA, with only a minority represented by the foetal cfDNA (also known as free foetal DNA, ffDNA). The ffDNA is reliably detectable from around the seventh week of pregnancy and continues to increase with gestation as the foetus and placenta increase in size, and therefore contributes a greater proportion of cfDNA. WGS is used to sequence the cfDNA and very sensitive analysis pipelines are run to detect if there is a relative abundance of chromosome 21 sequence. To put this into perspective, if the Down syndrome foetal fraction is 5% then 5% of the cfDNA has three copies of chromosome 21 (the remaining 95% cfDNA is maternal, contributing two copies of chromosome 21). The ratio of chromosome 21 reads, wild-type karyotype: Down syndrome karyotype would be 2: 2.05 (1:1.025). Highly sensitive techniques are required to reliably detect this level of difference. As NIPT is only a screening test (it is not diagnostic) invasive prenatal sampling is required to confirm the result.

The next point in the UK Foetal Anomaly Screening Programme (UKFASP) is through maternal serum screening which looks at analytes in the maternal blood (bHCG and PAPP-A), the nuchal translucency measurement (NT), gestational age, and maternal age to calculate a risk for Down syndrome (and also for Patau and Edwards syndrome). This is known as the **combined test** and is normally performed between 12 and 14 weeks gestation. For women who present later the **quadruple test** can be performed between 14 and 20 weeks—this looks at four analytes in the maternal blood. A high-risk result is considered to be greater than 1 in 150. In addition to the serum screening tests, ultrasound scanning of the foetus is also normally performed as part of the dating scan (end of first trimester), and foetal well-being scan (second trimester) and can identify the presence of foetal abnormalities which may be associated with Down syndrome (or other chromosomal abnormalities).

Invasive testing is offered for pregnancies with a high-risk screening result to determine if the foetus is affected with Down syndrome or not. Invasive testing involves direct testing of foetal tissue, which is obtained through a surgical procedure: a chorionic villus biopsy of the placenta for first trimester pregnancies, or an amniocentesis for later gestation pregnancies. Both of these procedures are associated with an increased risk of miscarriage for the pregnancy, and therefore patients are counselled before making a decision. If an invasive sample is taken it will be sent to the laboratory for testing. Rapid aneuploidy testing by QF-PCR is performed which can identify trisomy 21 and be reported within 3 days whilst cultures are set up and grown for karyotyping which takes around 10–14 days. Even though QF-PCR accurately detects trisomy 21, it does not provide any structural information for the chromosomes. In 95% of cases the additional chromosome 21 is present as a completely additional chromosome (e.g. with a 47,XY,+21 karyotype) which carries a low risk of recurrence. However, in ~5% of cases the additional chromosome 21 is structurally abnormal and attached to another acrocentric chromosome, termed a

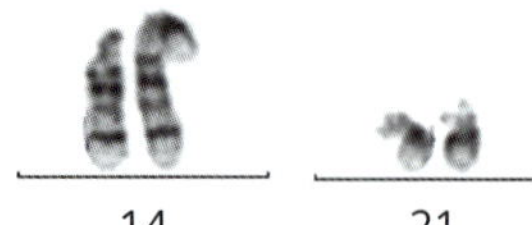

FIGURE 6.16
Robertsonian translocation of chromosomes 14 and 21 with an additional chromosome 21.

Robertsonian translocation, which may carry an increased risk of recurrence if inherited from a carrier parent. For a non-homologous Robertsonian translocation (i.e. with a different acrocentric chromosome—see **Figure 6.16** which shows a Robertsonian translocation involving chromosomes 14 and 21) the recurrence risk is up to 15%. However, for a homologous Robertsonian (both chromosome 21s fused together) the risk of an abnormal conception is 100% as they segregate together at meiosis (resulting in either trisomy 21, or monosomy 21 at fertilization). Genetic counselling can be offered to carriers of Robertsonian translocations, to discuss their options for prenatal diagnosis in any future pregnancies as well as offering testing to other at-risk family members.

Down syndrome may also be suspected at birth due to the baby's phenotype. The mother may not have undergone prenatal screening, was not identified as high risk, or the baby did not present with any visible abnormalities prenatally. In this case urgent trisomy testing is normally performed (by either FISH or QF-PCR) to quickly confirm the trisomy which will then be followed up by a karyotype analysis.

SELF-CHECK 6.3

1. Why do we need to confirm a NIPT result?
2. Why do we need to karyotype rather than array to test for Down syndrome?

CASE STUDY 6.2 Diagnosis of cystic fibrosis

Cystic fibrosis is an autosomal recessive genetic condition with clinical symptoms including obstructive pulmonary disease, recurrent respiratory infections, pancreatic insufficiency, and male infertility. It is caused by compound heterozygous or homozygous pathogenic variants in the *CFTR* (cystic fibrosis transmembrane conductance regulator) gene. These variants are more common in northern Europeans, with an approximate carrier frequency of 1/250. More than 1500 pathogenic variants have been described to date. However, most are very rare, and the majority of CF cases are caused by a small number of pathogenic variants. The *CFTR* gene codes for a cell membrane ion channel and pathogenic variants lead to either reduced or complete loss of function, resulting in chloride imbalance. This affects the osmotic potential and results in thick mucus which is difficult to expel and is, therefore, more prone to infection. Milder pathogenic variants are associated with CFTR-related disease such as isolated bronchiectasis, pancreatitis, and male infertility due to congenital bilateral absence of the vas deferens (CBAVD). Early detection and intervention can lead to improved health outcomes in patients with CF with half of patients living into their 40s. As such, newborn screening is part of the NHS Newborn Blood Spot screening programme. **Figure 6.17** shows the flow of newborn CF testing.

The newborn screening begins with a heel-prick blood sample when the baby is 5 days old. This is tested for a range of rare conditions, including CF, in which early intervention leads to improved outcomes for patients and families. The blood spot is first tested by the IRT assay (immunoreactive trypsinogen). IRT is raised (at or above 99.5th centile) in patients with CF due to blockage of pancreatic ducts leading to abnormal enzyme drainage. Babies with a positive result then undergo genetic testing for the four most common pathogenic variants in the UK population, which detects more than 80% of the disease-causing variants (**Figure 6.18**). If two pathogenic variants are identified, CF is suspected and

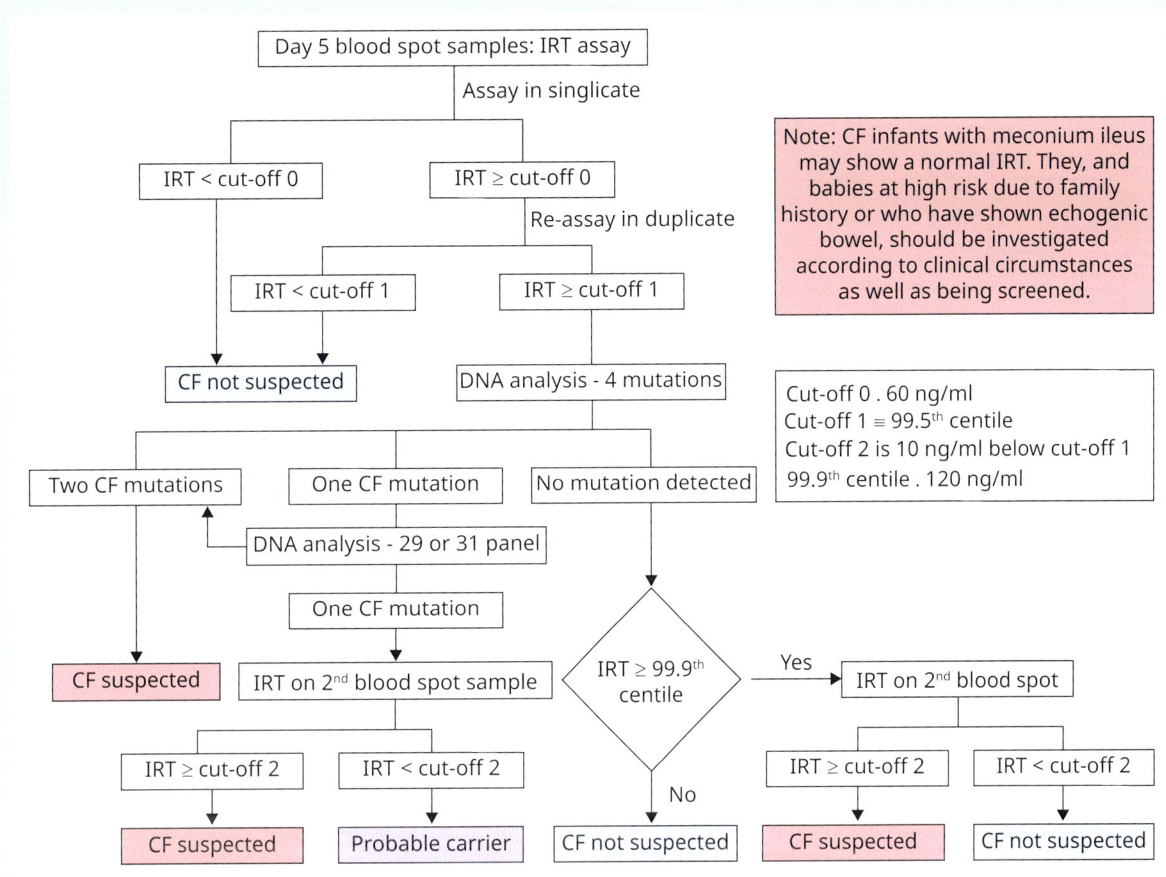

FIGURE 6.17
CF newborn screening flow chart.

the patient is referred immediately to a CF specialist. If only a single pathogenic variant is identified then DNA analysis for more mutations is undertaken to try and identify a second pathogenic variant. A larger panel of 31 pathogenic variants is used, which detects over 90% of pathogenic variants known to be in the UK population. If a second pathogenic variant is still not found, IRT can be performed on a second blood spot sample and, if positive, CF is suspected. Sequencing of the whole gene may be undertaken to look at the whole *CFTR* coding sequence for any rare pathogenic variant. Sequencing the entire gene is also performed in cases where no pathogenic variant is identified by the common variant or extended panel, where IRT is positive on both the first and second blood spot samples.

Until a clinical diagnosis is made, CF is only suspected with positive test results which warrant a full clinical assessment and also sweat testing. This is due to the heterogeneous nature of the pathogenic variants—even though some mutations are more likely to be associated with classical CF there is still a phenotypic range, and therefore a clinical evaluation and functional evaluation by the sweat test is required for a clinical diagnosis. Patients with CF have an increased concentration of salt in their sweat due to loss of function of the transmembrane ion channels. A special pad soaked with pilocarpine is placed on the child's arm or leg and a small electric current (painless) is applied which stimulates sweat production. The pad is removed, and the skin is washed before a sweat collector is placed over the area. Patients with the presence of two pathogenic variants and a positive sweat test have a confirmed diagnosis of CF. The absence of two CF-causing variants does not exclude a diagnosis of CF and further investigation is required. Patients with a single CF-causing variant and a negative sweat test are likely carriers of CF but not affected.

Once a clinical diagnosis is confirmed management and therapy can begin. This can include special infant formulas

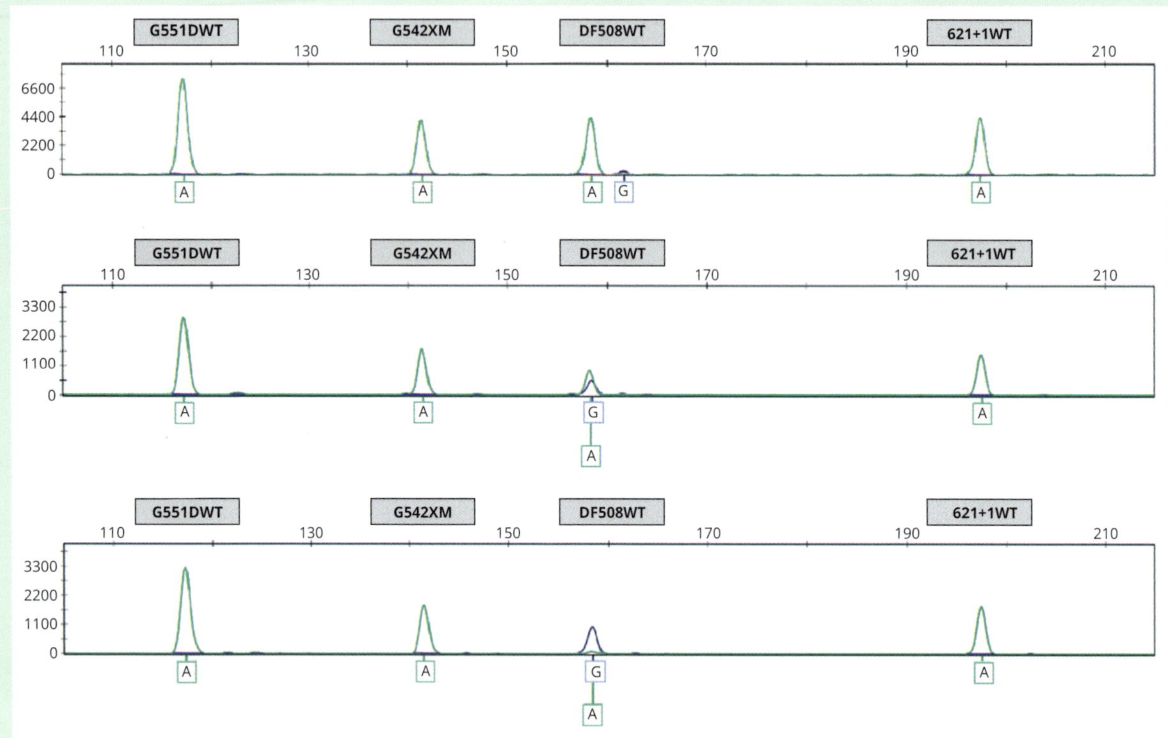

FIGURE 6.18

Targeted mutation testing for the 4 most common CF-causing pathogenic variants in the UK. The top image shows a normal result, the middle image shows the common DF508 deletion (heterozygote), and the bottom image shows a DF508 deletion (homozygote). Green peaks indicate the normal (wild-type) allele; the blue peak indicates the pathogenic variant.

to help with nutrition and weight gain in infants, pancreatic enzyme replacement, airway clearance techniques, antibiotics to prevent chronic airway infection, and assisted reproductive technologies (ART) for male infertility. Routine surveillance is also carried out to monitor for progression of the disease.

Genetic counselling will be recommended to discuss the findings of the pathogenic variants in the child and to offer cascade testing for other at-risk family members. Genetic testing will be offered to the parents of the affected child in the first instance to confirm their carrier status (with the assumption both parents are each carriers for a single pathogenic variant), with counselling offered more widely to other family members at risk of being carriers and testing offered as appropriate. Carrier parents have a 25% chance of having a further affected child, a 50% chance of having an asymptomatic carrier, and a 25% chance of having an unaffected non-carrier. Prenatal diagnosis (including NIPD) can be offered for future pregnancies to either plan and prepare for an affected birth or to make decisions about termination of the pregnancy.

CF may be indicated prenatally from ultrasound examination if high-grade foetal echogenic bowel or meconium ileus are present. If the parents have not had previous genetic testing then they will be tested first to identify if they are carriers, before offering prenatal diagnosis.

SELF-CHECK 6.4

What types of prenatal testing could be offered for CF?

Chapter summary

- Diagnostic laboratories are required to be accredited to high standards.
- Genetic conditions have a range of causes, not only due to the underlying DNA sequence, but also due to modification of gene expression, such as imprinting.
- Genetic testing can be complex, with different testing methods required for different disorders.
- Results of genetic tests can have implications for the whole wider family.
- Informed consent is critical to ensure patients understand the significance of potential findings which can include results of uncertain clinical significance and secondary findings.
- Genetic counselling is critical for discussion of the implications of positive test results.

Discussion questions

6.1 Why do we offer genetic testing?

6.2 What are the different types of genetic abnormalities?

6.3 What is the difference between cytogenetics and molecular genetics? Are there any similarities or overlap?

6.4 How do we decide which is the most appropriate genetic test to apply?

6.5 How are the results of genetic testing different to other types of diagnostic tests such as blood biochemistry or radiography?

Further reading

- Gardner RJ, Amor DJ (2018) ***Gardner and Sutherland's Chromosome Abnormalities and Genetic Counselling.*** 5th Edition, Oxford University Press, New York.

  A detailed textbook on cytogenetic abnormalities, and also the clinical consequences and genetic counselling.

- Strachan T, Read A (2019) ***Human Molecular Genetics.*** 5th Edition, CRC Press, Boca Raton.

  A detailed textbook for molecular genetics including background theory and testing.

- Arsham MS, Barch MJ, Lawce HJ (2017) ***The AGT Cytogenetics Laboratory Manual.*** 4th Edition, Wiley-Blackwell, Oxford.

  The complete resource for cytogenetic laboratory techniques.

- Carvalho, CMB, Lupski, JR (2016) ***Mechanisms underlying structural variant formation in genomic disorders***. *Nature Reviews Genetics*, 17(4), 224.

- Colvin K, Yeager M (2017) ***What people with Down Syndrome can teach us about cardiopulmonary disease.*** *European Respiratory Review*, 26, 160098. doi: 10.1183/16000617.0098-2016

- El Mecky J, *et al.* (2019) ***Reinterpretation, reclassification, and its downstream effects: challenges for clinical laboratory geneticists***. *BMC Medical Genomics*, 12(1), 170.
- Hart MR, *et al.* (2019) ***Secondary findings from clinical genomic sequencing: prevalence, patient perspectives, family history assessment, and health-care costs from a multisite study***. *Genetics in Medicine*, 21(5), 1100–1110.
- Hastings RJ, *et al.* (2016) ***Guidelines for cytogenetic investigations in tumours***. *European Journal of Human Genetics*, 24(1), 6–13.
- Horton R, *et al.* (2019) ***Direct-to-consumer genetic testing***. *BMJ*, 367, l5688.
- Hung ECW, Chiu RWK, Lo YMD (2009) ***Detection of circulating fetal nucleic acids: a review of methods and applications***. *Journal of Clinical Pathology*, 62(4), 308–13.
- Johnson JA, Cavallari LH (2015) ***Warfarin pharmacogenetics***. *Trends Cardiovasc Med*, 25(1), 33–41. doi:10.1016/j.tcm.2014.09.001
- Johnson SB, *et al.* (2020) ***Rethinking the ethical principles of genomic medicine services***. *European Journal of Human Genetics*, 28(2), 147–54.
- Lord J, *et al.* (2019) ***Prenatal exome sequencing analysis in fetal structural anomalies detected by ultrasonography (PAGE): a cohort study***. *The Lancet*, 393(10173), 747–57.
- Ramsden SC, *et al.* (2010) ***Practice guidelines for the molecular analysis of Prader-Willi and Angelman syndromes***. *BMC Medical Genetics*, 11(1), 70.
- Richards S, *et al.* (2015) ***Standards and guidelines for the interpretation of sequence variants: a joint consensus recommendation of the American College of Medical Genetics and Genomics and the Association for Molecular Pathology***. *Genetics in Medicine*, 17(5), 405–23.
- Riggs ER, *et al.* (2020) ***Technical standards for the interpretation and reporting of constitutional copy-number variants: a joint consensus recommendation of the American College of Medical Genetics and Genomics (ACMG) and the Clinical Genome Resource (ClinGen)***. *Genetics in Medicine*, 22(2), 245–57.
- Robson SC, *et al.* (2017) ***Evaluation of array comparative genomic hybridisation in prenatal diagnosis of fetal anomalies: a multicentre cohort study with cost analysis and assessment of patient, health professional and commissioner preferences for array comparative genomic hybridisation***. *Efficacy and Mechanism Evaluation*, 4(1), 1–104.
- Schwarz M, *et al.* (2008) ***Testing Guidelines for molecular diagnosis of Cystic Fibrosis***. Clinical Molecular Genetics Society, UK.
- Smith K, *et al.* (2015) ***General genetic laboratory reporting recommendations***. Association for Clinical Genetic Science, Birmingham, UK.
- Thornhill AR, *et al.* (2005) ***ESHRE PGD Consortium Best practice guidelines for clinical preimplantation genetic diagnosis (PGD) and preimplantation genetic screening (PGS)***. *Human Reproduction*, 20(1), 35–48.
- Wakeling EL, *et al.* (2017) ***Diagnosis and management of Silver–Russell syndrome: first international consensus statement***. *Nature Reviews Endocrinology*, 13(2), 105.
- Yohe S, Thyagarajan B (2017) ***Review of clinical next-generation sequencing***. *Archives of Pathology & Laboratory Medicine*, 141(11), 1544–57.

Useful websites

- Genereviews. **https://www.ncbi.nlm.nih.gov/books/NBK1116/**

  A very detailed online resource for the more common genetic disorders. Includes clinical phenotypic information, aetiology, testing, and management of patients.

- National Genomic Test Directory.**https://www.england.nhs.uk/publication/national-genomic-test-directories/**

  Lists the range of genetic testing available on the NHS based on referral groups and requesting speciality.

7

The Genetics and Genomics of Cancer

Heather Williams and Zoe Hatt

Heredity
The passing of traits from parents to offspring, which may include genetic predispositions to certain diseases like cancer.

Environmental factors
External factors such as exposure to chemicals, radiation, and lifestyle choices that can influence the development of cancer.

Malignant
Refers to cancerous cells that can spread and invade other parts of the body.

Ductal carcinoma *in situ* (DCIS)
A non-invasive breast cancer where abnormal cells are found in the lining of a breast duct but have not spread outside the duct.

Invasive ductal carcinoma (IDC)
A common type of breast cancer that starts in the milk ducts and invades surrounding tissues.

As the second leading cause of death in western populations, the factors responsible for the development of cancer are numerous; not only are there **heredity** and **environmental factors**, but chance also contributes to oncogenesis.

These unique drivers of the **malignant** process diverge cellular pathways which generate wide variation between cancers (e.g., rare cancers of the blood (leukaemia) with fewer than 15 cases per 100,000 people per year compared to the most common, such as breast cancer, with nearly 1.3 million cases per year; US National Cancer Institute Surveillance, Epidemiology, and End Results Program (SEER)) and within cancer types. For example, two different breast cancers: **ductal carcinoma *in situ* (DCIS)** or stage 0 breast cancer, a non-invasive only potentially malignant process with death an extremely rare outcome *versus* **invasive ductal carcinoma (IDC)**, an aggressive malignant process of the ducts with invasion of the stromal area which accounts for 75% of breast cancers. Both are diseases of the breast, with very different possible outcomes for patients.

Yet, there are commonalities in the process of oncogenesis that allow us to understand the process at the most basic level. In this chapter we will explore how the acquisition of genetic mutation(s) provides the aberrant cell an advantage over normal cells within the **microenvironment**. This often leads to a growth advantage providing the cell the ability to continue the malignant process unhindered by natural checkpoints designed to prevent disease. We will look at how heredity, the environment, and chance influence the genetics and genomics of oncogenesis and provide the basis of human cancer development.

The widespread advancement of genomic sequence data and the ubiquitous adoption of molecular methodologies have provided a peek into the underlying complexity of malignancy. Within the aberrant cell, many genes may be altered in their structure or copy number, whereas others may have inappropriate expression, with each change resulting in a unique clinical phenotype. The development of cancer has been linked to several **familial cancer genes**. However, the role, if any, of many **low penetrance variants** (or in some cases, polymorphisms) to the risk of sporadic cancer development remains elusive. Great discoveries have been made about the molecular pathogenesis of cancer, which has generated molecular classifications of disease. In this chapter we will look at how the widespread adoption of advanced detection methodologies in the clinical laboratory have enhanced and expanded these classifications, allowed for timely detection of disease, and provided the basis of targeted treatments of disease.

Microenvironment
The immediate environment surrounding a cell, including other cells, molecules, and blood vessels, that influences its behaviour and function.

Familial cancer genes
Genes that, when mutated, can increase the risk of cancer within a family due to inheritance.

Low penetrance variants
Genetic variations that do not always result in disease but may increase the risk of disease under certain conditions.

Learning Objectives

After studying this chapter, you should confidently be able to:

- List common cancers
- State the basic stages of cancer development
- Describe the difference between inherited (hereditary) and sporadic cancers
- Identify tumour suppressor genes and oncogenes critical in the malignant process
- Describe several types of variants (mutations) and how they lead to loss and gain of function in the disease process
- Explain how Knudson's hypothesis (two-hit theory) provides a framework for understanding cancer development
- Describe the role of the clinical genetics and genomics laboratory in cancer diagnosis
- Describe the clinical utility (and limitations) of genomic testing platforms.

7.1 Hereditary versus sporadic cancer

Broadly, the development of cancer requires genetic events termed mutation(s) or changes to the genetic material to produce disease. The inheritance of change in the gene (mutation or **variant**) may increase an individual's susceptibility to neoplasia. Look at **Table 7.1** for a list of hereditary predisposition cancer syndromes and notice the wide range of cancers associated. Some carriers of a mutation never develop cancer due to reduced **penetrance**. In some instances, two or more mutations produce only a benign precursor lesion, and other random (sporadic) genetic events are all that are necessary for the rapid and dramatic onset of **oncogenesis**.

These situations provide the basis of the **Knudson's 'two-hit' hypothesis**, a unifying model of carcinogenesis that is discussed later in this chapter.

In **Table 7.1**, hereditary cancer predisposition syndromes are indicated in the first column. The diseases associated with inheritance of variants (mutations) in the gene responsible are indicated in the second column. The official Human Genome Organisation (HUGO) Gene Nomenclature Committee (HGNC) approved names and gene symbols, in italic as per human genes nomenclature conventions, are presented in the third column. The chromosomal location or locus of the gene is in the fourth column.

Variant (mutation)
A change in the DNA sequence that can lead to changes in protein function and may contribute to diseases such as cancer.

Penetrance
Percentage of individuals with a certain genetic variant who present with the related trait.

Oncogenesis
The process of tumour formation or development of cancer.

Knudson's hypothesis (two-hit theory)
A theory that suggests cancer is the result of accumulated variants (mutations) in a cell's genes, with a first 'hit' being a genetic predisposition variant and a second 'hit' being an acquired variant (mutation).

SELF-CHECK 7.1

Have you noticed that the inheritance of a genetic mutation (variant) can predispose to multiple tumours throughout an individual's lifetime? Can you recall the names of some of the hereditary cancer predisposition syndromes in which this happens?

TABLE 7.1 A list of hereditary cancer predisposition syndromes

Syndrome	Disease manifestation/Tumour type(s)	Gene(s)	Chromosomal Location
Ataxia telangiectasia (AT)	Neurological manifestations, ocular telangiectasia, immunodeficiency, cancer susceptibility (100x increased risk) and radiation sensitivity	ATM serine/threonine kinase (*ATM*)	11q22.3
Bloom syndrome	Predisposition to development of numerous cancers (e.g., leukaemia; lymphoma)	BLM RecQ like helicase (*BLM*)	15q26.1
Hereditary breast and ovarian cancer (HBOC)	Development of breast cancer; ovarian cancer (increased risk vs. *BRCA2*)	BRCA1 DNA repair associated (*BRCA1*)	17q21.31
	Development of breast cancer; ovarian cancer	BRCA2 DNA repair associated (*BRCA2*)	13q13.1
Familial adenomatous polyposis (FAP)	Development of colorectal and intestinal carcinoma; papillary thyroid carcinoma.	APC regulator of WNT signalling pathway (*APC*)	5q22.2
Hereditary non-polyposis colon cancer (HNPCC/Lynch Syndrome)	Development of colorectal carcinoma; uterine carcinoma; gastric carcinoma, and others	See special section on Lynch syndrome (Section 7.1.4)	
Hereditary papillary renal cell carcinoma (PRCC)	Development of papillary renal cell carcinoma (kidney)	MET proto-oncogene, receptor tyrosine kinase (*MET*)	7q31
Hereditary retinoblastoma	Development of multiple primary cancers early in life (e.g., bilateral retinoblastoma, pineoblastoma (intracranial), melanoma (skin), sarcoma (bone)).	RB transcriptional corepressor 1 (*RB1*)	13q14.2
Li-Fraumeni syndrome (LFS)	Development of multiple primary cancers early in life (e.g., acute leukaemia, breast carcinoma, sarcoma (bone)).	Tumour protein p53 (*TP53*)	17p13.1
Multiple endocrine neoplasia (MEN) syndromes: MEN1, MEN2 (includes: MEN2A, MEN2B, familial medullary thyroid carcinoma (FMTC), MEN4)	Development of multiple endocrine tumours (e.g., pancreatic endocrine neoplasm; pituitary adenoma): 'MEN1' or 'Werner Syndrome'.	Menin 1 (*MEN1*)	11q13
	Development of multiple endocrine tumours (e.g., medullary thyroid cancer; parathyroid adenoma; phaeochromocytoma (adrenal gland)): 'MEN2A' or 'Sipple syndrome'; 'MEN2B' or 'Gorlin syndrome'.	Ret proto-oncogene (*RET*)	10q11.21
	Development of multiple endocrine tumours (e.g., pancreatic endocrine neoplasm; pituitary adenoma): 'MEN4'.	Cyclin dependent kinase inhibitor 1B (*CDKN1B*)	12p13
Von Hippel-Lindau Syndrome (vHL)	Development of multiple vascular tumours (e.g., retinal (eye) and central nervous system haemangioblastomas; clear cell renal carcinoma (kidney); phaeochromocytoma (adrenal gland)).	von Hippel-Lindau tumour suppressor (*VHL*)	3p25.3

7.1.1 Tumour suppressor genes

A normal cellular gene whose functional role often involves the inhibition of cell proliferation, whose aberrant inactivation results in a loss of this cell cycle 'break' function and is known to drive cancer development, is called a **tumour suppressor gene**. These tumour suppressors are critical regulators of the cell cycle, differentiation, and apoptosis. Mutations in tumour suppressors disrupt cellular homeostasis and initiate malignant transformation. Tumour suppressor genes often lose their function during cancer via mutation, (typically) by annihilating the functions of the encoded proteins, which may result in loss of regulatory oversight of protein function and/or the mutation alters the protein's affinity for a substrate. **Missense mutations** (i.e., a point mutation; a single nucleotide change, which results in a codon substitution) account for approximately 90% of mutations in the tumour suppressor gene 'Tumour protein p53' (*TP53*). Mutations in the *TP53* gene occur preferentially within the DNA-binding domain (i.e., a 'mutation hot spot') of the p53 protein between amino-acid (aa) residues 102–292 (out of 393 aa in the full length protein), which can alter protein conformation, subsequently resulting in its unfolding, as occurs with many missense mutations, or result in producing no protein at all; both changes disrupt normal cellular regulation. The latter, which produce no protein, are referred to as '**loss of function (LOF)**', as they lead to loss of the protein's native function within the cell. LOF mutations within the *TP53* gene account for 10% of mutations, and result from various mechanisms such as nonsense, frameshift, or deletion mutations. Located in the *TP53* hot spot, the p53 R175H missense mutation (i.e., 'p.R175H'—the p53 protein (p.) has a codon substitution, which results in a change from Arginine (R) to Histidine (H) at aa position 175) has the highest occurrence in cancer patient somatic (acquired) testing (TCGA and International Agency of Research on Cancer (IARC) *TP53* databases). [For more information on mutation (variation) nomenclature see Human Genome Variation Society (HGVS) nomenclature. *NB.* Somatic (acquired) mutations are often referred to using the protein 'p.' nomenclature (however, clinical reporting should always include the genomic location as well for clarity), whereas germline (constitutional) mutations are generally reported using both the protein 'p.' and coding DNA reference sequences (c.) nomenclature.] Interestingly, the p53 R175H mutation results in both loss of native p53 transactivating functions due to a defective DNA binding interface, and consequently, a **gain of function (GOF)** through several mechanisms, e.g., interactions with transcription factors to enhance or suppress target gene expression. Comparatively, the Brazilian founder missense germline (inherited) mutation in *TP53*, p53 p.R337H (i.e., 'p.R337H'—the p53 protein (p.) has a codon substitution, which results in a change from Arginine (R) to Histidine (H) at aa position 337) leads to compromised functionality (loss of function), which is observed in the extremely rare malignancy Paediatric Adrenocortical Tumour (ACT). Often attributed to the founder mutation, and the resulting impaired p53 activity, this cancer has a high incidence in South and Southeast Brazil.

Tumour suppressor genes
Genes that normally help to prevent uncontrolled cell growth. When these genes are mutated, cells can grow uncontrollably and become cancerous.

Missense mutation
A type of genetic mutation where a single nucleotide change results in a codon that codes for a different amino acid.

Loss of function (LOF) mutations
Mutations that result in the loss of a gene product.

Gain of function (GOF) mutation
Mutations that result in a new or enhanced activity of a gene product, which can lead to abnormal cellular behaviour, such as uncontrolled growth.

Nonsense mutations (i.e., another type of codon substitution), alter the DNA sequence to create a stop signal that prematurely results in termination of the encoded protein; the resulting abnormal protein may be non-functional, function incorrectly, or be inappropriately targeted for destruction within the cell. Observed in head and neck squamous cell carcinoma, the *TP53* mutation p53 Q167*, (i.e., 'p.Q167*'—the p53 protein (p.) has a codon substitution, which results in a change from Glutamine (Q) to termination (*)), which results in truncation and loss of protein function. Other mutations, such as the *TP53* R209Kfs*6 (i.e., p.R209Kfs*6, a deletion of the DNA codon c.626_627delGA ((transcript NM_000546.5) the coding DNA reference sequences (c.) has a deletion of two nucleotides at nucleotide positions 626 to 627 (delGA)), that creates a substitution in the protein (p.) of Arginine (R) to Lysine (K) that results in a translational frameshift (fs) that shifts the reading of the sequence (to a different reading frame)

Nonsense mutation
A type of genetic change that creates a stop signal resulting in the premature termination of the encoded protein.

and subsequently results in termination at 6 residues downstream), are observed in multiple cancer types (e.g., breast, oesophageal).

Translocation
A type of chromosomal abnormality where a chromosome segment is moved from one location to another.

In Anaplastic Large Cell Lymphoma (ALCL), the t(2;5)(p23;q35), i.e., **translocation** (t) of the short arm (p23) of chromosome 2 with the long arm (q35) of chromosome 5, results in the *NPM1::ALK* (previously denoted as NPM1-ALK) fusion, which leads to constitutive activation of the ALK protein. This uncontrolled GOF mutation induces several signalling cascade pathways, which initiate and maintain malignant transformation. Mutations such as the *NPM1::ALK* fusion, have become crucial targets of therapy, e.g., crizotinib, a specific ALK inhibitor, inhibits NPM-ALK protein phosphorylation leading to cell death, and is used as a chemotherapy for relapsed or refractory, systemic ALK-positive ALCL.

Homozygous
Concerning both allelic copies.

Heterozygous
Concerning only one allelic copy.

Sometimes a mutation in a tumour suppressor can create a cellular state similar to **homozygous** loss (no copies of the normal gene) even in the presence of a single **heterozygous** mutation.

Although the earliest discoveries of a tumour suppressor gene involve retinoblastoma and the *RB1* gene (look at **Section 7.1.3** for an extensive history of this discovery), it is a mutation of the tumour suppressor *TP53* (also known as the 'guardian of the genome') that occurs in many cancers including breast, colorectal, oesophageal, ovarian, and leukaemia.

Wild-type gene
A gene in its natural, unmutated form.

Loss of heterozygosity (LOH)
Loss of one copy of a segment of DNA.

Discovered in 1979, the *TP53* gene-encoded protein p53 was named only for its molecular weight, as its function(s) were unknown (Lane and Crawford 1979; Linzer and Levine 1979). However, investigations of the functions of *TP53* remained elusive as attempts by scientists to further characterize the p53 protein quickly ran into issues, as many cloned p53 proteins were inconsistent between laboratories, and it was subsequently realized that different genetic sequences caused these inconsistencies. In fact, the wild type *TP53* sequences were actually mutants that had been inadvertently identified as **wild type** because scientists had not yet discovered that the mutation of *TP53* is highly prevalent in human cancer, occurring in approximately 50% of all cancers. It would take the analysis of genetic losses common in colorectal cancers by Muleris and colleagues in 1985 before the location of *TP53* was established using karyotypic observation of recurrent loss of the short arm of chromosome 17. Yet, a further five years (and many more experiments) would pass before these colon cancer observations were combined and summarized to generate our understanding of how **loss of heterozygosity (LOH)** or the loss of one allele coupled with secondary mutations in the other *TP53* allele could work in tandem to cause cancer. In these experiments, upon comparison of colon cancer and normal tissues it was found that the secondary mutations were absent and had instead been acquired (known as somatic change) during cancer development. It is the inappropriate inactivation of the p53 protein that causes loss of cell checkpoint controls and inappropriate growth and division, which can allow for the rapid development of cancer. *TP53* acts as the 'guardian of the genome'; however, when lost, genome integrity is compromised and cells with damaged DNA are allowed to proliferate thus driving tumorigenesis.

Somatic mutations
Genetic alteration acquired by a cell.

In breast cancer, **somatic mutations** in the *TP53* gene occur in approximately 40% of patients whereas germline mutations are associated with Li–Fraumeni syndrome (LFS) and result in an increased risk of cancer development (not limited to breast cancer) at a young age (<30 years) (look at **Table 7.1** for the many other cancers patients with LFS are at an increased risk of developing).

7.1.2 Oncogenes

Proto-oncogene
The unmutated version of an oncogene.

A normal cellular gene whose aberrant functions are known to drive cancer development when mutated is called a **proto-oncogene**. These proto-oncogenes are often highly evolutionarily conserved between species and critical cell growth and differentiation regulators. Mutations in proto-oncogenes that disrupt cellular pathways that can initiate malignant transformation, convert the genes to **oncogenes**. Oncogenes are characterized by mutations that do not wipe out their function, but rather change (typically by increasing, i.e. a gain of) the functions of the encoded proteins. Typically, these mutations are activating, which may result in loss of regulatory oversight of protein function and/or the mutation altering the protein's affinity for a substrate.

Oncogenes
Genes that normally help cells grow. When mutated or expressed at high levels, they can turn a normal cell into a cancerous cell.

The earliest identification of genes responsible for cancer resulted from the discovery of oncogenes in cancer inducing retroviruses. The experiments of Rous (1911) involved attempts to transmit chicken sarcomas between animals using cell-free tumour extracts. After extracting fibrosarcomas (tumours of the connective tissue) from chickens, Rous passed the supernatant through specialized filters that retained the smallest of bacteria, and subsequently injected healthy chickens. As most of the healthy chickens went on to develop sarcomas, it was later established that these extracts contained the causative agent for the sarcoma—the Rous sarcoma virus (RSV). RSV is a retrovirus now known to utilize **reverse transcriptase** to integrate into a host genome, effectively generating an oncogene that drives cancer development. However, it would take until the mid/late 70s before Bishop and Varmus would establish that retroviral causes of avian cancer result from retroviral genes that are simply variations of the normal genes within the chicken genome. Specifically, RSV contains the viral form of the Src protein, which results from a mutant form of the proto-oncogene *SRC* (SRC proto-oncogene, non-receptor tyrosine kinase). The inappropriate constitutive activation of the viral Src protein causes inhibited and inappropriate phosphorylation of targeted proteins which can induce the rapid development of cancer.

Reverse transcriptase
An enzyme that synthesizes DNA from an RNA template.

In breast cancer, mutation of the proto-oncogene *ERBB2* (erb-b2 receptor tyrosine kinase 2), commonly identified and previously named HER2, contributes to the development of disease in a variety of ways. Occurring in ~20–30% of breast cancers, the overexpression of HER2 is associated with an aggressive disease, higher rates of recurrence, and shortened survival. These mutations vary but include acquired (somatic) mutations of *ERBB2 (HER2)* (some of which result in amplification of the oncogene), and over-expression of the HER2 protein. Through aberrant activation of HER2 transmembrane receptor, there is inappropriate stimulation of several intracellular pathways involved in transcriptional activity, which drives uncontrolled cell growth and division leading to cancer development (look at **Figure 7.10** for an example of a laboratory test for *ERBB2* (HER2) copy number changes).

Key Points

Cancer predisposition genes can be divided into two groups: oncogenes and tumour suppressor genes.

SELF-CHECK 7.2

How was a chicken sarcoma critical to our understanding of cancer development?

7.1.3 Two-hit hypothesis

First proposed by Geneticist Alfred Knudson in 1971, the 'two-hit' hypothesis serves as a basis for the role of loss of function mutations in tumour suppressor genes in cancer development. A result of studies performed by Knudson and colleagues over 20 years between the mid 1940s and 1960s on retinoblastoma, a retinal cancer which develops in young children, this hypothesis used statistical methods to explain why some children could inherit a germline mutation and develop retinoblastoma, whereas others did not. Study **Figure 7.1**. Notice how those who

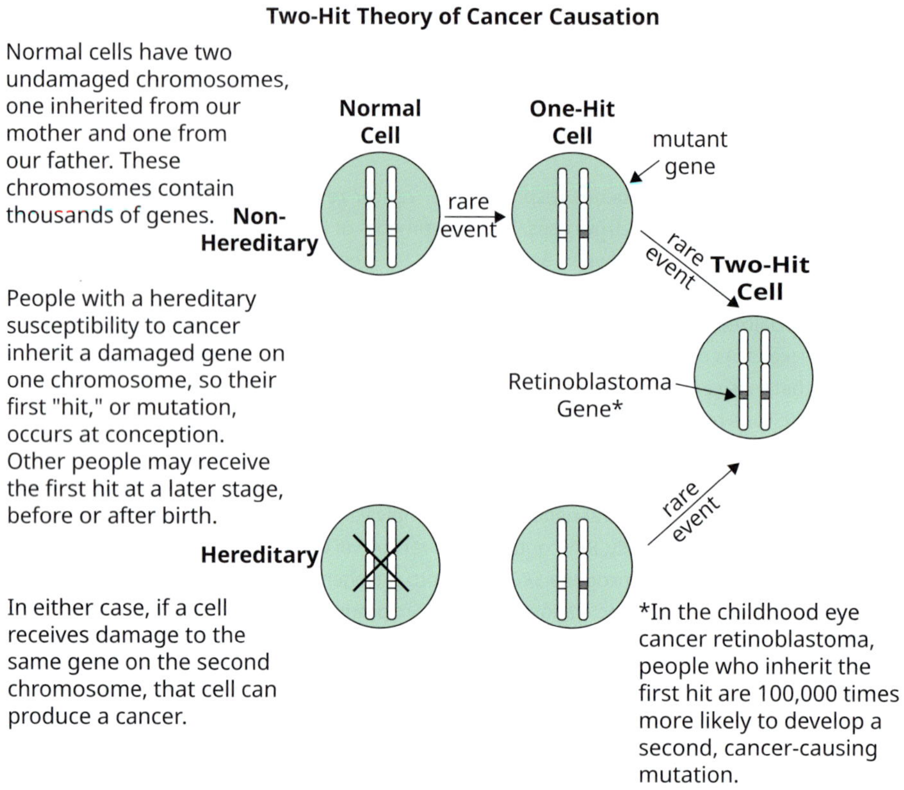

Two-Hit Theory of Cancer Causation

Normal cells have two undamaged chromosomes, one inherited from our mother and one from our father. These chromosomes contain thousands of genes.

Non-Hereditary

Normal Cell

rare event

One-Hit Cell

mutant gene

rare event

Two-Hit Cell

Retinoblastoma Gene*

People with a hereditary susceptibility to cancer inherit a damaged gene on one chromosome, so their first "hit," or mutation, occurs at conception. Other people may receive the first hit at a later stage, before or after birth.

Hereditary

rare event

In either case, if a cell receives damage to the same gene on the second chromosome, that cell can produce a cancer.

*In the childhood eye cancer retinoblastoma, people who inherit the first hit are 100,000 times more likely to develop a second, cancer-causing mutation.

FIGURE 7.1
Illustration of Knudson's two-hit hypothesis.

inherit one causative mutation are approximately one hundred thousand times more likely to develop retinoblastoma due to a sporadic mutation.

Although the gene responsible for retinoblastoma, *RB1* (RB transcriptional corepressor 1) on chromosome 13, would not be discovered until 1986, Knudson's studies reviewed patient family history, tumour location (one or both eyes), and number of tumours within the eye to characterize childhood populations from clinical records. Using this information, Knudson concluded the disease results from two mutations that occur in each allele (copy) of a tumour suppressor gene; those individuals who inherit mutations in the tumour suppressor gene would develop the disease at a younger age with more than one tumour likely to develop. The RB protein is now known to be involved in numerous cellular processes including inhibiting cell proliferation and division, and regulation of apoptosis. Inherited in an autosomal dominant manner, germline (constitutional) mutations in *RB1* are associated with an increased risk of retinoblastoma with ~60% of affected individuals developing unilateral disease by 24 months of age and ~40% with developing bilateral disease by 15 months.

The two-hit hypothesis is a fundamental underpinning to our scientific understanding of the inheritance patterns of genetic mutations in hereditary cancer syndromes. It is nearly unthinkable to discuss the development of cancer without mention of the Knudson two-hit hypothesis. However, with the discovery of **haploinsufficiency**, and its application to the model of carcinogenesis, we now understand that two hits are not always required and that the loss of a single copy of a gene can result in malignancy. In this situation, tumours that are haploinsufficient for a tumour suppressor gene product are unable to produce the level of a protein required to complete the normal cellular functions and this 'single hit' is sufficient for disease. For example,

Haploinsufficiency
A situation in which heterozygosity for a gene mutation leads to insufficient protein product.

haploinsufficiency of the p53 protein resulting from mutation in the *TP53* gene, in haematopoietic cells of patients with Li–Fraumeni syndrome (LFS) can promote the development of leukaemia with loss of a single functional gene copy. Adding to the complexity, **dominant negative effects** create yet another exception to the 'two-hit' hypothesis of disease. In this situation, the 'single hit' is not loss of the wild type gene copy; instead, a mutation results in decreased functional ability of the abnormal protein that acts to sabotage the role of the normal protein within the cell often through the formation of non-functional complexes. For example, dominant negative effects of the APC protein resulting from mutation in the *APC* gene, in colon cells of patients with Familial Adenomatous Polyposis (FAP), can promote the development of colorectal cancer due to disruption of the normal APC protein's ability to interact with other proteins.

Dominant negative effect
A type of genetic mutation that results in a protein that interferes with the co-expressed wild-type protein.

Finally, **pleiotropy**, the process whereby a single gene product can impact several cellular factors can contribute to the development of (multiple) cancers. That is, when pleiotropy occurs there are inconsistent effects of the loss of a gene product on the gene product's cellular functions and this could result in enhancement and/or inhibition. For example, the pleiotropic role of BRCA1 protein resulting from mutation in the *BRCA1* gene in breast cells of patients with hereditary breast and ovarian cancer (HBOC), which can promote aberrations of both cell checkpoint activation (thus stalling development) and DNA repair pathways (thus inhibiting repair that is necessary for DNA fidelity) that results in discrepancies across the genotype–phenotype spectrum of breast cancer observed in families with HBOC.

Pleiotropy
The phenomenon whereby a gene can affect multiple traits.

SELF-CHECK 7.3

What childhood cancer helped uncover our understanding of cancer development?

7.1.4 Hereditary (inherited) cancer syndromes

Although most cancers are sporadic (the individual has no inherited genetic change that would increase their risk of cancer development), approximately 5–15% of all cancers arise in individuals with heightened susceptibility given the inheritance of a genetic mutation associated with specific cancer(s). [*NB.* Advances in sequencing technology (see **Table 8.3**) continue to facilitate the discovery of cancer predisposition genes (CPGs) and although previous large sample studies indicated ~5–15% of patients with cancer carry a CPG, post sequencing-era data (mid 2000s to present) has emerged, which indicates that the incidence of hereditary cancer *may* be as high as 17.5%; e.g., a 2015 study from St. Jude Children's Research Hospital (Memphis, Tennessee, USA) indicated that ~4.4% of patients (of 1120 patients total) with leukaemia carried germline mutations, specifically, CPG mutations; however, only 23% had a family history suggestive of cancer predisposition. Studies such as this are changing screening and diagnostic guidelines in oncology care pathways.]

These cancer predisposition genes are inherited in a variety of patterns including dominant, recessive, X-linked, and in some cases with variable penetrance within families. Today, great efforts are being made in cancer surveillance and prevention programmes aimed at reducing cancer risks in this special population.

For example, Lynch syndrome (or hereditary non-polyposis colorectal carcinoma syndrome, HNPCC) is an autosomal dominant genetic condition associated with increased risk of colorectal, endometrial, gastric, and other cancers. One of the most prevalent hereditary cancer syndromes in humans, Lynch syndrome, like most familial cancer syndromes, is associated with a

TABLE 7.2 MMR gene mutations in Lynch syndrome

Gene Name	HGNC Symbol*	Chromosomal Location
mutL homolog 1	*MLH1*	3p22.2
mutS homolog 2	*MSH2*	*2p21–p16.3*
mutS homolog 6	*MSH6*	*2p16.3*
PMS1 homolog 2, mismatch repair system component	*PMS2*	*7p22.1*

Microsatellite
Repeated sequences of DNA (mononucleotide, dinucleotide, or higher-order nucleotide repeats such as $(A)_n$ or $(CA)_n$) that are prone to mutations and are used in genetic studies, especially in the context of cancer research.

Microsatellite instability (MSI)
A condition of genetic hypermutability that results from impaired DNA mismatch repair, associated with certain types of cancer like Lynch syndrome (LS).

Chromosomal location
The specific location of a gene on a chromosome, often identified by numbers and letters that indicate its position.

Germline mutation
Genetic alteration inherited by a cell.

Immunohistochemical (IHC) staining
A laboratory technique used to visualize specific proteins in tissue samples, often used to diagnose diseases like cancer.

Polymerase chain reaction (PCR)
A laboratory method used to make multiple copies of a specific DNA segment, enabling detailed analysis.

Next-generation sequencing (NGS)
A high-throughput method of DNA sequencing that allows for the rapid and comprehensive analysis of genetic material.

spectrum of tumours in mutation carriers. The identification of patients with Lynch syndrome is of clinical importance given they have an increased lifetime risk of colorectal cancer of nearly 80%, and nearly 60% for endometrial cancer. However, the spectrum of risk differs between mutations in genes associated with Lynch syndrome.

Lynch syndrome is characterized by defective mismatch repair (MMR) pathways resulting in cells that are unable to repair defects in short repetitive sequences (**microsatellites**), which is referred to as **microsatellite instability (MSI)**. The DNA MMR pathway plays a fundamental role in the maintenance of genomic stability to prevent mutation. MMR functions through the repair of base-to-base mismatches and mispairs from deletions/insertions following replication and recombination events. Germline mutations in the MMR genes *MLH1*, *MSH2*, *MSH6*, and *PMS2* are the most frequent cause of Lynch syndrome. However, mutations in only the *MLH1* and *MSH2* genes account for more than 90% of all mutations identified in Lynch syndrome families. Given the heterogeneity of genes associated with Lynch syndrome, the lifetime risks of cancer development are significantly higher in individuals with *MSH2* and *MLH1* mutations compared to those with *MSH6* or *PMS*—a difference that is thought to be related to the functional redundancy of the MSH6 (with *MSH3*) and PMS2 (with *MLH3* and *PMS1)* MMR proteins that are critical to genomic stability maintenance.

Look at **Table 7.2** for more information on the locus or **chromosomal location** of these genes. The most common mismatch repair genes associated with Lynch syndrome (first column). The official Human Genome Organisation (HUGO) Gene Nomenclature Committee (HGNC) approved symbol (second column). The chromosomal location of the gene (third column). Although loss of function results from a **germline mutation** (inherited) in one allele of the MMR genes, the second allele must be inactivated, generally through a somatic (acquired) mutation, for cancer development. Two methodologies for screening in Lynch syndrome include **immunohistochemical staining** and microsatellite instability testing (often performed via **polymerase chain reaction (PCR)** based assays; however, novel molecular-based methods have emerged such as various custom **next-generation sequencing (NGS)** approaches).

SELF-CHECK 7.4

What is the difference between inherited and sporadic cancer?

SELF-CHECK 7.5

What are the most common cancers associated with Lynch syndrome?

7.2 The development of cancer

Although the earliest written documentation of growths we would now consider cancer comes from Ancient Egypt (3000 to 2500 BCE), we have since that time struggled to recognize, classify, and effectively treat this disease. In fact, the early written evidence (known as the Edwin Smith Papyrus, named for the American dealer of antiquities whose death resulted in a donation of the scroll to The New York Historical Society) provided insight into several cases of early breast cancer in which a single case report reported the diagnosis as: 'very cool and bulging to thy hand with swellings which have spread all over the breast.' However, fundamental changes in the classification of diseases proposed by Sydenham and Boissier de Sauvages in the 1600s and 1700s, respectively, combined with the industrial revolution of the 1800s generated great scientific interest in changing the broad classification of 'cancer' known for millennia, into attempts to identify the underlying cause(s) of cancer. Combined with advances in medical laboratory sciences in the 1900s, literature on cancer blossomed and began to describe characteristic symptoms and natural histories of unique subtypes of cancer.

Today, the molecular biology discoveries of the past 40 years have expanded our understanding of cancer considerably, including identification of detectable underlying genetic abnormalities. However, more research is needed; for example, breast cancer alone continues to cause over a half million deaths globally each year.

The development of cancer is a complex process resulting from disruptions of the exquisitely regulated process of cell growth and division. Cancer development can occur in any of the cells within the body, typified by the numerous forms of cancer. However, tumours that result from transformation processes following any abnormal proliferation event can be benign or malignant. It is the malignant tumour, which results from aberrant differentiation, development, and invasion of normal tissue, which ultimately spreads through the body (metastasis). The multistep process of cancer development often results from the accumulation of genetic variation; this explains why so many cancers have higher incidence in older age given the cells take years to develop these abnormalities.

Cancer development can result from many different factors, including the inheritance of a mutation (as discussed above) or from exposure to carcinogens. Of note, it is naïve to discuss the development of cancer as the result of exposure to a single agent, given the multistep process; however, there are established links between carcinogens and disease. For example, tobacco smoking is a known health hazard and contains carcinogens responsible for the development of lung cancer. The use of tobacco remains the most preventable cause of death with ~85% of lung cancers attributed to the nearly 4000 chemicals within tobacco smoke known to be harmful to human health, including arsenic, benzene, formaldehyde, and toluene. Other environmentally associated cancers include skin cancers, which have the highest increased incidence (significantly faster than any other cancer) in fair-skinned populations in Europe, North America, and Oceania during the last four decades. Within the UK, where skin cancer is the fifth most common cancer, approximately 86% of all melanomas (a skin cancer) result from excessive UV radiation (or light) exposure. Specifically, it is the intermittent exposure to high-intensity sunlight (e.g., prolonged sunbathing at the lido) rather than chronic sunlight exposure (e.g., routine daily exposure through work) in the environment, which is most closely associated with cancer development.

A variety of other substances in the environment can sometimes result in cancer when humans are inappropriately exposed. These may be rare events, such as the largest-ever radiation accident at the Chernobyl Nuclear Power Plant in Ukraine on 26 April 1986, which has been associated with increased development of cancer in children exposed to radiation residing in

Belarus, the Russian Federation, and Ukraine. The repeated screening of a Ukrainian-American cohort over the last four decades following the accident has consistently demonstrated a significant elevated risk of radiogenic thyroid cancer. In studies of patients exposed to radiation as young children in the disaster, targeted sequencing of their papillary thyroid carcinomas (PTCs) has revealed that the *BRAF* (B-Raf proto-oncogene, serine/threonine kinase) p.V600E mutation (i.e., 'V600E'—the B-Raf protein (p.) has a codon substitution, which results in a change from Valine (V) to glutamate (E) at aa position 600), which represents over 90% of all *BRAF* mutations in human cancers, has been associated with the natural course of these environmental carcinogen-based cancers. In particular, the presence of the *BRAF* p.V600E has been associated with an older age at the time of the disaster with an extended period of latency between the disaster and cancer diagnosis, and older age at thyroidectomy operation. This is a great example of how sequencing information is aiding our understanding of disease.

7.2.1 Initiation

All cells contain protective signalling mechanisms that form a tightly regulated system of activation and inhibition of genes responsible for growth and division. In healthy cells, the balance between these states is maintained and prevents further errors as the cell is blocked from continuing through the cell cycle.

However, after a cell acquires a genetic aberration, in true Darwinian fashion, it is those mutations that provide a selective advantage to the cell. These cells may have acquired a mutation that alters gene expression to override cell checkpoint controls. As described by Hanahan and Weinberg (2011), the 'Hallmarks of Cancer' include the acquisition of specific biological capabilities, i.e., sustained proliferative signalling, evasion of growth suppression, resistance to cell death (apoptosis), enabled replicative immortality, induction of angiogenesis, activation of invasion and metastasis, reprogrammed energy metabolism, and evasion of immune destruction mechanisms. Updates to the Hanahan and Weinberg (2011) 'Hallmarks of Cancer' were proposed by Hanahan (2022), which suggest additional characteristics, such as non-mutational **epigenetic reprogramming**, avoiding immune destruction, tumour-promoting inflammation, polymorphic microbiomes, inducing or accessing vasculature, senescence, genome instability and mutation, deregulating cellular metabolism, and unlocking phenotypic plasticity, also lead to disease.

Epigenetic reprogramming
The process by which gene expression is altered without changing the underlying DNA sequence.

Figure 7.2 shows that cancer development can result from the disruption of several regulatory and maintenance cellular processes.

These hallmarks result from an underlying genome instability that has allowed for the accumulation of genetic diversity (including mutations) within cells. Quite often these mutations accumulate within genes with critical roles in the cycle cell: tumour suppressor and oncogenes (look at **Section 7.1.3** for an in-depth review of how these genes result in disease).

It is the acquisition of one or multiple biological capabilities that allows aberrant cells to undergo rapid growth and division to overtake restraints in the cellular environment resulting in the initiation of cancer.

7.2.2 Progression

The cells that have acquired mutations soon become dominant cells in the tumour microenvironment given their success in clonal selection due to their unique advantage of outcompeting others for survival. It is the accumulation of unique cellular mutational profiles, i.e., genomic instability and a concomitant genomic heterogeneity (diversity of these genomic

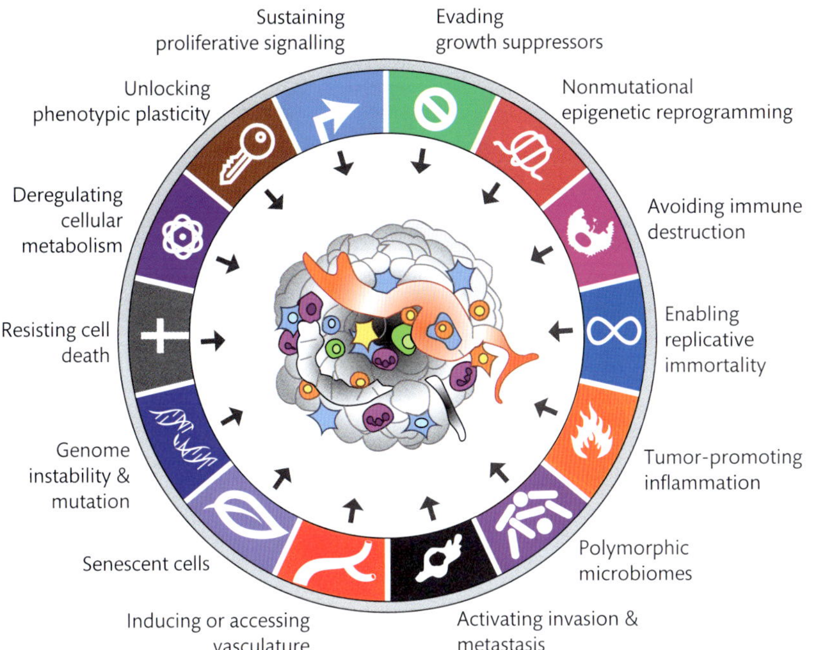

FIGURE 7.2
'The Hallmarks of Cancer' Hanahan (2022) adapted from Hanahan and Weinberg (2011).

changes), that in combination provide this competitive advantage to initiate malignant transformation, subsequently serving as the basis of progression and development of cancer. This genomic heterogeneity contributes to disease causation in several ways, as proposed by McClellan and King (2010): (1) individually rare mutations collectively play a critical role in complex illnesses; (2) the same gene may harbour many (hundreds or even thousands) of unique rare severe mutations in unrelated affected individuals; (3) the same mutation may lead to different clinical manifestations (phenotypes) in different individuals; and (4) mutations in different genes in the same or related pathways may lead to the same disease. In fact, in cancer, it is the analysis of not only the diagnostic specimen for prognosis, but routine monitoring of disease for progression (through the detection of additional aberrations, i.e., genomic heterogeneity), which may indicate the presence of subclone(s) that are genetically distinct from the primary tumour that confer different biological properties and therapeutic sensitivities—thus altering disease course and prognosis. Sequencing technologies are now embedded into routine clinical care and are providing valuable insights into the phylogenetic relationship of primary tumours, subclones, and metastatic development through the documentation of somatic tumour mutation profiles throughout the disease course in many cancers—which have added to the already expansive cytogenomic information discovered in previous decades.

As this clonal selection process occurs through multiple generations, the tumours acquire additional mutations that override the regulatory mechanisms, resulting in uncontrolled proliferation (see **Figure 7.3**). For example, a mutation that results in sustained proliferative signalling within a cell may drive the process of cell growth and division only to be stopped by cell checkpoint control mechanisms. However, the progression of the malignant transformation process occurs when this same cell acquires a second change, conferring resistance to cell death (apoptosis). The cell has a selective advantage and can succeed in the progression of the disease because it can grow and divide and is not stopped by internal controls that have activated the cell safety signals, which would normally signal an aberrant to activate cell-suicide pathways

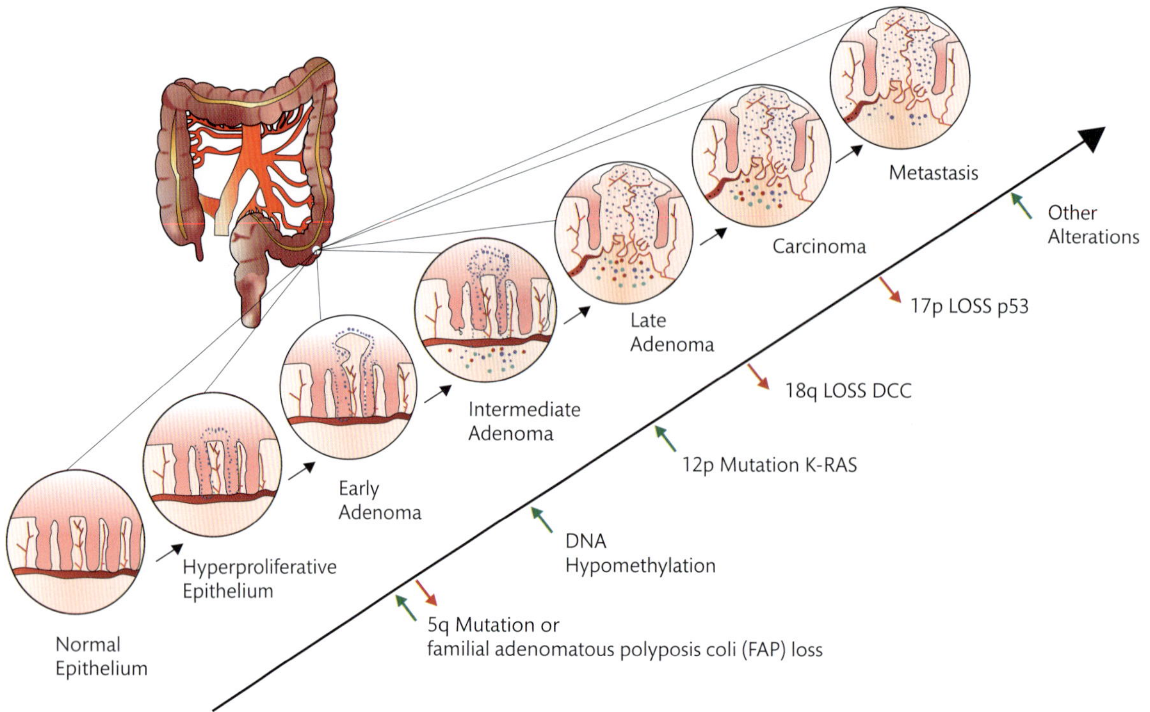

FIGURE 7.3

'A genetic model for colorectal tumourigenesis' Menter et al. (2019) adapted from Fearon and Vogelstein, *Cell*, 61, 759–767, 1990.

(cellular apoptosis) and ultimately, there is cell death. In **Figure 7.3**, a series of genetic aberrations contributes to tumourigenesis. As described in the figure, early aberrations contribute to disease development, such as alterations to chromosome 5q or adenomatous polyposis coli (*APC*) loss of function mutation associated with familial adenomatous polyposis coli (FAP). These mutations combine with alterations in oncogenes (e.g., KRAS proto-oncogene, GTPase (*KRAS*)) and tumour suppressor genes (particularly those on chromosomes 5q, 17p, and 18q). Subsequent events, such as DNA methylation (e.g., hypomethylation, i.e., loss of the methyl group), followed by additional alterations, e.g., loss of 18q or mutations in the DCC netrin 1 receptor gene (DCC) (formerly referred to as 'deleted in colorectal cancer'), follow and then 17p loss (alteration to the p53 protein) contribute to disease; subsequent mutations may precede the development of metastasis.

Aneuploidy
The presence of an abnormal number of chromosomes in a cell.

The cellular lineage affected by the aberrant activity can influence the cancer subtype and whether the disease is rapidly progressive. For example, acute leukaemia in young infants, which affects cells of the bone marrow, results from genetic aberrations including **aneuploidy**, chromosomal rearrangements, and gene mutations. In these children (< 12 months of age) with leukaemia, chromosomal rearrangements involving the *KMT2A* (Lysine (K) methyltransferase 2A) (formerly MLL (Mixed-Lineage Leukemia 1)) locus are believed to be the primary alteration required for initial cancer development. In fact, these *KMT2A* rearrangements occur *in utero* during foetal development, which disrupts cell cycle regulation and allows for the acquisition of additional mutations needed to drive rapid progression of leukaemia.

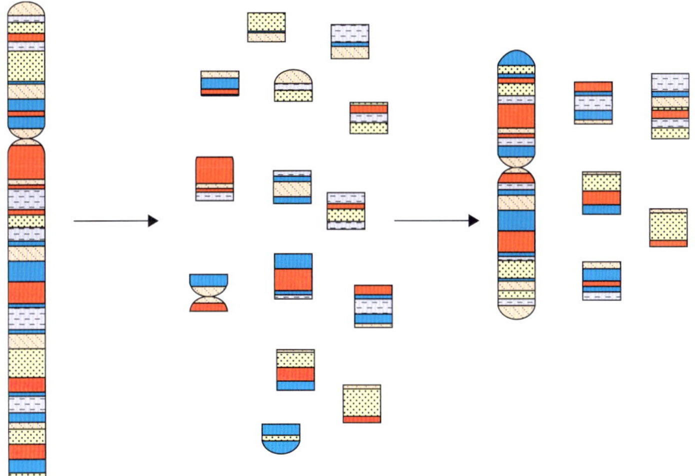

FIGURE 7.4
A schematic diagram of the chromothripsis process.

7.2.3 Chromothripsis

Chromothripsis is a cellular phenomenon that generates complex structural genomic rearrangements from a single catastrophic event. The name chromothripsis comes from the Ancient Greek 'chroma' or colour (so named to indicate the stained chromosomes when they were originally viewed under a microscope) and 'thripsis' for 'shattering into pieces'. During chromothripsis, chromosomes are shattered into many pieces, and subsequently reassembled to form mosaic chromosomes. Although a genetic event long known to occur, Stephens and colleagues first suggested the term 'chromothripsis' in 2011.

Chromothripsis
A phenomenon where chromosomes are shattered into many pieces and then reassembled incorrectly, leading to complex genetic rearrangements associated with cancer.

This finding challenged the textbook model of cancer development, as chromothripsis does not involve the stepwise progression of disease through the accumulation of multiple genetic changes before malignant transformation. Instead, this single event generates tens to hundreds of genomic rearrangements and copy number alterations at once. Interestingly, although many theories predicted that cells suffering from chromothripsis might be susceptible to signals that induce apoptosis, this reconfiguration of the genome may in reality allow for selective growth advantages for daughter cells, for example aberrations of the *MYC* proto-oncogene in the paediatric malignant brain tumour, medulloblastoma. *MYC* aberrations promote cell proliferation and highly aggressive tumours frequently harbour *MYC* gene amplification and/or protein overexpression. Look at **Figure 7.4** and notice how the normal chromosome example is utterly unrecognizable after this synthetic visual depicting a chromothripsis event.

The chromosomes are shattered and then placed back together incorrectly resulting in multiple copy number changes, which results in the gain and loss of genetic material.

7.3 Translational genomics

The utilization of genomic technologies in clinical practice has increased our understanding of the underlying molecular mechanisms through the identification of factors that have a role in the diagnosis, prognosis, and treatment of disease. Globally, specialized genetic services are provided by trained Medical Geneticists, Genetic Counsellors and/or Nurses, Laboratory Genetic Technologists, and Clinical Laboratory Geneticists to create a patient-focused care team. Although the first two groups are often involved in direct patient care, it is the work of the latter two, of equal importance, often hidden away in the laboratory, that generate genomic test results. There are ongoing global efforts to support the training and recognition of professionals in genetic medicine.

These specialized services are often performed at regional centres within networks of care providers. These centres use precision medicine-based oncology for personalized care through the application of advanced diagnostics, often created via an interface with the most recent advances in translational research.

7.3.1 Diagnostic methodology and technology

Due to the rapid development and presentation of many cancers, immediate and accurate diagnoses are necessary to manage these patients. This is highly dependent upon appropriate genetic classification in many cancers for appropriate diagnosis, prognosis, and treatment.

The detection of genetic changes can include gross chromosomal changes (i.e. copy number gains/loss, translocations, inversions), large deletion or insertion mutations (i.e. loss or duplication of more than one exon), point mutations (i.e. the substitution of a base) (and more) is completed in the clinical genomics laboratory using a variety of techniques discussed herein.

See **Table 7.3** for a summary of these technologies, their limitations, and application in diagnostic testing.

TABLE 7.3 Test methodologies in the clinical genomics laboratory

Methodologies	Technique(s)	Limitations	Application	Diagnostic Example
Cytogenetics (CG)	Karyotyping via G-banding	Detects large chromosomal changes at a low resolution (range of 5–10 Mb)	Detection of changes in chromosome number (copy number) and spatial arrangement (translocations, inversions)	Detects the t(9;22)(q34;q11) rearrangement involving the 'Philadelphia chromosome' in Chronic Myelogenous Leukaemia (CML)
Fluorescence *in situ* Hybridization (FISH)	Target testing of specific genetic sequence using fluorescent labels	Detects only the genomic loci under evaluation (range of 100–200 kb)	Microdeletions, cryptic changes below the resolution of karyotyping	Detects loss of *TP53* (17p13) in Chronic Lymphocytic Leukaemia (CLL)
Microarray Array CGH (aCGH)	Whole genome testing using thousands of DNA probes on an array	Detects copy number changes (<5 kb); unable to detect balanced changes	Detection of changes in chromosome number (copy number), cryptic changes below the resolution of karyotyping	Detects (using aCGH SNP) copy neutral-loss (cn-LOH) of heterozygosity of *TET2* (4q24) in Myelodysplastic Syndrome (MDS)
Polymerase Chain Reaction (PCR)	Targeted testing via amplification of specific sequence	Detects single base pair changes; unable to amplify large genes (>100kb)	Detects changes in DNA sequence (e.g., small insertions/deletions) of varying size, with very little template specimen	Detects (using real time PCR) large deletions in the *VHL* gene in Von Hippel-Lindau Syndrome (vHL)
Sequencing: Sanger and Next Generation Sequencing (NGS)	Sequencing of loci of interest or whole genomes (e.g., targeted gene panels, Whole Exome Sequencing (WES) or Whole Genome Sequencing (WGS))	Sanger: Very low error rate but limited by low throughput NGS: Higher error rate (>0.1%) but with high throughput	Detection of changes (e.g., translocations, large deletions) across the genome; coverage can range from 10 to >1000×.	WES analysis may identify rare pathogenic mutations (variants) in familial colon cancer without a previously known genetic cause. Sanger sequencing is used to confirm the variants identified by NGS.

7.3.2 Genome approaches

Conventional cytogenetic methods (i.e., **karyotyping**) via Giemsa staining was the original whole genome technique and remains a powerful diagnostic test. To visualize and identify chromosomes, a special combination of protein digestion and staining produces distinctive patterns for all chromosomes. This test involves the analysis of chromosomes under a light microscope, most often with a magnification of 1000× (modern laboratories utilize an automated microscope to capture hundreds of metaphase images for analysis!), to produce a karyogram, a display of the patient's karyotype.

Look at **Figure 7.5**, which shows an assembled **karyogram** produced from the analysis and pairing of the entire chromosome complement of a cell stopped at metaphase demonstrating normal (**Figure 7.5(a)**) and abnormal (**Figure 7.5(b)**) karyotypes.

Although, FISH can be used as a whole genome screening technique (via spectral karyotyping (SKY) and multiplex FISH (mFISH)), it is routinely used as a targeted technique in clinical diagnostics. FISH is performed via selection of probe(s) that are complementary to the target genomic sequence of interest: the probe is applied to a glass slide with patient material, and is subsequently denatured, hybridized, and washed prior to analysis. This test involves the analysis of interphase or metaphase chromosomes under a fluorescent microscope, most often with a DAPI (4′,6-diamidino-2-phenylindole) counterstain to enhance the visualization of colourful probe signals! **Figure 7.6** shows an interphase FISH result.

Karyotyping
A laboratory technique that visualizes chromosomes under a microscope to detect abnormalities in chromosome number and/or structure.

Karyogram
A diagram or a photograph of the chromosomes in a cell, arranged in homologous pairs and numbered sequence.

Fluorescence *in situ* hybridization (FISH)
A molecular cytogenetic technique that uses fluorescent probes to detect specific DNA sequences.

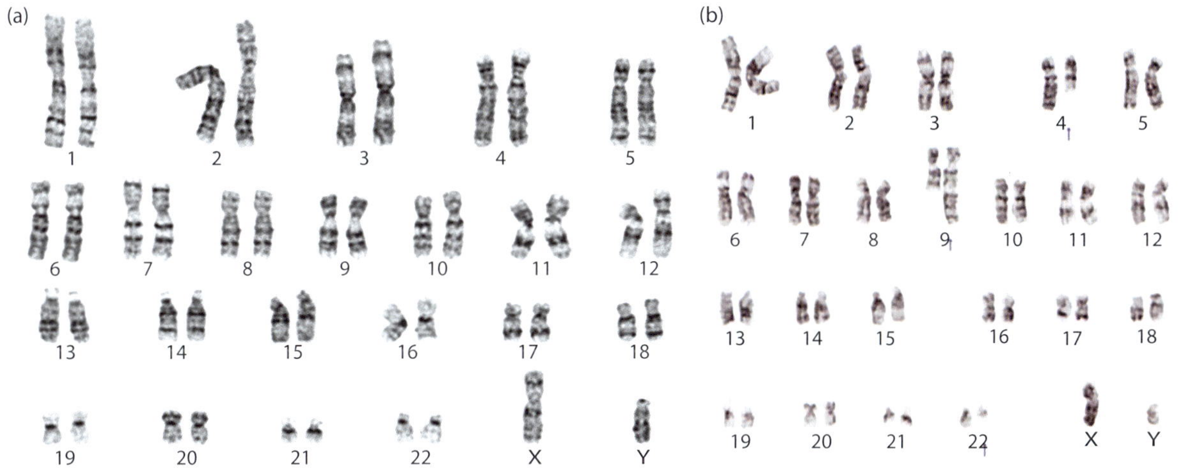

FIGURE 7.5
Normal karyogram (a) and abnormal karyogram (b), showing a four-way translocation. (a) Example karyotype prepared with the Giemsa-stained following Trypsin Treatment (GTG) procedure. The karyotype for this male individual according to the International System for Human Cytogenetic Nomenclature (ISCN) is 46,XY[20]; which indicates that all 20 cells analysed demonstrated a normal male chromosome complement. (b) Karyotype showing the abnormal chromosomes from a male with CML prepared with the GTG procedure ISCN: 46,XY,t(8;9;22;20)(p22;q34;q11.2;p13)[10]; which indicates that the 10 cells analysed contained a four-way translocation between the short arm of chromosome 8 (p22), the long arms of chromosomes 9 (q34) and 22 (q11.2), and the short arm of chromosome 20 (p13), leading to production of a Philadelphia chromosome t(9;22)(q34;q11) diagnostic of CML, which was subsequently confirmed to involve the *BCR::ABL1* (BCR activator of RhoGEF and GTPase:: ABL proto-oncogene 1, non-receptor tyrosine kinase, formally 'BCR-ABL1' or breakpoint cluster region-v-abl Abelson murine leukemia viral oncogene homolog 1) translocation via **fluorescence *in situ* hybridization (FISH)**.

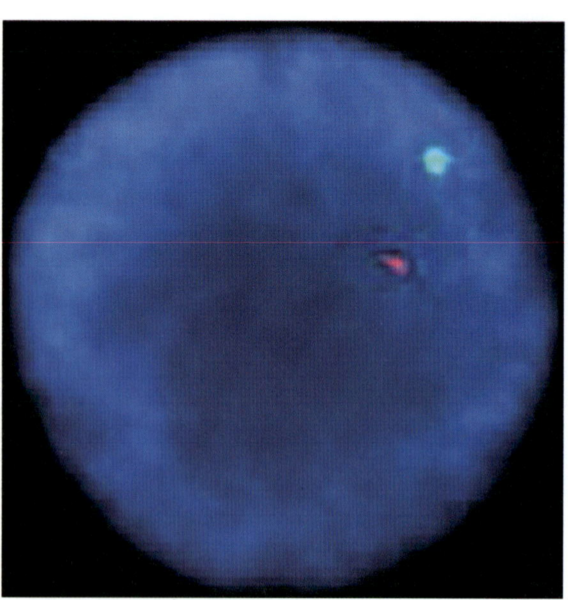

FIGURE 7.6
Interphase FISH analysis. Copy number FISH using the Abbott Molecular (Chicago, IL, USA) TP53/CEP17 probe indicates 1 red (R) *TP53* signal and 1 green (G) CEP17 signal. This is monosomy of chromosome 17 (which includes a loss of *TP53*). A normal cell would have a 2R2G pattern; two reds (one for each *TP53*) and two greens (one for each chromosome 17 centromere).

Advanced molecular techniques in clinical use can provide higher resolution whole genome analysis as compared to karyotyping. Introduced in the 1990s, microarray technologies advanced cytogenetic techniques by allowing for copy number and sequence analysis while addressing some of the limitations of G-banding (e.g., required cultured cells) and FISH (e.g., higher resolution but limited to selected loci). These techniques are based on the comparison of differences between a specimen and reference specimen(s) to provide relevant clinical information. **Array comparative genomic hybridization (aCGH)** and microarray with single nucleotide polymorphism (SNP-A) are commonly used in the clinical genomics laboratory.

Array comparative genomic hybridization (aCGH)
A technique that analyses the entire genome for copy number aberrations by comparing the tumour DNA to a reference genome.

The aCGH technology is based on slides ('arrays') that contain millions of short sequences of DNA linked to their surface for 'target' sequences of interest (i.e., normal human genetic sequences) in a precise grid, often referred to as an array chip. The resolution of the array is the ability to detect aberrations of a given size, e.g., resolution of 25 kb indicates that the system is capable of detecting 'target' aberrations from the size of 25 kb. Several factors contribute to determining the resolution of arrays, and the number of target probes is only one of these factors. It is the spacing between these target probes, not just the number, that help determine resolution, i.e., many array platforms do not space the target probes uniformly, but instead concentrate the probes within regions of greatest interest (e.g., mutation hotspots) and in other platforms, these target probes may be replicated to increase diagnostic confidence when an abnormality is detected. As the space between the target probes on the array decreases, the array is able to detect smaller aberrations, such as micro- deletion and amplifications of the genome. To analyse a specimen, DNA from the patient is 'digested' (i.e., cut up into small fragments), and in combination with a reference DNA specimen (from an individual or pool of individuals with no genetic abnormalities), these digested fragment 'DNAs' are labelled with unique fluorescent dyes and hybridized to the array. After washing and staining, a scanner measures the intensity of the fluorescence at each position comparatively between patient and reference, to identify relative gains or losses at a given locus. Specialized software then displays these changes using a log2 ratio derived from the differences (log2 ratio = $\log2_{sample} - \log2_{reference}$).

Polymorphisms
Common variations in the DNA sequence that may or may not influence health or contribute to disease.

Copy number variations (CNVs)
Alterations in the number of copies of a particular gene or DNA segment in the genome.

Both single nucleotide **polymorphism** arrays (SNP-A) and aCGH also have the ability to detect **copy number variations (CNVs)**, but SNP-A has the added ability to detect copy-neutral loss of heterozygosity (cn-LOH), a frequent event in the development of malignancy. In SNP-A,

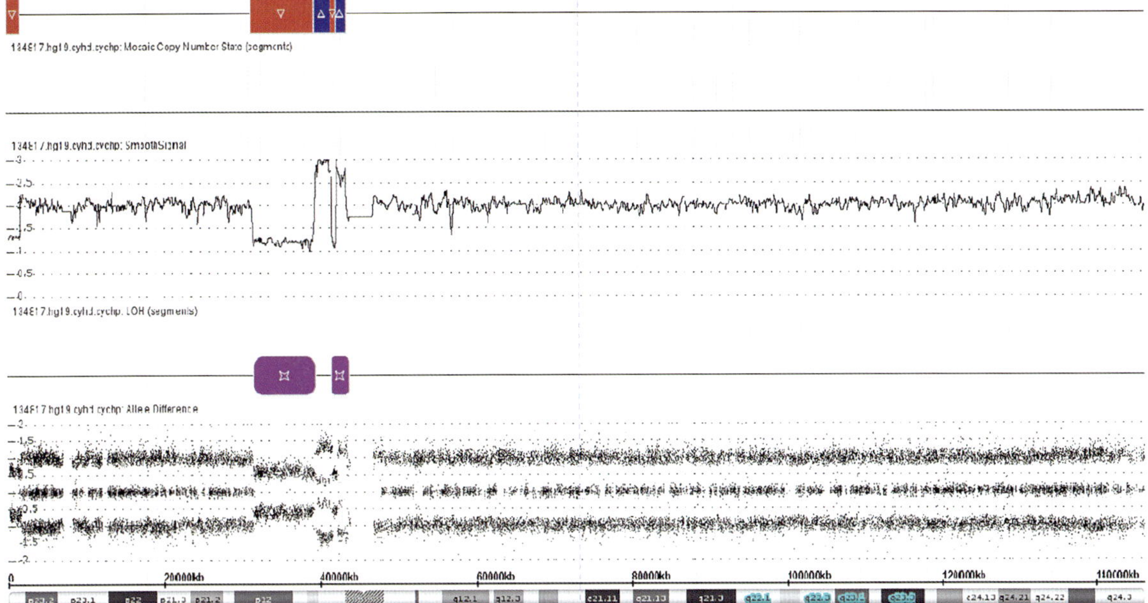

FIGURE 7.7

Copy number changes of chromosome 8. Look at the blue (gains) and red (losses) boxes generated by the Chromosome Analysis Suite Software (ChAS) (Affymetrix/ThermoFisher Scientific) that correlate with the respective changes in the log2 ratio. Deletions are associated with loss of signal intensity consistent with a decrease in log2 ratio. Gains are associated with gain of signal intensity consistent with an increase in log2 ratio. Log2 transformation makes it easier to visualize the data as it compresses the range of ratios by centering the data around zero. Courtesy of Dr. Azim Mohamedali (King's College Hospital, Laboratory for Molecular Haemato-Oncology (LMH)).

genomic DNA (gDNA) must be isolated, digested, and ligated to adapters to perform a multiplex PCR reaction. These PCR products are purified and fragmented; the DNA product is end labelled with a modified biotinylated base and hybridized to the array. Again, washing, staining, scanning, and analysis follows with specialized software; however, the copy number evaluation of each allele is generated through comparison of the signal intensity to a known reference. This signal intensity in SNP-A generates two output values: the SNP (e.g., A, B, or AB) and copy number status (using log2 ratio).

Figure 7.7 shows the results of SNP-A. This display is only of copy number changes on chromosome 8.

SELF-CHECK 7.6

What is a limitation of aCGH when compared to karyotyoping and vice versa?

Introduced by Mullis in 1987, polymerase chain reaction (PCR) is a widely used technique to amplify specific sequences of DNA. This technique relies on the complementary base pairing, double-stranded structure, and melting temperature of DNA to amplify the target sequence via an enzymatic reaction. Firstly, a template is required along with primers (complementary to the sequence of interest), nucleotides Adenine (A), Cytosine (C), Thymine (T), and Guanine (G), as 2′-deoxynucleotide triphosphates (dNTPs), and DNA polymerase combined in a mixture. Using a specially designed thermocycler, the mix is cycled through three stages via alternating the temperatures between 90–97°C, 50–65°C, and 75–80°C: 1) denaturation (melting), 2) annealing, 3) elongation (replication) stages. During denaturation, the heat results in disruption of the hydrogen bonds that hold together the two strands of DNA, the lowering of the temperature during annealing follows as the complementary bases between template and primer combine, and finally, as the temperature increases the activity of the polymerase does so in tandem to allow for DNA replication. During this period, new DNA strands are synthesized as DNA polymerase assembles the nucleotides in the 3′ to 5′ direction to produce the result: two complete complementary strands of DNA. This new DNA is identical to the starting template DNA and will be recycled as the basis of additional cycles of PCR to generate millions of copies of the target. As these cycles continue, the amplification of DNA increases exponentially until no template reagents remain in the reaction. The results of a PCR reaction are often visualized via gel electrophoresis, a methodology that uses an electrical field across agarose gel to separate negatively charged nucleic acids by size (i.e., negatively charged DNA will move based on weight towards the positively charged electrode, e.g., smaller DNA fragments will move quickly through the gel when compared to larger DNA fragments, which results in arrangement of the nucleic acids by size that can be selected for from the gel).

The first generation of sequencing technologies, Sanger sequencing, introduced in 1977, a now modernized technique (we no longer use radioactive labelling systems), continues to be used in the laboratory. This technique involves the selective incorporation of chain terminating deoxynucleotides (which results in chain termination) by DNA polymerase during *in vitro* DNA replication. Firstly, Sanger sequencing starts with the denaturation of double-stranded DNA (dsDNA) into single-stranded DNA (ssDNA). This ssDNA is subsequently annealed to oligonucleotide primers, using a mixture of necessary nucleotides Adenine (A), Cytosine (C), Thymine (T), and Guanine (G), as 2′-deoxynucleotide triphosphates (dNTPs) and chain terminating (because they lack the 3′ hydroxyl group needed for DNA chain extension) fluorescent marked 2′,3′-dideoxynucleotide triphosphates (ddNTPs), a new dsDNA is built by DNA polymerase (e.g., ddATP to the A reaction). The reaction continues with extension of the DNA structure using dNTPs until a ddNTP is attached. During this process, varying lengths of created DNA sequence are generated. Today, an automated capillary electrophoretic machine with a laser can detect these fluorescent intensities as a 'trace' or 'peak'.

Key Points

Previously, Sanger sequencing was performed with four separate reactions (one for each base) using radioactively labelled products that were separated based on molecular mass using a polyacrylamide gel.

Generally, A is green, C is blue, T is red, and G is black in the display of these trace results on the electropherogram. The presence of a homozygous mutation (variant) will result in replacement of the expected fluorescent colour by the new base pair colour; in the presence of a heterozygous mutation (variant) there will be two peaks with equal intensity. However, this sequence assay is labour intensive and cannot be scaled up when there are a high number of genomic targets. **Figure 7.8** shows the results of Sanger sequencing with a mutation.

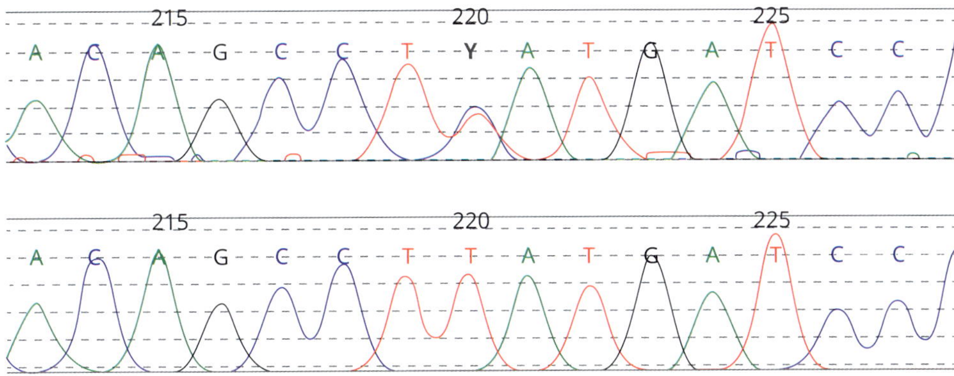

FIGURE 7.8

Sanger sequencing for variant confirmation. A sequence of concern visualized on the Mutation Surveyor software (SoftGenetics, State College, PA, USA). The forward sequencing reaction (top) from the patient versus the wild type result (bottom) are shown. Note the heterozygous variant (C/T) (noted above with a grey Y) in the patient versus the T in the reference. In this case, there are different nucleotides (C and T) in each allele.

Instead, the successor to Sanger sequencing, Next Generation Sequencing (NGS) or (massively parallel sequencing) platforms allow for the generation of large amounts of DNA sequence data that can be produced rapidly and are cost effective. NGS tests include disease-targeted multigene panels (where a select group of genes known to be associated with a disease are tested e.g., *ASXL*, *EZH2*, *GATA2*, *TP53* in haematological malignancies of the myeloid lineage), whole exome sequencing (the coding regions of the genome are tested) and **whole genome sequencing** (the coding and non-coding regions of the genome are tested, with some exceptions due to technical limitations of sequencing technology) in the clinical laboratory.

Whole genome sequencing
A method for determining the complete DNA sequence of an organism's genome, used to identify genetic variations associated with diseases.

One of the most popular NGS platforms—Illumina—uses the incorporation of reversible dye terminators that allow for the identification of single bases during incorporation into DNA strands. The procedure starts with fragmented genomic DNA (~200–600 bps). There is ligation of adapters containing terminal sequences, primer binding sites, and index sites to both ends of the DNA (each fragment has two different adapters attached). These created adapter-ligated DNA molecules are washed across onto a flow cell on the sequencing machine, and the terminal attached sequences subsequently bind to immobilized oligonucleotides on flow cell surface. Using a localized PCR, the DNA attached to the flow cell is replicated to create DNA clusters. Unlabelled nucleotide bases and DNA polymerase are added to the replicated sequences to create longer DNA that can create bridges of dsDNA between dsDNA and primers on the flow cell surface. After heating, the dsDNA is denatured into ssDNA, and primers and the fluorescently labelled ddNTPs (A, T, C, and G, chain terminators) are added. DNA polymerase binds to the primer and subsequently incorporates bases; however, once a fluorescently labelled terminator nucleotide is added, strand extension ceases until this base is removed from the DNA strand. Lasers are subsequently passed over the flow cell to activate fluorescence signals with unique colours for each base. After the signals are read and recorded by the computer, dye and terminator are chemically removed from the DNA strand, and the cycle repeats to generate millions of base-by-base sequences. Following the alignment of the data to a reference sequence via bioinformatic processing, scientists look for changes in the sequenced product. **Figure 7.9** shows the results of next generation sequencing, showing a genetic mutation.

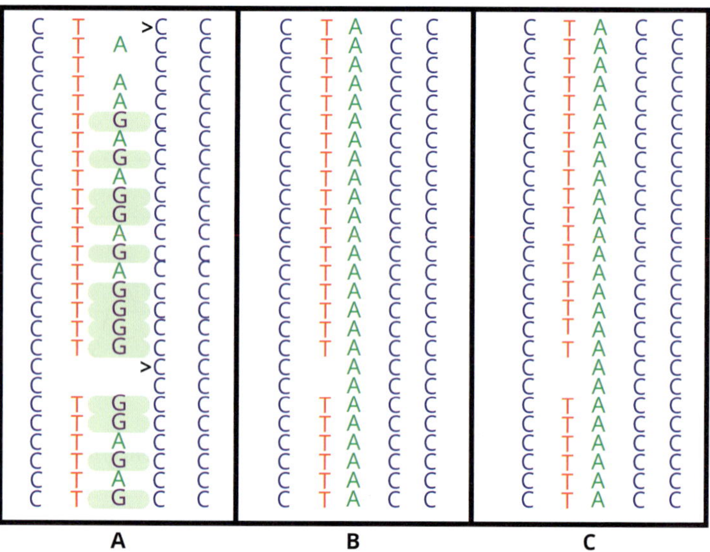

FIGURE 7.9
Sequence alignment showing a mutation. The next generation sequencing of the proband (patient) (left, A). Next generation sequencing of the mother (middle, B). Next generation sequencing of the father (right, C). Note the heterozygous change (A/G) (highlighted) in the proband versus the A in both parents. In this case, the proband carries a heterozygous mutation (variant), that is not detected in the mother or father.

SELF-CHECK 7.7

What whole genome techniques are routinely used clinically?

7.3.3 Prognosis and therapeutics

Several diagnostic methodologies are available to inform on the disease course including predicted outcomes (prognosis) and the response to specific therapeutic agents. In diseases with heterogeneity, identifying and classifying patients into subgroups can be strong predictive markers critically guiding treatment decisions.

In CLL the presence of specific chromosomal aberrations (del(11q), del(13q), del(17p), and trisomy 12) and specific gene mutations (*IGHV, TP53, NOTCH1*) can be used to generate subgroups of patients and to select therapy. For example, karyotyping analysis may reveal the three or more chromosomal aberrations, referred to as a 'complex karyotype' and this has been associated with poor patient outcomes. In addition, the *IGHV* (immunoglobulin heavy chain variable) region mutation status affects prognosis of the disease. Patients with unmutated *IGHV* often have an aggressive disease course as compared to those with mutated *IGHV*, which is associated with an indolent course. Finally, *TP53* mutations in CLL are associated with chemotherapy-refractory disease as patients generally respond poorly to the standard first-line chemotherapy—as such this has led to guidance that chemoimmunotherapy should no longer be considered a standard therapy for patients harbouring a *TP53* mutation. Instead, mutation status may guide therapy selections as patients may be offered other drugs such as Venetoclax, which acts by preventing one of the hallmarks of cancer development, the acquired resistance to cell death (apoptosis). It is therapies such as this aimed at circumventing cancers' attempts to hijack cellular processes that show promise to effectively treat the disease.

CASE STUDY 7.1 Solid tumour

A 27-year-old woman detected a lump in her right breast; she visited her GP who ordered imaging studies (i.e., a mammogram), which detected a tumour. The tumour was removed from the breast. The excision specimen was reviewed by a pathologist and classified as a stage II breast cancer. Specimen was sent to the clinical genomic laboratory for fluorescence *in situ* hybridization (FISH) studies to assist in the molecular classification of breast cancer.

Figure 7.10 shows interphase nuclei of the tumour specimen. Notice how the red signals appear 'cloud-like' with few green signals; this is amplification of HER2.

Q1 Can you make a preliminary classification of the molecular HER2 phenotype? What are the abnormalities?

Q2 What gene(s) might be requested for sequencing in this patient?

Q3 Why is family history information critical for this patient?

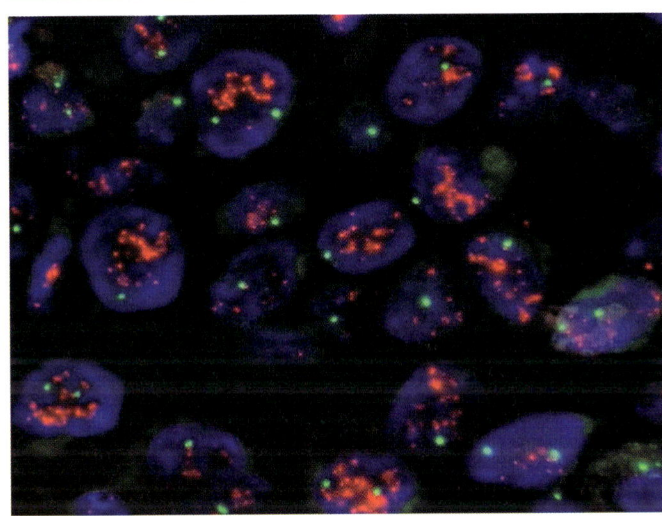

FIGURE 7.10
FISH using the Metasystems (Altlußheim, Germany) XL ERBB2 (HER2)/CEP17 probe indicates ~20 red HER2 signals and 1–2 green CEP17 signals.

CASE STUDY 7.2 Haematological cancer

A 71-year-old man with recent involuntary weight loss, fatigue, and fever visited his GP. The GP examined him, noticed his enlarged lymph nodes, and immediately ordered a full blood count. The FBC indicated lymphocytosis, the blood film indicated the presence of mostly small lymphocytes with few larger cells with a prominent nucleolus (prolymphocytes). A specimen was sent to the clinical genomics laboratory for SNP-A and sequencing studies to assist in the molecular classification of Chronic Lymphocytic Leukaemia (CLL) (**Figure 7.11**).

Q1 What catastrophic cellular event might the frequent oscillations of gene copy number in chromosome 15 detected in the array indicate?

Q2 What gene(s) might be requested for sequencing in this patient?

Q3 Why is sequencing of *TP53* critical for therapy selection for the patient?

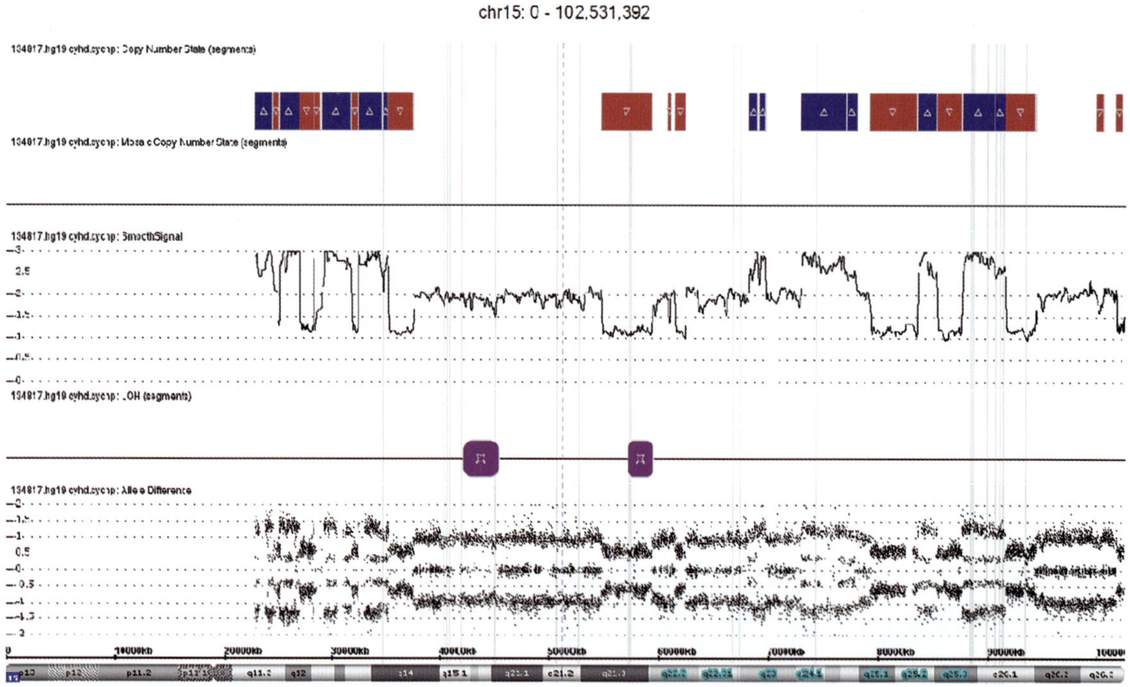

FIGURE 7.11
Chromosomal Microarray: aCGH SNP analysis in CLL. Multiple copy number changes of chromosome 15. Blue (gains) and red (losses) boxes generated by the Chromosome Analysis Suite Software (ChAS) (Affymetrix/ThermoFisher Scientific) correlate with the respective changes in the log2 ratio.

Chapter summary

- Cancer is a genomic disorder characterized by dysregulation of cellular proliferation.
- Cancer has been described for millennia but we are just starting to unravel all the processes that can result in malignant transformation.
- Cancer results from germline (inherited) and somatic (acquired) changes.
- Cancer development is a multi-step process selective for unchecked proliferation, survival, invasion, and metastasis.
- The inheritance of a germline mutation may result in increased susceptibility to single or multiple cancers.
- The 'two-hit' hypothesis is a unifying genetic model that can be used to explain both heritable and sporadic cancers.

- The 'two-hit' hypothesis has evolved and includes many exceptions that result in cancer including haploinsufficiency, dominant negative effects, and pleiotropy.
- 'Shattering' of the genome or 'chromothripsis' can lead to catastrophic consequences including malignancy.
- The use of genomic platforms allows for the molecular stratification of many cancers.

Discussion questions

7.1 Why is the clinical genomics laboratory important in investigating cancer? Describe the types of tests performed in this laboratory.

7.2 What are the 'Hallmarks of Cancer'?

7.3 What is an oncogene? Find one in the chapter and explain how it is involved in the development of a given cancer.

7.4 What is a tumour suppressor gene? Find one in the chapter and explain how it is involved in the development of a given cancer.

7.5 Describe how genomic instability plays a role in the development of cancer.

Further reading

- Berger MF, Mardis ER (2018) ***The emerging clinical relevance of genomics in cancer medicine***. *Nat Rev Clin Oncol.*, 15(6), 353–365. doi:10.1038/s41571-018-0002-6

  'Presents a very thorough account of the role of genomics-based assays in the characterization of germline and somatic changes in tumour specimens.'

- Chen HZ, Bonneville R, Roychowdhury S (2019) ***Implementing precision cancer medicine in the genomic era***. *Semin Cancer Biol.*, 55, 16–27. doi:10.1016/j.semcancer.2018.05.009

  'Presents a very thorough account of how the identification of actionable genomic alterations is directing precision oncology care.'

- Vogelstein B, Kinzler KW (2004) ***Cancer genes and the pathways they control***. *Nat Med.*, 10(8), 789–799. doi:10.1038/nm1087

  'Presents a very thorough account of genetic events and the resulting aberrant pathways that contribute to the development of cancer.'

- Vogelstein B, Papadopoulos N, Velculescu VE, Zhou S, Diaz LA Jr, Kinzler KW (2013) ***Cancer genome landscapes***. *Science*, 339(6127), 1546–1558. doi:10.1126/science.1235122

  'Reviews the recently revealed genomic landscapes including lists of commonly mutated genes.'

- Nowell PC, Rowley JD, Knudson AG Jr (1998) ***Cancer genetics, cytogenetics—defining the enemy within***. *Nat Med.*, 4(10), 1107–1111. doi:10.1038/2598

  'Presents the history of fundamental discoveries in genetics.'

- Anderson MA, Deng J, Seymour JF, *et al*. (2016) ***The BCL2 selective inhibitor venetoclax induces rapid onset apoptosis of CLL cells in patients via a TP53-independent mechanism***. *Blood*, 127(25), 3215–24. doi:10.1182/blood-2016-01-688796
- Aoki K, Taketo MM (2007) ***Adenomatous polyposis coli (APC): a multi-functional tumor suppressor gene.*** *J Cell Sci.,* 120(Pt 19), 3327–35. doi:10.1242/jcs.03485
- Bagacean C, Tempescul A, Ternant D, *et al*. (2019) ***17p deletion strongly influences rituximab elimination in chronic lymphocytic leukemia***. *J Immunother Cancer*, 7(1), 22. doi:10.1186/s40425-019-0509-0
- Balmain A, Gray J, Ponder B (2003) ***The genetics and genomics of cancer***. *Nat Genet*, 33 Suppl, 238–44. doi:10.1038/ng1107
- Bellanger M, Zeinomar N, Tehranifar P, Terry MB (2018) ***Are global breast cancer incidence and mortality patterns related to country-specific economic development and prevention strategies?*** *J Glob Oncol*, 4, 1–16. doi:10.1200/JGO.17.00207
- Ben-Skowronek I, Kozaczuk S (2015) ***Von Hippel--Lindau Syndrome***. *Horm Res Paediatr*, 84(3), 145–52. doi:10.1159/000431323
- Bien SA, Peters U (2019) ***Moving from one to many: insights from the growing list of pleiotropic cancer risk genes***. *Br J Cancer*, 120(12), 1087–9. doi:10.1038/s41416-019-0475-9
- Bodmer WF (2006) ***Cancer genetics: colorectal cancer as a model.*** *J Hum Genet*, 51(5), 391–6. doi:10.1007/s10038-006-0373-x
- Bose R, Kavuri SM, Searleman AC, *et al*. (2013) ***Activating HER2 mutations in HER2 gene amplification negative breast cancer***. *Cancer Discov*, 3(2), 224–37. doi:10.1158/2159-8290.CD-12-0349
- Breasted JH, editor (1930) ***The Edwin Smith Surgical Papyrus***. The University of Chicago Press, Chicago, Illinois. Special edition, 1984.
- Bunz, F, editor (2016) ***Principles of Cancer Genetics***. 2nd Edition, Springer, New York.
- Campo E, Cymbalista F, Ghia P, et al. (2018) ***TP53 aberrations in chronic lymphocytic leukemia: an overview of the clinical implications of improved diagnostics***. *Haematologica*, 103(12), 1956–8. doi:10.3324/haematol.2018.187583
- Chernoff J (2017) ***Alfred G Knudson Jr, MD PhD: In Memoriam (1922–2016)***. *Cancer Res*, 77(4), 815–6. doi:10.1158/0008-5472.CAN-16-3547
- Correa H (2016) ***Li–Fraumeni Syndrome***. *J Pediatr Genet*, 5(2), 84–8. doi:10.1055/s-0036-1579759
- Cortés-Ciriano I, Lee JJ, Xi R, et al. (2020) ***Comprehensive analysis of chromothripsis in 2,658 human cancers using whole-genome sequencing [published online ahead of print, 2020 Feb 5]***. *Nat Genet*. 10.1038/s41588-019-0576-7. doi:10.1038/s41588-019-0576-7
- Cooper GM (2018) ***Cancer***. In: *The Cell: A Molecular Approach*. 8th Edition, Oxford University Press, New York.
- Dharmawardana PG, Giubellino A, Bottaro DP (2004) ***Hereditary papillary renal carcinoma type I***. *Curr Mol Med*, 4(8), 855–68. doi:10.2174/1566524043359674

- Dutzmann CM, Vogel J, Kratz CP, Pajtler KW, Pfister SM, Dörgeloh BB (2019) **Ein Update zum Li-Fraumeni-Syndrom [Update on Li-Fraumeni syndrome]**. *Pathologie*, 40(6), 592–9. doi:10.1007/s00292-019-00657-y
- Ebenazer A, Rajaratnam S, Pai R (2013) **Detection of large deletions in the VHL gene using a Real-Time PCR with SYBR Green**. *Fam Cancer*, 12(3), 519–24. doi:10.1007/s10689-013-9606-2
- El-Khoury R, Hajj M, Khraibani J, Audi E, Monsef C, Farra C (2019) **Novel pleiotropic BRCA2 pathogenic variants in Lebanese families**. *Cancer Genet*, 231–232, 32–5. doi:10.1016/j.cancergen.2018.12.005
- Esteban-Jurado C, Vila-Casadesús M, Garre P, et al. (2015) **Whole-exome sequencing identifies rare pathogenic variants in new predisposition genes for familial colorectal cancer**. *Genet Med.*, 17(2), 131–42. doi:10.1038/gim.2014.89
- Evans DG, Birch JM, Narod SA (2008) **Is CHEK2 a cause of the Li-Fraumeni syndrome?** *J Med Genet*, 45(1), 63–4. doi:10.1136/jmg.2007.054700
- Foucault M (1970) **The Order of Things: An Archaeology of the Human Sciences.** Tavistock Publications, London.
- Gomez A, Korf B. (2018) **Genetic Testing Techniques in Pediatric Cancer Genetics**. 1st Edition, Elsevier, St. Louis, Missouri.
- Goss KH, Groden J (2000) **Biology of the adenomatous polyposis coli tumor suppressor**. *J Clin Oncol*, 18(9), 1967–79. doi:10.1200/JCO.2000.18.9.1967
- Haas NB, Nathanson KL (2014) **Hereditary kidney cancer syndromes**. *Adv Chronic Kidney Dis*, 21(1), 81–90. doi:10.1053/j.ackd.2013.10.001
- Hagemann I (2015) **Overview of Technical Aspects and Chemistries of Next Generation Sequencing in Clinical Genomics**. Elsevier, London.
- Hammond EC, Horn D (1954) **The relationship between human smoking habits and death rates: a follow-up study of 187,766 men**. *J Am Med Assoc*, 155(15), 1316–28. doi:10.1001/jama.1954.03690330020006
- Hegde M, Ferber M, Mao R, Samowitz W, Ganguly A (2014) **Working Group of the American College of Medical Genetics and Genomics (ACMG) Laboratory Quality Assurance Committee. ACMG technical standards and guidelines for genetic testing for inherited colorectal cancer (Lynch syndrome, familial adenomatous polyposis, and MYH-associated polyposis)**. *Genet Med*, 16(1), 101–16. doi:10.1038/gim.2013.166
- Hogan AJ (2013) **Locating genetic disease: the impact of clinical nosology on biomedical conceptions of the human genome (1966–1990)**. *New Genetics and Society*, 32(1), 78–96. doi:10.1080/14636778.2012.735855
- Hucklenbroich P (2014) **'Disease entity' as the key theoretical concept of medicine**. *J Med Philos*, 39(6), 609–33. doi:10.1093/jmp/jhu040
- Huszno J, Grzybowska E (2018) **TP53 mutations and SNPs as prognostic and predictive factors in patients with breast cancer.** *Oncol Lett*, 16(1), 34–40. doi:10.3892/ol.2018.8627
- Inoue K, Fry EA (2017) **Haploinsufficient tumor suppressor genes**. *Adv Med Biol*, 118, 83–122.
- Janitz, M. editor (2008) **Next-Generation Genome Sequencing: Towards Personalized Medicine**. 1st Edition, Wiley-Blackwell, Germany.

- Kalkat M, De Melo J, Hickman KA, et al. (2017) ***MYC deregulation in primary human cancers.*** ***Genes (Basel)***, 8(6), 151. doi:10.3390/genes8060151
- Kamihara J, Bourdeaut F, Foulkes WD, et al. (2017) ***Retinoblastoma and neuroblastoma predisposition and surveillance***. *Clin Cancer Res*, 23(13), e98–e106. doi:10.1158/1078-0432.CCR-17-0652
- Kleinerman RA, Schonfeld SJ, Tucker MA (2012) ***Sarcomas in hereditary retinoblastoma***. *Clin Sarcoma Res*, 2(1), 15. doi:10.1186/2045-3329-2-15
- Kratz CP, Achatz MI, Brugières L, et al. (2017) ***Cancer screening recommendations for individuals with Li-Fraumeni Syndrome***. *Clin Cancer Res*, 23(11), e38–e45. doi:10.1158/1078-0432.CCR-17-0408
- Kumar P, Gill RM, Phelps A, Tulpule A, Matthay K, Nicolaides T (2018) ***Surveillance screening in Li-Fraumeni Syndrome: raising awareness of false positives***. Cureus, 10(4), e2527. doi:10.7759/cureus.2527
- Lakhtakia R (2014) ***A brief history of breast cancer: Part I: Surgical domination reinvented***. *Sultan Qaboos Univ Med J*, 14(2), e166–e169.
- Lawrenson K, Kar S, McCue K, et al. (2016) ***Functional mechanisms underlying pleiotropic risk alleles at the 19p13.1 breast–ovarian cancer susceptibility locus***. *Nat Commun*, 7, 12675. https://doi.org/10.1038/ncomms12675
- Libby W (1922) ***The History of Medicine in its Salient Features***. Houghton Mifflin Company, Boston/New York.
- Liehr T, Carreira IM, Balogh Z, et al. (2019) ***Regarding the rights and duties of Clinical Laboratory Geneticists in genetic healthcare systems; results of a survey in over 50 countries***. *Eur J Hum Genet*, 27(8), 1168–74. doi:10.1038/s41431-019-0379-4
- Lorenz TC (2012) ***Polymerase chain reaction: basic protocol plus troubleshooting and optimization strategies***. *J Vis Exp*, 63, e3998. doi:10.3791/3998
- Luzón-Toro B, Fernández RM, Villalba-Benito L, Torroglosa A, Antiñolo G, Borrego S (2019) ***Influencers on thyroid cancer onset: molecular genetic basis***. *Genes (Basel)*, 10(11), 913. doi:10.3390/genes10110913
- Luzzatto L, Pandolfi PP (2015) ***Causality and chance in the development of cancer***. *N Engl J Med*, 373(1), 84–8. doi:10.1056/NEJMsb1502456
- Lynch HT, Snyder CL, Shaw TG, Heinen CD, Hitchins MP (2015) ***Milestones of Lynch syndrome: 1895–2015***. *Nat Rev Cancer*, 15(3), 181–94. doi:10.1038/nrc3878
- Ma X, Shao Y, Tian L, et al. (2019) ***Analysis of error profiles in deep next-generation sequencing data***. *Genome Biol*, 20(1), 50. doi:10.1186/s13059-019-1659-6
- Makki J (2015) ***Diversity of breast carcinoma: histological subtypes and clinical relevance***. *Clin Med Insights Pathol*, 8, 23–31. doi:10.4137/CPath.S31563
- Marcozzi A, Pellestor F, Kloosterman WP (2018) ***The genomic characteristics and origin of chromothripsis***. *Methods Mol Biol*, 1769, 3–19. doi:10.1007/978-1-4939-7780-2_1
- McKusick VA (1966) ***Mendelian Inheritance in Man; Catalogs of Autosomal Dominant, Autosomal Recessive, and X-Linked Phenotypes***. Johns Hopkins Press, Baltimore.

- Miranda-Filho A, Piñeros M, Ferlay J, Soerjomataram I, Monnereau A, Bray F (2018) **Epidemiological patterns of leukaemia in 184 countries: a population-based study**. *Lancet Haematol*, 5(1), e14–e24. doi:10.1016/S2352-3026(17)30232-6

- Mitri Z, Constantine T, O'Regan R (2012) **The HER2 receptor in breast cancer: pathophysiology, clinical use, and new advances in therapy**. *Chemother Res Pract*, 2012, 743193. doi:10.1155/2012/743193

- Mohamedali AM, Smith AE, Gaken J, et al. (2009) **Novel TET2 mutations associated with UPD4q24 in myelodysplastic syndrome**. *J Clin Oncol*, 27(24), 4002–4006. doi:10.1200/JCO.2009.22.6985

- Muleris M, Salmon RJ, Zafrani B, Girodet J, Dutrillaux B (1985) **Consistent deficiencies of chromosome 18 and of the short arm of chromosome 17 in eleven cases of human large bowel cancer: a possible recessive determinism**. *Ann Genet*, 28(4), 206–13.

- Narod SA, Iqbal J, Giannakeas V, Sopik V, Sun P (2015) **Breast cancer mortality after a diagnosis of ductal carcinoma in situ**. *JAMA Oncol*, 1(7), 888–896. doi:10.1001/jamaoncol.2015.2510

- Peterson JF, Aggarwal N, Smith CA, *et al.* (2015) **Integration of microarray analysis into the clinical diagnosis of hematological malignancies: How much can we improve cytogenetic testing?** *Oncotarget*, 6(22), 18845–62. doi:10.18632/oncotarget.4586

- Raghavendra P, Pullaiah T (2018) **Pathogen Identification Using Novel Sequencing Methods in Advances in Cell and Molecular Diagnostics**. 1st Edition. Elsevier, London.

- Rednam SP, Erez A, Druker H et al. (2017) **Von Hippel-Lindau and hereditary pheochromocytoma/paraganglioma syndromes: clinical features, genetics, and surveillance recommendations in childhood**. *Clin Cancer Res*, 23(12), e68–e75. doi:10.1158/1078-0432.CCR-17-0547

- Rekhtman N, Baine M, Bishop J (2019) **Tumor Syndromes.** In: Rekhtman N, Baine M, Bishop J (eds), *Quick Reference Handbook for Surgical Pathologists*. 2nd Edition, Springer, Switzerland, pp. 133–44. doi.org/10.1007/978-3-319-97508-5_12

- Roberts KG (2018) **Genetics and prognosis of ALL in children vs adults**. *Hematology Am Soc Hematol Educ Program*, 2018(1), 137–45. doi:10.1182/asheducation-2018.1.137

- Rothblum-Oviatt C, Wright J, Lefton-Greif MA, McGrath-Morrow SA, Crawford TO, Lederman HM (2016) **Ataxia telangiectasia: a review**. *Orphanet J Rare Dis*, 11(1), 159. doi:10.1186/s13023-016-0543-7

- Roy R, Chun J, Powell SN (2011) **BRCA1 and BRCA2: different roles in a common pathway of genome protection**. *Nat Rev Cancer*, 12(1), 68–78. doi:10.1038/nrc3181

- Santana Dos Santos E, Lallemand F, Burke L, et al. (2018) **Non-coding variants in BRCA1 and BRCA2 genes: potential impact on breast and ovarian cancer predisposition**. *Cancers (Basel)*, 10(11), 453. doi:10.3390/cancers10110453

- Schedler KJ, Traine PG, Lohmann DR, Haritoglou C, Metz KA, Rodrigues EB (2016) **Hereditary diffuse infiltrating retinoblastoma**. *Ophthalmic Genet*, 37(1), 95–97. doi:10.3109/13816810.2014.921315

- Shendure J, Balasubramanian S, Church GM, ***et al.*** (2017) **DNA sequencing at 40: past, present and future [published correction appears in Nature, 2019, 568(7752), E11]**. *Nature*, 550(7676), 345–53. doi:10.1038/nature24286

- Souchkevitch GN (1996) ***Main scientific results of the WHO International Programme on the Health Effects of the Chernobyl Accident (IPHECA)***. *World Health Stat Q*, 49(3–4), 209–12.
- Soussi T (2010) ***The history of p53. A perfect example of the drawbacks of scientific paradigms***. *EMBO Rep*, 11(11), 822–6. doi:10.1038/embor.2010.159
- Steinke V, Engel C, Büttner R, Schackert HK, Schmiegel WH, Propping P (2013) ***Hereditary nonpolyposis colorectal cancer (HNPCC)/Lynch syndrome***. *Dtsch Arztebl Int*, 110(3), 32–38. doi:10.3238/arztebl.2013.0032
- Stephens PJ, Greenman CD, Fu B, ***et al.*** (2011) ***Massive genomic rearrangement acquired in a single catastrophic event during cancer development. Cell***, 144(1), 27–40. doi:10.1016/j.cell.2010.11.055
- Susswein LR, Marshall ML, Nusbaum R, ***et al.*** (2016) ***Pathogenic and likely pathogenic variant prevalence among the first 10,000 patients referred for next-generation cancer panel testing*** [published correction appears in ***Genet Med, 2016, 18(5), 531–2]***. *Genet Med*, 18(8), 823–832. doi:10.1038/gim.2015.166
- Toufektchan E, Toledo F (2018) ***The guardian of the genome revisited: p53 downregulates genes required for telomere maintenance, DNA repair, and centromere structure***. *Cancers (Basel)*, 10(5), 135. doi:10.3390/cancers10050135
- Vogelstein B, Fearon ER, Kern SE, ***et al.*** (1989) ***Allelotype of colorectal carcinomas***. *Science*, 244(4901), 207–211. doi:10.1126/science.2565047
- Wadsten C, Garmo H, Fredriksson I, Sund M, Wärnberg F (2017) ***Risk of death from breast cancer after treatment for ductal carcinoma in situ***. *Br J Surg*., 104(11), 1506–13. doi:10.1002/bjs.10589
- Welch BM, Kawamoto K, Drohan B, Hughes KS (2014) ***Clinical Decision Support for Personalized Medicine.*** In: Greenes R. (ed.), *Clinical Decision Support. The Road to Broad Adoption*. 2nd Edition, Academic Press (Elsevier), London, pp. 383–413. doi.org/10.1016/C2012-0-00304-3
- Win AK, Lindor NM, Winship I, ***et al.*** (2013) ***Risks of colorectal and other cancers after endometrial cancer for women with Lynch syndrome***. *J Natl Cancer Inst*,105(4), 274–9. doi:10.1093/jnci/djs525
- Winters AC, Bernt KM (2017) ***MLL-rearranged leukemias: an update on science and clinical approaches***. *Front Pediatr*, 5, 4. doi:10.3389/fped.2017.00004
- Wu Y, Zhang N, Yang Q (2017) ***The prognosis of invasive micropapillary carcinoma compared with invasive ductal carcinoma in the breast: a meta-analysis***. *BMC Cancer*, 17(1), 839. doi:10.1186/s12885-017-3855-7
- Yarden Y (2001) ***Biology of HER2 and its importance in breast cancer***. *Oncology*, 61 Suppl 2, 1–13. doi:10.1159/000055396

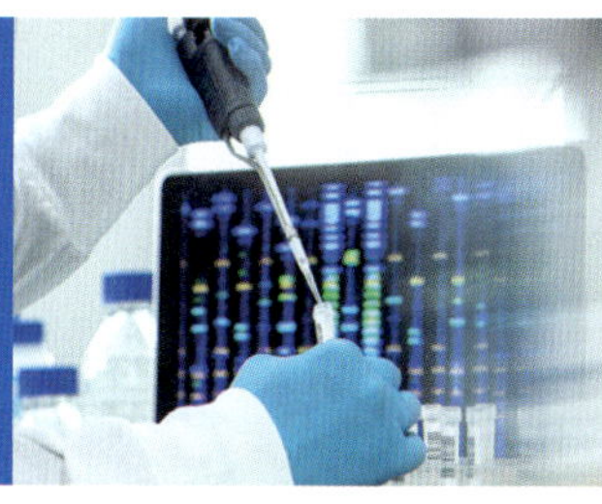

8

Clinical Microbiology in the Genomic Era

Vincenzo Torraca

One of the greatest challenges that clinical microbiologists face in the twenty-first century is the need for the rapid detection and identification of microorganisms and their attributes from clinical samples. The SARS-COV-2 pandemic of 2019 underscored the potential of emerging pathogens to rapidly spread worldwide. The diagnostic tools for infectious diseases have been greatly improved in the past 50 years. Cultivation, microscopy, and serology are no longer the only methods that we can apply to detect infectious agents. Molecular diagnostics tools have emerged that allow us to detect very small amounts of microbial nucleic acids in clinical samples. These techniques have revolutionized modern clinical microbiology. Polymerase chain reaction (PCR)-based techniques allow us to detect specific agents in a matter of hours. PCR and other amplification techniques also allow for the determination of particular features of the microorganisms, such as drug resistance or virulence, aiding, therefore, the choice of the most suitable therapeutic approaches. Molecular diagnostics are widely adopted for qualitative analysis; however, they have also been successfully adopted for quantitative purposes and in-depth characterization. For example, the implementation of quantitative viral load testing provides clinicians with useful information to monitor the response of patients to antiviral therapy.

In most clinical diagnostic approaches, specific PCR primers are designed for each microorganism/group of microorganisms or a genetic feature of interest (i.e., an antimicrobial resistance gene, a virulence factor, a serotype determinant). Based on the history and the symptoms, the patient can be screened for a panel of pre-selected putative microorganisms. Genomics can also be extremely useful in the case of unknown pathogens or pathogen features. For example, whole DNA sequencing allows the detection and characterization of microorganisms without any initial sequence information. Molecular techniques are also largely applied to track epidemiologically relevant isolates. In fact, sequence-based identification and analysis of specific markers can be exploited in epidemiology to monitor and control diseases spreading in the population.

In this chapter, we will dissect how nucleic acid detection and sequencing methods have been adopted to identify, monitor, and characterize pathogens to diagnose or help treat infectious diseases. It is important to emphasize that most molecular diagnostics techniques highlighted here find application beyond microbiology diagnostics and can be applied to study host-related conditions, for example, genetic diseases and the host response to a variety of physiological and pathological stimulations.

Learning Objectives

After studying this chapter, you should confidently be able to:

- Appreciate the large variety of infectious agents within the groups of viruses, bacteria, parasites, and fungi
- Describe the advantages and limitations of molecular techniques versus classical diagnostic approaches
- Explain how nucleic acid detection techniques can be used to detect and qualitatively characterize infectious agents
- Discuss how nucleic acid-based methods can be also applied to quantitatively measure the titre of a pathogen
- Describe how nucleic acid-based diagnostic tools can be applied to inform the most suitable therapeutic approaches for a patient
- Outline how nucleic acid-based techniques can be helpful to public health to monitor and manage infectious diseases circulating in the population.

8.1 Infectious agents and the microbiome

Microbes are categorized into three major groups: microbes with an acellular organization, microbes with a **prokaryotic** cell organization, and microbes with a **eukaryotic** cell organization. **Bacteria** are prokaryotes (i.e. they have a simplistic cell structure that lacks a nucleus and membrane-lined organelles). Other microorganisms, such as **fungi** and **protozoa**, are eukaryotes (i.e. they have a more complex cell structure, which includes a nucleus and several membrane-lined organelles). **Viruses** are instead acellular particles and are neither classified as prokaryotes nor eukaryotes. They are defined instead as *obligate parasites*, as they rely on cellular organisms for their replication. Most microbes have a genome consisting of double-stranded DNA. The exception to this is represented by several viruses, which have an **RNA** genome. The size and architecture of the microbial genomes also vary dramatically, from about 1 kbp for some viruses to a few Gbp for some eukaryotic microbes. Look at **Figure 8.1**, which gives an overview of the genome size for different microbial classes.

microbes
Microscopic organisms, including bacteria, viruses, fungi, and protozoa, that can be found in various environments.

prokaryotic
Referring to cells that lack a nucleus and other membrane-bound organelles, characteristic of bacteria and archaea.

eukaryotic
Referring to cells that have a nucleus and other membrane-bound organelles, characteristic of animals, plants, fungi, and protists.

bacteria (singular bacterium)
Unicellular prokaryotic microorganisms that propagate by cellular fission and are a major cause of disease.

fungi
A kingdom of eukaryotic organisms that includes yeasts, moulds, and mushrooms, which decompose organic material and can cause infections.

protozoa
Single-celled eukaryotes that can be free-living or parasitic, causing diseases such as malaria and giardiasis.

viruses
Infectious agents composed of a protein coat and nucleic acid (DNA or RNA) that replicate inside living host cells.

ribonucleic acid (RNA)
A nucleic acid present in all living cells that plays a role in coding, decoding, regulation, and expression of genes.

8.1.1 Viruses

Viruses range from about 20 nm to 300 nm in diameter. Whilst conventional microscopes are used to visualize bacterial cells, viruses (apart from the largest viruses, e.g. poxviruses) can only be seen using specialized electron microscopes. Viruses are obligate endoparasites of cells, therefore they cannot be cultured on their own. For this reason, molecular techniques are largely exploited for the diagnosis of viral infections, as they allow the detection of a virus directly from a clinical sample.

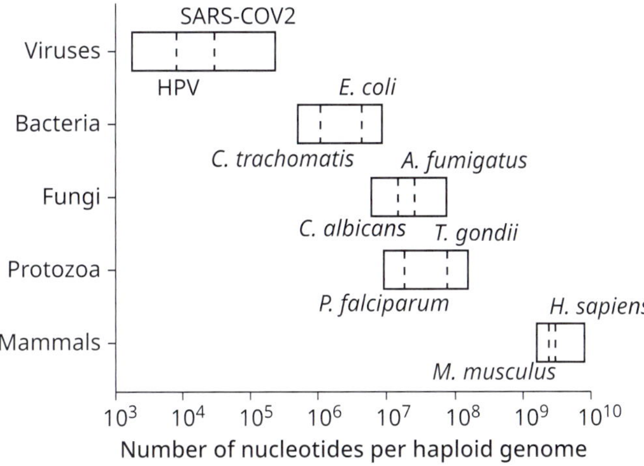

FIGURE 8.1
Illustration of the range of genome sizes (in number of base pairs for haploid genome) for different groups of microbes, and their comparison to the size of mammalian genomes.

Viruses are categorized based on their genome architecture. DNA viruses have a DNA genome, while RNA viruses have an RNA genome. Differently from other microbes, mature viruses only have one type of nucleic acid, not both.

Both the genomes of DNA and RNA viruses can be single-stranded or double-stranded (i.e. the nucleic acid consists of a single or a double filament), linear or circular (i.e. the nucleic acid molecules can have free extremities, or these can be fused), and segmented or non-segmented (i.e. consisting of single or multiple molecules).

Additionally, single-stranded viral genomes may either have a positive (+) sense (i.e. the genomic ssRNA has the same polarity as the mRNA and can be translated into proteins) or a negative (–) sense (i.e. the genomic ssRNA has the opposite polarity to the mRNA and is first used as a template to produce the complementary mRNA that is then translated into proteins). Some single-stranded viruses are also ambisense (i.e. they contain a genome that is a mixture of + and – sense RNA). **Figure 8.2** shows how viruses can be classified based on their genome architecture.

Notice how different viruses exploit different processes to convert their genetic information into **messenger RNA (mRNA)**. Group 1 and Group 2 viruses have a DNA genome. However, Group 1 viruses have a dsDNA genome that can be transcribed directly into mRNA, while Group 2 viruses have a ssDNA genome that is first used as a template to synthesize a complementary DNA strand. The resulting dsDNA is then used for mRNA transcription. Groups 3 to 6 viruses all have an RNA-based genome. Group 3 viruses have a dsRNA genome that can be used directly for mRNA transcription. Group 4 viruses instead have a positive sense ssRNA genome. This means that the genome sequence in the ssRNA is identical to the intended mRNA sequence. Therefore, their genome needs to be first transcribed into a complementary (negative sense) ssRNA which is then used for the transcription of mRNA molecules. Group 5 viruses have a negative sense ssRNA genome, so this filament can be used directly for mRNA transcription. Group 6 viruses, also known as RNA **retroviruses**, have a negative sense ssRNA genome like Group 4 viruses, but their genome is not directly used as a template for the transcription of mRNA molecules. Instead, the viral genome is first converted into a dsRNA, then into dsDNA by the enzyme **reverse transcriptase**, and the dsDNA is finally used as a template for the transcription of mRNA molecules. Group 7 viruses are dsDNA viruses but differ from Group 1 viruses because their genome is not directly used as a template for mRNA transcription. Instead, it is first

messenger RNA (mRNA)
A type of RNA that carries genetic information from DNA to the ribosome, where it is used to synthesize proteins.

retroviruses
A family of viruses that replicate by reverse transcribing their RNA into DNA, integrating into the host genome (e.g., HIV).

reverse transcriptase
An enzyme used by retroviruses to convert their RNA genome into DNA.

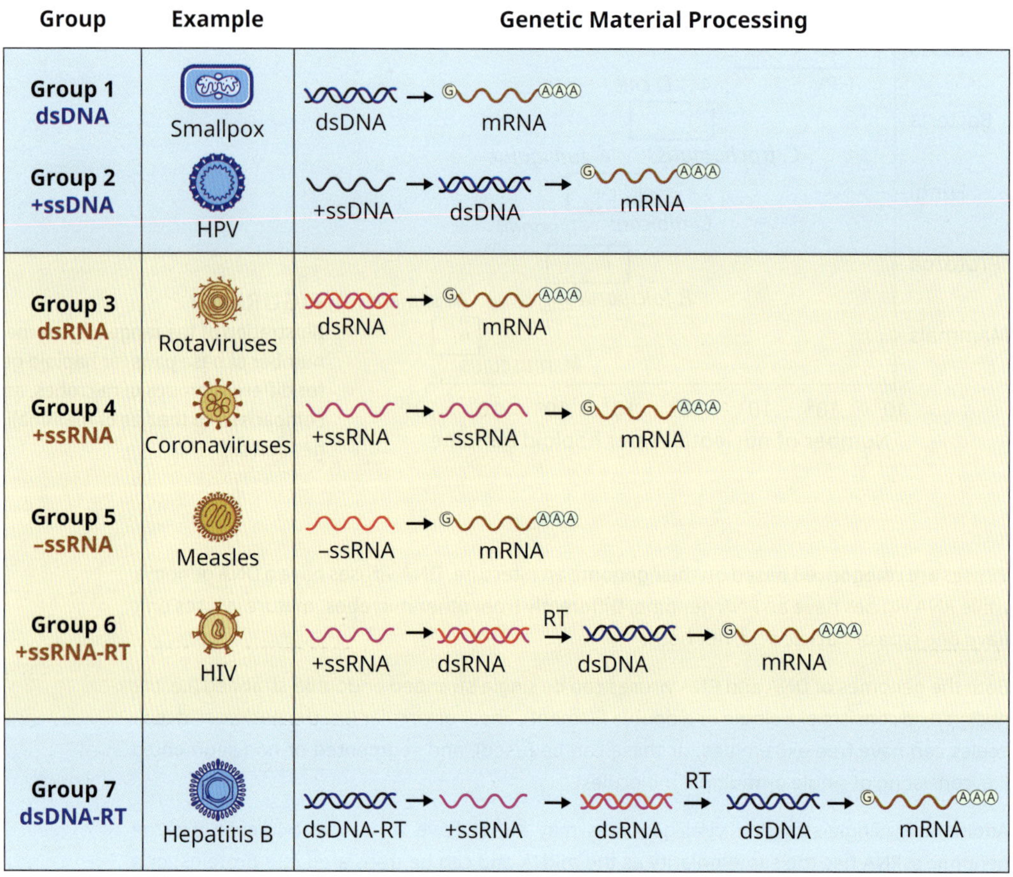

Group	Example	Genetic Material Processing
Group 1 dsDNA	Smallpox	dsDNA → mRNA
Group 2 +ssDNA	HPV	+ssDNA → dsDNA → mRNA
Group 3 dsRNA	Rotaviruses	dsRNA → mRNA
Group 4 +ssRNA	Coronaviruses	+ssRNA → -ssRNA → mRNA
Group 5 -ssRNA	Measles	-ssRNA → mRNA
Group 6 +ssRNA-RT	HIV	+ssRNA → dsRNA →(RT) dsDNA → mRNA
Group 7 dsDNA-RT	Hepatitis B	dsDNA-RT → +ssRNA → dsRNA →(RT) dsDNA → mRNA

FIGURE 8.2

Classification of viruses based on their genome. Viruses are classified into seven groups based on their genome composition (RNA or DNA), structure (double- or single-stranded genome, and positive or negative polarity), and the mechanism of genetic material processing (i.e. dependence/independence from reverse transcriptase and presence/absence of different intermediary steps of conversion between RNA and DNA). See also the main text for further information.

transcribed into positive sense ssRNA, complemented as dsRNA, and retrotranscribed back to dsDNA by the enzyme reverse transcriptase, before it can be used for mRNA transcription.

Viruses do not have **ribosomes** of their own, nor do they contain other elements of the cellular machinery, such as many essential enzymes and energy stores. Therefore, they cannot reproduce and amplify the information in their genomes without the assistance of the cellular components of their host.

As highlighted before, a particular group of viruses is represented by retroviruses. These viruses have an RNA genome and utilize the viral enzyme reverse transcriptase to retrotranscribe their RNA genome into DNA. This DNA can be integrated into the host genome and used to produce viral mRNAs. An example of retroviruses includes lentiviruses, such as **Human Immunodeficiency Virus (HIV)**. The human genome also contains many integrated **Human Endogenous Retroviruses (HERVs)**. These are reminiscent of retroviruses that had originally infected the germline. The genetic material of HERVs duplicates with the host genome as the

ribosome

A cellular organelle that translates mRNA into proteins, found in both prokaryotic and eukaryotic cells.

Human Immunodeficiency Virus (HIV)

The virus that causes AIDS, attacking the immune system and making the body vulnerable to infections and certain cancers.

host cells divide. Therefore, HERVs are contained in every cell of the host. Most endogenous retroviruses are defective and are unable to form new viral particles that can infect new cells. Therefore, they can only proliferate during host genome duplication.

Human Endogenous Retrovirus (HERV)
Retroviral sequences in the human genome that originated from ancient viral infections of germ cells.

Importantly, viral genomes are remarkably smaller than other microbial genomes and can mutate particularly rapidly. In most cases this is because the viral polymerases involved in the genome duplication generally lack proofreading activity.

Several examples of viruses of clinical importance are reported in **Table 8.1**.

TABLE 8.1 Examples of clinically relevant viruses

Virus name	Envelope	Virion size (nm)	Genome size (kbp)	Genome type	Genome segmentation	Disease provoked
Adenovirus	Non-enveloped	70–90	~26–45	Linear dsDNA	Non-segmented	Respiratory infections, conjunctivitis, gastroenteritis, etc.
Astrovirus	Non-enveloped	28–30	~6.8–7.9	Linear, positive-sense ssRNA	Non-segmented	Gastroenteritis
Cytomegalovirus (CMV)	Enveloped	150–200	~235	Linear dsDNA	Non-segmented	Cytomegalovirus infections
Dengue virus	Enveloped	40–60	~10–11	Linear, positive-sense ssRNA	Non-segmented	Dengue fever
Ebola virus	Enveloped	970×80	~18–19	Linear, negative-sense ssRNA	Non-segmented	Ebola virus disease
Epstein–Barr virus (EBV)	Enveloped	180–200	~170	Linear dsDNA	Non-segmented	Infectious mononucleosis, Burkitt's lymphoma, nasopharyngeal carcinoma
Hepatitis B virus (HBV)	Enveloped	42–45	~3.2	Circular dsDNA	Non-segmented	Hepatitis B
Hepatitis C virus (HCV)	Enveloped	55–65	~9.6	Linear, positive-sense ssRNA	Non-segmented	Hepatitis C
Hepatitis D virus (HDV)	Enveloped	36–37	~1.7	Circular, negative-sense ssRNA	Non-segmented	Hepatitis D
Hepatitis E virus (HEV)	Non-enveloped	27–34	~7.2	Linear, positive-sense ssRNA	Non-segmented	Hepatitis E
Herpes simplex virus (HSV)	Enveloped	120–300	~152	Linear dsDNA	Non-segmented	Herpes infections (cold sores, genital herpes)
Human Immunodeficiency Virus (HIV)	Enveloped	120	~9.2–9.6	Linear, positive-sense ssRNA	Non-segmented	Acquired Immunodeficiency Syndrome (AIDS)
Human papillomavirus (HPV)	Non-enveloped	55	~8	Circular dsDNA	Non-segmented	Cervical cancer, genital warts

(*Continued*)

TABLE 8.1 *Continued*

Virus name	Envelope	Virion size (nm)	Genome size (kbp)	Genome type	Genome segmentation	Disease provoked
Influenza virus	Enveloped	80–120	~13.5	Linear, negative-sense ssRNA	Segmented	Influenza (flu)
Measles virus	Enveloped	150–300	~16	Linear, negative-sense ssRNA	Non-segmented	Measles
Mumps virus	Enveloped	150–200	~15	Linear, negative-sense ssRNA	Non-segmented	Mumps
Norovirus	Non-enveloped	38	~7.5	Linear, positive-sense ssRNA	Non-segmented	Gastroenteritis (stomach flu)
Poliovirus	Non-enveloped	25–30	~7.5	Linear, positive-sense ssRNA	Non-segmented	Poliomyelitis (polio)
Poxvirus	Enveloped	200–450	~130–360	Linear dsDNA	Non-segmented	Smallpox, monkeypox, cowpox, etc.
Rabies virus	Enveloped	180	~12	Linear, negative-sense ssRNA	Non-segmented	Rabies
Respiratory syncytial virus (RSV)	Enveloped	150–300	~15.2	Linear, negative-sense ssRNA	Non-segmented	Respiratory tract infections, bronchiolitis, pneumonia
Rotavirus	Non-enveloped	70–75	~18.5	Linear dsRNA	Segmented	Gastroenteritis (in infants and young children)
SARS-CoV-2	Enveloped	50–200	~29.8–29.9	Linear, positive-sense ssRNA	Non-segmented	COVID-19
Varicella zoster virus (VZV)	Enveloped	150–200	~125	Linear dsDNA	Non-segmented	Chickenpox, shingles (Herpes zoster)
Yellow fever virus	Enveloped	40–60	~11	Linear, positive-sense ssRNA	Non-segmented	Yellow fever

Key Points

Viruses exhibit a wide range of genome architectures, including single-stranded or double-stranded DNA or RNA, with variations in strand polarity, segmentation, and circularity.

patho-adaptation
The process by which a pathogen evolves to adapt to its host environment, often leading to increased virulence.

non-culturable bacteria
Bacteria that cannot be grown in standard laboratory culture media, often detected by molecular methods.

8.1.2 Bacteria

Bacteria are prokaryotic microorganisms with a typical size of a few μm. Many bacteria are free-living and contain all the essential elements of the cellular machinery to metabolize nutrients, grow, and replicate. Bacterial cells are often characterized by a rigid cell wall that provides shape to them. They have cytoplasmic membranes, cytosol, and ribosomes. However, bacteria do not have a nucleus and their genetic material is free in the cytosol, rather than being compartmentalized by a nuclear membrane. The human host can be infected by a large variety of bacterial pathogens. Some of them are extracellular pathogens and can live associated with the body surfaces, such as skin and mucosae. Others are intracellular pathogens. These can invade the host cell and grow intracellularly. During the evolutionary process of **patho-adaptation**, some bacteria have lost their free-living capabilities and underwent a process of genome streamlining. Much like viruses, these bacteria cannot be cultured on their own and require the host environment for nutrient supply. Molecular techniques are especially useful to detect these **non-culturable bacteria** (i.e. bacteria that cannot be isolated and studied with conventional culture-based techniques).

The human host also carries a vast community of non-pathogenic microbes. Some of them are **commensals** and might become pathogenic in particular circumstances. Others are beneficial for the host such as most microorganisms living in the human gut. The microbial communities present in the human host represent the **microbiome**. Different tissues of the host might have very different compositions of the microbiome. For example, the skin and the gut have very different microbiomes. While the human microbiome also contains viruses and eukaryotic microorganisms, these are much less characterized, while the bacterial microbiome is the subject of substantial investigations.

Bacterial genomes are always made of double-stranded DNA. As you can see in **Figure 8.3**, most bacteria have only one essential molecule of DNA, which is also referred to as the **bacterial chromosome**. In most bacteria, this consists of a circular molecule of DNA.

A few bacteria with linear chromosomes exist, however. The spirochaete *Borrelia burgdorferi* is an example of a bacterium with linear chromosomes. Many bacteria also contain accessory molecules of circular DNA, much smaller than the bacterial chromosome. These DNA sequences are named **plasmids** and often encode for factors that provide an advantage in certain contexts. For example, plasmids often encode **antimicrobial resistance genes** and are responsible for antibiotic resistance spreading from one pathogen to another. Several bacterial species also have virulence plasmids that encode major **virulence factors**. For example, ***Shigella*** strains represent a human-adapted pathovar of *E. coli*. *Shigella* strains have acquired a large virulence plasmid, also known as pINV or plasmid of invasion, which contains an array of virulence determinants. Bacteria can also be infected by a special class of viruses, called **bacteriophages** (or phages). These viruses have a DNA genome and can integrate into the bacterial genome. Although not human pathogens on their own, bacteriophages are relevant for human health because, similar to plasmids, they can equip the host bacterium with genes encoding virulence factors that make the bacterium pathogenic for the host. This is the case for ***Vibrio cholerae***, as the cholera toxin is encoded by a phage that often infects this pathogen.

Unlike eukaryotes, bacterial genomes can contain a wide range of genomic **GC contents**, from 25% to 75%. The genomic architecture of bacteria is much more simplistic than the eukaryotic genome. For example, bacterial genes do not contain **introns** and do not have large sections of non-coding DNA. Bacterial promoters are small and they drive the expression of **polycistronic mRNA** molecules. This means that sets of co-expressed and functionally related genes are often found in immediate proximity on the bacterial genome and are transcribed as a single mRNA molecule containing in tandem the coding sequences of the individual genes. Different sections of this single mRNA are then translated into different protein products. For these reasons, during genome sequencing, bacterial genomes are much easier to assemble, study, and annotate than eukaryotic genomes. On the other hand, bacterial genomes have large *within-species diversity*. This is due to different mechanisms of **horizontal gene transfer** that are particularly effective in bacteria. Horizontal gene transfer is a process by which genes

commensals
Organisms that live on or within another organism (the host) without causing harm, often referring to the normal microbiota.

microbiome
Community of microorganisms (also called microbiota) that live in a particular environment, such as the human gut.

bacterial chromosome
The main DNA molecule that contains the genetic information of a bacterium.

plasmid
A small, circular DNA molecule found in bacteria that can replicate independently of the chromosomal DNA.

antimicrobial resistance genes
Genes that provide bacteria with the ability to resist the effects of antibiotics and other antimicrobial agents.

virulence factors
Molecules produced by pathogens that contribute to the pathogenicity and facilitate the disease process.

Shigella
A Gram-negative bacterium that causes dysentery and shigellosis, a form of severe diarrhoea.

bacteriophages
A virus that infects and replicates within bacteria.

Vibrio cholerae
A Gram-negative bacterium that causes cholera, a severe diarrhoeal disease.

GC contents
The percentage of guanine (G) and cytosine (C) bases in a DNA molecule, often used as a measure of genomic composition.

intron
A non-coding region of a gene that is removed during RNA splicing.

polycistronic mRNA
An mRNA molecule that encodes multiple proteins, typically found in prokaryotes.

horizontal gene transfer
The movement of genetic material between organisms other than by descent from parent to offspring.

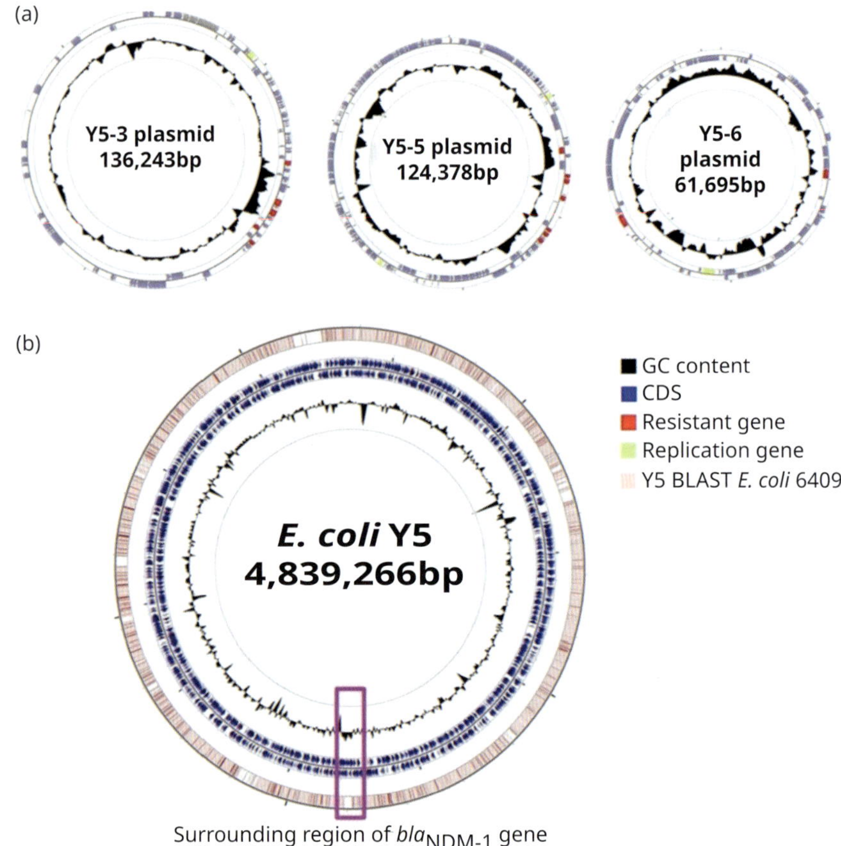

FIGURE 8.3
Representative genome of an *E. coli* strain (*E. coli* Y5) and its three plasmids. Blue arrows denote coding sequences, red arrows denote resistance genes (mostly carried by the bacterial plasmids), and replication genes are denoted by green arrows. GC-content is the inner circle in black. *E. coli* strain Y5 is a multidrug-resistant strain, originally isolated from a human urine sample. Amongst the other antibiotic resistance determinants, this strain also encodes a metallo-beta-lactamase (*bla*NDM-1), a gene that has disseminated rapidly throughout the world and poses an urgent threat to public health. Interestingly, *bla*NDM-1 is normally carried by virulence plasmids, but in this strain, the gene was integrated stably into the chromosomal genome. The region surrounding the *bla*NDM-1 gene is highlighted with a purple frame. Shen, P., Yi, M., Fu, Y., Ruan, Z., Du, X., Yu, Y., and Xie, X. (2016). Detection of an *Escherichia coli* ST167 strain with two tandem copies of bla NDM-1 encoded in the chromosome. *Journal of Clinical Microbiology*. 55. JCM.01581-16. 10.1128/JCM.01581-16.

conjugation
A process by which one bacterium transfers genetic material to another through direct contact.

transduction
The transfer of genetic material from one bacterium to another by a bacteriophage.

transformation
The uptake of free DNA from the environment by a bacterial cell.

can be passed from one bacterium to another that is not directly related to it. Many plasmids, for example, can be transferred from one bacterium to another via a process of **conjugation**. Plasmids that have this capability are also called conjugative plasmids. Horizontal gene transfer can also occur via bacteriophage infections (a process called **transduction**) or by the acquisition of free DNA from the environment, for example DNA fragments released by dead bacteria (a process called **transformation**).

Several bacteria examples of clinical importance are reported in **Table 8.2**.

TABLE 8.2 Examples of clinically relevant bacteria

Bacteria Species	Gram staining	Cell shape	Cell size (μm)	Genome size (Mbps)	%GC content	Chromosome architecture	Disease provoked
Acinetobacter baumannii	Negative (Variable)	Bacillus	1.0–1.5×1.5– 2.5	~4.0	~39%	Circular	Sepsis, urinary tract infection, pneumonia
Borrelia burgdorferi	Negative	Spiral	10–30×0.2–05	~1.5	~29%	Linear (always with at least 17 additional linear and circular plasmids)	Lyme disease
Bordetella pertussis	Negative	Coccobacillus	0.4–0.8	~4.0	~67%	Circular	Pertussis (whooping cough)
Campylobacter jejuni	Negative	Spiral	0.5–5×0.2–0.9	~1.7	~21%	Circular	Campylobacteriosis
Chlamydia trachomatis	Negative	Coccoid	0.2–0.7	~1.0	~40%	Circular (always with an additional virulence plasmid)	*Chlamydia* infection
Clostridioides difficile	Positive	Bacillus	3–8×0.5–0.6	~4.3	~28%	Circular	*Clostridioides difficile* infection
Clostridium perfringens	Positive	Bacillus	3–8×0.4–1.2	~3.3	~28%	Circular	Gas gangrene
Escherichia coli	Negative	Bacillus	1–2×0.2–1	~4.6	~50%	Circular	Gastroenteritis, urinary tract infection
Haemophilus influenzae	Negative	Coccobacillus	0.3–1	~1.8	~38%	Circular	Meningitis, pneumonia, sinusitis
Helicobacter pylori	Negative	Spiral	2–4×0.5–1	~1.6	~39%	Circular	Peptic ulcers, gastritis, stomach cancer
Klebsiella pneumoniae	Negative	Bacillus	06–6×0.3–1	~5.5	~57%	Circular	Pneumonia, urinary tract infection
Legionella pneumophila	Negative	Bacillus	2–20×0.3–0.9	~3.3	~39%	Circular	Legionnaires' disease
Listeria monocytogenes	Positive	Bacillus	0.5–4×0.5–2	~2.9	~38%	Circular	Listeriosis
Mycobacterium tuberculosis	Resistant	Bacillus	2–4×0.3–0.5	~4.4	~65%	Circular	Tuberculosis
Mycoplasma genitalium	Non applicable (no cell wall)	Flask	0.6–0.7×0.3–0.4	~0.6	~32%	Circular	Urethritis, pelvic inflammatory disease

(*Continued*)

TABLE 8.2 *Continued*

Bacteria Species	Gram staining	Cell shape	Cell size (μm)	Genome size (Mbps)	%GC content	Chromosome architecture	Disease provoked
Neisseria gonorrhoeae	Negative	Cocci	0.6–1	~2.1	~51%	Circular	Gonorrhoea
Neisseria meningitidis	Negative	Cocci	0.6–1	~2.2	~51%	Circular	Meningococcal meningitis
Pseudomonas aeruginosa	Negative	Bacillus	1–5×0.5–1	~6.5	~66%	Circular	Respiratory infection, urinary tract infection
Salmonella enterica	Negative	Bacillus	2.5–0.5×1.5	~4.6	~52%	Circular	Salmonellosis
Shigella species	Negative	Bacillus	1–6×0.3–1	~4.6	~50%	Circular (always with an additional large virulence plasmid)	Shigellosis
Staphylococcus aureus	Positive	Cocci	0.5–1	~2.8	~32%	Circular	Skin and soft tissue infection
Streptococcus pneumoniae	Positive	Cocci	0.6–1.25	~2.0	~39%	Circular	Pneumonia, meningitis, ear infection
Streptococcus pyogenes	Positive	Cocci	0.6–1	~1.8	~38%	Circular	Streptococcal pharyngitis, skin infection
Treponema pallidum	Negative	Flat wave	6–15×0.1–0.2	~1.1	~53%	Circular	Syphilis
Vibrio cholerae	Negative	Curved bacillus	2.7–3.5×0.3–0.4	~4.0	~47%	Circular (2 chromosomes)	Cholera

Key Point

Bacterial genomes are typically organized in a single circular chromosome, which can be accompanied by smaller circular DNA molecules called plasmids. Their genomes lack introns and extensive non-coding regions.

8.1.3 Fungi and protozoa

Fungi and protozoa represent an extremely diversified group, which includes species with unique life cycles, morphological specializations, and a range of nutritional requirements. Fungi and protozoa are also responsible for diseases of great public health relevance. Malaria, for example, is a protozoal disease and is a major cause of death worldwide.

Fungi and protozoa have a general eukaryotic cell architecture, like human cells. These micro-organisms can be unicellular or pluricellular and their cells have a clearly defined eukaryotic nucleus, as well as membrane-bound organelles such as mitochondria and the endoplasmic reticulum. These structures are lacking in the prokaryotic cell.

Unicellular microbial fungi are also called **yeasts**, while pluricellular microbial fungi are called **moulds**, and are characterized by a filamentous organization of their cells, known also as *hyphae* or *mycelium*. Some microbial fungi, such as ***Candida albicans***, are dimorphic and can transition from a yeast to a hyphal form. All fungi are also characterized by a rigid cell wall, made of the polysaccharide chitin.

Protozoa is an informal term used to refer to many eukaryotic microorganisms, usually single-celled. Differently from fungi, protozoa are restricted to aquatic or moist environments, they do not grow in hyphal filaments, and do not have a cell wall made of chitin. Protozoa also exhibit animal-like traits, such as the ability to move independently through the medium. Many protozoa are free-living; however, this group also includes many parasitic species of pathological interest. Eukaryotic microorganisms have a vast diversity of structures and a comprehensive description of the different groups of eukaryotic microorganisms is beyond the scope of this chapter.

Fungal genomes are among the smallest genomes of eukaryotes and range from less than 10 Mbp to hundreds of Mbp. Due to their compact size, fungal genomes can be sequenced with fewer resources than most other eukaryotic genomes.

Protozoan genomes span a wide size range. The freshwater protozoan *Polychaos dubium* (previously described as *Amoeba dubia*) holds the record for the largest genome size known to date (670 billion base pairs). Its genome is 231-fold larger than the human genome. The genomes of parasitic protozoa, however, are in general much smaller. For example, the human parasite *Babesia microti* has a nuclear genome of approximately 6.5 Mbp and the genomes of the different ***Plasmodium*** species range between 20 and 25 Mbp.

The genome of both fungi and protozoa is organized in linear chromosomes. Like in animals and plants, the number of chromosomes in the **karyotype** varies from species to species. Different fungal and protozoan organisms exist as stable **haploid**, **diploid**, or **polyploid** cells. In some species, the ploidy changes during different phases of the life cycle or in response to environmental conditions. Additionally, in many protozoa, the exact number of chromosomes and ploidy are still unclear, because their nuclei are extremely small and difficult to study or because their chromosomes do not condense during cell division.

Several fungi and protozoa of clinical importance are reported in **Table 8.3**.

yeast
A type of unicellular fungus.

mould
A type of fungus that grows in multicellular filaments called hyphae, often causing food spoilage and health issues.

Candida albicans
A species of yeast that can cause infections in humans, particularly in immunocompromised individuals.

Plasmodium
The genus of protozoa that causes malaria in humans.

karyotype
The number and appearance of chromosomes in the nucleus of a eukaryotic cell.

haploid
Having a single set of chromosomes, typically found in gametes (sperm and egg cells).

diploid
Having two sets of chromosomes, one from each parent, typical of most somatic cells.

polyploid
Having more than two sets of chromosomes, common in some plants, animals, and protozoa.

Key Points

Fungi and protozoa exhibit diverse life cycles, morphological adaptations, and nutritional requirements. They possess eukaryotic cell structures, such as a defined nucleus and membrane-bound organelles, and can be unicellular or multicellular. These pathogens are responsible for a wide range of diseases, from fungal infections like candidiasis to protozoal diseases like malaria, making them significant contributors to public health challenges.

SELF-CHECK 8.1

Discuss the different groups of infectious agents and contrast unique features within each group, with special attention to differences in genome architecture.

TABLE 8.3 Clinically relevant fungi and protozoa

Species name	Type	Cell size (μm)	Haploid genome size (Mbps)	Number of haploid chromosomes	Disease provoked
Acanthamoeba castellanii	Protozoan	15–40	~42	Unclear	Acanthamoeba keratitis
Aspergillus fumigatus	Fungus	2–4	~30	8	Aspergillosis
Candida albicans	Fungus	5–7	~15	8	Candidiasis (thrush)
Cyclospora cayetanensis	Protozoan	2–7	~44	Unclear	Cyclosporiasis
Cryptococcus neoformans	Fungus	8–11	~19	14	Cryptococcosis
Cryptosporidium parvum	Protozoan	4–6	~9	8	Cryptosporidiosis
Entamoeba histolytica	Protozoan	10–60	~24	14	Amebiasis (amoebic dysentery)
Fusarium oxysporum	Fungus	3–54	~52	At least 11	Keratitis, onychomycosis
Giardia lamblia	Protozoan	9–21	~12	5	Giardiasis
Leishmania donovani	Protozoan	2–6	~32	36	Leishmaniasis
Pneumocystis jiroveci	Fungus	2–5	~8	20	Pneumocystis pneumonia
Plasmodium falciparum	Protozoan	1–2	~23	14	Malaria
Toxoplasma gondii	Protozoan	7–10	~65	13	Toxoplasmosis
Trypanosoma cruzi	Protozoan	15–30	~40	Variable	Chagas disease (trypanosomiasis)

SELF-CHECK 8.2

What is the microbiome?

SELF-CHECK 8.3

What is reverse transcriptase, and how is it utilized by retroviruses?

SELF-CHECK 8.4

What are the mechanisms of horizontal gene transfer in bacteria, and how do they contribute to bacterial diversity?

8.1.4 Collection and processing of clinical samples for molecular biology analysis

Clinical samples, such as swabs, blood, and urine, are collected for the diagnosis of infections, the monitoring of the patient, or the screening of the general population for epidemiological surveillance practice. Collection procedures vary according to the specimen. It is possible to extract nucleic acids directly from the diagnostic samples. This is generally sufficient for all techniques involving amplification. However, for some applications that require high quality and quantity of nucleic acids (for example, pulse field gel electrophoresis or whole genome DNA–DNA hybridization), isolation and culturing of the microbe in large amounts might still be necessary.

The discussion in this section focuses on the most collected specimens and the methods for direct extraction of nucleic acids from these samples. Further guidance should be found in specialized books and laboratory guidelines. When handling samples from a potentially infected patient, safety should be carefully considered, and preventive measures should be applied to minimize the risk of infection of the operator.

After collection, nucleic acids are extracted, purified, and/or concentrated. **Method Box 8.1** and **Figure 8.4** within the box show a general pipeline for nucleic acid extraction from clinical samples. The **gold standard** for nucleic acid extraction is the *phenol: chloroform extraction* method. However, a variety of commercial methods are now available and often preferred for their rapidity and safety. Depending on the application, the nucleic acids might also be probed for quality and yield before further processing. Absorbance measurements at 260 and 280 nm and the *A260/280 ratio* are widely adopted and provide an estimate of purity from protein and other chemicals of the extracted samples. Gel electrophoresis might also be used to assess nucleic acid integrity.

Blood should be collected by a **phlebotomist** in appropriate tubes, containing additives for the specific application. For example, anticoagulants or RNase inhibitors might be added. Before

gold standard
The best available diagnostic test or benchmark against which other tests or procedures are measured.

phlebotomist
A healthcare professional trained to draw blood from patients for clinical or medical testing, transfusions, donations, or research.

METHOD BOX 8.1 Extraction of nucleic acid from clinical samples

Figure 8.4(a) is a representative nucleic acid extraction process from a swab. The sample collected is resuspended in an appropriate buffer. Depending on the nature of the sample and the intended applications, the resuspension might require additional mechanical or enzymatic steps to lyse the cells/tissues and ensure that the nucleic acids are released in solution. Treatment with DNase (to extract RNA only) or with RNase (to extract DNA only) might be applied at this stage as well. The addition of phenol and chloroform enables the separation of the sample into two phases, an aqueous upper phase, and a lower phenolic phase. Nucleic acids are more soluble in the aqueous phase, while proteins are more soluble in the phenolic phase or form a precipitate at the interphase. The upper phase enriched in nucleic acids is transferred into a new tube and the nucleic acids are precipitated with an alcohol (e.g. isopropanol or ethanol). The pellet is dried and resuspended in an adequate medium, commonly nuclease-free water or Tris-EDTA (TE) buffer.

Moving on to **Figure 8.4(b)**, for some applications, the yield and purity of the nucleic acid might be relevant. In these cases, spectrophotometric measurements are widely adopted. Nanodrop instruments can be used to measure concentrations and purity of nucleic acids from microlitres of solutions and can detect concentrations as little as a few ng/μl. Nucleic acids have a peak of absorbance at 260 nm; hence the concentration can be estimated by the level of absorbance of the sample at this wavelength. Reading the absorbance at multiple wavelengths allows us to estimate the purity as well. For example, proteins have absorbance peaks at 280 nm, hence 260/280 ratios can be used to estimate the level of residual proteins in the sample. High concentrations of salts can be inferred by checking the 230 nm absorbance.

Figure 8.4(c) shows how the quality of prokaryotic RNA preparation can be estimated by gel electrophoresis. Since ribosomal RNA represents the most abundant RNA molecule, the bands corresponding to the large (23S rRNA) and small (16S rRNA) prokaryotic ribosomal subunits would be particularly visible, and a band corresponding to the 5S tRNA might be also detectable. Some smeared RNA material might be visible. This is expected since mRNA molecules

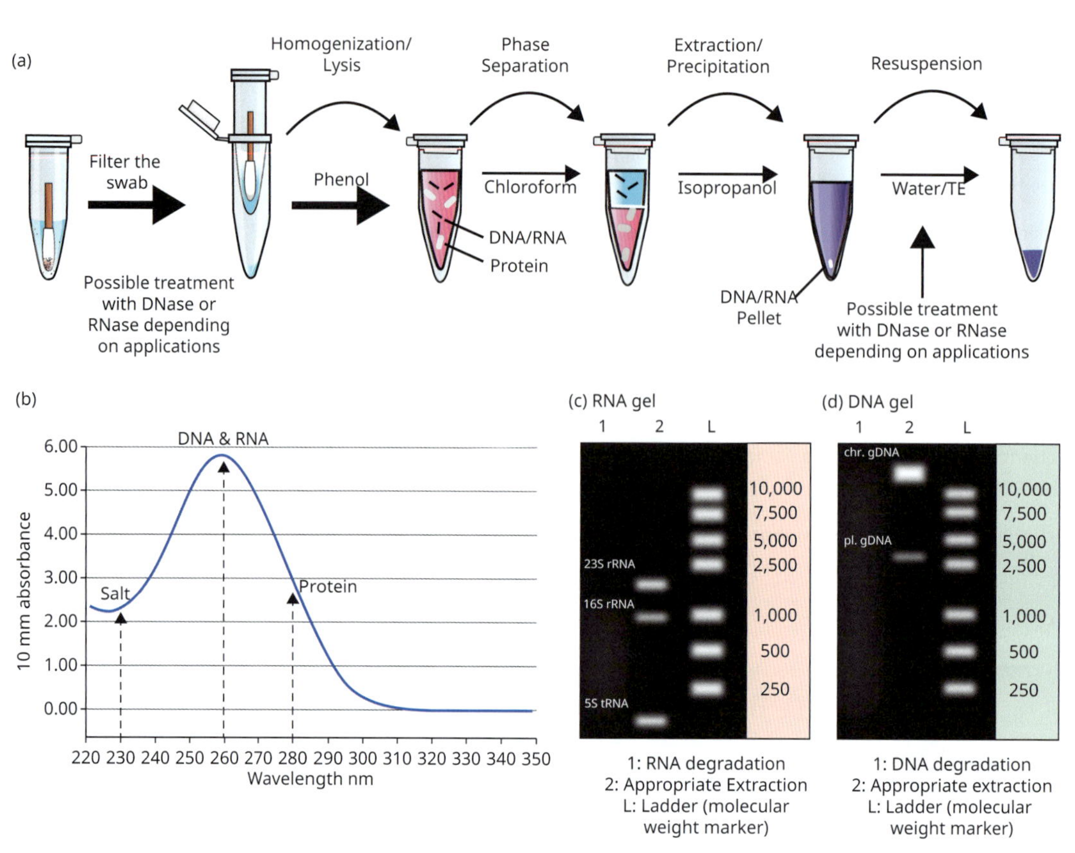

FIGURE 8.4

Extracting nucleic acid from clinical samples. (a) Representative nucleic acid extraction process from a swab. (b) Spectrophotometric measurements of nucleic acid yield and quality. (c) Representative gel electrophoresis of prokaryotic RNA. (d) Representative gel electrophoresis of prokaryotic gDNA.

have a whole spectrum of molecular weights. If the 23S and 16S bands are not detected in a whole RNA preparation, the sample has been likely degraded.

Figure 8.4(d) shows how the quality of prokaryotic gDNA preparation can be estimated by gel electrophoresis. Since the vast majority of prokaryotes have only one large chromosome, a single band at a very high molecular weight is expected (chr. gDNA). However, the exact molecular weight of the prokaryotic chromosome (i.e. several Mbp) is too high to be estimated on conventional gel electrophoresis systems. In some cases, other discrete bands might be detectable, for example, if the bacterial strain has plasmids (pl. gDNA). These normally have sizes of a few kbp and might be estimated on a gel. Smeared material in a gDNA gel might be visible, which might be due to partial degradation of the gDNA or contamination with RNA. If a high molecular weight band is not detected, the sample was likely degraded. Note that schematics in **Figures 8.4(c)** and **(d)** represent typical results that can be obtained from a pure culture of a prokaryote. If the nucleic acids are directly extracted from a clinical sample, bands corresponding to the host nucleic acids (and possibly a variety of other microbes) might also be detected. In many cases the nucleic acids might not be sufficiently concentrated to detect them with conventional gel-based methods; however, the sample might still be suitable as a template for many molecular applications, where the nucleic acids can be detected by amplification (for example in PCR).

processing for molecular biology analysis, blood is often fractioned to separate the cell/corpuscular components present in the samples. For some applications, the **serum** (the non-cellular part of the blood but containing the coagulation factors) or the **plasma** (the non-cellular part of the blood, from an anticoagulant-treated sample) might be needed. Anticoagulants are often used during blood sampling. The choice of anticoagulant is very important, as some anticoagulants can interfere with the downstream molecular techniques. For example, citrate and heparin are commonly used anticoagulants, but they can inhibit the **polymerase chain reaction (PCR)**. Additionally, nucleic acids in blood samples are very susceptible to enzymatic degradation, especially RNA. The addition of commercially available **RNase/DNase inhibitors** can be applied to preserve RNA integrity.

Swabs and saliva contain exfoliated epithelial cells as well as possible pathogens associated with the epithelial layers. Self-collection of swabs or saliva is safe and convenient. The collected sample generally contains a sufficient amount of nucleic acids for amplification-based techniques. A large variety of pathogens can be diagnosed from swabs. For example, buccal/nasopharyngeal swabs are widely adopted to screen for respiratory infections (e.g. **influenza virus** and **Severe Acute Respiratory Syndrome Coronavirus 2 (SARS-CoV-2)**), while rectal swabs can be applied to diagnose gastroenteric infections (e.g. **Astrovirus**, **Adenovirus**, **Rotavirus**, and **Norovirus**).

Urine is an ideal diagnostic sample because it can be collected easily and non-invasively. Urine nucleic acid tests find applications in diagnosing *urinary tract infections* (UTI) such as ***E. coli***, ***Klebsiella***, and ***Enterococcus***. It can be also applied to screen for certain sexually transmitted infections (STI), such as infections caused by ***Neisseria*** and ***Chlamydia***. A variety of viruses (including viruses that are non-tropic to the urinary tract) are directly shed in the urine to levels that are detectable with nucleic acid-based approaches. Hence, screening of urine samples for several viruses, such as **Cytomegalovirus (CMV)**, has been proposed. Urine is often collected by the patient and contamination during collection is an important limitation.

Stool samples can be collected non-invasively from patients and the patient can directly perform sampling without specialized training. As for urine samples, the likelihood of contamination during collection is a limitation. Stool samples can be used to screen for a wide variety of gastroenteric pathogens. For example, gastrointestinal multiplex diagnostic tools are available that can be used to detect over 20 different pathogens in a single assay, spanning across viruses, bacteria, and protozoa (e.g. Astrovirus, Norovirus, Rotavirus, *Shigella*, ***Campylobacter***, ***Salmonella***, ***Cryptosporidium***, ***Entamoeba***, and ***Cyclospora***).

Cerebrospinal fluid (CSF) is a colourless liquid found in the spinal cord and brain. During infections, some pathogens that infect the brain or the spinal cord (such as pathogens causing meningitis and encephalitis) might be detectable in the spinal fluid. Detection of pathogen nucleic acids in the CSF has been successfully applied for **Herpes simplex virus (HSV)** and ***Toxoplasma gondii***. Direct detection in the CSF also informs that the pathogen is localized in the nervous system. However, due to the invasiveness of the sampling method, CSF sampling is only limitedly applied, and less invasive approaches are often preferred.

Biopsies are solid tissue samples. Many collection techniques, such as rapid freezing or collection in commercial solutions that prevent nucleic acid degradation, do not aim at preserving the native tissue architecture. The whole nucleic acids can be extracted from the tissue and used for downstream applications, such as PCR. If the sample is fixed in formalin and mounted on coverslips, this may be used to preserve the tissue structure and detect pathogens *in situ*, for example by hybridization of fluorescent probes with the pathogen nucleic acids.

serum
The liquid part of blood that remains after clotting, used in diagnostic tests and research.

plasma
The liquid component of blood that contains water, electrolytes, proteins, and other molecules, but without the cells.

polymerase chain reaction (PCR)
A technique used to amplify small segments of DNA by copying them repeatedly.

RNase inhibitor
A substance that inhibits the activity of ribonucleases (enzymes that degrade RNA).

DNase inhibitor
A substance that inhibits the activity of deoxyribonucleases (enzymes that degrade DNA).

influenza virus
The virus that causes influenza (flu), a contagious respiratory illness.

Severe Acute Respiratory Syndrome Coronavirus 2 (SARS-CoV-2)
The virus responsible for COVID-19, a contagious respiratory illness.

Astrovirus
A family of viruses that cause gastroenteritis, especially in children.

Adenovirus
A group of viruses that can cause respiratory, gastrointestinal, and eye infections.

Rotavirus
A virus that causes severe diarrhoea, primarily in infants and young children.

Norovirus
A highly contagious virus that causes gastroenteritis, leading to vomiting and diarrhoea.

E. coli
A Gram-negative bacterium found in the intestines of humans and animals, some strains of which can cause various illnesses, including foodborne diseases and urinary tract infections.

Klebsiella
A genus of Gram-negative bacteria that can cause pneumonia, bloodstream infections, and other infections.

Enterococcus
A genus of Gram-positive bacteria that are normal inhabitants of the human intestine but can cause serious infections.

Neisseria
A genus of Gram-negative bacteria that includes species causing gonorrhoea and meningitis.

Chlamydia
A genus of Gram-negative bacteria that cause chlamydia, a common sexually transmitted infection.

Cytomegalovirus (CMV)
A common virus that can cause serious illness in immunocompromised individuals and congenital infections.

stool
Solid waste discharged from the intestines through the rectum, used often as a diagnostic sample.

Campylobacter
A genus of Gram-negative bacteria that is a common cause of food poisoning, leading to diarrhoea.

Salmonella
A genus of Gram-negative bacteria that causes foodborne illnesses, such as typhoid and enteric fever.

Cryptosporidium
A genus of protozoa that causes cryptosporidiosis, a diarrhoeal disease.

Key Points

Clinical samples, including swabs, blood, urine, stool, cerebrospinal fluid, and biopsies are collected for various purposes such as infection diagnosis, patient monitoring, and epidemiological surveillance. Application of nucleic acid-based techniques to these clinical samples can provide valuable information for diagnosing a wide range of infectious agents.

SELF-CHECK 8.5

What are the advantages and limitations of using swabs and saliva samples for molecular diagnostics? Provide examples of pathogens commonly diagnosed from these specimens.

SELF-CHECK 8.6

Why is urine considered a valuable diagnostic sample? What types of infections can be detected using urine nucleic acid tests?

8.2 Advantages and limitations of nucleic acid-based techniques

Some advantages and limitations for target organisms are reported in **Table 8.4**. The pace at which nucleic acid-based techniques have advanced, their universality of application, and the speed at which these analyses can be performed made them increasingly more adopted in diagnostic laboratories. Although nucleic acid-based detection holds great promise for enhancing the current diagnostic potential, several limitations also exist and restrict its use for routine clinical diagnoses. Widely recognized limitations include, for example, the occurrence of false-negative and false-positive results. We are additionally still far from being able to predict a variety of phenotypic properties (e.g. virulence, antibiotic susceptibility) exclusively on the grounds of genotypic information. Therefore, despite the advances in nucleic acid detection and genome analysis, infectious organisms are still often identified by other means such as culture, proteome, bacteriophage, or biochemical profiles. In this section, we will discuss in detail the advantages and limitations of nucleic acid-based techniques.

8.2.1 Advantages

Molecular diagnostics in clinical microbiology provide a series of advantages compared to classical techniques.

Speed

Compared to culture-based techniques that normally require several days/weeks for the isolation and characterization of a pathogen, molecular techniques generally can allow identification within a few hours/days.

TABLE 8.4 Advantages and limitations of PCR-based versus conventional diagnostic methods for the detection of three target organisms

Target organism	Conventional diagnostic method			PCR-based method		
	Method	Pros	Cons	Method	Pros	Cons
M. tuberculosis	Culture	- Allows susceptibility testing - High specificity with nucleic acid-based identification	- Inadequate sensitivity - Prolonged time to result (>2 weeks) - Requires further identification after positive culture - High cost (direct and indirect) associated with delayed diagnosis	PCR (TB Amplicor)	- High sensitivity (93%)/ specificity (98–100%) - Rapid detection time - No transport requirement - Allows detection from non-invasive specimens - Best for diagnosis and screening	- Potential contamination - Unable to assess viability - Limited ability for genotype and susceptibility testing
	Acid-fast stain	- Rapid detection	- Inadequate sensitivity/ specificity			
C. trachomatis	Culture	- High specificity (100%) - Detects only viable organisms - Allows further genotype or susceptibility testing	- Low sensitivity (70–80%) - Invasive specimen collection - Prolonged time to result - Cold transport - High cost	PCR (Amplicor)	- Rapid detection time - Moderate to high sensitivity (80–90%) with high specificity (95–100%) - Cost effective	- Limited ability for multidrug-resistance testing - Currently used as an adjunctive test
	Antigen detection method	- High specificity (98–99%) - Allows assessment of specimen adequacy - Rapid detection time - No transport requirement	- Requires technical expertise - Moderate sensitivity (80–90%) - Requires high expertise			
Enterovirus	Viral isolation	- High specificity (100%)	- Prolonged time to result (10–14 days) - Low sensitivity (65%–75%) - Multiple cell lines needed for isolation - Serotyping time-consuming, labour-intensive, and costly	RT-PCR (Amplicor EV)	- Higher sensitivity (98%) - High specificity (94%) - Rapid detection time - More adaptable for serotyping - Cost effective	

Entamoeba
A genus of amoebas, some species of which cause amebiasis, an intestinal illness.

Cyclospora
A genus of protozoa that causes cyclosporiasis, an intestinal infection leading to diarrhoea.

cerebrospinal fluid (CSF)
The clear fluid found in the brain and spinal cord, which protects and nourishes the central nervous system.

Herpes simplex virus (HSV)
A virus that causes herpes infections, including oral and genital herpes.

Toxoplasma gondii
A protozoan parasite that causes toxoplasmosis, which can be serious for pregnant women and immunocompromised individuals.

Trypanosoma
A genus of protozoa that causes diseases such as African sleeping sickness and Chagas disease.

Specificity

Molecular techniques can be designed to be extremely specific against a certain microbe, while most non-molecular techniques require multiple assays to confirm a certain infectious agent and exclude other similar ones.

Sensitivity

Nucleic acid-based techniques that involve amplification (such as PCR) allow the detection of pathogens at very low titres. Direct observation methods, serological and culturing techniques are unable to reach the same level of sensitivity. PCR-based detection is particularly useful when a culture appears negative.

Direct identification potential

The application of molecular diagnostics allows the direct detection from complex samples, without the necessity to purify or culture the live microorganism. This is particularly useful for the detection of viruses, slow-growing bacteria (e.g. ***Mycobacterium tuberculosis***), fastidious or non-cultivable bacteria (e.g. *Chlamydia*, ***Mycoplasma***, **Borrelia**), certain fungi (e.g. *Candida*, ***Cryptococcus***, ***Aspergillus***), and certain protozoa (e.g. *Toxoplasma*, ***Plasmodium***, ***Trypanosoma***). These are organisms for which no alternative diagnostic methods are available or for which the current alternatives to nucleic acid amplification methods (e.g., culture or serology) are slow or laborious.

Stability of nucleic acids

While RNA is notoriously quite susceptible to degradation, DNA is very stable. This is also exemplified by its use in forensics and archaeological research. Except for RNA viruses, all other microbes have a DNA genome. Therefore, DNA stability is of great advantage when processing the majority of clinical samples. If appropriately handled and stored, RNA can also be preserved for a long time. Additionally, RNA may be transcribed into cDNA, which has higher stability.

Universality of approach

All microbes contain nucleic acids and can be detected and characterized by nucleic acid-based techniques. No other diagnostic approach can be universally adopted for the detection and characterization of all microbes.

Automation potential

Nucleic acid-based techniques can be automated, and data analysis computerized.

8.2.2 Challenges and limitations

Nucleic acid-based techniques are not free from limitations.

Occurrence of false-positive results

Amplification-based nucleic acid techniques are prone to false-positive errors. This is mostly caused by the fact that amplification from nucleic acids is highly efficient and thus prone to detect contaminations. The most likely sources of contamination are other samples handled at the same time or products from previous amplifications. Processing of negative samples along with each extraction batch can control for contamination by sample carryover.

Occurrence of false-negative results

Nucleic acids can be subject to degradation by nucleases. Treatments of samples with nuclease inhibitors or nuclease inactivation by denaturation can mitigate the risk of false negatives due to degradation. Another cause of false-negative errors is the inhibition of enzymatic activities used in the nucleic acid-based technique. For example, DNA polymerase activity may be inhibited by chemicals present in the sample. Because of the complex composition of clinical samples, inhibitors of enzymatic reactions can sometimes be present. Notorious PCR inhibitors include anticoagulants and immunoglobulin G (e.g. in plasma preparations), haemoglobin and lactoferrin (e.g. in erythrocytes and leukocytes). Purification methods, however, exist to mitigate this risk. Additionally, an artificial target (with an amplification product that can be differentiated from that of the real target) may be added to amplification reactions as a positive control. False-negative results could also be caused by suboptimal reaction conditions. For example, in a PCR reaction, the operational parameters might need to be optimized empirically.

Limited quantitative applications

Quantification is important for diagnostic purposes, especially for samples where the sole presence of an organism is not sufficient to diagnose an infectious disease. This is the case of organisms that might be present in low numbers as commensals. For example, in a urine sample, the number of organisms present is normally used to assess a UTI and whether this requires therapeutic intervention. Several approaches for quantification based on nucleic acid exist, such as the quantitative polymerase chain reaction (qPCR, see also **Section 8.5.1**) but a reproducible quantitative amplification assay is very difficult to develop in many clinical scenarios and its application remains limited.

Difficulty in inferring phenotype from gene presence/absence

Even when the whole genome of the microbe has been sequenced, it can be challenging to infer several phenotypic features from the genotype alone. For example, a PCR test would be able to detect the presence of an antibiotic resistance gene, even if this is constitutively suppressed or defective, and therefore the infectious agent might be genotypically, but not phenotypically, positive. Similarly, antibiotic resistance to the same antibiotic can be achieved with a variety of mechanisms, requiring different genes, and different reactions would be needed to cover each mechanism of resistance. Multiplex PCRs (see also **Section 8.3.1**) have been implemented but these still only partly solve the issue.

Exclusive specificity of detection

Due to the high specificity of the nucleic acid methods, each assay is limited to the detection of a single infectious species or genus. However, clinical samples can contain different pathogens that lead to very similar symptoms. Specific investigation of all these individual pathogens would require the use of numerous molecular assays, which can be impractical and expensive. However, the presence of conserved genomic regions throughout the bacterial kingdom or within some eukaryotic groups (e.g. fungi) allows scientists to design certain molecular assays that can be used to obtain an amplification product from any member of a certain pathogen group (see also **Section 8.3.5**).

High costs and need for specialized expertise

Compared to most other diagnostic techniques, nucleic acid-based techniques can be costly as they rely on expensive/complex reagents (enzymes, nucleotides, molecular-grade quality chemicals), infrastructure (thermocyclers, sequencers), and data analysis software. It also requires highly qualified expertise to run the procedures safely and accurately.

SELF-CHECK 8.7

What are some advantages of nucleic acid-based techniques over classical methods in clinical microbiology?

SELF-CHECK 8.8

Why are nucleic acid techniques considered at risk of false-positive and false-negative occurrences? Give some examples of measures that can be taken to mitigate these risks.

8.3 Detection and identification of microorganisms

Molecular techniques can be applied to the detection of virtually any microbial pathogen. While for several microorganisms (especially bacteria) culturing and biochemical identification methods are still widely adopted and represent the gold standard for the diagnosis of infectious disease, the detection of viruses is now mostly performed by molecular methods. The most widely adopted molecular technique is the polymerase chain reaction (PCR) which allows the amplification and detection of a specific fragment of DNA. Many variants of PCR exist that improve sensitivity (e.g. nested PCR), allow simultaneous detection of different targets (e.g. multiplex PCR) or allow detection of RNA templates (e.g. reverse transcriptase PCR). New amplification tests, such as the **ligase chain reaction (LCR)**, have greater specificity than conventional PCR-based amplification techniques and have been recently licensed in some countries for the diagnosis of some infectious diseases, such as gonorrhoea and chlamydial infection from urine samples. Alternative identification methods include detection by **hybridization** of **nucleic acid probes** to the target DNA/RNA template (e.g. Southern/Northern blotting). This section summarizes these technologies and reports examples of widely adopted assays used in clinical microbiology diagnostics.

ligase chain reaction (LCR)
A variation of the PCR technique that amplifies DNA using a ligase enzyme for detecting specific sequences.

hybridization
The process of combining complementary nucleic acid strands to form a double-stranded filament.

nucleic acid probes
Short strands of DNA or RNA used to detect the presence of complementary sequences by hybridization.

8.3.1 Polymerase chain reaction (PCR)

Conventional PCR can be done using DNA templates prepared from different samples, including blood, skin swabs, saliva, and cultured microbes. Only small amounts of DNA template are necessary for PCR to generate an amplicon that can be analysed using conventional laboratory methods, for example by gel electrophoresis. **DNA polymerase** is the enzyme responsible for the polymerization of individual nucleotides to form a PCR product. DNA polymerase is unable, however, to initiate the copy of a single-stranded DNA molecule on its own and it requires a primer sequence. **Primers** are short sequences of ssDNA that are complementary to the sequence adjacent to the DNA sequence to be amplified. A conventional PCR requires a set of two primers that are designed to be specific for the desired target. The PCR product, also called an **amplicon**, consists of billions of molecules of double-stranded DNA molecules spanning the sequence between the two primers. The PCR reaction mix (containing the DNA polymerase, the single deoxynucleotides (A, T, G, C), the primers, the template DNA, and the PCR buffer) is assembled and incubated in a thermocycler. Conventional PCR reactions often require under two hours to complete.

DNA polymerase
An enzyme that synthesizes DNA molecules from nucleotide building blocks.

primer
Short sequences of nucleotides that provide a starting point for DNA synthesis during PCR.

amplicon
A piece of DNA or RNA that is the product of amplification events (such as in PCR).

Historically, the most widely adopted PCR approach includes loading and running the sample on an **agarose gel** via **electrophoresis**, as shown in **Figure 8.6(b)**. The PCR product would form a discrete band on the gel at the expected molecular weight, which can be estimated by comparison with a molecular weight marker. The detection of the DNA on a gel is generally obtained by staining with dyes that intercalate the DNA, such as ethidium bromide or SYBR™ Safe. An alternative detection method involves **capillary electrophoresis**. In this setting, the PCR product is still run on gel support but within a capillary. This method has the advantage of increasing sensitivity and resolution. It can distinguish, for example, PCR bands that differ from each other by only a few nucleotides. PCR products can also be detected in solution and without electrophoretic runs, for example by the use of fluorescent dyes that increase their fluorescence upon intercalating the DNA in solution (i.e., SYBR™ Green) or via the use of amplification primers or probes (i.e. DNA sequences complementary to the DNA) that were conjugated to fluorescent dyes. In this case, the fluorescence is detected as a readout of successful amplification. Notice that *in solution* detection methods do not allow assessment of specificity by the visualization of a band unless the sample is still run on a gel matrix post-amplification. However, a **melting curve** analysis can be performed, as in **Figure 8.5**. In this approach, the amplicon is progressively melted by small increases in temperature. This allows the measurement of the **temperature of melting (Tm)**, at which 50% of the amplicon is in dsDNA form and 50% is in ssDNA form. The Tm is very specific for a certain amplicon, and it depends on its exact sequence and molecular weight. Another alternative way to analyse a PCR product is via sequence confirmation by **Sanger sequencing**. This could be particularly useful for the **genotyping** of variants in the sequence of a certain amplicon.

Further sensitivity and specificity in a PCR reaction may be obtained by the *nested PCR* technique, whereby the DNA is amplified in two steps. Notice from **Figure 8.6(a)** that the first step of a nested PCR is a conventional PCR reaction. However, the nested PCR differs from the conventional PCR because of a second amplification step, whereby the product obtained from the first step is used as a template with two inner primers.

Figure 8.6(b) exemplifies how nested PCR may be particularly useful when the number of DNA copies is very small (i.e. a low titre of the infectious agent). In this experiment, a dilution series was performed on a DNA template. A nested PCR was performed using each dilution, demonstrating the sensitivity of detection of as little as 1 femtogram (fg) of template DNA. The nested PCR also increases specificity because of the use of a second, internal primer set in the second step. Similar to conventional PCR, nested PCR can also be combined with capillary electrophoresis, which further increases the sensitivity of the approach. Whilst nested PCR can allow the detection of very small amounts of template, capillary systems allow the detection of smaller concentrations of the produced amplicon.

Conventional PCR and its variants are very sensitive techniques. It is possible to obtain a sensitivity of one DNA molecule in a clinical sample. While this is generally a strength, it can also represent a drawback. For example, even a very small amount of contamination is sufficient to generate a false-positive test. Moreover, a positive PCR result can be difficult to interpret. In fact, this does not necessarily mean the presence of an infectious disease. This is a problem of relevant concern, for example, for the detection of viruses that can also be **latent**, such as CMV. Latent CMV genomes may be detected in the blood of healthy (i.e. asymptomatic) individuals, although these carriers do not require any treatment for CMV infection. PCR-based techniques also do not distinguish between viable and non-viable microbes and they might be able to continue to detect the residual presence of the pathogen's nucleic acids for several days or weeks after the viable microbes have been destroyed by therapy or upon remission. For example, PCR-based diagnostic tools often detect pathogen-specific nucleic acids for a longer period, compared to other techniques such as **lateral flow tests** (i.e. devices which detect microbes

agarose gel
A gel matrix used in electrophoresis to separate DNA or RNA fragments by size.

electrophoresis
A technique used to separate nucleic acids or proteins based on their size and charge by applying an electric field.

capillary electrophoresis
A technique that separates ions based on their electrophoretic mobility using capillary tubes.

melting curve
A graph showing the denaturation of nucleic acids as temperature increases, used to analyse PCR products.

temperature of melting (Tm)
The temperature at which half of the DNA strands are in the double-helical state and half are in the 'melted' single-stranded state.

Sanger sequencing
A method of DNA sequencing that uses labelled chain-terminating nucleotides.

genotyping
The process of determining differences in the genetic make-up (genotype), by examining their DNA sequence.

latent
Referring to a pathogen that is present in the body but not currently causing symptoms.

lateral flow tests
A simple diagnostic device used to confirm the presence or absence of a target analyte, such as a virus, in a sample.

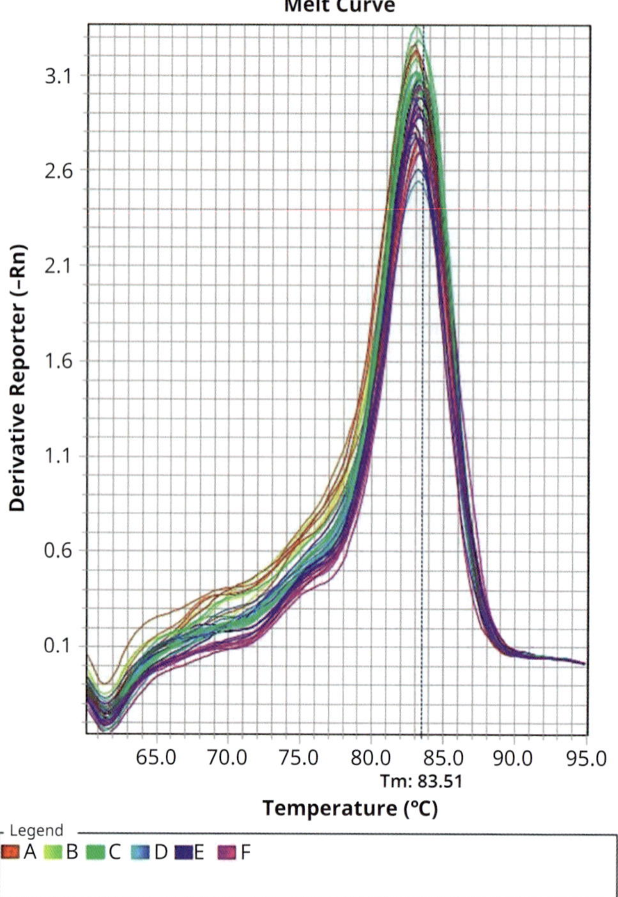

FIGURE 8.5
Melting curve analysis of a PCR reaction in a thermocycler. The melting point is dependent on the sequence and size of the amplicon so it can be used as a confirmatory tool, just like the size of the band on a gel.

based on the recognition of a specific antigen via an antibody). Despite these drawbacks, PCR is widely used to diagnose viral infections, especially as the assays are becoming less expensive and highly automated systems are readily available.

Multiplex PCR
A variant of PCR that allows simultaneous amplification of multiple targets in a single reaction.

Multiplex PCR is characterized by the use of multiple primer sets within a single PCR reaction to simultaneously produce multiple amplicons (of varying sizes to enable distinction among them), each specific to a different DNA sequence investigated. By targeting multiple genes at the same time, more information may be obtained from a single test, which would otherwise require several different PCR runs. Although in most cases 4–7 primer sets are used, up to 19 different microorganisms have been detected at once in a multiplex assay. **Figure 8.7(a)** shows how this method can be used to detect different types of viruses in a clinical sample.

Notice how each lane of the gel in **Figure 8.7(a)** contains an internal control amplicon (used as a positive control). As expected, this is amplified from each clinical sample, confirming good sample quality. The other bands are specific for the detection of different viruses. Notice that multiplex controls (M) were also run. These are created by combining nucleic acids from all the pathogens that are meant to be detected simultaneously, to confirm that co-detection is possible. Indeed, both multiplex controls provided simultaneous amplification for all the possible targets. The gel also has a negative control (N). This is a sample from a healthy individual, where only the internal amplification control can be detected. All the other patients have amplification only of one or more viral-specific amplicons, and the nature of the amplicon can be inferred by molecular weight comparison. Also note that in some cases (patients 2, 3, 7, 8, 9,

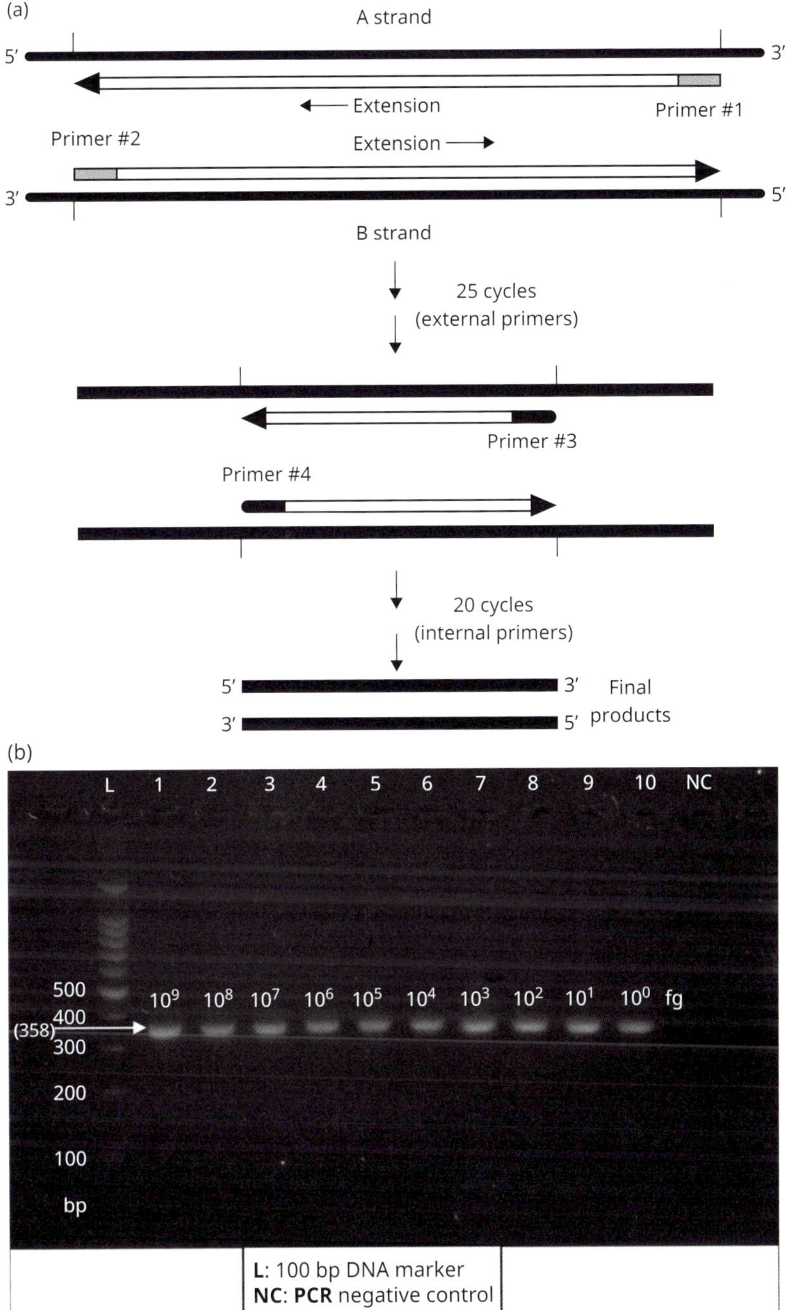

FIGURE 8.6

(a) Diagram representing the nested PCR amplification approach. In a nested PCR, a first primer pair is used to amplify a sequence of interest. This is then further amplified using a second pair of inner primers. This increases the sensitivity and specificity of the method compared to a regular PCR. (b) Gel image of sensitivity assay for a nested PCR for *Leishmania donovani* infection, done using a dilution series of a positive control of known concentration. Sizes of relevant molecular weight markers are reported, including the expected band size of 358 bp. The amount of DNA template used in the reaction is reported in fg on the corresponding lane. NC: negative control. Deepachandi B, Weerasinghe S, Soysa P et al. A highly sensitive modified nested PCR to enhance case detection in leishmaniasis. *BMC Infect Dis* 19, 623 (2019). https://doi.org/10.1186/s12879-019-4180-3

and 11), two distinct viral typing products could be detected at the same time, which could indicate the presence of two different viral types. As **Figure 8.7(b)** shows, multiplex PCR can be combined with capillary electrophoresis, to increase sensitivity and promote standardization.

Multiplex PCR has been particularly adopted for the precise diagnosis of pathogens that share similar symptomatic manifestations. Common applications include screening of respiratory pathogens (e.g. influenza A, influenza B, adenovirus, **respiratory syncytial virus (RSV)**), STIs/

respiratory syncytial virus (RSV)
A virus that causes respiratory infections, especially in young children and the elderly.

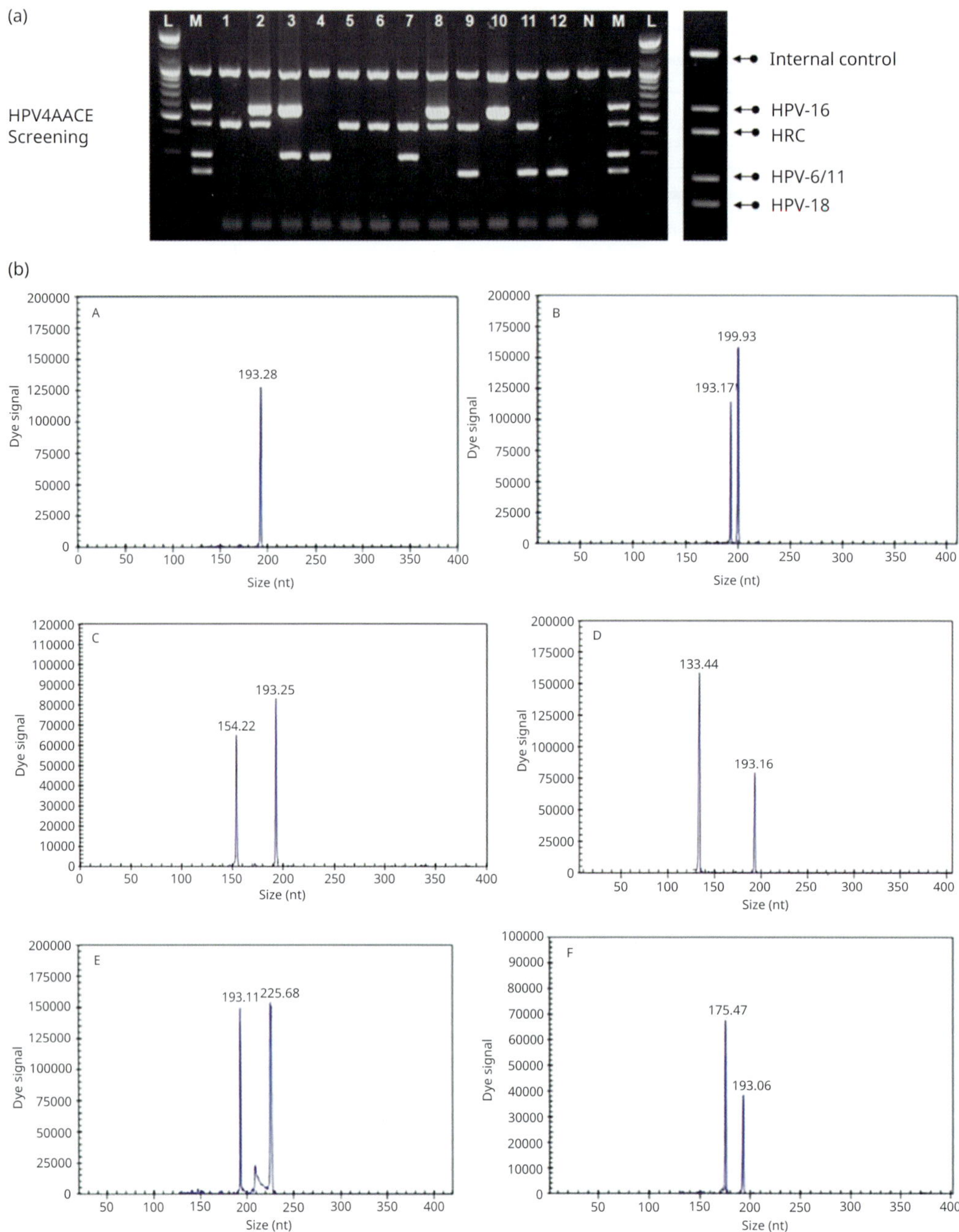

FIGURE 8.7

(a) Representative multiplex PCR for viral screening via analysis on a gel. L: ladder (molecular weight marker); M: multiplex all-positive control; N: multiplex all-negative control (except the internal amplification control); Lanes 1–12: different clinical samples, where one or more target products were detected. The all-positive control is magnified on the right. HPV-16: Human papillomavirus type 16; HPV-18: Human papillomavirus type 18; HPV-6/11: Human papillomavirus type 6 or 11; HRC: HPV high-risk common 11 types. Shim HS, Noh S, Park AR et al. Detection of sexually transmitted infection and human papillomavirus in negative cytology by multiplex-PCR. *BMC Infect Dis* 10, 284 (2010). https://doi.org/10.1186/1471-2334-10-284 (b) Representative multiplex PCR detection of eight avian influenza A viral subtypes via capillary electrophoresis. Panels (A–I) show the individual target gene amplification results. Red peaks in J indicate the position of the molecular weight markers. K represents an all-positive control. Li M, Xie Z, Xie Z et al. Simultaneous detection of eight avian influenza A virus subtypes by multiplex reverse transcription-PCR using a GeXP analyser. *Sci Rep* 8, 6183 (2018). https://doi.org/10.1038/s41598-018-24620-8

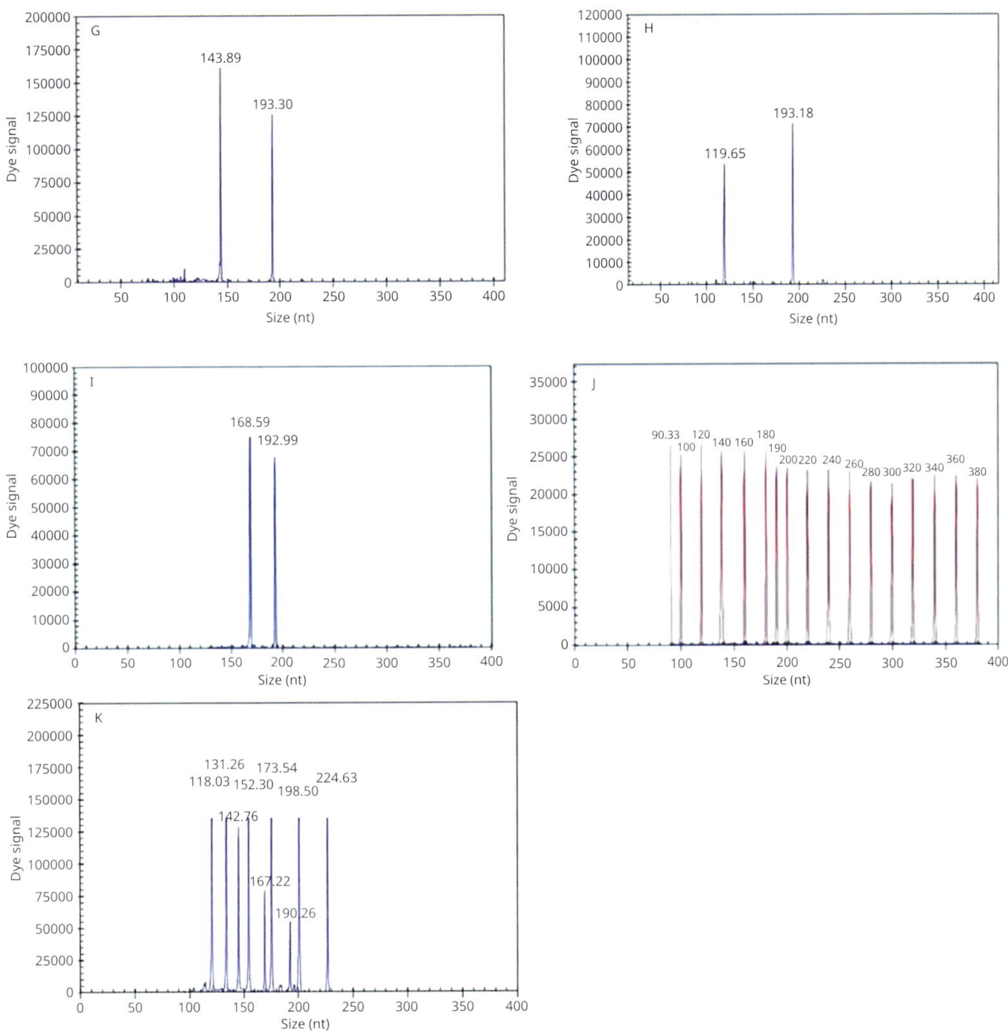

FIGURE 8.7 (Continued)

UTIs (e.g. *Mycoplasma* species, *Neisseria gonorrhoeae, Chlamydia trachomatis,* uropathogenic *E. coli*), gastrointestinal pathogens (e.g. adenovirus, astrovirus, norovirus, rotavirus, *Shigella*, *Salmonella*) and pathogens causing meningitis/encephalitis (e.g. HSV, CMV, **Epstein–Barr virus (EBV)**, **varicella zoster virus (VZV)**).

Single-stranded RNA molecules cannot be directly amplified by PCR. However, to identify viral nucleic acid of RNA viruses (e.g. HIV) in clinical specimens, an additional step can be undertaken before the PCR cycles are initiated. This procedure is termed **reverse transcriptase PCR (RT-PCR)**. This technique is further illustrated in **Figure 8.8**. Notice how, during RT-PCR, the RNA sequence is first converted to a single-stranded DNA sequence (complementary DNA or cDNA) by using a reverse transcriptase enzyme from a retrovirus. The reverse transcriptase also requires a DNA primer annealed to the template RNA to produce its complementary DNA. A specific primer can be used for the virus of interest; alternatively, random hexamer primers are often used to create first a cDNA library (i.e., all the RNA molecules in the sample are used as a template to produce cDNA). In case the target RNA is polyadenylated (i.e. numerous adenines are added at the end of the RNA sequence), a polyT complementary

Epstein–Barr Virus (EBV)
A virus that causes infectious mononucleosis and is associated with certain types of cancer.

varicella zoster virus (VZV)
The virus that causes chickenpox and shingles.

reverse transcriptase PCR (RT-PCR)
A technique that combines reverse transcription of RNA into DNA and PCR amplification of the resulting DNA target.

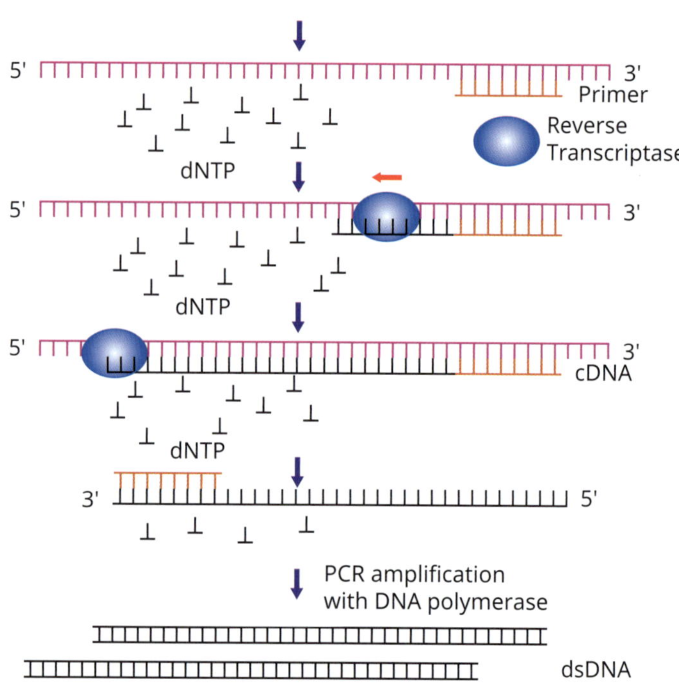

FIGURE 8.8
Diagram representing the process of RT-PCR for the detection of viral RNA. In RT-PCR, the viral RNA template is first converted into DNA. This can be done in multiple ways. For example, if the pathogen of interest is known, a single primer that is complementary to the viral RNA sequence can be used to guide the reverse transcriptase enzyme. The reverse transcriptase will produce copied DNA (cDNA) by incorporating dNTPs according to the complementarity of nucleotides. This produces a single-stranded DNA template which can be used in a PCR amplification reaction.

CASE STUDY 8.1 Detection of viral respiratory infections

Respiratory infections often present themselves with very similar symptoms and signs in the patient, although the pathogen responsible for the disease can differ. In this case study, a multiplex PCR test was developed to screen for an array of respiratory bacterial/viral pathogens and markers. The test includes: Influenza A Virus (flu A), Human Adenovirus (HAdV), Human Bocavirus (HBoV), Human Rhinovirus (HRV), Influenza A virus subtype 2009H1N1 (2009H1N1), Human Parainfluenza Viruses (HPIV), *Chlamydia* species (Ch), Human Metapneumovirus (HMPV), Influenza B Virus (flu B), *Mycoplasma pneumoniae* (Mp), Influenza A virus subtype H3 (H3), Human Coronaviruses (HCoV), and Human Respiratory Syncytial Virus (HRSV). Peaks for a human RNA control (huRNA), a human DNA control (huDNA), and an internal control (IC) were also detected.

Look at **Figure 8.9** which represents the multiplex PCR peaks obtained from running the PCR products on a capillary electrophoresis support. **Figure 8.9A–E** represents individual patients. **Figure 8.9F** represents the combined positive control for each product. **Figure 8.9G** indicates the molecular weight markers.

Q1 What is the diagnosis for patients A–E?

Q2 Is it possible to detect co-infections from the data? Are there patients with triple infections?

Q3 Based on your study on this case, what are the advantages of the multiplex PCR technique, compared to conventional PCR methods?

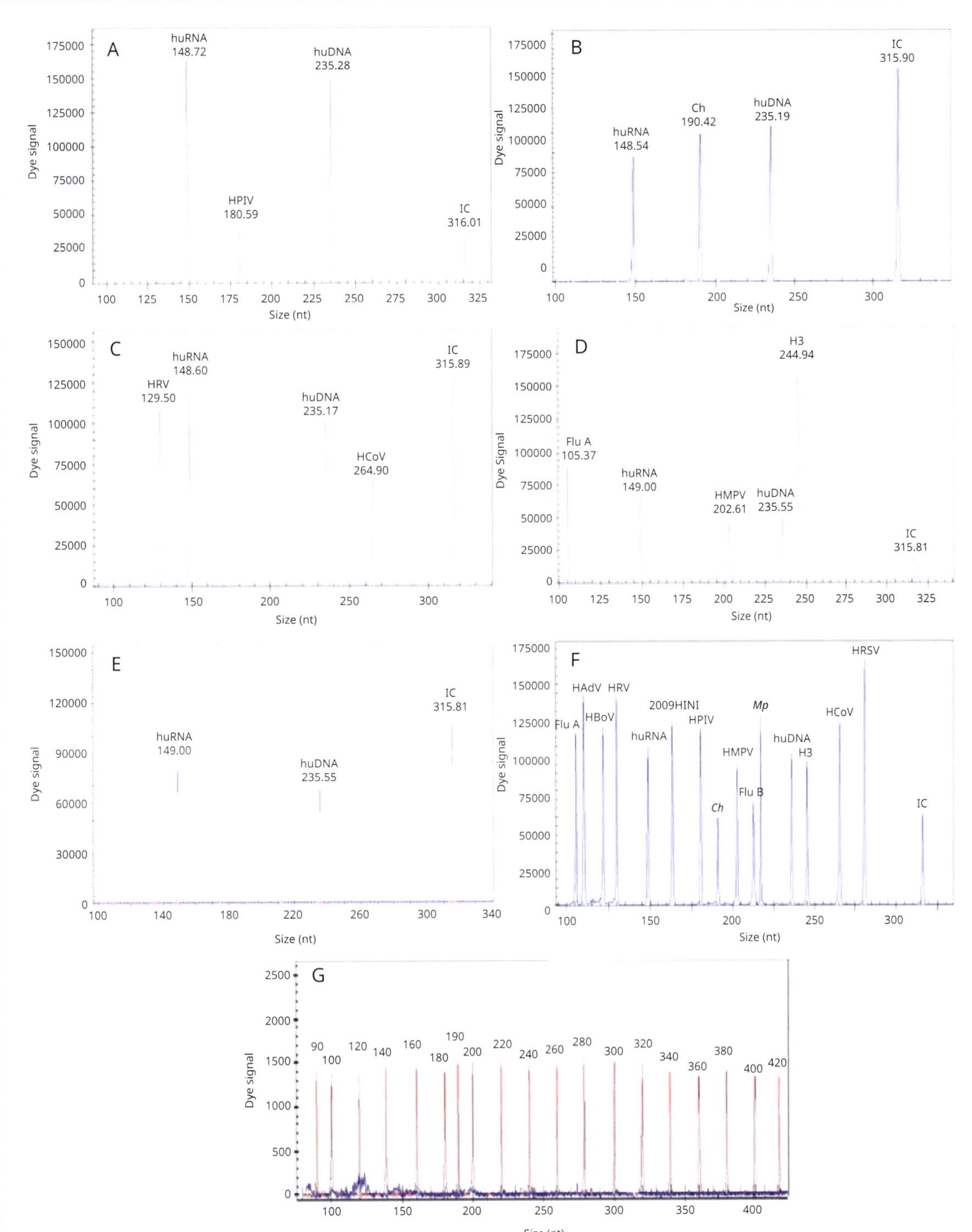

FIGURE 8.9

Multiplex PCR peaks obtained from running the PCR products on a capillary electrophoresis support. Elsevier

primer might also be employed. Using either primer source, the reverse transcriptase synthesizes a cDNA strand by adding bases complementary to the RNA template. The template RNA is then degraded using an RNase, leaving intact the single-stranded cDNA generated in the reaction. This cDNA sequence can then be amplified using conventional PCR or any of its variants previously described.

Key Points

Polymerase chain reaction (PCR) is a fundamental molecular biology technique used to amplify DNA sequences from diverse samples. The process involves thermal cycling, producing billions of copies of the target DNA sequence within hours. PCR products can be detected through various methods including gel electrophoresis, capillary electrophoresis, fluorescent detection, and melting curve analysis. Nested PCR enhances sensitivity and specificity, while multiplex PCR allows simultaneous detection of multiple DNA targets. Reverse transcriptase PCR (RT-PCR) couples the PCR with reverse transcription, allowing the detection of RNA molecules. PCR and its variants have revolutionized diagnostic microbiology, offering rapid and sensitive detection of pathogens from various clinical samples.

8.3.2 Strand displacement amplification (SDA)

SDA is another DNA polymerase-based amplification technique, and it is known for its high specificity in detecting DNA targets. The main difference between PCR and SDA is that SDA is an isothermal technique, meaning that it does not require a thermocycler and the whole reaction is conducted at a constant temperature. Like a PCR, in SDA a DNA polymerase is used to amplify the target DNA sequences. However, the technique involves the use of multiple primer sets that hybridize to the target DNA and initiate the amplification process. The DNA polymerase extends the primers and displaces the downstream DNA strands, creating single-stranded DNA regions that are further targeted by additional primers for further extension and displacement. Since SDA is conducted at a constant temperature, this technique eliminates the need for thermal cycling equipment, making it simpler, more cost-effective, and ideal for applications requiring rapid and on-site detection. SDA has been successfully used in the diagnosis of *Mycobacterium tuberculosis*, as well as sexually transmitted infections such as *Chlamydia*. However, SDA is not as widely used as PCR due to limitations in sensitivity and real-time monitoring compared to PCR-based methods.

8.3.3 Ligase chain reaction (LCR)

LCR is a method of DNA amplification that differs from PCR because it amplifies a probe molecule (consisting of a short molecule of DNA), rather than producing an amplicon through the polymerization of nucleotides. As **Figure 8.10** shows, in LCR four probes are used (two probes for each DNA strand). Each probe is complementary to the probe binding to the opposite strand and anneals to the DNA template right next to the probe that hybridizes on the same strand. The two probes annealing to the same DNA strand are ligated together by the ligase, to form a single probe complementary to the single probe formed

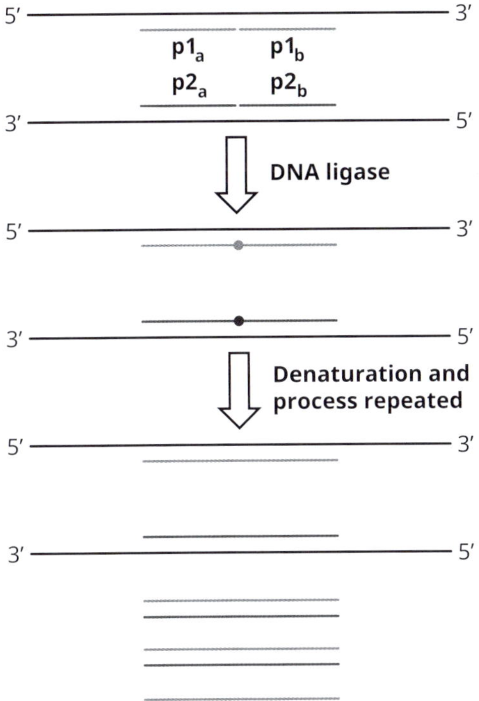

FIGURE 8.10
Diagram illustrating the process of ligation chain reaction (LCR). In LCR, two sets of DNA probes are used. In the diagram, probes p1a and p1b are designed to anneal to the forward DNA strand, next to each other. Probes p2a and p2b are designed to anneal to the reverse strand, next to each other. Additionally, p1a is complementary to p2a and p1b is complementary to p2b. If the templates are present, p1a will bind next to p1b and p2a will bind next to p2b. This allows the ligase enzyme to ligate together the neighbouring probes (p1a with p1b and p2a with p2b). Upon multiple cycles of denaturation, annealing, and ligation, a product will be amplified which is composed of the 4 probes ligated and annealed together.

by the ligation of the other two probes on the opposite strand. Like in PCR, LCR requires a thermocycler to drive the reaction and each cycle results in the doubling of the target nucleic acid sequence. LCR can have greater specificity than PCR, but it requires that the exact sequence in the region being amplified is known. LCR has been recently applied to diagnose *N. gonorrhoea* and *Chlamydia* infections from urine samples. The possibility to test urine samples provides the opportunity to rapidly and non-invasively screen for sexually transmitted infections (STIs) in community-based settings, such as clinics, schools, and other institutions.

8.3.4 Southern and Northern blotting

Southern and Northern blotting are methods used for the detection of a specific DNA or RNA sequence, respectively, within a non-amplified DNA sample. As shown in **Figure 8.11(a)**, these methods involve the separation of the nucleic acid fragments in a sample by gel electrophoresis and then transfer to a membrane by blotting. The membrane is subsequently incubated with one or multiple probes for fragment detection by probe hybridization. The probes will hybridize with the samples, based on base complementarity and sequence homology. Normally, the sample is prepared to separate the two strands of dsDNA molecules and to eliminate higher-order structures of ssRNA molecules, by a process of denaturation that involves exposing the nucleic acid sample to high temperature, low salt concentration, or by the addition of various chemicals. After the membrane with the immobilized sample is incubated with the probes, the denaturing conditions can be reversed to favour probe binding (hybridization). For detection, the probe is generally labelled with biotin, digoxigenin, or fluorescent dyes. Historically,

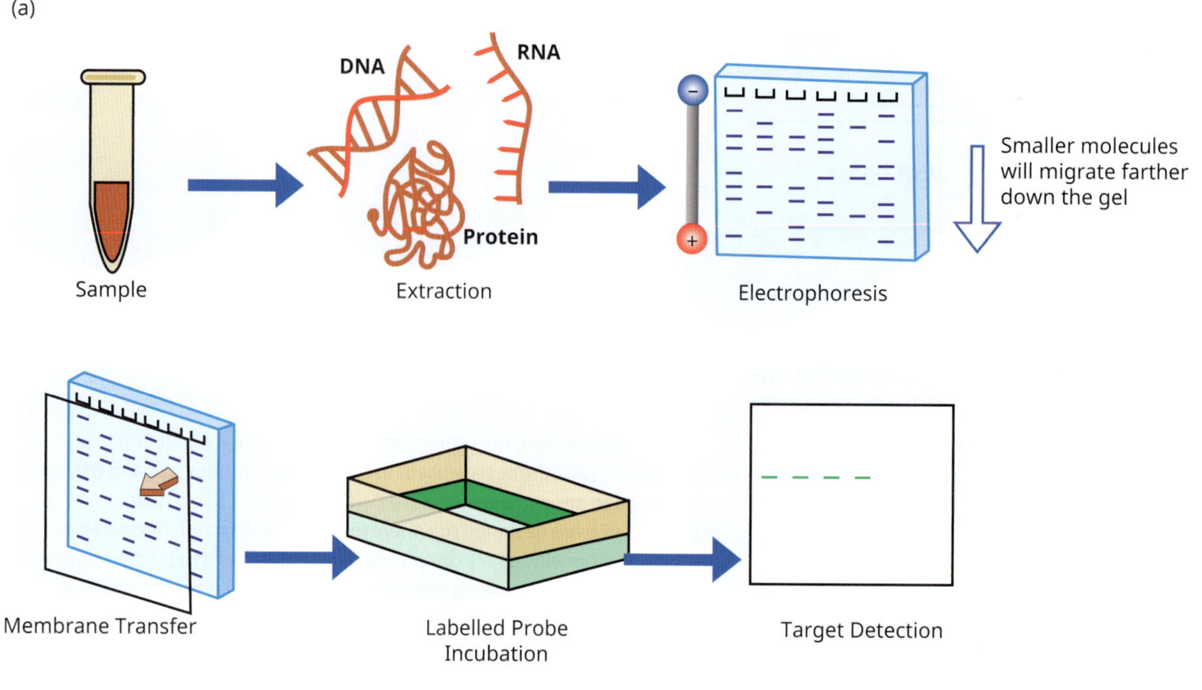

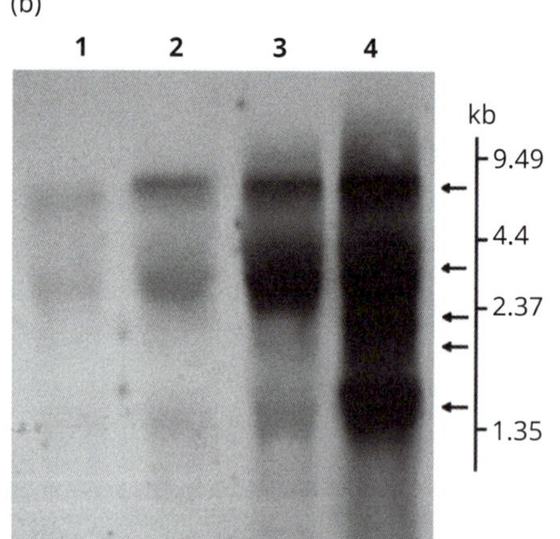

FIGURE 8.11

(a) Workflow of Southern/Northern blotting procedure. A sample containing nucleic acids (DNA for Southern blotting, RNA for Northern blotting) is loaded on a gel and separated by electrophoresis. Nucleic acids are then transferred to a nylon membrane and detected by hybridization of a complementary and labelled probe. (b) Northern blot analysis of SARS-CoV total RNA. The total RNA of SARS-CoV was extracted from SARS-CoV-infected Vero E6 cells (a cell line, largely used to maintain a variety of viruses). The RNA was separated in a 1% denaturing agarose gel containing 6.29% formaldehyde. Afterwards, the RNA was transferred to a positively charged nylon membrane and hybridized with digoxigenin-labelled PCR fragments specific for the 1b (lane 1), S (lane 2), M (lane 3), and N (lane 4) genes of the virus. The vertical bar on the right shows the molecular size reference. Arrows indicate the transcripts hybridized with the N probe. The signals were analysed by chemiluminescence. Hui, R., Zeng, F., Chan, C., Yuen, K., Peiris, J., and Leung, F. (2004). Reverse Transcriptase PCR Diagnostic Assay for the Coronavirus Associated with Severe Acute Respiratory Syndrome. *Journal of Clinical Microbiology.* 42. 1994–9. 10.1128/JCM.42.5.1994-1999.2004.

radioisotopes (e.g. ^{32}P, ^{35}S, and ^{125}I) were used. **Figure 8.11(b)** shows a representative Northern blot gel, performed to detect SARS-COV total RNA.

While Southern/Northern blotting may also allow the quantification of DNA/RNA present in the sample, it is often found that the sensitivity of these techniques does not reach the sensitivity of techniques that involve amplification and is comparable with other non-nucleic acid-based diagnostic methods, such as serology methods.

8.3.5 16S/18S ribosomal RNA (rRNA) gene sequencing

The **ribosomal RNA (rRNA)** genes are highly conserved components of the transcriptional machinery of all life forms (as we discussed earlier, this does not include viruses). As shown in **Figure 8.12**, bacterial and fungal rRNA genes contain both highly conserved and hypervariable segments. This allows the design of pan-bacterial or pan-fungal PCR primer sets that can be used to amplify DNA fragments from a diversity of microorganisms, rather than just from one species. Universal primers can be designed in conserved areas that flank a hypervariable and genus-/species-specific area. Sequencing of these products can then be performed to identify the microbe by sequence homology analysis. For bacterial species, the 16S rRNA gene is generally used as a target, while for fungal species the 18S rRNA gene is used. Sequencing of these gene targets has the potential to allow the direct identification of an unknown pathogen from a clinical sample. It has to be noted, however, that, due to the universality of the primers used in 16S/18S rRNA sequencing, this approach can be difficult if multiple microbial species are present in the clinical sample. For example, 16S/18S rRNA gene primers would also amplify sequences derived from commensals and other species present in the host microbiome. Clonal amplicons (i.e. from a single microorganism) can be sequenced with a conventional Sanger sequencing approach; however, amplicons derived from multiple co-existing microbes can only be distinguished by **next generation sequencing (NGS)** approaches, where individual sequences are read. Because of these properties, 16S/18S rRNA sequencing is commonly used also for the identification, classification, and quantitation of microbes within complex biological mixtures such as environmental and microbiome samples. In fact, reads belonging to different microorganisms can be also counted and their frequency over the total number of reads can be used to assess the differential abundance of different microbial species.

ribosomal RNA (rRNA)
A type of RNA that makes up ribosomes and is essential for protein synthesis.

next generation sequencing (NGS)
Advanced sequencing technologies that allow rapid sequencing of large amounts of DNA.

Table 8.5 reports some example results of how 16S rRNA sequencing has been instrumental in reaching a diagnosis in cases where conventional culture results failed to detect a pathogen. It is worth highlighting that the identification of viruses is not possible via rRNA sequencing strategies due to their lack of viral rRNA genes. Although theoretically possible, identification of protozoa via rRNA sequencing has also been applied with scarce success, as protozoa do not share widely conserved rRNA sequences that can be efficiently and broadly used as priming sites.

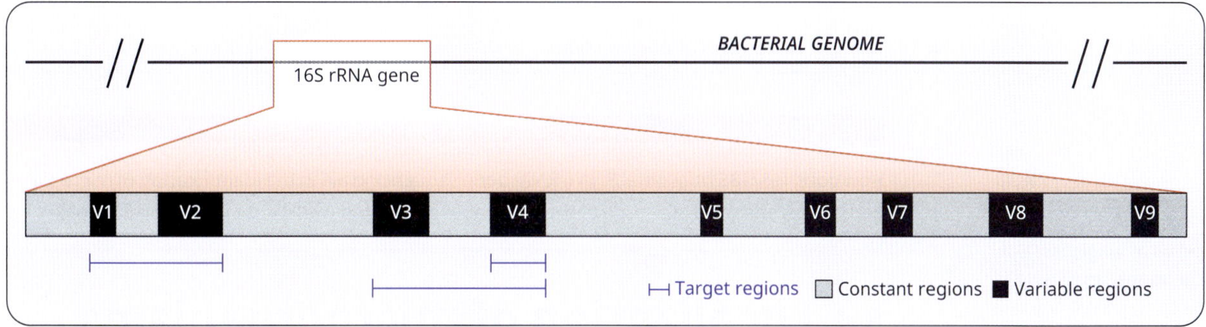

FIGURE 8.12

Representation of the 16S rRNA gene on a bacterial genome. The different variable regions (V1–9) are indicated, in the context of the neighbouring constant regions. Regions V1–2 and V3–4 are the most targeted regions for 16S rRNA sequencing. By designing primers in constant regions, the inner variable regions are amplified and then sequenced. Since this sequence is hypervariable, it is often possible to determine the genus and sometimes the exact species present in the sample.

TABLE 8.5 Application of 16S rRNA sequencing in diagnostics

Specimen type	Conventional culture result	16S rRNA sequencing result	Antimicrobial regimen after 16S rRNA sequencing result	Clinical diagnosis
Bronchoalveolar lavage	No growth	*Candida parapsilosis*	Amphotericin/ Meropenem/Linezolid	Candidal pneumonia
Bone—Mastoid	No growth	*Streptococcus pneumoniae*	Ceftriaxone	Mastoid abscess
Bone—Vertebral body/ spinal biopsy	No growth	*Kingella kingae*	Ceftriaxone	Vertebral osteomyelitis
Cerebrospinal fluid	No growth	No bacterial DNA detected	None	Aseptic meningitis
Joint fluid	No growth	*Streptococcus sanguinis*	Linezolid	Septic arthritis
Joint fluid—Elbow aspirate	No growth	*Kingella kingae*	Amoxicillin/Clavulanic acid	Septic arthritis, osteomyelitis
Joint fluid—Hip	No growth	No bacterial DNA detected	None	Bilateral hip effusion
Joint fluid—Hip aspirate	No growth	*Kingella kingae*	Amoxicillin/Clavulanic acid	Septic arthritis
Joint fluid—Hip fluid	No growth	*Propionibacterium acnes*	Cephalexin	Septic arthritis
Lung biopsy	No growth	No bacterial DNA detected	None	Lung nodule
Lymph node	No growth	No bacterial DNA detected	None	Reactive lymphadenitis
Pleural fluid	No growth	*Streptococcus pneumoniae*	Amoxicillin/Clavulanic acid	Complicated pneumonia
Pleural fluid	No growth	*Streptococcus pneumoniae*	Ceftriaxone	Complicated pneumonia
Pleural fluid	No growth	No bacterial DNA detected	None	Pleural effusion
Pleural fluid	No growth	*Streptococcus pneumoniae*	Ceftriaxone/ Clindamycin	Complicated pneumonia
Pleural fluid	No growth	*Streptococcus pyogenes*	Amoxicillin—clavulanic acid	Complicated pneumonia
Pleural fluid	No growth	*Streptococcus pneumoniae*	Ampicillin	Lung abscess

SELF-CHECK 8.9

How does 16S/18S ribosomal RNA (rRNA) gene sequencing aid in the identification of microbial pathogens, and what are some of the challenges associated with its use in clinical diagnostics?

8.4 Localization of microorganisms

In most nucleic acid-based techniques, the approach used for the DNA/RNA extraction destroys the tissue and the histopathological architecture of the collected sample. However, it can be very informative to localize the microbial nucleic acid in the context of its original tissue section. Hybridization-based detection methods for nucleic acids can be adapted to identify the presence of a microorganism *in situ*, for example on a fixed tissue sample.

Aspergillus
A genus of fungi that includes species causing infections, particularly in immunocompromised individuals.

Staphylococcus
A genus of Gram-positive bacteria that includes species causing various infections, such as skin infections and pneumonia.

8.4.1 Direct *in situ* hybridization (ISH)

Typically, ISH is used when it is relevant to identify the pathogen in the context of intact cells or tissues. The information provided by this technique can confirm the role of the organism investigated in the disease and give valuable information on the pathogen distribution and burden. This application is used, for example, when looking for EBV-driven tumours in solid organ transplant recipients. Direct hybridization of nucleic acid probes has also been applied for the localized detection of HPV, ***Aspergillus*** *species*, ***Staphylococcus*** species, ***Streptococcus*** species, ***Helicobacter pylori***, and ***Legionella***. In ISH, a tissue sample is fixed, sliced, and attached to a microscopy glass support. The sample is permeabilized, with detergents and/or by treatment with proteases to expose the nucleic acids. High-temperature treatment is performed to denature the nucleic acids. A labelled probe is applied and the temperature is reduced to allow the hybridization of the probe with the target nucleic acids in the samples. The signal associated with the label of the probe can be finally detected to visualize the microbial nucleic acids directly in the context of the histopathological structure of the sample. **Figure 8.13** exemplifies how this approach can be used to diagnose infection of foetal membranes with *Staphylococcus* or *Streptococcus* during pregnancy.

Streptococcus
A genus of Gram-positive bacteria that includes species causing strep throat, pneumonia, and other infections.

Helicobacter pylori
A Gram-negative bacterium that causes chronic gastritis, peptic ulcers, and is linked to stomach cancer.

Legionella
A genus of Gram-negative bacteria that causes Legionnaires' disease, a severe form of pneumonia.

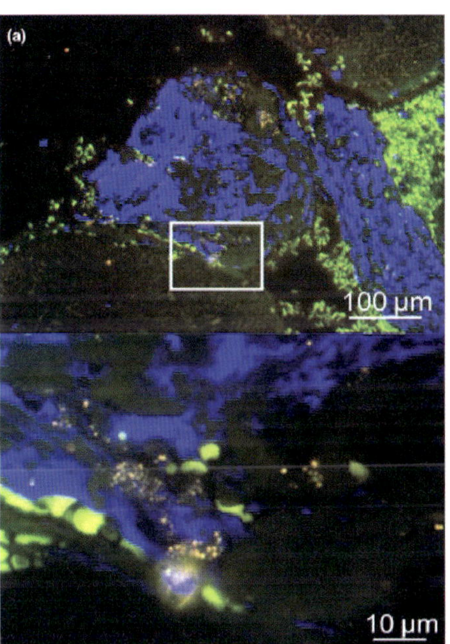

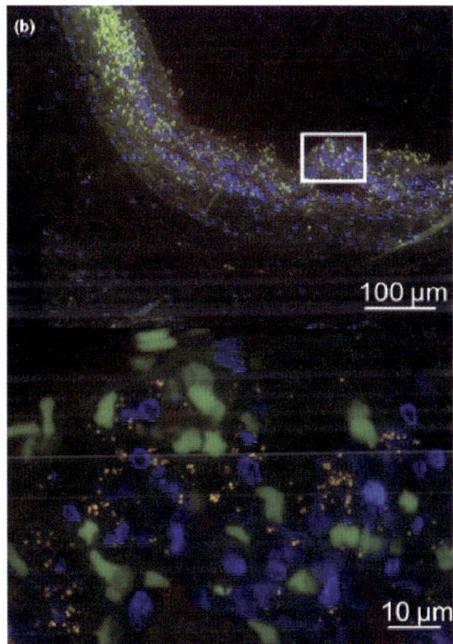

FIGURE 8.13
Visualization of *Staphylococcus aureus* **(a)** and *Streptococcus mitis/Streptococcus oralis* group **(b)** in foetal membranes. **(a)** Top: overview. Bottom: higher magnification of the inset. Bacterial nucleic acids were hybridized simultaneously with the probes STAPHYFITC (green, which is specific for *Staphylococcus* species), SAUCy3 (orange, which is specific for *S. aureus*), and NONEUBCy5 (magenta, a non-sense probe to test for unspecific probe binding, no signals detected), and stained with 4′,6-diamidino-2-phenylindole (DAPI) (blue, an unspecific nucleic acid stain). Bacteria are shown scattered within the tissue of the foetal membranes. All bacteria detected are positive both for the *Staphylococcus*-specific probe and the *S. aureus*-specific probe, thus identifying the microorganisms as *S. aureus*. **(b)** Top: overview. Bottom: higher magnification of the inset. Bacterial cells were hybridized simultaneously with the probe EUB338FITC (green, which is specific for most bacteria), STREP1/2Cy3 (orange, a *Streptococcus mitis/Streptococcus oralis* group-specific probe), and NONEUBCy5 (magenta, no signals detected). Nucleic acids were non-specifically stained with the fluorochrome DAPI (blue). Single *Streptococcus mitis/Streptococcus oralis* group cells are shown scattered in the superficial tissue layers of the foetal membranes. Schmiedel D et al. Fluorescence in situ hybridization for identification of microorganisms in acute chorioamnionitis. *Clinical Microbiology and Infection*, 20(9), O538–O541.

8.5 Quantification of microbial load

Nucleic acid amplifications, such as PCR, have been available for over a decade and have been applied in many variants to detect microbes. Traditionally, PCR products are amplified in a commercial thermocycler followed by visualization of PCR products either on a gel-based system or via fluorescence of probes associated with amplified products. However, a drawback of PCR detection is that it is difficult to quantify the presence of a pathogen, based on the PCR amplification. The development of 'real-time' PCR has added a huge advantage to traditional PCR. This technique allows for the real-time quantification of PCR products as they are produced during consecutive PCR amplification cycles. Quantitative analysis is possible during the early stages of the PCR assay, where reagents are in excess, and the amount of amplified product is small. Real-time PCR allows the estimation of the amount of DNA present in a sample, by working in the early (exponential) phase of amplification. In recent years, new quantitative molecular methods have emerged, including Transcription-Mediated Amplification (TMA) and the branched DNA (bDNA) assay. These methods are summarized in the following sections. As summarized in **Table 8.6**, microbial load monitoring by molecular techniques is widely adopted for viral infections. Due to the ability to produce highly sensitive and rapid results, this strategy has also been adopted to monitor the load of slow-growing or unculturable bacteria and parasites.

TABLE 8.6 Examples of pathogens for which molecular monitoring has been implemented to inform patient treatment

Pathogen	Type of pathogen	Disease
Human Immunodeficiency Virus (HIV)	RNA virus	AIDS
Hepatitis C virus (HCV)	RNA virus	Hepatitis C
Influenza virus	RNA virus	Influenza
Severe Acute Respiratory Syndrome Coronavirus 2 (SARS-CoV-2)	RNA virus	COVID-19
Hepatitis B virus (HBV)	DNA virus	Hepatitis B
Cytomegalovirus (CMV)	DNA virus	CMV infection
Human papilloma virus (HPV)	DNA virus	HPV infection
Herpes simplex virus (HSV)	DNA virus	Herpes infection
Mycobacterium tuberculosis	Bacterium	Tuberculosis
Chlamydia trachomatis	Bacterium	*Chlamydia* infection
Neisseria gonorrhoeae	Bacterium	Gonorrhoea
Plasmodium species	Protozoan	Malaria
Trypanosoma species	Protozoan	Chagas disease, sleeping sickness
Leishmania species	Protozoan	Leishmaniasis

8.5.1 Real-time quantitative PCR (qPCR)

Real-time **quantitative PCR** is a tool now widely used in clinical diagnostics as it allows quantification and monitoring of the microbial load in a clinical sample. Real-time qPCR follows the same principles as conventional PCR. However, detection is allowed by the incorporation of fluorescent dyes during the amplification process. The thermocycler in which the PCR occurs is connected to a detector that detects the increase of fluorescence as the PCR progresses. The method can be exploited quantitatively because samples with different concentrations of nucleic acids will require a different number of PCR cycles to reach a certain level of fluorescence. Look at **Figure 8.14** which shows the amplification curve for a dilution series of nucleic acid from a sample. Notice that each sample is run in technical duplicates (i.e., samples A–B, samples C–D etc. are technical repeats). Technical repeats are very important in qPCR because they allow control for pipetting errors. Due to the exponential nature of the PCR reaction, even small pipetting errors may result in large differences in amplification rate. Notice also that the most concentrated samples (samples A–B) reach the desired level

quantitative polymerase chain reaction (qPCR)
A variant of PCR that enables quantification of the amount of DNA in a sample in real time.

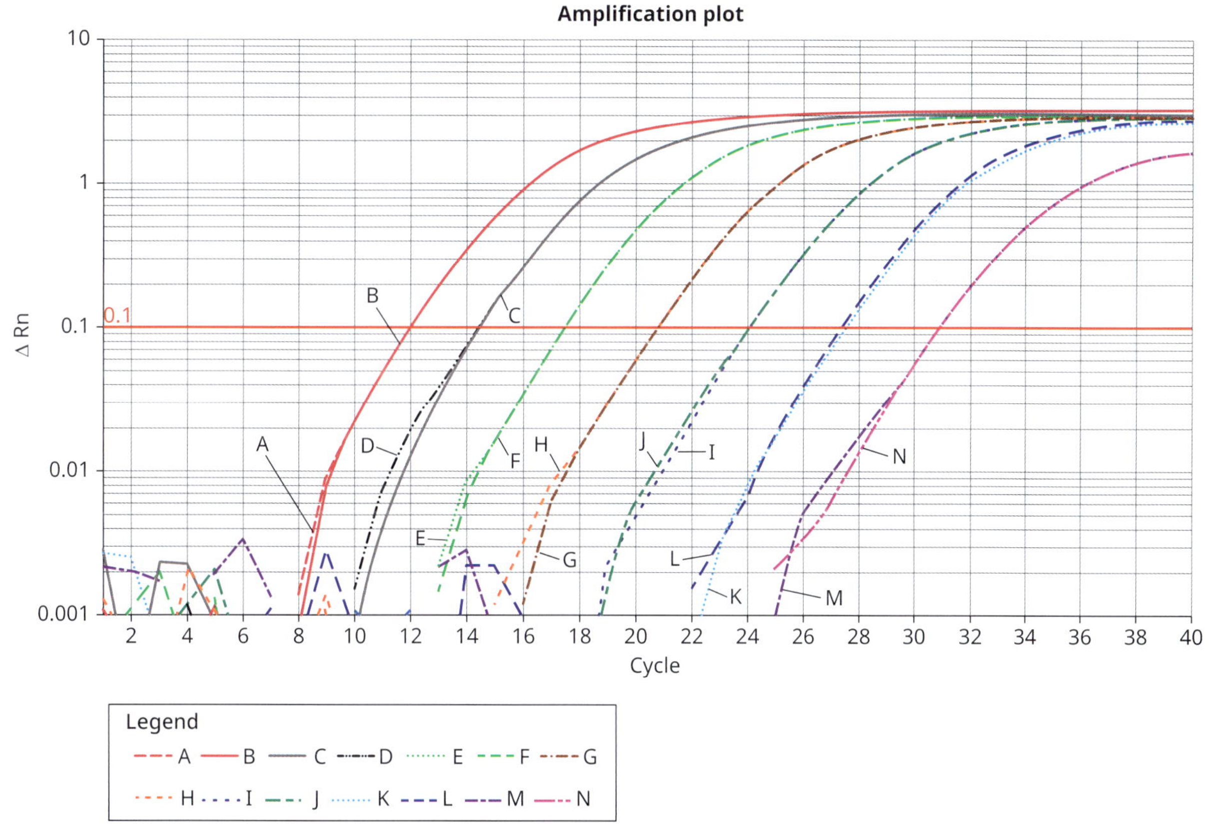

FIGURE 8.14
Example of real-time qPCR amplification plot. The number of PCR cycles is reported on the *x*-axis, while the *y*-axis reports the changes in normalized reporter value (ΔRn). The threshold of the fluorescent signal was set to ΔRn = 0.1. Samples labelled. Experiment was run in two technical repeats (A and B, C and D, E and F, G and H, I and J, K and L, M and N represent technical replicate pairs). A gradient of progressively lower concentrations was detected from samples A,B to samples M,N, which can be evidenced by the fact that more diluted samples require more PCR cycles to reach the designated signal threshold. The cycle at which the threshold is reached is called the threshold cycle or Ct. Note that no amplification is detected for samples O,P.

of amplification (indicated by the red line) with the smallest number of cycles, while the most diluted sample (samples O–P), requires the greatest number of cycles to reach the same level of amplification. The number of cycles required to reach a certain level of amplification is also referred to as the *threshold cycle*. There are two main variants of real-time qPCR. In one version a short probe, specific to the DNA of interest, is used together with the PCR primers. A fluorescent dye and a quencher are attached at the two extremities of the probe. Due to the physical proximity of the fluorescent dye with its quencher, no fluorescence is emitted by the dye. During the extension phase, the DNA polymerase encounters the quenched probe hybridized to its target DNA. The polymerase utilized for this version of real-time qPCR has an exonuclease activity and degrades the probe, therefore separating the quencher and fluorescent dye, which are released freely in the solution. In this state, the fluorescent dye can fluoresce.

Another version of real-time qPCR takes advantage of fluorescent intercalating agents, such as SybrGreen. These molecules have an increased fluorescence when they accumulate in the dsDNA of the amplicon produced by the PCR. The qPCR technique can be adapted to also detect RNA molecules. In this case, a reverse transcriptase step is included to convert the RNA molecules into cDNA.

multiplex real-time qPCR
A variant of PCR that allows simultaneous amplification of multiple targets in a single reaction.

Multiplex real-time qPCR is also possible, for example by combining primers and probes that are unique for different products and that are conjugated to different fluorescent dyes. For example, multiplex real-time qPCR for EBV, CMV, and adenovirus has been developed to measure the viral loads of these viruses in immunocompromised patients, such as solid organ transplant recipients and haematological-oncology patients.

Key Points

PCR-based techniques also find quantitative applications with significant impact on the management of certain infectious diseases. Real-time quantitative PCR (qPCR) enables accurate quantification of microbial load in clinical samples. The method monitors fluorescence levels elicited by the occurrence of the PCR reaction in real time, allowing quantification based on the number of PCR cycles required to reach a threshold level of fluorescence.

8.5.2 Transcription-mediated amplification (TMA)

As shown in **Figure 8.15**, TMA is a method that allows the amplification of RNA. TMA involves a reverse transcription step in which an RNA template is retro-transcribed into DNA by a reverse transcriptase enzyme. Subsequently, a DNA-dependent RNA polymerase is used to generate numerous RNA transcripts. Upon amplification, the RNA copies are hybridized with a complementary probe to detect them via a chemiluminescent label. Like a PCR, TMA is also run in cycles and yields 100–1,000 copies of RNA per cycle, which results in a 10-billion-fold increase in 15–30 min. TMA is becoming very popular in clinical microbiology labs, and a variety of commercial tests based on this technology are already available. TMA is widely used for the detection of RNA-based viruses like HIV and **hepatitis C virus (HCV)**, contributing to early diagnosis and monitoring of viral load. During the **COVID-19** pandemic, TMA-based tests were also employed for Severe Acute Respiratory Syndrome Coronavirus 2 (SARS-CoV-2) detection.

hepatitis C virus (HCV)
A virus that causes liver infection, which can lead to chronic liver disease and liver cancer.

COVID-19
The disease caused by the SARS-CoV-2 virus, characterized by respiratory illness and, in severe cases, pneumonia.

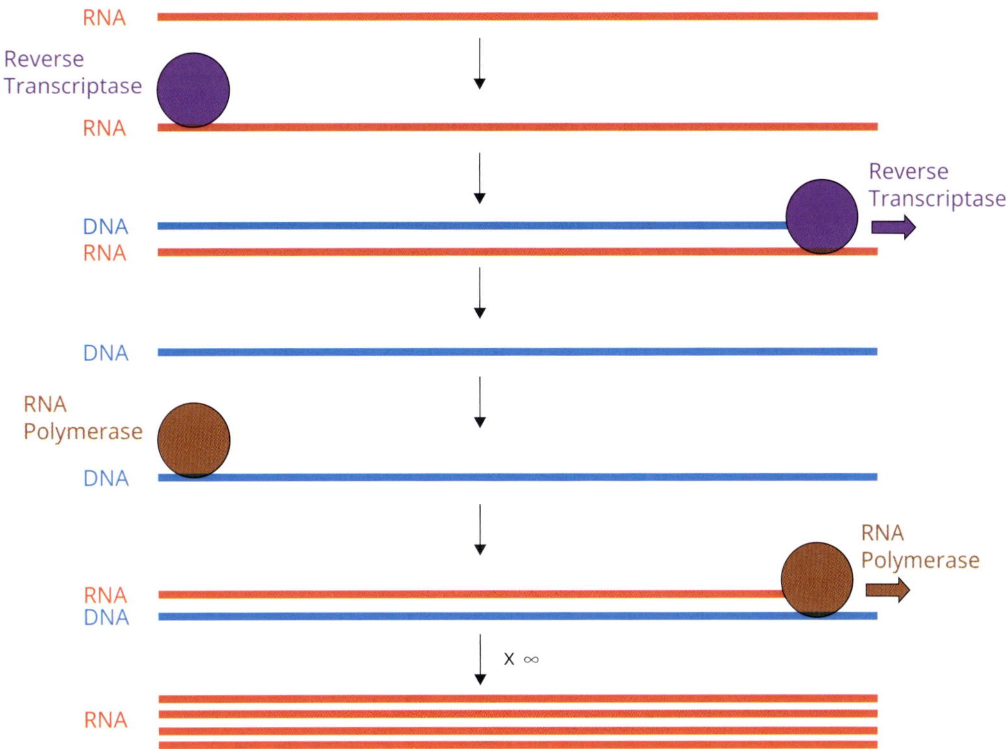

FIGURE 8.15
Diagram representing the process of transcription-mediated amplification (TMA). TMA is commonly applied for the detection of viral RNA. The RNA is first used as a template to synthesize a cDNA, with the same procedure previously described for RT-PCR. Next, an RNA polymerase is used to produce numerous RNA molecules complementary to the cDNA. These RNA copies are hybridized with a complementary probe to detect them via a chemiluminescent label.

8.5.3 Loop-mediated isothermal amplification (LAMP)

LAMP is a nucleic acid amplification technique used to amplify specific DNA sequences rapidly and efficiently. Unlike traditional PCR, LAMP operates at a constant temperature, eliminating the need for thermal cycling equipment and providing a simplified and cost-effective approach to molecular diagnostics. Additionally, LAMP utilizes a set of four to six primers that recognize distinct regions within the target DNA sequence, resulting in a robust and rapid amplification process. The amplification products are abundant and can be easily quantified with several fluorescent dyes. Positive reactions can also be detected by the naked eye, making LAMP particularly suitable for point-of-care testing in resource-limited settings. Due to its high sensitivity, specificity, and user-friendly nature, LAMP has found widespread applications in detecting various pathogens, including bacteria, viruses, and parasites. Its isothermal nature, simplicity, and potential for real-time monitoring make LAMP a promising tool in various fields, especially in remote and underserved areas where access to sophisticated laboratory equipment may be limited. In several countries, LAMP technology has been largely adopted during the SARS-COV-2 pandemic.

8.5.4 Nucleic acid sequence-based amplification (NASBA)

NASBA is another isothermal nucleic acid amplification technique designed for RNA targets. It offers high sensitivity and rapid turnaround time, making it suitable for early detection of infections. However, it is limited to RNA targets and cannot be used to directly detect DNA. Like TMA, NASBA requires specialized reagents and expertise for assay development. Similar to TMA and LAMP, NASBA is often used for the detection and monitoring of viral load for RNA viruses like HIV and HCV.

8.5.5 Branched DNA assay (bDNA)

Fundamentally different from target amplification methods such as PCR, the bDNA assay directly measures nucleic acid molecules (either DNA or RNA) at physiological levels using signal amplification rather than by replicating target sequences as the means of detection. The signal amplification in a bDNA assay is obtained by using a combination of synthetic oligonucleotide probes. As shown in **Figure 8.16**, the bDNA assay is based on a series of specific hybridization reactions, followed by a chemiluminescent detection of hybridized probes. In the simplest

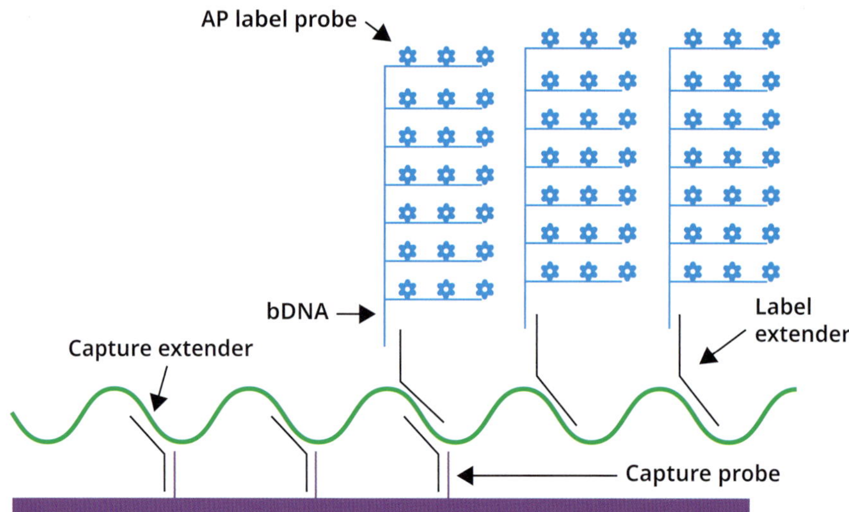

FIGURE 8.16
Diagram representing the bDNA assay. The bDNA assay is unique amongst the quantitative techniques because it does not rely on sequence amplification for the detection of the nucleic acids; rather it detects these directly based on chemiluminescent systems of signal amplification. Briefly, oligoprobes called capture probes are linked to the bottom of microtitre wells. These are hybridized with partly complementary capture extenders. Upon application of the sample containing the nucleic acid, this is captured by the capture extender branches by complementarity. The capture of the target nucleic acid is then quantified using hybridization with a label extender and a bDNA probe amplifier, which is conjugated with alkaline phosphatase (AP) activity.

versions of the method, DNA oligonucleotides with a specific nucleotide sequence, called 'capture' probes, are attached to the surface of a microwell. The capture probes bind to a series of capture 'extenders' which are partly complementary to the capture probe and the target nucleic acid molecule. After hybridization with the nucleic acids in a sample, the target nucleic acid will be attached to the extenders and immobilized on the microwell surface, while the rest of the sample is washed away. Detection of the target nucleic acid and amplification of the signal is accomplished through another series of hybridizations, involving label extenders that are partly complementary to the target nucleic acid and to 'amplifier' probes. The latter is generally linked to the alkaline phosphatase (AP) enzyme. The unhybridized amplifier probe is washed away and the substrate for AP is added. Conversion of the substrate by AP leads to the development of a chemiluminescent signal. Providing a reliable means for direct quantification in clinical specimens, the bDNA assay has been applied successfully in many areas, including the diagnosis and monitoring of patients with viral diseases, such as HIV, CMV, HCV, and **hepatitis B virus (HBV)** infections. It provides high sensitivity due to signal amplification rather than target amplification. More recent variants of the bDNA assay also allow multiplexing for the simultaneous detection of different targets. However, bDNA requires specific probes consisting of branched DNA molecules, which makes it more complex and costly. Although powerful, the bDNA assay is still less commonly used than PCR and other nucleic acid amplification techniques.

hepatitis B virus (HBV)
A virus that causes liver infection, which can lead to chronic liver disease and liver cancer.

METHOD BOX 8.2 Molecular quantification of HIV for viral load monitoring

Treatment of HIV-positive patients has advanced greatly through the development of a real-time qPCR-based technique for the routine measurement of the amount of HIV in the patient's bloodstream. The recommended lower limit of detection for the optimal monitoring of HIV patients is fifty copies/ml of blood. For this, a large sample volume (>1 ml of plasma) is generally taken and processed.

The HIV viral load is used to predict how long an infected person will stay healthy, or how rapidly the disease might progress to AIDS. A viral load greater than 100,000 copies/ml of blood normally means that the patient is likely to develop AIDS within the following five years. A viral load of fewer than 10,000 copies/ml of blood in the early stages of the disease indicates instead a reduced risk of developing AIDS.

While receiving effective antiretroviral therapy (ART) HIV-positive patients can have a viral load that is so low to be difficult to detect, even by using conventional molecular amplification methods. Patients with continual undetectable viral load are not cured of HIV, and in fact, if antiviral therapy is suspended, the virus might become detectable again. Additionally, even during effective ART therapy, viral load measurements may occasionally be able to detect a low-level increase of the virus (usually less than 500 copies/ml blood). Often, on repeat testing, these values return to an undetectable level. These temporary states of detectable viraemia are also called 'blips' and do not mean that the virus is developing resistance to the drug used in the therapeutic regime. Blips might be elicited by co-infections with other viruses, such as colds and influenza. In fact, they are more common in the winter months when these infections are more common. Many other factors have been investigated as possible causes of transient blips. In contrast, high levels (more than 500 copies/ml blood) and a prolonged increase in the viral load are often related to the development of drug resistance and/or viral mutations. In this case, modifications of the ART regime are often necessary.

CASE STUDY 8.2 Quantification of viral load during routine HIV patient monitoring

A 40-year-old male with HIV infection and on antiretroviral therapy (ART) is being assessed prior to his periodic HIV monitoring appointment at a local healthcare clinic. He has been successfully on ART therapy for several years and the values of his test results were within the normal reference range at the last two monitoring appointments six months and one year ago (i.e. undetected viraemia, normal blood cell counts for all cell populations, and normal bilirubin levels). He recently complained of some fatigue and lethargy, but no other major symptoms. The patient also confirmed consistent adherence to the ART regime since the last tests. The new results for the patient are summarized in **Table 8.7**.

Q1 Consulting also other resources, can you make a preliminary assessment of the patient's status based on these results?

Q2 What is the most likely explanation for the symptoms of the patient?

Q3 Should the patient be advised to modify his therapeutic regime?

Q4 What are the next diagnostic steps?

TABLE 8.7 Test results for the patient of Case study 8.2

Test	Technique	Result	Reference range	Units of measurement
HIV viral load	Real-time reverse transcriptase qPCR	100	Undetected	HIV RNA copies per ml
Red blood cell count	Flow cytometry	4.1	4.5–6.1	million cells per mm^3
White blood cell count	Flow cytometry	5.1	4.3–10.8	thousand cells per mm^3
CD4 cell count	Flow cytometry	8.2	5.9–11.2	hundred cells per mm^3
CD8 cell count	Flow cytometry	4.4	3.3–7.9	hundred cells per mm^3
Bilirubin blood test	Van den Bergh's reaction	0.5	0.1–1.2	mg/dl

8.6 Characterization and typing of pathogens

Molecular techniques are widely used for the in-depth characterization of a pathogen, for example, to assess its virulence, antimicrobial resistance, or similarity to an epidemiologically relevant cluster. These clues can be key for the management of individual patients as well as for the management of the whole population at risk of infection, as they help optimize therapies and public health interventions. Multiple techniques discussed so far, for example PCR, can be used to detect genetic traits of the microbe. Whole genome sequencing analysis has also become widely used for this purpose, as it provides comprehensive genotypic information on the specific microbial isolate. Nucleic acid hybridization techniques, as well as polymorphic mapping, are also adopted, although these approaches are generally only used to provide information on the relatedness of a specific isolate with other isolates of interest, based on the hybridization efficiency and/or the similarity in the pattern of **polymorphisms**.

polymorphism
A common variation in the DNA code.

8.6.1 Detection of antimicrobial resistance and virulence genes

Antimicrobial resistance continues to increase worldwide. For example, antibiotic resistance has been increasing over the past decades, with some bacterial strains that have been reported to be resistant to >15 antimicrobial agents at once. Quick and accurate determination of susceptibility of a clinical isolate to an antimicrobial can be useful in informing the patient's therapy, as the presence of resistance markers can help to infer drug susceptibility from the genotype. Similarly, some pathogens become particularly dangerous if they carry specific virulence determinants. For example, different non-pathogenic *E. coli* strains can reside within a healthy human gut microbiome. However, detection of determinants associated with pathogenic *E. coli* strains—such as the genes *Stx1* and *Stx2*, indicative of Shiga toxin-producing *E. coli* (STEC), and the gene *eaeA*, indicative of Enteropathogenic *E. coli* (EPEC)—could help distinguish commensal colonization from infection with pathogenic strains. A range of virulence factors is produced by pathogens, which can allow an infectious agent to establish infection or evade host defences. Besides the virulence factors characterizing the pathovars of *E. coli*, other broadly relevant examples include capsule formation determinants (for example in *Streptococcus pneumoniae*), gene-encoding toxins (for example in *Vibrio cholerae*, *Staphylococcus aureus*, and *Streptococcus pyogenes*), and gene-encoding adhesins (for example in *N. gonorrhoeae*). In some microorganisms, such as *V. cholerae*, the disease is entirely due to the secretion of a single virulence factor (a powerful enterotoxin in the case of *V. cholerae*).

The use of molecular diagnostic methods for the detection of drug resistance and virulence determinants is evolving rapidly and has become a routine practice in many diagnostic settings.

The use of nucleic acid-based methods to determine resistance and virulence factors can be broadly applied to bacteria, viruses, fungi, and protozoa. However, the full potential of molecular diagnostics for virulence or drug resistance testing in microbiology is yet to be reached. For example, the application of antifungal and antiprotozoal molecular testing is currently an area under development. There is still a lot to learn about the genetic determinants and mechanisms that cause virulence/resistance, especially in eukaryotic pathogens, because the processes involved are more complex than those present in viruses or bacteria. For example, evaluation of the gene expression levels (rather than solely gene presence/absence) has been particularly important in assessing fungal/protozoan infections and informing treatments. Until the full potential of drug-resistant and virulence markers is better comprehended, molecular antimicrobial testing should still be flanked by traditional culturing methods.

whole genome sequencing (WGS)
A laboratory process that determines the complete DNA sequence of an organism's genome at a single time.

Haemophilus influenzae
A Gram-negative bacterium that can cause respiratory infections and, in severe cases, meningitis.

phylogenetic tree
A branching diagram that represents the evolutionary relationships among strains, lineages, or species based on their genetic characteristics.

antimicrobial susceptibility testing (AST)
Laboratory testing to determine the susceptibility of bacteria to various antibiotics.

METHOD BOX 8.3 PCR methods for pathogen characterization

Rapid PCR-based methods have the advantage that they do not rely on extended incubations or different expression levels that might depend on the incubation or culturing conditions. Thus, viable suggestions for therapy can be made early in the diagnosis. This is especially relevant with globally important and slow-growing organisms such as *Mycobacterium tuberculosis*, especially considering that multidrug-resistant (MDR) and extremely drug-resistant (XDR) *M. tuberculosis* have been identified. PCR detection techniques have been successfully adopted to detect many antibiotic resistance genes in bacterial isolates, including resistance genes against β-lactams, aminocyclitols, aminoglycosides, chloramphenicol, fluoroquinolones, glycopeptides, isoniazids, macrolides, rifampin, sulfonamides, tetracyclines, trimethoprim, and others. A well-documented example is the detection of methicillin-resistant *Staphylococcus aureus* (MRSA). A single factor, the *mecA* gene, mediates methicillin resistance in most MRSA. The use of rapid PCR tests plays an important role in identifying carriers, for example in hospital settings, and helps to prevent and control the spreading of MRSA.

The identification of viral mutations associated with drug resistance has also been studied. Examples include mutations leading to resistance against polymerase and protease inhibitors (e.g. for HIV), acyclovir and penciclovir (e.g. for HSV), ganciclovir (e.g. for CMV), famciclovir and lamivudine (e.g. for HBV) and amantadine (e.g. for IAV). Genotypic resistance testing is nowadays widely used to manage infections with life-threatening viruses, such as HIV.

An important limitation of PCR detection of antibiotic resistance is that the same antibiotic resistance can be achieved with a variety of mechanisms, requiring different genes. For example, beta-lactam resistance can be a result of beta-lactamase production, altered penicillin-binding proteins, alteration of porins controlling membrane permeability, and other mechanisms. Several primer pairs would be necessary to cover all the potential resistance determinants. An alternative approach could be **whole genome sequencing (WGS)** as discussed in Method Box 8.4. WGS provides a comprehensive representation of the microorganism and therefore can resolve the need to design multiple primer sets.

METHOD BOX 8.4 Whole genome sequencing (WGS) for pathogen characterization

In 1981 Sanger and his colleagues published the sequence of DNA from the first viral genome. In 1995, scientists at The Institute for Genomic Research (TIGR) described the first two complete DNA sequences of bacterial genomes, those of ***Haemophilus influenzae*** and *Mycoplasma genitalium*. Since then, the genome sequences of many human pathogens have been collected. These now represent an important resource for pathogen detection and diagnostics. Comparative genomics (for example between an emerging isolate and isolates that were previously described) can pinpoint differences in virulence and host adaptations between isolates.

In recent years, there has been an increase in the use of next generation sequencing (NGS) in the diagnosis of infectious diseases. NGS is an umbrella term that describes several non-Sanger sequencing technologies, where huge numbers of reads are sequenced. Several manufacturers nowadays produce NGS instruments. For a more in-depth discussion of the principles of next generation sequencing, please see **Chapter 5**.

NGS has received both praise and criticism, with common themes around the cost of testing and complexity in interpretation. Being the most comprehensive genome-based approach, whole genome sequencing can be used for the most accurate identification of a microorganism, as well as for its in-depth characterization. Look for example at **Figure 8.17**, which shows a **phylogenetic tree** reconstruction of different *E. coli* isolates based on whole genome sequencing. Each isolate is represented by a node. Two nodes at a time are connected by branches and the depth of the branches represents the time distance that can be estimated from the sequence differences in WGS from the different isolates. The figure also presents on the right a panel of antibiotic-resistant genes, colour-coded based on the antibiotic class. The nodes at which the antibiotic resistance gene was detected by WGS are represented with filled squares.

Whilst studies have illustrated how WGS can be used for bacterial identification, its current cost and speed are unlikely to see it replace other identification methods in routine laboratories, especially in circumstances where exact microbial species identification is not even necessary to inform treatment. However, as more rapid sequencing techniques are developing, this situation may rapidly change in the future.

Rather than for pathogen detection, WGS is widely adopted for the qualitative characterization of microbes. WGS is a technique that can effectively detect a wide variety of characteristics of a pathogen, including its virulence and antibiotic resistance pattern.

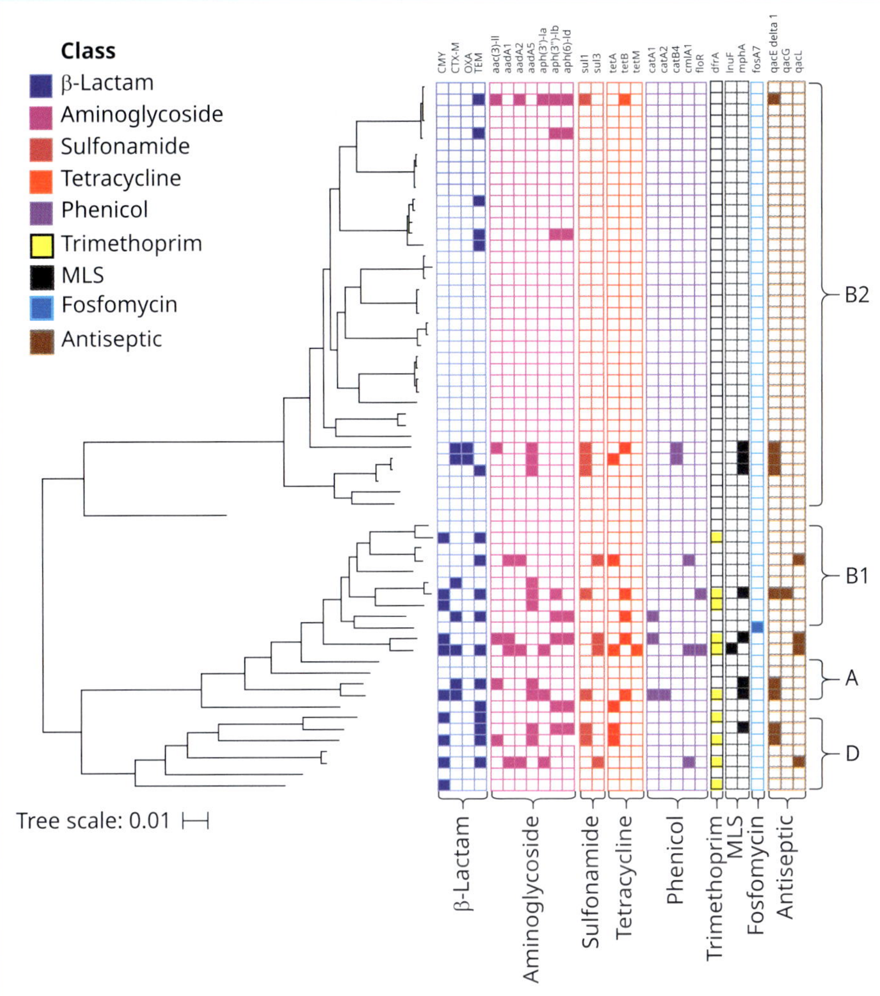

FIGURE 8.17

WGS-based phylogeny and antimicrobial resistance gene predictions in *E. coli* strains. Midpoint-rooted core genome phylogenetic tree of *E. coli* isolates with antimicrobial resistance gene predictions. Each column corresponds to the antimicrobial-resistant gene listed along the top, with colours corresponding to the antibiotic class to which that gene confers resistance. A filled box indicates the detection of that gene.

The pathogen WGS will likely play a very important role in the future in helping to prevent the emergence, transmission, and spread of infectious agents. In 2017, a report by the European Committee on Antimicrobial Susceptibility Testing (EUCAST) reviewed the present status of WGS for **antimicrobial susceptibility testing (AST)** in bacteria. The report concluded that WGS, as a tool to infer antimicrobial susceptibility, is still in its infancy. Significant research is still required before WGS can be implemented universally across a wide variety of microbes and antimicrobial resistance factors. Despite this, a large amount of data is being continuously collected on this theme, and our knowledge is expanding at a rapid pace.

For antibiotic resistance, studies have reported a 95–98% match of WGS with culture-based AST, across a panel of commonly used antibiotics and in frequently diagnosed antimicrobial-resistant pathogens such as *S. aureus*, *E. coli*, *Salmonella*, and *Klebsiella*. There are, however, accuracy problems in predicting susceptibility, especially for certain classes of antibiotics and for antibiotics that are less commonly applied, such as meropenem, vancomycin, amikacin, and tobramycin. For example, AST and resistance predictions from WGS data have a comparability of less than 90% for meropenem and amikacin. For the most part, this is because of the current lack of comprehensive knowledge about resistance mechanisms to these antibiotics. Frequent discrepancies have also been reported for the *S. aureus* resistance to mupirocin, a commonly used topical antibiotic to eradicate *S. aureus* (including MRSA) from the nasal cavity. Another problem associated with poor detection of antimicrobial resistance via WGS is that the resistance may not be due to the presence of a gene, but simply due to a change in permeability of the bacterial cell to the antibiotic. Such permeability changes may be, for example, a result of altered expression of several surface proteins. Predictions of altered expressions are still quite difficult to achieve from WGS data alone. In ***Acinetobacter baumannii***, for example, resistance to beta-lactamases can occur due to insertion elements that alter gene expression. Among the different NGS sequencing methods, short reads-based methods are believed to have the most significant limitations in this respect as they can only poorly detect repetitive insertion sequences. This specific case exemplifies how WGS needs to evolve and highlights the importance of standardization of methods and techniques, as different sequencing approaches might provide different results. For these reasons, culturing is still widely considered the gold standard for antimicrobial susceptibility testing.

Another important area to which WGS is being applied is the detection of virulence-associated genes. WGS of *S. aureus* isolates has been applied for the *in silico* detection of virulence genes such as *eta*, *etb*, and *etd* (which code for exfoliative toxins) and *sel(u–w)* (which code for production of enterotoxins by which *S. aureus* can cause food poisoning). WGS has also been widely applied to the characterization of *E. coli* pathogenic isolates, for example for the identification of Shiga toxin genes in *E. coli* O157:H7 strains, which have been responsible for outbreaks of bloody diarrhoea and haemolytic uremic syndrome (HUS) worldwide.

Overall, compared to PCR-detection methods, WGS technology has the potential to help characterize an infectious agent in depth. However, there are still limitations to this approach. In particular, the WGS technology only allows the detection (i.e. presence/absence) of genes. We still lack effective tests that enable us to quantify the expression level of the pathogen's genes during infection and how control of gene expression might influence pathogenicity and inform therapeutic interventions.

CASE STUDY 8.3 Monitoring of SARS-COV-2 outbreak

SARS-CoV-2 (Severe Acute Respiratory Syndrome Coronavirus 2) is a recently emerged coronavirus. It was first identified in December 2019 in Wuhan, China. SARS-CoV-2 was responsible for the COVID-19 pandemic, which had a profound impact on global health and society. The virus primarily spreads from person to person through respiratory droplets and can cause a range of symptoms, from mild respiratory problems to severe illness and death. In February–March 2020, SARS-CoV-2 began to spread in Lebanon. Eight cases of COVID-19 were investigated from Lebanese individuals during the beginning of the pandemic in Lebanon. Data were collected by real-time reverse transcriptase qPCR. The patients' information and detection level of the virus (expressed as Ct values) are reported in **Table 8.8**.

For the cases where the virus was detected (S1–6), the whole genome was also sequenced, and a phylogenetic tree was constructed as depicted in **Figure 8.18**. Representative genomes that were already available from other countries were also included.

TABLE 8.8 Summary of patients' demographics for Case study 8.3

Sample	Gender	Age (years)	Location of exposure	Sample type	Ct	Patient status	Signs and Symptoms
S1	Male	56	Egypt	Sputum/PBS	14.78	Hospitalized: deceased	Early stages: flu-like symptoms followed by severe dyspnoea and severe ADRS. Chest X-ray: patchy bilateral upper lobe consolidation
S2	Male	63	Local, community acquired	Nasopharyngeal VTM	26.15	Hospitalized: released	Asymptomatic
S3	Female	33	United Kingdom	Nasopharyngeal VTM	34.67	Hospitalized: released	Rhinorrhoea and headache
S4	Male	42	Iran	Nasopharyngeal VTM	16.11	Hospitalized: released	Asymptomatic
S5	Female	74	Local, community acquired	Nasopharyngeal VTM	33.8	Hospitalized: released	Dyspnoea
S6	Female	25	United Kingdom	Nasopharyngeal VTM	34	Hospitalized: released	Asymptomatic
S7	Male	36	Local, community acquired	Sputum/PBS	-	Hospitalized: released	Asymptomatic
S8	Male	22	Italy	Nasopharyngeal VTM	-	Hospitalized: released	Early stages: dysuria, fever, and flu-like symptoms followed by severe dyspnoea and ARDS. Chest CT: bilateral infiltrates with ground glass appearance

Dark blue circles on the tree branches represent the confidence of the branch (i.e. the bootstrap). Different clades are represented by different colours. Domestic cases in Lebanon (S1–6) are highlighted with stars of different colours according to the lineages they belong to. The figure reports, in brackets after the sample name, if the patient had recently travelled abroad or is likely to have acquired the infection locally.

Q1 From the data, can you estimate which patients were likely to have the highest viral loads at the time of isolation and whether there is a link between viral load and prognosis?

Q2 Based on the phylogenetic tree, how many distinct clusters of SARS-CoV-2 circulated in Lebanon at the time of the analysis?

Q3 Are any of the cases strictly related (i.e. do you suspect any direct patient-to-patient transmission amongst the patient cases reported here)?

Q4 Was there any evidence of domestic transmission of SARS-CoV-2 in Lebanon at the time of the analysis?

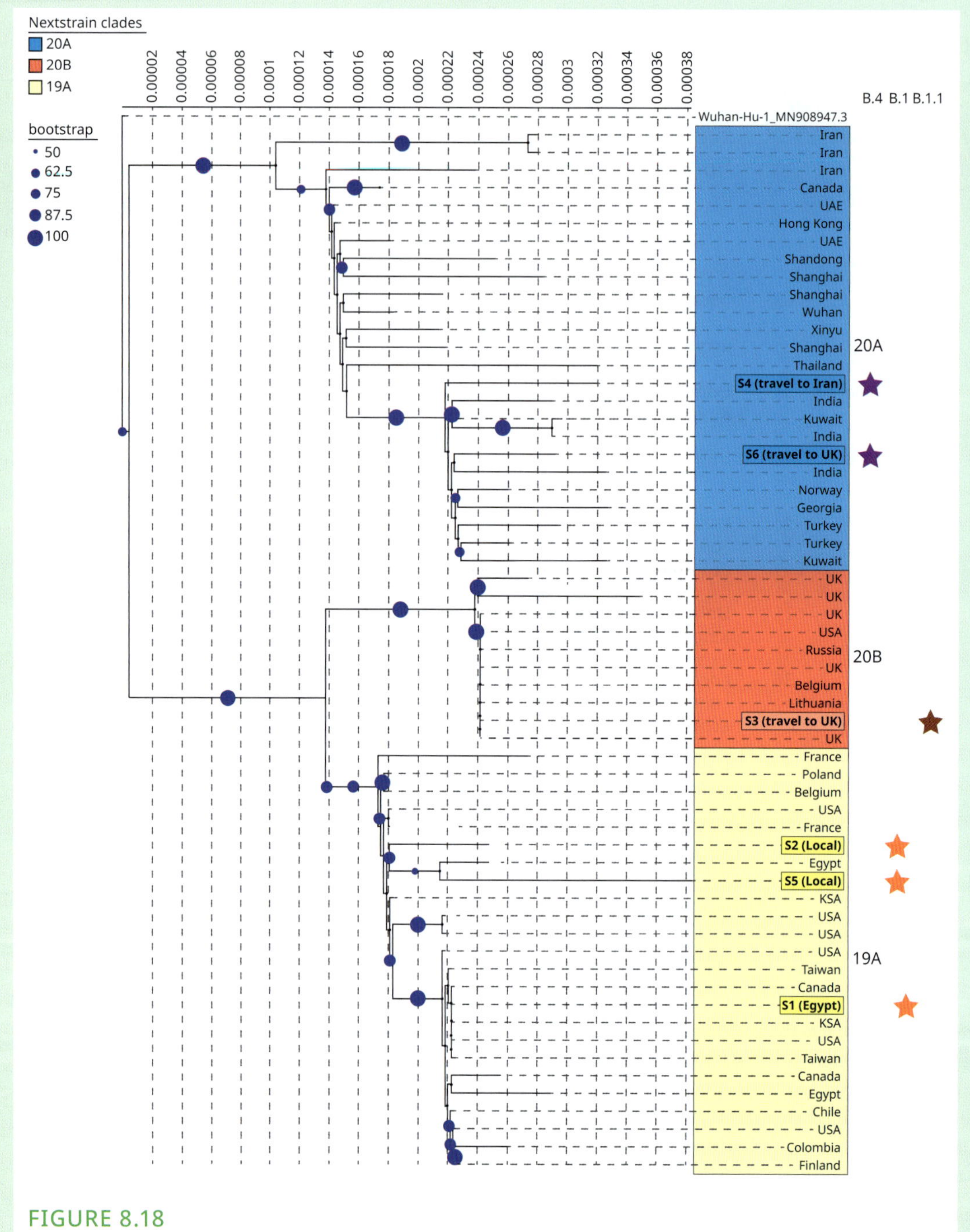

FIGURE 8.18
Phylogenetic tree of SARS-CoV-2 genomic sequences isolated from Lebanon.

8.6.2 Determination of relatedness of clinical isolates (typing)

Acinetobacter baumannii
A Gram-negative bacterium that causes a range of infections, often resistant to antibiotics, and associated with hospital-acquired infections.

Typing of microbes provides information regarding relatedness of different strains of the same microbe. Traditionally, typing methods are based on phenotypes, such as the biochemical profile, antibiogram (i.e. the antibiotic resistance profile), serotype (i.e. the agglutination reaction when treated with a certain serum), or phage type (i.e. the array of phages that a bacterial isolate carries). However, in many cases, traditional typing only provides limited differentiation as any typing technique must have sufficient discriminatory power to highlight any similar, but epidemiologically unrelated, strains investigated. Also, typing techniques must be rapid, inexpensive, easy to perform, and reproducible. Different nucleic acid-based techniques have been useful for these applications and are summarized in the sections below.

Genome–genome hybridization

Nucleic acid–nucleic acid hybridization, especially DNA–DNA hybridization (DDH), has been used by microbiologists since the 1960s to assess how related or distant two strains or isolates are. This approach is still important in the taxonomic definition of microbial species. The method has had enormous relevance, especially for the classification of prokaryotes. As shown in **Figure 8.19**, genome–genome hybridization is based on the fact that a positive hybridization

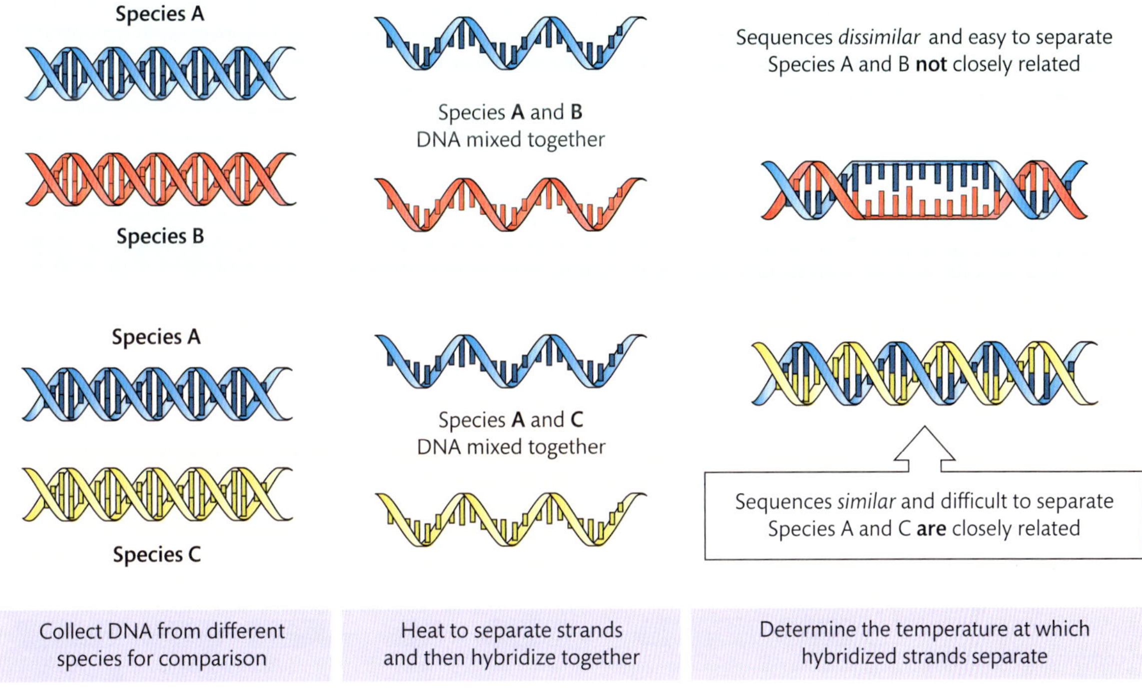

FIGURE 8.19
Diagram illustrating DNA–DNA hybridization to determine if two DNA samples are similar enough to be considered derived from isolates/strains of the same species. See also the main text for more details.

reaction between nucleic acids derived from two different sources indicates sequence homology and therefore genetic relatedness between the two isolates.

Hybridization assays require that a nucleic acid strand is derived from a known microbe while the other strand is derived from the microbe to be characterized. Different alternative methods have been proposed to evaluate the degree of relatedness of two genomes and are based on either measuring the degree of hybrid reassociation or the thermal stability of the hybrids. In the simplest form of DDH, the DNA of one microbe is labelled and then mixed with the unlabelled DNA to be compared. The mixture is incubated to allow homoduplex DNA strands to dissociate and then cooled to form heteroduplex (hybrid) DNA. Hybridized sequences with a high degree of similarity will bind strongly and more energy will be required to separate them. To assess the DNA homology, the hybrid dsDNA is bound to a column and heated using small temperature increments. At each step, the column is washed. If the hybrid dsDNA melts it will become single-stranded and washed off the column. Hence, the temperatures at which labelled DNA is eluted from the column reflect the similarity between the sequences. A self-hybridization sample control serves as a reference. For prokaryotes, a similarity value greater than 70% and ≤5°C in the difference between the melting temperature of the hybrid and non-hybrid DNA normally indicates that the compared strains belong to the same species. DDH has often been criticized as cumbersome and obsolete. For these reasons, DDH methods are being progressively replaced by alternative approaches based on genome sequencing and genome-wide sequence comparisons.

Multilocus sequence typing (MLST)

MLST is a method that characterizes isolates of microbial species using the DNA sequences of internal fragments of multiple housekeeping genes. The approach involves PCR amplification followed by DNA sequencing: 450–500 base pair internal fragments are generally selected, for each of the housekeeping genes used. For different species, the housekeeping genes dissected can vary. Nucleotide differences between strains can be found in a variable number of genes depending on the degree of differentiation required. However, in most MLST typing schemes, 7 housekeeping genes are being evaluated, like in the example in **Figure 8.20**.

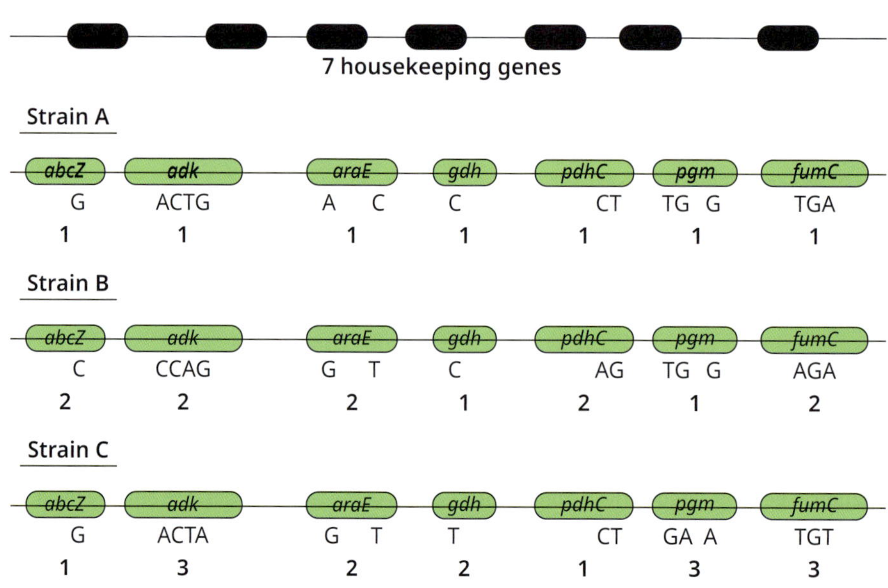

FIGURE 8.20
An example of MLST typing. In this typing scheme, 7 genes are considered, namely *abcZ*, *adk*, *araE*, *adh*, *pdhC*, *pgm*, and *fumC*. For each of these genes, different genotypes (i.e. SNPs or small indels) have been reported for the species of interest. Each genotype is given a number, hence a certain combination of alleles for the 7 genes of interest can be characterized by a 7-digit code. Strains that share an identical pattern have the same sequence type and are likely closely related.

MLST can be used readily during outbreaks of infection and has been validated for the typing of a wide range of important fungi and bacteria, such as *Candida albicans*, *Escherichia coli*, *Salmonella enterica*, *Campylobacter jejuni*, *Enterococcus faecalis*, *Staphylococcus aureus*, *Streptococcus pneumoniae*, *Haemophilus influenzae*, *Listeria monocytogenes*, and *Bordetella pertussis*.

Polymorphic typing

Systems that use different types of gels are used for the characterization of microbes, for example during outbreaks. Several methods to determine the clonal types present in microbial populations have been investigated. These include pulse-field gel electrophoresis (PFGE), restriction fragment length polymorphism (RFLP), and amplified fragment length polymorphism (AFLP).

As displayed in **Figure 8.21**, PFGE uses genomic DNA isolated from strains of target organisms and digested by restriction enzymes that cut the genome into 10–40 fragments that are then visualized into fingerprint patterns upon gel electrophoresis in an apparatus that switches the direction of the current in a predetermined manner. This apparatus is used to increase the resolution of differences between high molecular weight DNA fragments. As exemplified by **Figure 8.21**, PFGE can be useful for the rapid comparison of isolates and enables the identification of clonal subgroups of isolates. PFGE has been the gold standard for the rapid determination of how bacterial strains are related and has been applied in the analysis of outbreaks of many pathogens including MRSA and vancomycin-resistant strains of *E. faecium*. The technique is highly discriminatory and reproducible between laboratories. Public health laboratories have standardized PFGE for its application in outbreak investigations. Digital imaging has provided the possibility for cross-gel comparative analyses as well. However, the approach is relatively complex and not practical in many laboratory settings. Additionally, as it utilizes the whole genome from a pathogen, it does require culturing of the pathogen for quantitative extractions of high-quality genomic DNA.

RFLP is a technique in which microbes may be differentiated by analysing the pattern derived from the cleavage of their DNA. The fragments produced and detected are generally smaller than those obtained in PFGE. Additionally, in the RFLP method, the DNA fragments are

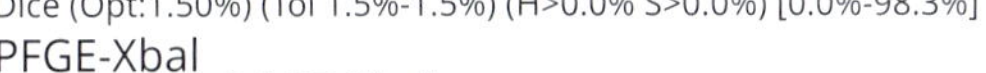

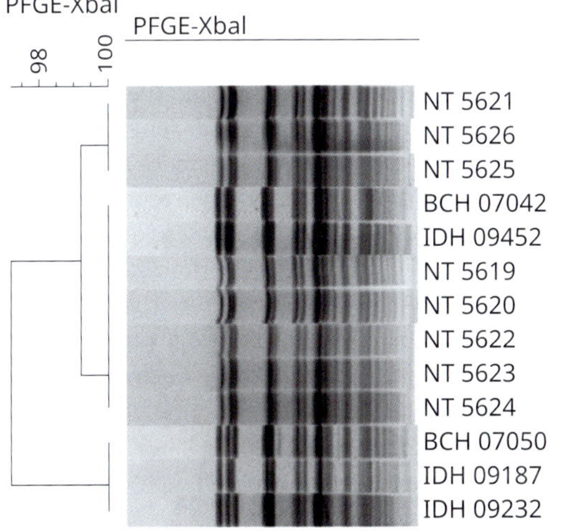

FIGURE 8.21
Pulsed-field gel electrophoresis (PFGE) analysis of XbaI-digested genomic DNA of *Shigella sonnei* strains isolated during the outbreak along with the strains isolated in the surveillance study in Kolkata, India during 2016. Similar band patterns on the gel indicate closely related strains. The electropherograms of the individual strains have been rearranged according to the closeness of the relationship between characterized strains. For example, strains NT 5621, 5626, and 5625 show a near-identical band pattern. Debnath, F., Mukhopadhyay, A., Chowdhury, G., Saha, R., and Dutta, S. (2018). An outbreak of foodborne infection caused by *Shigella sonnei* in West Bengal, India. *Japanese Journal of Infectious Diseases*. 71. 10.7883/yoken.JJID.2017.304.

transferred to a membrane after the electrophoretic run and the alleles of interest are detected via specific probes, by Southern blot.

AFLP is based on the PCR amplification of restriction fragments from a digest of genomic DNA. Fingerprints are produced without prior sequence knowledge, for example by using a limited set of primers that can frequently bind to the genome in multiple sites. This approach combines the robustness and reliability of RFLP and the power of a PCR technique. The amplified fragments can be separated on a gel or by capillary electrophoresis.

Comparative genomics

single nucleotide polymorphisms (SNPs)
Variations at a single position in a DNA sequence among individuals.

Comparative genomics can be performed by sequence alignments. If WGS is performed, the sequence can be compared to online databases to identify the relatedness to previously described strains. A large variety of algorithms have been developed for this purpose and help the construction of phylogenetic trees. Broadly adopted methods are based on sequence alignment of a set of specific genes, and calculation of the genetic distance based on **single nucleotide polymorphisms (SNPs)** as well as *K-mer comparison* methods. In the latter, the whole genome is computationally fragmented in sequences of a K-mer size. The sequence profile of the isolate is compared with the profiles from previously described isolates to infer similarity.

Microarrays

microarray
An assay used to simultaneously detect the expression of thousands of genes by hybridizing them to known nucleic acid sequences fixed on a solid surface.

Microarrays are small solid supports onto which the sequences of thousands of different nucleotide sequences are immobilized, or attached, at fixed locations. These are created by robotic instruments. The supports usually consist of glass microscope slides but may also be silicon chips or nylon membranes. The DNA is spotted or synthesized directly onto the support. Gene sequences in a microarray are arranged on the support in an ordered way since the analysis devices use the location of each spot to index all the different sequences. Whilst whole genome sequencing is based on sequencing by synthesis, microarrays are based on hybridization using fluorescently labelled probes. Microarrays are simple to employ and construct. The limitations of this approach include cost and impracticality, especially in clinical diagnostic settings that have limited resources.

In a typical microarray experiment, the microarray support is hybridized competitively with two DNA samples, for example, a test and a control, each fluorescently labelled with different dyes, usually with Cy5 (red) and Cy3 (green). The fluorescently labelled probes within the two samples hybridize competitively with the immobilized target DNA. The unbound probes within the microarray are then washed off and the chip is read in a scanner. The fluorescent labels are excited by a laser, and a digital image of the array is obtained. The red-to-green fluorescence ratio is calculated. The intensity of the red and green fluorescence at each spot is proportional to the amount of Cy5 and Cy3-labelled probe that has hybridized to the DNA at that spot. Microarrays are commonly used in epidemiological investigations, as genomic variations can be characterized and differentiation between strains can be performed. The use of DNA arrays for typing has been demonstrated for example for HIV and *M. tuberculosis*. Various analysers are being developed that employ DNA microarray technology for the identification of bacterial isolates. These include solutions where the pathogen DNA can be detected directly

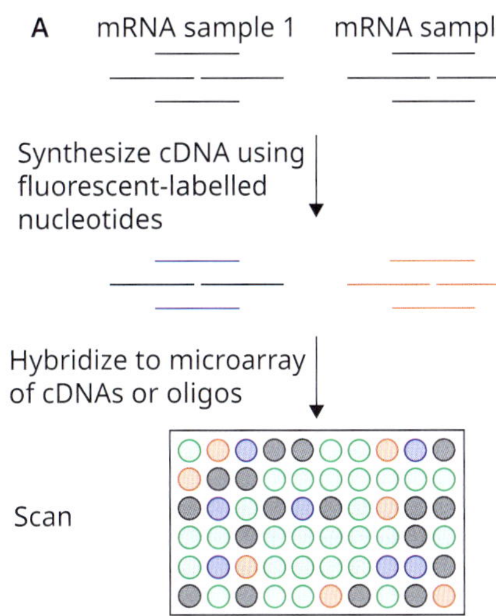

FIGURE 8.22
Use of microarrays to study gene expression. In this example, 60 data spots show the hybridization efficiency of 60 different cDNAs (representing 60 genes) from two different samples (mRNA sample 1, for which green fluorescently labelled cDNA was synthesized and mRNA sample 2, for which red fluorescently labelled cDNA was synthesized). A green signal indicates high gene expression in sample 1, while a red signal indicates high gene expression in sample 2. A yellow signal (resulting from equal levels of sample 1 and sample 2 cDNA hybridizations) indicates equivalent gene expression in both samples. Grey (no signal) indicates lack of gene expression in both samples.

from samples. It is therefore likely that such systems will become more embedded in clinical microbiology diagnostic laboratories in the future.

Importantly, microarrays can be applied also for expression analysis. **Figure 8.22** shows the workflow of a microarray experiment used for transcriptional studies. In this case, a cDNA library is constructed from an original RNA sample. The application of expression microarrays to clinical microbiology holds great promise for providing a tool to quantify the gene expression of the pathogen during infections of the host. However, their application in this context is still very limited.

SELF-CHECK 8.10

How are molecular techniques, such as PCR and whole genome sequencing, utilized for the characterization and typing of microorganisms?

8.6.3 Characterization of microbial diversity

Most molecular assays can target only a limited set of microbes using primers or probes. **Metagenomics** represents a unique method that allows us to collect information from all DNA or RNA molecules present in a sample. This allows the analysis of the composition and characteristics of the host microbiome. The capacity to detect virtually all potential pathogens has great potential utility for diagnosing infectious diseases.

metagenomics
The study of genetic material recovered directly from environmental samples, allowing the analysis of microbial communities without the need for culturing.

So far, several studies have provided examples of the potential use of metagenomics in clinical and public health settings. For example, metagenomics was used to diagnose neuroleptospirosis in a critically ill child with meningoencephalitis and allowed clinicians to provide the patient with appropriate treatment and eventually clear the infection.

Even if metagenomics is collecting some recent success (especially when applied for research purposes), its clinical diagnostic applications have remained quite limited. A complex interplay of microbial and host determinants influences human health, and it is sometimes uncertain if a microorganism detected by metagenomics represents a contaminant, a commensal, or a disease-causing pathogen. Moreover, clinical labs follow universal reference standards and well-established methods to ensure quality. Considering that metagenomics is still an emerging technology, standardized metagenomic assays are still missing. However, the breadth and potential clinical utility of these applications are likely to transform the field of diagnostic microbiology in the future.

Figure 8.23 represents the workflow for *untargeted metagenomics* of clinical samples. This is probably the most comprehensive approach for the characterization of the microbial community in a sample and the diagnosis of a potential infection. Theoretically, all pathogen groups can be identified in a single experiment. However, no microbiome-based tests have been clinically validated so far for the diagnosis of disease or to inform treatment.

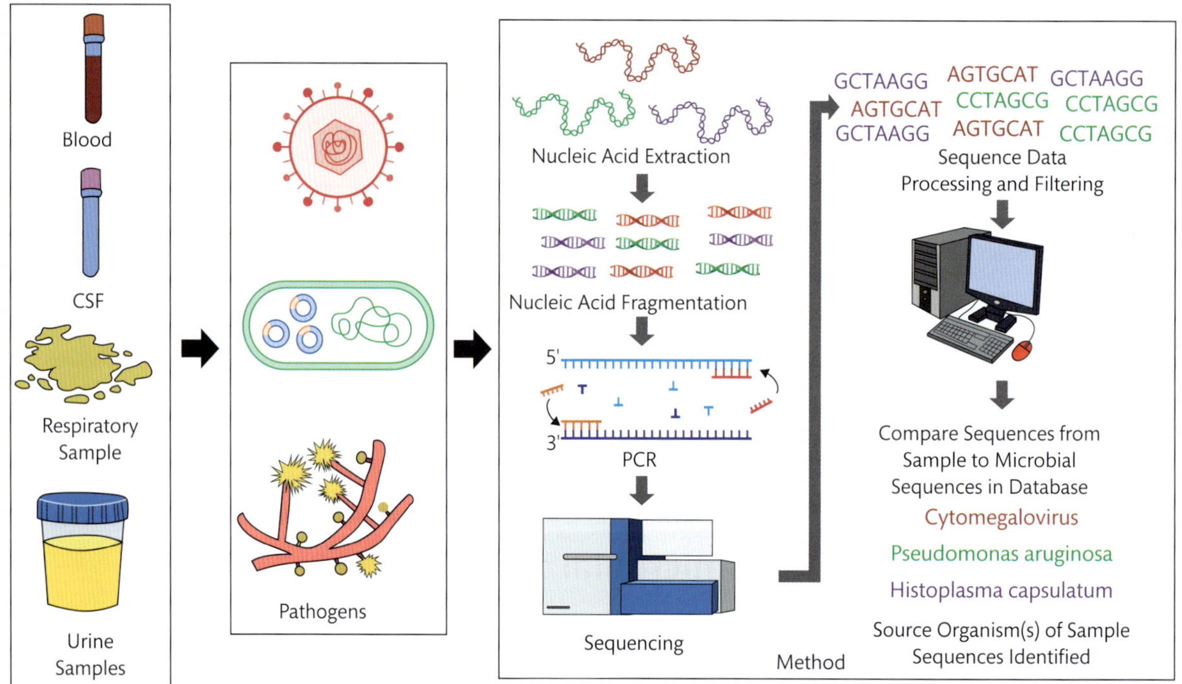

FIGURE 8.23

Workflow for untargeted metagenomics. Nucleic acids can be extracted from a variety of biological samples which may contain a community of microbes (i.e. viruses, bacteria, fungi, and parasites). The mixed nucleic acids from these complex sources are amplified and sequenced. The sequences generated are mapped to genomes available in sequence databases. This allows the identification and relative quantification of the different microbes present in the clinical sample.

A future clinical application of microbiome analysis may be to manage and treat ***Clostridioides difficile*** infections. *C. difficile* is an opportunistic bacterium that can infect the gut. *C. difficile* infection occurs only if the microbiome has been altered, for example after a course of broad-spectrum antibiotics. The relevance of the microbiome for *C. difficile* colonization is emphasized by the fact that faecal transplants have an 80–90% effectiveness in treating the infection. The application of metagenomics to analyse the microbiome has also helped to propose bacterial probiotic formulations that can be delivered to prevent or treat *C. difficile* infections.

Clostridioides difficile
A Gram-positive bacterium that causes severe diarrhoea and colitis, often associated with antibiotic use. Previously named *Clostridium difficile*.

Another use of metagenomics is for the analysis of microbial diversity, which can provide information on the patient's health and risk for several conditions. Alteration of the microbiome is known as **dysbiosis** and can be related to many conditions, spanning from obesity to diabetes mellitus and inflammatory bowel disease. Therapeutic interventions into the microbiome may represent a new avenue to develop treatments for these conditions, and metagenomics may represent a valuable monitoring system to assess the need and evaluate the effectiveness of these approaches.

dysbiosis
An imbalance in the microbial community, often in the gut, that can be associated with disease.

Key Points

Metagenomics offers a unique approach to characterizing microbial diversity by analysing all DNA or RNA molecules present in a sample, providing insights into the composition and characteristics of the entire host microbiome. This method has potential utility in diagnosing infectious diseases by detecting all potential pathogens present. While metagenomics has shown success in research settings, its clinical diagnostic applications are still limited due to the lack of standardized assays and methodologies. However, its breadth and potential clinical utility suggest it could transform diagnostic microbiology in the future. Metagenomics may also have future applications in managing conditions such as *Clostridioides difficile* infections and assessing microbial diversity to evaluate health and risks for various diseases.

8.7 Impact of molecular detection on pathogen evolution

Our ability to detect and identify microorganisms through the application of molecular methods has had huge positive impacts on our ability to control the spread of pathogens and improve patient outcomes. However, microbial genomes are not fixed entities and evolve continuously. This section makes some considerations on how the application of molecular methods can lead to the selection of pathogen variants that can elude these detection systems, highlighting the need for vigilance when deploying molecular diagnostic methods.

8.7.1 Elusion of molecular diagnostics by *Chlamydia trachomatis*

The bacterium *Chlamydia trachomatis* is the most common cause of bacterial sexually transmitted infections (STIs) in industrialized countries such as the United Kingdom and the United States. Several countries, including the United Kingdom, have therefore implemented screening programmes for *C. trachomatis*, based on nucleic acid amplification tests.

Most nucleic acid-based commercial tests to detect *C. trachomatis* target either conserved regions of the 16S rRNA, the *Chlamydia* gene *omp1,* or the *Chlamydia*-specific 7.5 kbp plasmid, also called a cryptic plasmid. A non-nucleic acid-based method is also commonly adopted, which detects the *Chlamydia* lipopolysaccharide (i.e. the *Chlamydia* rapid test).

Notably, a *C. trachomatis* variant with a deletion in the cryptic plasmid was identified in 2006 in Sweden, after an apparent 25% decrease in *C. trachomatis* infections was noted, due to failure of the detection test. Look at **Figure 8.24**, showing the map of the mutant and wild-type versions of the *C. trachomatis* plasmid.

Notice how the new variant has a 377 bp deletion in the CDS1 region of the plasmid. This area is targeted by two commercial tests, the M2000 (Abbott Laboratories, Abbott Park, IL, USA) and Amplicon CT/NG PCR (Roche Diagnostic Systems, Branchburg, NJ, USA) tests, which were widely adopted before the discovery of this variant. Due to this deletion, detection of the *C. trachomatis* variant with these tests has yielded false-negative results. You might also notice that another modification is detectable on this plasmid, which is a 44 bp duplication of a region in CDS3. However, this mutation did not impact the detection per se, as the area was not targeted by any commercially available detection assays.

The *Chlamydia* cryptic plasmid variant remained detectable by molecular assays based on the detection of the 16S rRNA (Gene probe Aptima Combo 2), the *omp1* gene (Artus RealArt CT PCR) or different unaffected areas of the cryptic plasmid (BD Biosciences ProbeTec ET). It is

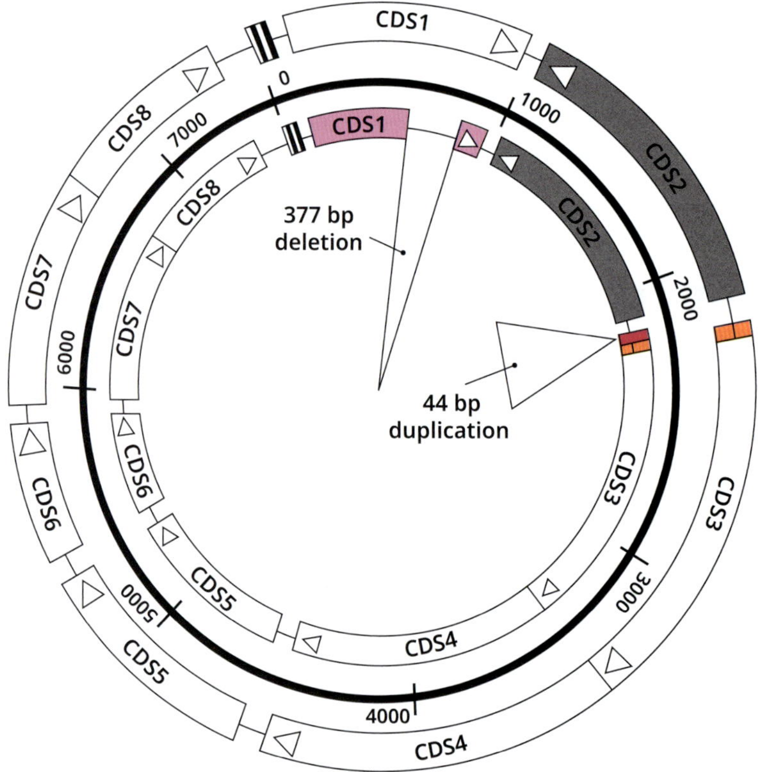

FIGURE 8.24
Comparison of *Chlamydia* cryptic plasmid variant (inner circle) with the wild-type version of the plasmid (outer circle). The mutant plasmid carries a 377 bp deletion in CDS1 and a 44 bp duplication immediately upstream of CDS3 (shown in red).

additionally still detectable by the non-molecular *Chlamydia* rapid test, based on lipopolysaccharide detection. It has been hypothesized that this mutation might have been positively selected as it conferred an advantage to that *Chlamydia* variant as being less likely to be detected, and therefore treated. This case also exemplifies that clinicians and biomedical scientists should be alerted to the possibility that some pathogens might develop variants that are not detectable by widely adopted tests. It also emphasizes the importance of alternative detection routes, when possible, to confirm a negative result.

CASE STUDY 8.4 Detection of a novel pathogen variant

A 28-year-old heterosexual African man attended the Ambrose King Centre (AKC) at the Royal London Hospital in December 2006 complaining of dysuria over a three-week period. The patient reported pain while passing urine and revealed that he had engaged in unprotected sex with four partners in the United Kingdom and mainland Europe during the previous three months. A genital examination revealed urethral discharge. Tests for *Neisseria gonorrhoeae*, *Treponema*, HIV, and *Mycoplasma genitalium* were all negative. The urine sample collected was also negative for *C. trachomatis* by the ProbeTec ET assay (a routine molecular assay used to detect *C. trachomatis*).

The sample was further tested with both the *C. trachomatis* lipopolysaccharide rapid test and the Amplicon CT/NG PCR assay, and it was found to be reactive in the first test but again negative in the second test. To examine whether the lipopolysaccharide rapid test was a false positive, a RealArt CT PCR assay was applied, to detect the chromosomally encoded gene *omp1*. An Aptima Combo 2 test was also used to detect 16S ribosomal RNA. The sample was positive for both *omp1* and *Chlamydia* 16S rRNA (**Table 8.9**). A set of in-house PCRs were set up to detect different regions of the *Chlamydia* cryptic plasmid, and they all appeared negative (**Figure 8.25**). In contrast, an in-house PCR test for *omp1* was positive.

Q1 What is the most likely explanation for the data reported in this case?

Q2 Is it possible that the patient carries a Swedish variant of *C. trachomatis*, with the 377 bp deletion in the cryptic plasmid?

Q3 What next steps would you recommend in term of recommendations to the patient, patient treatment, and public health reporting?

TABLE 8.9 Summary of test results for the patient of Case study 8.4
C. trachomatis tests performed on the urine sample of the proband

Test manufacturer or developer	Test name	Test target	Result
Commercial tests			
BD Biosciences	ProbeTec ET	Plasmid	-
Roche	Amplicor CT/NG PCR	Plasmid	-
Artus	RealArt CT PCR	*omp1*, chromosomally encoded	+
GenProbe	Aptima Combo 2	16S ribosomal RNA	+
University of Cambridge	*Chlamydia* rapid test	Lipopolysaccharide	+

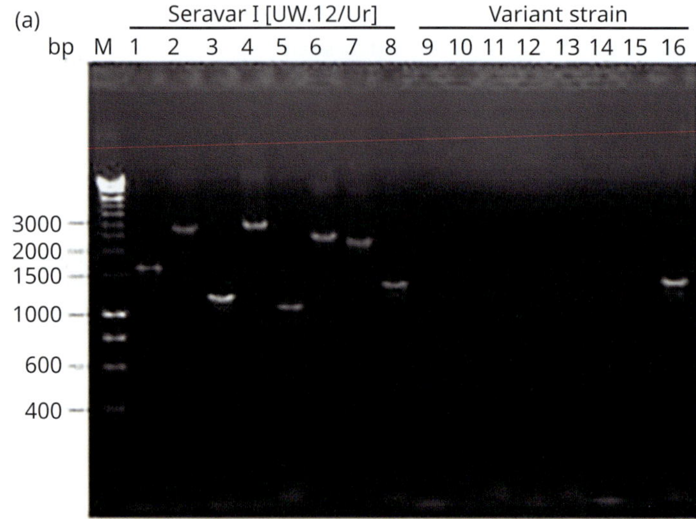

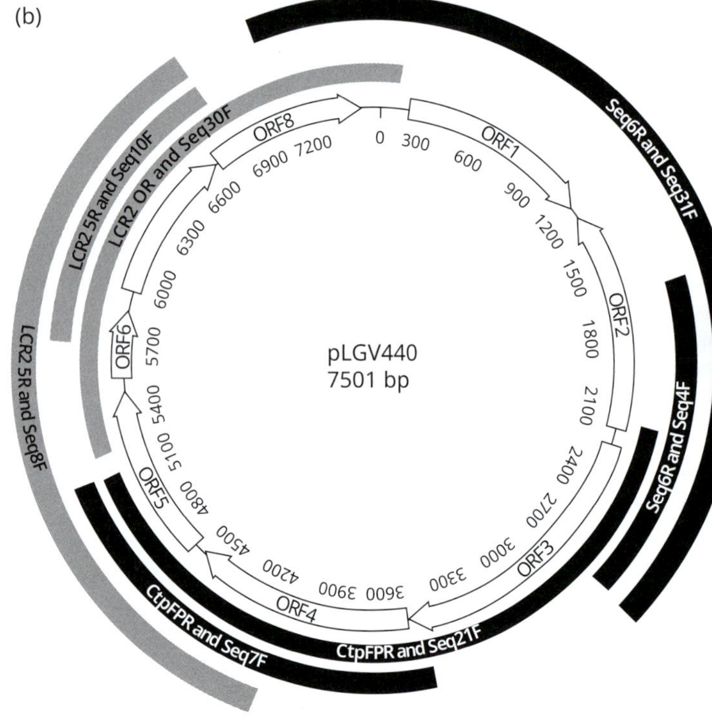

FIGURE 8.25

PCR analysis of the *C. trachomatis omp1* and cryptic plasmid sequences for the case study. **(a)** A PCR analysis was performed using a WT strain of *C. trachomatis* (lanes 1–8) as a control and DNA extracted from the clinical sample of the case study (lanes 9–16) as templates. PCR reactions were performed with seven primer pairs specific for various regions of the 7.5 kbp cryptic plasmid (lanes 1–7 and 9–15) or with primers that amplify the *omp1* sequence (lanes 8 and 16). The PCR products were separated by electrophoresis in a 1% agarose gel and stained with ethidium bromide. Lane M represents the molecular size markers. The primers and expected sizes of the corresponding amplification products are as follows: lanes 1 and 9, CtpFPR and Seq7F (1680 bp); lanes 2 and 10, CtpFPR and Seq21F (3037 bp); lanes 3 and 11, Seq6R and Seq4F (1228 bp); lanes 4 and 12, Seq6R and Seq31F (3221 bp); lanes 5 and 13, LCR25R and Seq10F (1098 bp); lanes 6 and 14, LCR25R and Seq8F (2643 bp); lanes 7 and 15, LCR20R and Seq30F (2368 bp); and lanes 8 and 16, *omp1* primers (1327 bp). The WT strain produced bands for each primer set at the expected molecular weight. In contrast, the variant strain failed to produce a band with any primer set for the plasmid but amplified *omp1*. Magbanua, J.P., Goh, B.T., Michel, C.E., Aguirre-Andreasen, A., Alexander, S., Ushiro-Lumb, I., Ison, C., and Lee, H. Chlamydia trachomatis variant not detected by plasmid based nucleic acid amplification tests: molecular characterization and failure of single dose azithromycin. *Sex Transm Infect.* 2007 Jul;83(4):339–43. doi: 10.1136/sti.2007.026435. Epub 2007 Jun 13. PMID: 17567684; PMCID: PMC2598672. **(b)** Cryptic plasmid map, showing the open reading frames (ORFs) and regions covered by the various PCR products obtained from the WT strain with the primer pairs used in (a).

Chapter summary

- Viruses are obligate intracellular parasites, reliant on host cells for replication. They are further classified based on their genome architecture, which includes single-stranded or double-stranded DNA or RNA, with variations in strand polarity, segmentation, and circularity.
- Bacteria are prokaryotic microorganisms. With some exceptions, bacteria have a rigid cell wall containing peptidoglycan. Their genomes are always made of double-stranded DNA, with most having a single circular chromosome supplemented by smaller circular DNA molecules called plasmids.
- Fungi are eukaryotic organisms with unicellular (yeast) or multicellular (mould) forms. They possess a rigid cell wall primarily composed of chitin. Fungal genomes are organized in linear chromosomes of double-stranded DNA.
- Protozoa, also eukaryotic microorganisms, can be free-living or parasitic and vary widely in genome size and complexity. Parasitic protozoa tend to have smaller genomes compared to free-living species.
- The human microbiome is a complex community of microorganisms living in and on the body. It affects many aspects of health, including digestion, metabolism, and immune function.
- Clinical samples, including swabs, blood, urine, and stool, are collected for various purposes such as infection diagnosis, patient monitoring, and epidemiological surveillance.
- Nucleic acids can typically be extracted from diagnostic samples in amounts that are sufficient for techniques involving amplification. For some applications requiring high quality and quantity of nucleic acids, isolation and culturing of the microbe might be necessary.
- Nucleic acid-based techniques offer rapid identification, high specificity, and exceptional sensitivity.
- Despite their advantages, nucleic acid-based techniques are prone to false-positive and false-negative results, primarily due to contamination and sample degradation issues. Quantitative applications remain limited, and inferring phenotypic features solely from genotype presents challenges.
- Polymerase chain reaction (PCR) is a fundamental molecular biology technique. The process involves thermal cycling, producing billions of copies of the target DNA sequence within hours.
- PCR and its variants have revolutionized diagnostic microbiology, offering rapid and sensitive detection of pathogens from various clinical samples.
- Besides PCR, other amplification-based methods have emerged, such as strand displacement amplification (SDA) and the ligase chain reaction (LCR).
- 16S/18S rRNA gene sequencing targets conserved nucleic acid regions and can aid in identifying bacterial/fungal pathogens.
- Direct *in situ* hybridization (ISH) can be used to detect pathogens within intact tissues, preserving their histopathological architecture. The technique is based on the treatment of a sample with a probe targeting a nucleic acid.
- Real-time quantitative PCR (qPCR) can be used to quantify microbial load in clinical situations. Quantification is possible by tracking the generation of a fluorescent signal during amplification.

- Other nucleic acid-based approaches can be also used to quantify microbial loads, such as transcription-mediated amplification (TMA), loop-mediated isothermal amplification (LAMP), nucleic acid sequence-based amplification (NASBA), and branched DNA assay (bDNA).
- Molecular techniques, such as PCR, whole genome sequencing (WGS), and several typing techniques, are pivotal for characterizing pathogens, aiding in understanding virulence, antimicrobial resistance, and epidemiological relationships.
- WGS allows the reconstruction of the entire genome of a pathogen, offering comprehensive insights into its characteristics.
- Different typing methods exist to assess relatedness of different microbial isolates. These include multilocus sequence typing (MLST), pulse-field gel electrophoresis (PFGE), restriction fragment length polymorphism (RFLP), and amplified fragment length polymorphism (AFLP).
- Microarrays are solid supports onto which thousands of nucleotide sequences are immobilized. They can be applied for gene expression analysis and genomic variation characterization.
- Metagenomics represents a revolutionary method for analysing microbial diversity by examining all DNA or RNA molecules present in a sample.
- Advancements in molecular detection have improved our capacity to recognize and diagnose infectious diseases. Nonetheless, the dynamic nature of microbial genomes underscores the importance of vigilance when implementing molecular diagnostic techniques.

Discussion questions

8.1 What are the main differences in genome structure, size, and composition among the different microbial groups (i.e. bacteria, viruses, fungi, protozoa) and how do these impact the choice of the most suitable diagnostic techniques?

8.2 What are the advantages and limitations of different molecular techniques in identifying and characterizing microbial pathogens?

8.3 How have advancements in nucleic acid-based detection transformed our approach to controlling the spread of pathogens compared to other methods?

8.4 How have techniques such as 16S rRNA sequencing and metagenomics revolutionized our ability to detect and characterize microbial pathogens in clinical samples?

8.5 How do molecular techniques enable the detection and characterization of both AMR determinants and virulence factors, and what implications does this have for clinical management and public health interventions?

8.6 Discuss the applications of quantitative nucleic acid-based detection in clinical microbiology, with a particular focus on patient monitoring, and its significance in public health.

8.7 How does *in situ* detection of microorganisms enhance our understanding of an infection? Discuss the advantages and challenges associated with *in situ* detection techniques.

8.8 In what ways do dynamic microbial genomes pose challenges for molecular diagnostics?

Further reading

- Chiu CY, Miller SA (2019) ***Clinical metagenomics***. *Nature Reviews Genetics*, 20(6), 341–55. doi: 10.1038/s41576-019-0113-7.
- Ford M (2019) ***Medical Microbiology***. Third Edition, Oxford University Press, Oxford.
- Fournier PE, Dubourg G, Raoult D (2014) ***Clinical detection and characterization of bacterial pathogens in the genomics era***. *Genome Medicine*, 6(11), 114. doi:10.1186/s13073-014-0114-2.
- Goering R, Dockrell HM, Zuckerman M, Chiodini PL (2018) ***Mim's Medical Microbiology and Immunology***. Sixth Edition, Elsevier, Amsterdam.
- Goyal PA, Bankar NJ, Mishra VH, Borkar SK, Makade JG (2023) ***Revolutionizing medical microbiology: how molecular and genomic approaches are changing diagnostic techniques***. *Cureus*, 15(10), e47106. doi: 10.7759/cureus.47106.
- Hui RK, Zeng F, Chan CM, Yuen KY, Peiris JS, Leung FC (2004) ***Reverse transcriptase PCR diagnostic assay for the coronavirus associated with severe acute respiratory syndrome***. *Journal of Clinical Microbiology*, 42(5), 1994–9. doi: 10.1128/JCM.42.5.1994-1999.2004.
- Marlowe EM, Wolk DM (2006) ***Pathogen detection in the genomic era***. In: *Advanced Techniques in Diagnostic Microbiology*. Springer, Boston, pp.505–23. doi: 10.1007/0-387-32892-0_28.
- Microbiology Services, PHE (2018) ***UK Standards for Microbiology Investigations***. http://www.legislation.gov.uk/uksi/1999/3106/regulation/3/made
- Pfaller MA (2001) ***Molecular approaches to diagnosing and managing infectious diseases: practicality and costs***. *Emerging Infectious Diseases*, 7(2), 312–18. doi: 10.3201/eid0702.700312.
- Prabhakar PK, Lakhanpal J (2020) ***Recent advances in the nucleic acid-based diagnostic tool for coronavirus***. *Molecular Biology Reports*, 47(11), 9033–41. doi: 10.1007/s11033-020-05889-3.
- Sandhu GS, Kline BC, Stockman L, Roberts GD (1995) ***Molecular probes for diagnosis of fungal infections***. *Journal of Clinical Microbiology*, 33(11), 2913–9. doi: 10.1128/jcm.33.11.2913-2919.1995.
- Schmiedel D, Kikhney J, Masseck J, Rojas Mencias PD, Schulze J, Petrich A, Thomas A, Henrich W, Moter A (2014) ***Fluorescence in situ hybridization for identification of microorganisms in acute chorioamnionitis***. *Clinical Microbiology and Infection*, 20(9), O538–41. doi: 10.1111/1469-0691.12526.
- Tang YW, Procop GW, Persing DH (1997) ***Molecular diagnostics of infectious diseases***. *Clinical Chemistry*, 43(11), 2021–38. doi: 10.1093/CLINCHEM/43.11.2021.
- Vaneechoutte M, Van Eldere J (1997) ***The possibilities and limitations of nucleic acid amplification technology in diagnostic microbiology***. *Journal of Medical Microbiology*, 46(3), 188–94. doi: 10.1099/00222615-46-3-188.
- Vaught JB, Henderson MK (2011) ***Biological sample collection, processing, storage and information management***. *IARC Scientific Publications*, 163, 23–42.

- Yang S, Rothman RE (2004) ***PCR-based diagnostics for infectious diseases: uses, limitations, and future applications in acute-care settings***. *The Lancet Infectious Diseases*, 4(6), 337–48. doi: 10.1016/S1473-3099(04)01044-8.
- Yu AC, Vatcher G, Yue X, Dong Y, Li MH, Tam PH, Tsang PY, Wong AK, Hui MH, Yang B, Tang H, Lau LT (2012) ***Nucleic acid-based diagnostics for infectious diseases in public health affairs***. *Frontiers in Medicine*, 6(2), 173–86. doi: 10.1007/s11684-012-0195-5.

Useful websites

- UKHSA (UK Health Security Agency) Specialist and Reference Microbiology: Laboratory Tests and Services: **https://www.gov.uk/guidance/specialist-and-reference-microbiology-laboratory-tests-and-services**

9

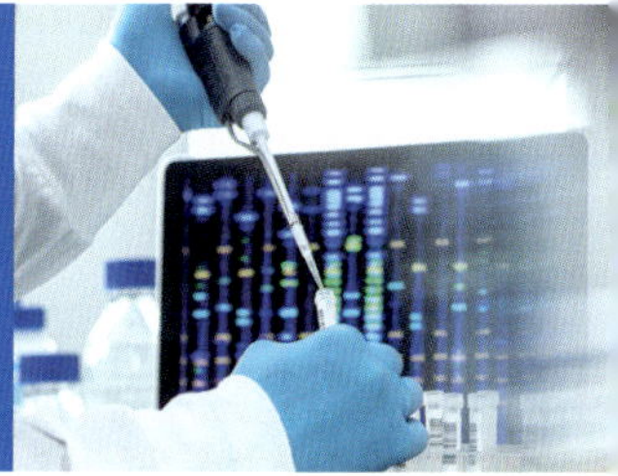

Ethics in Human Genetics Research and Clinical Genetics Practice

Angus Clarke and Nicola Taverner

This chapter discusses ethical issues in genetics and genomics, considering both human genetics/genomics research and clinical genetics practice in healthcare. Looking back over the development of these fields highlights the kind of issues that may arise, some of which have led to the publication of guidelines and legal requirements that are key for those working in genetics and genomics. Even with such guidelines in place, situations arise where it may be unclear how best to proceed, so it is also important to have an understanding of the major schools of moral philosophy to inform decisions about ethical practice. Difficult ethical issues tend to arise in the genetics clinic due to tensions between the wishes or interests of different people, sometimes including those who do not yet exist. Other issues may arise due to concern for the long-term welfare of children or future children, or to decisions about whether to participate in genetics research. Genetic counselling by medical genetics professionals (including doctors and genetic counsellors) enables patients to consider these issues, understand and adjust to their risk, and make the most appropriate decisions for them and their family about what they need to do.

Learning Objectives

After studying this chapter, you should confidently be able to:

- Describe key events in the development of human genetics research and of clinical genetics
- Be aware of key guidelines and legal obligations that are important for those working in human medical genetics and genomics
- Outline the key points of major schools of moral philosophy that can be used to consider ethical issues

- Discuss ethical issues that may arise in clinical genetics practice and describe how current best practice meets these challenges
- Discuss ethical issues in laboratory studies and research.

9.1 The historical context

When considering the ethics of human medical genetics and genomics, it is important to consider the historical context as the implications were so significant that they are still felt today.

9.1.1 Population eugenics

Within the genetics clinic, our focus is on the individual patient or family and enabling them to make informed decisions about whether they want to have genetic testing and what they may choose to do about their genetic information. Genetics can also be considered at a population level and in the past this has led to ethically problematic practices. In the nineteenth century in England, the inheritance of disease and non-disease traits in humans began to be widely discussed, with a focus on improving society through selective breeding. This involved judgements about which genetic traits are more or less favourable and who should be encouraged to have children, or not. As genetic traits are shared between individuals who are more closely related, this contributed to the making of judgements about which social groups should be encouraged to have children. Along with other studies of modern human populations and their supposed relation to human fossil remains, ideas developed about a hierarchy of 'races', with those in powerful positions considering their 'race' to be superior to others. A leading figure in these discussions was Francis Galton, who coined the term 'eugenics' for improving the genetic health of the population. In addition to identifying those with positive traits who were encouraged to breed, some were assessed as having negative traits and discouraged or prevented from breeding. These ideas gained popularity and were widespread across many countries, leading to practices such as the forced sterilization of those considered unfit to breed in the USA, Canada, Sweden, and, more recently, China. In Nazi Germany, eugenics took the form of both 'Race Hygiene' (maintaining the quality of the race) and 'Race Superiority' (asserting the superiority of the preferred 'race' over others), leading to the murder of those deemed unfit to survive on one or other of those grounds.

9.1.2 Ethics of research with human subjects

Ethical practices, both in research and clinical practice, have developed over time. Key events (some of which are covered below) have triggered some of these changes as well as more general social trends, with a shift away from paternalism in healthcare to a greater respect for individual autonomy.

Nuremberg Code

In addition to their programme of genocide, killing those of 'inferior races' in their 'ethnic cleansing' of Europe, in the Nazi era German physicians and anthropologists carried out medical experiments on thousands of concentration camp prisoners, resulting in death or permanent disability for many. They also selected children with inherited neurodegenerative disorders for killing, so that they could perform detailed neuropathological studies on them. These experiments were carried out without consent and, after the war was over, an American military tribunal was opened against these medically qualified and other researchers on charges related to these experiments, the selective killing of patients, and other activities associated with war crimes and crimes against humanity. This led to a significant step change in the conduct of research on human participants with the development of the Nuremberg Code (established in 1948), which stated 'The voluntary consent of the human participant is absolutely essential' for medical research and that the benefits of research must outweigh the risks. This was not enshrined in law, but these were the first international guidelines requiring participants to be willing and able to give their informed consent to participation in research.

Declaration of Helsinki

These principles were expanded in the Declaration of Helsinki, when the World Medical Association drew up recommendations for medical doctors carrying out research with human participants. Again, these are not legally binding in international law but are considered to be moral principles that physicians should follow. They include that informed consent is essential for research and that any risks should not exceed the potential benefits, as well as requirements for the researchers to be appropriately qualified and the research to be reviewed by an independent committee before it commences. The first Declaration was drafted in 1964 and it has been revised several times since.

Tuskegee syphilis study

Even with these guidelines and recommendations in place, unethical practices persisted for many years. An example of this is the Tuskegee syphilis study which ran from 1932 to 1972. This research project was carried out by the US Public Health Service and involved 600 low-income African American males, 400 of whom were infected by syphilis. They were given free medical care, but were not told about their disease and, when penicillin became available as a cure for syphilis in the 1950s, they did not receive treatment. Researchers even prevented treatment for some participants who were diagnosed by other physicians, and many participants died of syphilis. The study was only stopped when it was reported in the press, resulting in public outcry, and the panel set up to review the study concluded it was ethically unjustified and should be stopped. This and other examples have contributed to what is known as the '**Tuskegee effect**', a lack of trust from African Americans in government health services, so they are less likely to seek medical help or to engage with research.

Tuskegee effect
Describes the lower levels of participation and representation in research and in engagement with medical services by African Americans, attributed to the loss of trust which occurred as a result of the Tuskegee experiment. In the experiment, conducted between 1932 and 1972, African American men were denied treatment for syphilis in the name of research.

Asilomar conference: 1975 moratorium on recombinant DNA technology

A significant event in the development of ethics specifically around genetics research was the development of guidelines around the use of **recombinant DNA technology**. Scientists were starting to use viruses to introduce new genes into bacterial cells to study the effects, with the aim of moving on to do the same with mammalian cells. However, these researchers were concerned that this could be a risk to human health, for example there was concern that gene manipulation might cause cancer in humans. They proposed a moratorium on this research and established the international Asilomar conference in 1975 which brought together experts including scientists, lawyers, and government officials to debate whether the research should be allowed to continue. The conference recommendations were for there to be a risk assessment on the different types of experiment planned, and different safety measures used for different levels of risk. These included different degrees of physical containment and, for the

recombinant DNA technology
Laboratory-based techniques that involve manipulating DNA fragments from different sources to create new genetic combinations.

riskier experiments, the use of cells that had been engineered so they were not able to survive outside the laboratory. This enabled the research to be restarted, and to continue until today.

Moratorium on development of the human embryo in vitro *beyond 14 days*

in vitro

Scientific experiments or research conducted or produced in a test tube, culture dish, laboratory, or elsewhere outside a living organism.

Another important consideration in genetics research is the acceptability of work on human embryos. In 1978, the first baby was born by *in vitro* fertilization (IVF), demonstrating that it was possible to create viable human embryos in the laboratory, which could be implanted into the womb. Therefore, it was now possible to cultivate a human embryo ***in vitro***, at least for a short period. This led to much debate about human embryo research, and the Ethics Advisory Board of the US Department of Health, Education, and Welfare held a consultation resulting in a report which cautiously supported this research but put a stipulation in place that the embryos should not be kept alive for longer than 14 days after fertilization. This length of time was selected as, on the 15th day, the primitive streak forms, when the layers of germ cells start to become different, and the body plan begins to be established. In 1984, the UK Committee of Inquiry into Human Fertilisation and Embryology (the 'Warnock committee') adopted this 14-day limit, and it became part of UK law. This limit has been adopted by many different countries globally, including Australia, Canada, Japan, and the Netherlands, and is one of the most widely adopted rules in reproductive science. However, technology has now advanced sufficiently that it is possible to keep embryos alive *in vitro* for longer than 14 days, and ethical debate has begun as to whether this rule should remain in place, or whether research should be allowed to continue further into the development of the embryo.

Human gene therapy and the manipulation of the somatic genome

The ability to alter genes by insertion of other DNA sequences has led to considerable interest in gene therapy for genetic conditions. However, early attempts at gene therapy in human patients led to the insertion of the therapeutic DNA sequence into inappropriate sites in the genome, sometimes causing malignancy and even death. The importance of the thorough protection of human research participants was given shocking emphasis by the death of one patient in 1989, who had been given an experimental gene therapy treatment for carbamoyl transferase deficiency. He died from a strong immune response, and it became clear afterwards that he should never have been recruited to the trial, that the informed consent process was deeply flawed, and that commercial considerations had probably led to shortcuts to the proper clinical process. Several cases of leukaemia, and one death, resulted from gene therapy for severe combined immunodeficiency. Such cases led to a lengthy period in which gene therapy trials were halted in the USA and there have been further deaths in more recent gene therapy trials, with three deaths reported in 2020 in association with a trial of treatment for X-linked myotubular myopathy.

Human germline gene editing (hGGE)

In addition to enabling human embryos to survive longer within the laboratory, advances in technology mean that it is now possible to alter the DNA sequence within human cells, in a process called gene editing. It is hoped that this technique will provide new treatment opportunities for genetic conditions, by altering disease-causing sequence variants in specific cells. However, this opens up the possibility of altering the DNA sequence in sperm, eggs, or embryos so that the resulting child is genetically modified. Whilst this could be used to prevent the development of genetic disease, there is concern that this could lead to interest in 'enhancing' the child, altering the DNA sequence to their advantage. Therefore, there have been calls for a global moratorium on all clinical (i.e. human) uses of germline gene editing while an international framework can be drawn up to enable the technique to only be used in ways which are considered ethically acceptable to societies. However, this has not been established yet and there are also scientists, ethicists, and other interested parties who argue against such a moratorium, believing that it may be more helpful to work towards an international consensus for hGGE practice instead.

As these examples demonstrate, there have been many difficult ethical issues related to genetic and genomic research, many of which have not yet been fully resolved, and the situation is likely to evolve over time. Ethical issues also arise within the clinical genetics service, some of which are explored in **Section 9.3**.

Participation in genomic research and databases: representation of population groups

Most of the readily available data on the human genome, especially in combination with clinical information, has been gathered in the USA and Europe, with some also from Japan and Australia. Large numbers of people have had their genome sequenced in China too, although those data are not so available. Despite the inclusion of data from the USA of people with African heritage (notwithstanding the Tuskegee effect described above), there is a great under-representation of data from the populations of Africa. This is important because there is much more genomic variation in the African continent than anywhere else. The entire human species is of African origin but only a small proportion of that variation has been included in the diaspora, the groups that settled the rest of the world.

There are at least two very important consequences of this. First, it makes it much more difficult to interpret variants of uncertain significance (VUS) found in patients, as they might often be perfectly benign and data on African populations would help to establish that. Secondly, related to the first consequence, is that VUS are more likely to be found in patients of African ancestry and yet we are less likely to have the background population data to help interpret their significance. There is, therefore, not only an inequity of access to medical genomics services, with those of African descent in Europe and North America being disadvantaged by discrimination and the legacy of slavery, but when they succeed in gaining access to these services, they are less likely to be given clear and useful information.

Cell lines: the case of Henrietta Lacks

The final issue to be flagged here concerns the use of human tissue samples for research and for commercial gain. The outstanding example of this is that of Henrietta Lacks, a US black woman whose cells, taken originally for a diagnostic biopsy, were transformed to become a permanent cell line (**HeLa cells**) and, as such, have been commercialized and distributed around the world (to the profit of the original investigators). This occurred without consent from the patient or, once she had died, from her surviving relatives, and without any of the subsequent profits passing to her family. This has become a *cause célèbre*, especially since the Black Lives Matter movement has helped us all to ask whether this would have happened in quite the same way if Mrs Lacks had been white. It is now a legal case in the USA, with compensation sought by the relatives both because of how they feel they have been treated as a

HeLa cells
A strain of continuously dividing cancer cells cultured since its isolation in 1951 from a cervical carcinoma of the patient Henrietta Lacks.

family and because of the lack of return to them from the commercial exploitation of the cell line derived from Mrs Lacks.

There are parallels here too with the research and commercial exploitation of both DNA sequence information and clinical data obtained from and about patients in the clinic and participants in research projects. Similar issues arise, although the legal status of such information is often treated as different from the human tissue itself. Data is handled in the context of a long tradition of concern for privacy in general and respect for medical confidentiality in particular, reinforced recently in Europe (including the UK) through the General Data Protection Regulation (GDPR). The separate Human Tissue Act regulates the handling of tissue and what is done with it, including DNA extraction and sequencing, with specific attention to how and when it was obtained (especially as to before or after death, and with what sort of consent, if any, and from whom this was obtained). There are some anomalies that could perhaps be clarified in the future, as currently a spouse can decide to block further genetic analyses from being performed on a tissue sample after death, although it may be blood relatives who stand to benefit from such analyses. This requirement does not apply when the DNA has already been extracted from the tissue sample or a cell line has been produced. The potential for disagreement in this context is unhelpful, although such problems do not arise frequently. Note that the stipulations of the law in any one jurisdiction may not always coincide with the most compelling ethical position.

Key Points

Several critical occurrences in recent history have led to challenges and corrections of our practices with respect to the field of human medical genetics. Some of these challenges remain, as yet, unresolved. With the pace of advancement in research and technology ever increasing, these challenges will multiply. It is critical that we have informed, educated practitioners who can be mindful of past learnings as they navigate future challenges.

9.2 A framework for ethics in the medical genetics consultation

As **Section 9.1** shows, ethics has been important for the field of human medical genetics and genomics, both historically and today, and difficult issues also arise within a medical genetics consultation. Ethical frameworks can help to navigate these situations, enabling professionals to weigh up the different influencing factors and decide on the best course of action. Three of the main ethical traditions are outlined in **Box 9.1**.

There are many different theories within each of these traditions and different actions may be recommended by followers of the same tradition, depending upon their perspective. With the example of offering termination of pregnancy for Down syndrome within a deontological perspective, should the healthcare professional be considering the harm that would be done to the unborn child, or the harm that might be done to the parents and family unit, if they feel unable to cope with a child with additional needs? If we shift to the utilitarian tradition, the over-arching societal perspective may be seen to pre-judge the issue by having excluded the affected person from consideration as a valued member of the society. If the affected infant's interests are included alongside everyone else's, this may lead to a different outcome even from within the utilitarian tradition.

BOX 9.1 Ethical Traditions

Consequentialism: These theories focus on the consequences of an action: an action is believed to be right based on its outcome. An example of this is utilitarianism, where actions are morally right if they maximize the happiness of those they affect. Under a utilitarian perspective, offering termination of pregnancy for Down syndrome (which causes learning disability and a range of health conditions) could be argued to be morally right as it reduces the healthcare burden on society, maximizing the benefit for the population.

Deontology: These theories view actions as morally right or wrong, irrespective of the consequences. They consider that the healthcare professional has certain duties or obligations to fulfil, for example an obligation to do no harm. Under a deontological perspective, offering termination of pregnancy for Down syndrome could be argued as morally wrong as the professional is doing harm to the individual who would be born.

Virtue ethics: These theories focus on the individual agent, rather than the actions or their consequences. Under virtue ethics, the healthcare professional should embody appropriate character traits, such as justice and kindness, and will then make morally appropriate decisions. From a virtue ethics perspective, the healthcare professional should be demonstrating these virtues when deciding whether to offer termination of pregnancy for Down syndrome.

SELF-CHECK 9.1

Which of the following three ethical traditions focuses on the individual agent: consequentialism, deontology, or virtue ethics?

A good starting point to help resolve these difficulties in any particular case may be to consider the different stakeholders, and their competing interests, to enable a balancing exercise to be conducted before deciding upon a course of action. When carrying out such a balancing exercise, many healthcare professionals use the four principles of medical ethics, published by Beauchamp and Childress. These are:

- Autonomy: promoting the individual's right to choose how they wish to proceed
- Beneficence: doing good
- Non-maleficence: doing no harm
- Justice: fairness.

These principles were designed to enable those with different approaches to jointly discuss the merits and demerits of proposed solutions to a wide range of concrete problems, without needing to share a theoretical perspective, as illustrated in **Case study 9.1**. For a decision to be considered ethical, it should respect each of these principles.

However, in many situations within the medical genetics clinic, there is often not a clear-cut course of action as different stakeholders are involved. To return to the example of termination of pregnancy for Down syndrome, providing termination promotes the woman's autonomy as she has the right to choose to end the pregnancy if she wishes, but this removes the autonomy of the unborn child to choose to live. Similarly, there are different perspectives on doing good, doing no harm, and what is fair to all parties. How would you balance these principles? Do you believe that termination of pregnancy should be offered for Down syndrome? If you do not consider Down syndrome as sufficiently severe to warrant the termination of

CASE STUDY 9.1 Applying the four principles of ethics

A young couple are seen in the medical genetics clinic as the male partner has a pathogenic genetic variant which causes retinoblastoma, a tumour of the developing retina which occurs in children. The female partner is pregnant, and the medical genetics team explain that there is a 50% chance that the baby will inherit the variant, and therefore be at risk of developing retinoblastoma. The couple do not wish to have any genetic testing of the pregnancy, so the medical genetics team offer to test the baby at birth. If the baby has the variant, they will be offered regular eye examinations under anaesthesia until age six months, then less frequently until age three years, so that any **retinoblastoma** tumours that develop are detected early when they are small and treatable. If they do not have the variant, they do not need to have these procedures. The couple do not want the baby to have genetic testing, as they do not want the baby to have lots of interventions and prefer to take the chance of seeing how things develop.

In this situation, should the medical genetics team respect the decision of the parents, or should they continue to advocate for a genetic test?

The four principles of medical ethics can be applied to this situation:

- **Autonomy:** the baby is too young to make an autonomous decision for themselves, but the parents argue that they should have the right to choose whether their baby has the genetic test.
- **Beneficence:** if the baby carries the variant, they can be offered screening, which enables any tumours that develop to be picked up early when they are more treatable, giving the best outcome for the baby.
- **Maleficence:** if the baby does not carry the variant, the parents can be reassured, and the baby will not be put at risk by having anaesthesia.
- **Justice:** fairness can be considered in terms of use of healthcare resource. By having a genetic test, resources (such as eye examinations) can be allocated fairly to those who need them.

By balancing these principles, it is clear to the medical genetics team that they should continue to advocate for testing, although this does not promote the couple's autonomy, which is outweighed by the other principles.

retinoblastoma
A rare malignant tumour of the retina, affecting young children.

pregnancy, are there other conditions that you would recognize as sufficiently severe? Which of the ethical traditions do you feel is most appropriate to use for these kinds of decisions? This illustrates the limitations of this approach, which does not provide guidance about how to balance the principles. Healthcare professionals still need to consider the ethical traditions, and whether their arguments to support or reject a course of action are based on virtue ethics, deontology, consequentialism, or another prior commitment. In practice, the use of the framework may lend itself to lists of pros and cons, and the predominance of consequentialism. However, it gives a framework that can be used to consider the ethical issues explored in **Section 9.3**.

Key Points

There are a well-accepted and widely used set of three traditions and four principles to help researchers and practitioners conduct balancing exercises when needing to reach consensus on a best way forward.

9.3 Ethical issues in the medical genetics clinic

In this section, we consider some of the issues which arise in clinical genetics. As outlined earlier, many of these issues arise due to the patient's family and social context, as this affects their priorities and influences the decisions that they choose to make. When complex issues arise, it is imperative to have an ethical framework to consider the appropriate course of action, as outlined in **Section 9.2**.

9.3.1 Informed consent

As highlighted earlier in this chapter, informed consent has become a central doctrine in scientific research involving human participants, material (tissues, cells, DNA), or data (DNA sequence information or clinical data) derived from humans. Informed consent is also essential for all diagnostic procedures and treatments within healthcare, including genetic and genomic testing.

Informed consent for diagnostic testing

Genetic and genomic tests may achieve the diagnosis of a genetic condition, or confirm a diagnosis that was already suspected, but can also generate additional information for the individual and their family. For example, a condition may be diagnosed that has not yet begun to manifest: such information amounts to a predictive test and it is important to distinguish diagnostic tests from predictive tests as far as possible. Patients and their family need to understand what type of result might emerge from the proposed genetic investigation, so that they can decide whether or not to consent to the test and also be prepared for the possible results.

Informed consent for research in the genetics clinic

Case study 9.2 focuses on informed consent for genetic testing as part of healthcare. However, with so much still unknown about the genome, patients in the genetics clinic may also be offered the chance to participate in research, and the boundary between research and clinical practice may be blurred: if an uncharacterized variant is found, at what point do investigations

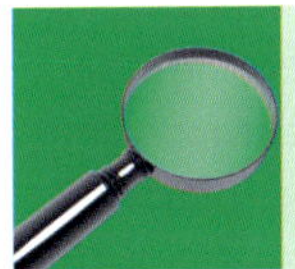

CASE STUDY 9.2 Informed consent and genetic testing

Susan, aged 48, has been diagnosed with breast cancer, which is 'triple negative' as the cancer cells do not contain receptors for oestrogen, progesterone, or human epidermal growth factor. As part of her journey through cancer diagnosis and treatment, she has had a number of tests including a biopsy and blood tests. She is aware that one of these blood tests is being performed to help decide about her treatment and is shortly afterwards informed that she has a variant in the *BRCA1* gene, which means that full mastectomy is recommended, rather than a lumpectomy. The surgery is discussed with her, and she agrees to go ahead. As the appointment is coming to a close, the healthcare professional tells her that this also means that she is at increased risk for ovarian cancer, but they will talk to her about that at some other point, and then ends the appointment.

Q1 How might Susan feel about this information?

Q2 How might this impact on her trust in the healthcare professionals?

Q3 Did she give her informed consent for this genetic test?

In order for Susan to give informed consent for her genetic test, she needs to have more information about what was being tested, why this is helpful, and what other health information this may provide. Finding out that she is at increased risk for ovarian cancer, with no prior warning and no discussion or explanation, is likely to be very challenging for her to accept, especially whilst in the middle of breast cancer treatment. With broader testing looking at a wider range of genes, it is often not possible to inform an individual of all the different risks that may be identified, depending on which gene contains a pathogenic variant, but it is important to discuss the type of health information that might be revealed, so that they are prepared for the possibilities that may arise.

With genetic testing, the results may have implications not only for the individual but also for their family members.

Susan is shocked and upset by the revelation that she is at increased risk of ovarian cancer. She is already fearful about surviving her breast cancer diagnosis, and she has heard that the outcomes are much worse in ovarian cancer, which is known as 'the silent killer'. She is anxious that she may already have ovarian cancer, and rings the clinical nurse specialist in a panic to ask about ovarian cancer screening. The clinical nurse specialist arranges a rapid appointment with her, when she explains about the implications of *BRCA* variants, including the risk of ovarian cancer. Susan feels reassured by this information and will think about the management options available to her. However, as part of this discussion, she learns that she may have passed this *BRCA* variant on to her two daughters, one of whom is currently pregnant with her first child.

Q1 How is this information about the risk to her children and future grandchild likely to affect Susan?

Q2 How is she likely to feel about the possibility that she has passed on a gene variant that increases their cancer risk?

Q3 How will it feel for Susan to tell her children about this risk?

In order for someone to give their informed consent to a genetic test, they need to be aware that the outcome may have implications for the health of other family members as well as themselves. For many individuals affected by a condition, finding out information for their family members is a major factor in their decision to have genetic testing, as they believe it is helpful for their relatives to know about their risk. However, if they are not aware of this possibility, it can come as a shock when they are in the middle of their treatment journey, and they may feel unable—for the moment—to engage with the risks to others in the family when they are so bound up with their own situation.

When an individual is manifesting a condition, the decision to have a diagnostic genetic test may be relatively straightforward, as it may be helpful to confirm the diagnosis or to inform management options. However, if an individual is at risk of a condition that is present in their family but is not yet showing symptoms, they may be able to have a presymptomatic genetic test to see if they carry the genetic variant that causes the condition in their family and find out whether they are likely to have the condition in the future. For conditions where there are screening or management options that can be put in place to manage the risk, the decision to have a genetic test may be more straightforward. However, for conditions where there are few or no management options, it may be more challenging for an individual to decide whether to have a genetic test.

Susan's daughter Lisa comes to the genetics clinic shortly after the birth of her son Cameron, as she wants to discuss genetic testing for the *BRCA1* variant identified in her mum. She was aware that this would lead to an increased risk of breast and ovarian cancer, but did not know how much the risk would be or what management options were available. She was particularly worried about the risk of ovarian cancer as a work colleague had recently been diagnosed with late-stage ovarian cancer and was not expected to live for long. Therefore, she was keen to access ovarian screening, and was shocked to find that there is no ovarian screening available that has been found to be effective. The main option to manage ovarian cancer risk is to have the ovaries removed, but this would not be offered until she was around age 40 and had completed her family. Lisa is currently aged 32 and is planning to have a couple more children.

Q1 How do you think Lisa would feel about the limited management options for ovarian cancer risk?

Q2 How do you think she would feel if she had genetic testing for the *BRCA1* variant and discovered that she is at increased risk of ovarian cancer before being informed about the management options?

Q3 Is it preferable for Lisa to know about the management options before or after the genetic test?

Lisa was shocked about the lack of management options for her potential ovarian cancer risk and was not sure whether she would want to know that she is at increased risk when she cannot access any treatment to mitigate this risk. By delaying the test, she felt able to live with some hope that she may not be at increased cancer risk, which would alleviate some of her anxiety around this. She planned to access breast screening for now and return for a pre-symptomatic genetic test when she had completed her family and was approaching age 40. Then, with the option of removing her ovaries if she was at risk, she felt she would want to have genetic testing to see whether she was at risk.

As this case study shows, informed consent involves many different aspects, not just about the possible outcomes of testing, but also what this may mean for the individual and their family and what options there may be to mitigate the risks. In addition, it encompasses the benefits, risks, and limitations of testing itself.

Susan's other daughter Sarah has had genetic testing which found that she carries the *BRCA1* variant identified in Susan. She is pregnant with her first child, and tells her midwife about the *BRCA1* variant, as she is worried that she may pass this down to her child. She would really like to know whether the child carries the variant so she can teach them about cancer awareness from a young age if needed. The midwife arranges for her to have the pregnancy tested by chorionic villus sampling. A needle is passed through the abdomen and a sample taken from the placenta, which arises from the fertilized egg. Genetic testing of these cells shows that the pregnancy does not have the *BRCA1* variant. However, Sarah has a miscarriage shortly after the procedure, which has a miscarriage rate of less than 1%. Sarah is very upset about losing the pregnancy and says that she was not aware that there was a miscarriage risk. She wishes that she had not had the procedure as she would have continued with the pregnancy even if the *BRCA1* variant was present.

Q1 Should Sarah have been offered prenatal testing for this variant when it was not going to affect the management of the pregnancy?

Q2 Does it sound like she gave informed consent for this procedure?

This last section of the case study illustrates the importance of understanding the purpose of the test as, if she would not terminate (end) a pregnancy with the *BRCA1* variant, then there was little reason to put the pregnancy at risk by carrying out the invasive prenatal testing procedure. Pre-symptomatic testing for a *BRCA* gene variant would not usually be offered until adulthood (see **Section 9.3.4**), so such testing would also not be offered for information, as the parents would then know their child's *BRCA* gene status without the child having had the option to request this information for themselves. We would also hope, if the child was tested, that the parents would pass this information to them in an appropriate way, perhaps 'staged' over some years. However, communication of information about genetic risk does not always work out within a family in this ideal fashion.

SELF-CHECK 9.2

Name the four principles of medical ethics that can usefully be applied when performing balancing exercises to reach a consensus about the best way to proceed.

to establish its pathogenicity become research? Those taking consent for research within the genetics clinic are required to undertake Good Clinical Practice (GCP) training, the international ethical, scientific, and practical standard to which all clinical research is conducted. If you look at **Box 9.2**, you will see the draft principles of Good Clinical Practice training produced by the International Council for Harmonisation of Technical Requirements for Pharmaceuticals for Human Use, that include informed consent as essential for ethical conduct. The UK Policy Framework for Health and Social Care Research requires researchers conducting clinical trials of investigational medicinal products to complete GCP training, and NHS Trusts also require this for staff consenting patients to research studies.

BOX 9.2 Good Clinical Practice Training

1. Clinical trials should be conducted in accordance with the ethical principles that have their origin in the Declaration of Helsinki and that are consistent with good clinical practice (GCP) and applicable regulatory requirement(s).
2. Clinical trials should be designed and conducted in ways that ensure the rights, safety, and well-being of participants.
3. Informed consent is an integral feature of the ethical conduct of a trial. Clinical trial participation should be voluntary and based on a consent process that ensures participants are well-informed.
4. Clinical trials should be subject to objective review by an institutional review board (IRB)/independent ethics committee (IEC).
5. Clinical trials should be scientifically sound for their intended purpose and based on robust and current scientific knowledge and approaches.
6. Clinical trials should be designed and conducted by qualified individuals.
7. Quality should be built into the scientific and operational design and conduct of clinical trials.
8. Clinical trial processes, measures, and approaches should be proportionate to the risks to participants and to the reliability of trial results.
9. Clinical trials should be described in a clear, concise, and operationally feasible protocol.
10. Clinical trials should generate reliable results.
11. Roles, tasks, and responsibilities in clinical trials should be clear and documented appropriately.
12. Investigational products used in a clinical trial should be manufactured in accordance with applicable Good Manufacturing Practice (GMP) standards and be stored, shipped, and handled in accordance with the product specifications and the trial protocol.

Ethical issues related to research are discussed in more detail in **Section 9.4**.

9.3.2 Family secrets and disclosure

As we share much of our genetic material with our relatives, this means that genetic information may have implications for them too. Therefore, a recurrent question that arises is when and how we can help and encourage (or even persuade) our patients to share their medical and genetic information with their close relatives. This type of information is usually passed to relatives if it has important health implications for them, especially if there are management options available to deal with an increased risk of serious medical problems. However, there are situations where this does not occur, as explored in **Case study 9.3**.

As can be seen from the case study, there is no duty to disclose risk information to relatives, and healthcare professionals must continue to weigh up the possible options to decide on the most appropriate strategy. It is worth noting that they may not be able to contact the relative, unless their personal details and contact information have been provided by the patient. In practice, genetics health professionals usually manage these situations by maintaining contact with the patient and continuing to encourage them to share the relevant information with their relatives, and it would be very unusual to force disclosure against an individual's wishes.

It can be argued that the details of a particular pathogenic variant that causes a genetic condition are not specific to an individual, as their relatives may have this same variant in their shared genetic material. Therefore, a distinction can be drawn between an individual's medical history, which must be kept confidential, and this genetic information, which (some argue) could be used to test relatives even without the consent of the original patient. This enables a

CASE STUDY 9.3 ABC vs St George's

A woman discovered that her father has Huntington's disease, a neurodegenerative disease that affects movement, cognition, and behaviour, after giving birth to her first child. She brought a claim in the UK court against her father's clinicians for 'wrongful birth' resulting from their failure to warn her about her father's condition, which had been diagnosed during the second (middle) trimester of her pregnancy. Clinicians had asked the father whether they could inform his daughters of their risk, but he refused his permission. She argued that she should have been told of the risk to her and her unborn child against her father's wishes, and that she would have terminated the pregnancy if she had known about the risk. The initial court ruling was to strike out the claim as the judge felt that an extension of a clinician's duty of care to inform a patient's relatives of their risk was not a step that he would be willing to take. The Court of Appeal reversed this decision and said that the case should go to court, to weigh up the competing interests of duty of care and duty of confidentiality.

Q1 Do you think that her father's clinicians should have disclosed the risk to her?

Q2 Would this apply in the same way, whatever the diagnosis might have been in the father, or with different conditions might the claim be stronger or weaker?

Q3 If someone is diagnosed with a genetic condition, who should be responsible for passing this information to their relatives?

Q4 If a patient is not willing to disclose their genetic information to a relative, should the healthcare professionals contact the relative to do so, if indeed they know enough about their patient's relatives to be able to contact them? If so, under what circumstances?

Sharing genetic information with relatives can be very difficult: it is often unwelcome news, and is particularly challenging when, as with Huntington's disease, there are no interventions to manage the risk. Relatives who are told that they are at risk but that there is no treatment available may respond with anger, fear, or resentment, so individuals may find it difficult to inform them. Sometimes, they may decide to wait until 'the right time', but then find that there is no 'right time' and the information is never passed on. Other barriers can also arise, such as geographical distance and family estrangement, and occasionally individuals may openly state that they do not wish to pass on the information, as in this case study.

If a relative may be at increased risk of developing a particular condition and can use this information, for example to access screening, to be vigilant about possible symptoms or complications of a condition, or to make reproductive decisions (as in this case study), it can be argued that the clinician has a duty of care to the relative. This is particularly acute when treatment options may save that relative's life. However, clinicians are also bound by a duty to keep their patient's information confidential if that is the patient's wish. Therefore, these duties must be considered in a careful balancing process. This process will vary with each case, depending on the condition, the options available to those at risk, and the competing agendas of the involved parties. These difficult cases are often discussed by groups of healthcare professionals who weigh up the options together and decide upon the most appropriate strategy. In this case study, the judge considered these conflicting duties and the process of weighing them up, as outlined below.

The woman had attended family therapy sessions with her father at the facility run by one of the defendants, and the judge ruled that the hospital therefore did have a duty of care towards her. However, this duty of care did not extend to telling her about the diagnosis, as this did not arise out of the therapy. The clinicians had consulted with others, including a clinical geneticist, and concluded that, on balance, disclosure should not take place. The judge was satisfied that the decision about whether or not to disclose had been considered appropriately and that, although other responsible clinicians would have disclosed, the decision not to disclose was supported by sufficient professional opinion. The judge also concluded that disclosure was unlikely to have led to a termination of the pregnancy, mainly due to the short timescales involved. Therefore, one of the defendants was found to have a duty of care to the woman which meant that they should carry out a balancing exercise about whether to disclose, and they had met this duty of care as this exercise had been carried out appropriately. It is interesting to note that, when the woman had found out about her father's diagnosis, she had made the decision not to tell her sister, who was also at risk and—at the time when ABC found out about her father's diagnosis—was pregnant.

Q1 Whilst this is an unusual case, are there lessons that can be learned that may be applicable to other cases?

Q2 What are your personal views on the balance between the duty of care and the duty of confidentiality?

situation where an individual is told that there is an inherited condition in their family and is offered genetic testing for the causative variant but is not told which relatives have been found to have this condition or carry the variant. In reality, they may be able to deduce which relatives have the condition from their personal knowledge of the family history and family structure, but this distinction may be useful to help consider where the boundaries of confidentiality can be drawn.

Key Points

When working cases of genetically inherited conditions a distinction can be made between an individual's medical history (which must be kept confidential) and the availability of genetic information that has become available because of them having been genetically tested.

9.3.3 Prenatal testing and disability

Prenatal testing raises many different ethical issues, especially whether, and in what circumstances, it is acceptable to terminate a pregnancy. Many patients are seen within the genetics clinic as they wish to have children but are considering whether they would end an affected pregnancy, as can be seen in **Case study 9.4**.

CASE STUDY 9.4 Prenatal testing and disability

A young couple were seen in the genetics clinic as the woman is a female carrier of the sex-linked disease Duchenne muscular dystrophy, a life-limiting condition which causes progressive muscle degeneration and weakness, with affected individuals usually living only until their twenties or thirties. Her brother, aged 18, is affected by the condition, and there is a 1 in 2 chance that any son that she has would be affected. She was in the early stages of pregnancy and the couple wanted to discuss their options for this pregnancy. They could have prenatal testing with the option to terminate if the foetus proved to be an affected male.

Q1 Does the offer of prenatal testing and termination convey a recommendation that the couple should choose this option?

Q2 What does the offer of termination say about the value of her brother's life? How might she feel about ending a pregnancy due to the condition which affects her brother?

The provision of prenatal testing and termination normalizes this as an acceptable route that the couple could consider. However, there is a risk that this could then be seen as the preferred option, with couples feeling that they are being told that they should not continue with an affected pregnancy. From a population perspective, prenatal screening with termination of affected pregnancies reduces suffering and improves the population's genetic health, whilst from a health economic perspective, it could be argued that this is more economical than providing long-term care for an individual with the condition, especially in a publicly funded health service. However, these perspectives fit with population eugenics as discussed in **Section 9.1.1**, which many individuals would reject. In the genetics clinic, the focus is on the couple and their family: how can we support them to make the best decision for their values, beliefs, and circumstances? This is complex, and there are no straightforward answers for the couple. Their decision will be influenced by their family experience, which may include her brother's views on termination of a pregnancy due to Duchenne muscular dystrophy, and genetics health professionals who strive to help them consider all their options and to make a choice as to which may be the least distressing or uncomfortable for them. If they opt for termination of an affected pregnancy, this would certainly be difficult, but it may be at least as troubling if they were to have a son affected with the same condition as his uncle, and might cause the uncle to feel great sadness and even some level of responsibility (however unfounded we might consider that to be).

Tied up in these decisions are thoughts about what termination of an affected pregnancy says about the value of another's life—if she is willing to terminate an affected pregnancy, does this imply that she feels her brother should not have been born? This is the sort of consideration that applies within a family, where the affected individual could sometimes (in some circumstances) take offence and feel hurt if a relative decides not to have a child with the same condition that affects them. The interplay between relationships within the family and decisions about reproduction can be very personal and highly specific to each family, with great variation between families as to what is discussed openly, what is acknowledged in other ways, perhaps with great sensitivity and mutual respect, and then—in contrast—what is kept private from each other as closely guarded secrets.

For some couples, embryo selection (or **pre-implantation genetic testing**) is a more acceptable option than prenatal testing and termination. In this process, the eggs and sperm are harvested, and fertilization takes place in the laboratory. After a few days, the resulting embryos are tested so that only those which do not have the causative gene variant are placed into the womb to develop. Whilst couples may choose this option, it is important for them to be aware of the complexities of the process, with various steps that need to be carried out and about a third of treatment cycles resulting in a baby. There is also concern that this embryo selection could lead to 'designer babies' selected for what are considered to be positive attributes rather than to avoid genetic conditions. Within the UK, the Human Fertilisation and Embryology Authority (HFEA) is responsible for deciding which conditions are sufficiently serious for embryo selection to be offered.

pre-implantation genetic testing
A specialized technique which combines *in vitro* fertilization (IVF) technology with genetic testing.

Population screening for genetic conditions

In contrast to considering a particular family, population screening for genetic conditions is a very different context, especially antenatal screening for a range of conditions. These include some conditions—some disorders—where the purpose of screening is to enable timely treatment and other conditions where the stated goal is to offer choice to the pregnant woman about whether to continue or terminate the pregnancy. Thus, treatment will be offered to the benefit of the foetus—the future baby—if the mother is found to be affected by syphilis or HIV, or if the foetus is affected by anaemia because the mother has reacted against surface proteins of foetal red blood cells, a few of which may have crossed the placenta into the mother. If that happens, factors inherited from the father may then trigger an immune response in the mother and the IgG anti-foetal red cell antibodies cross the placenta to cause anaemia in the foetus. Structural heart defects and other malformations are also often detected on ultrasound scans and may be managed more successfully after birth if they are detected before birth: appropriate plans can be made. This category of screening will sometimes detect conditions that are inevitably lethal or that are difficult to treat satisfactorily. Parents may then decide to terminate the pregnancy rather than continue all the way through, if they feel the outcome is likely to be bad for the baby. The driving purpose, however, is better treatment.

Different from this are those elements of the antenatal screening programme that seek to identify conditions where there is no prospect of useful treatment for the baby but where the whole purpose is to enable parents to decide to terminate the pregnancy if the foetus is affected. These are the conditions through which affected individuals can feel rejected by society: why else would society have decided to offer screening tests but to prevent them being born? While less personal than decisions made about you by family members, the sense of rejection by society can also be powerful. What motivates society's investment in these screening programmes? Is it a eugenic wish to reduce the number of individuals born with costly, burdensome conditions? The programmes did begin with that rationale, although most would argue that this is no longer the purpose. And does it devalue those with the condition?

These issues have been debated in relation to individuals with Down syndrome, which is routinely screened for during pregnancy. Screening tests are made available for Down syndrome (trisomy 21) and the other, less frequent autosomal trisomies (Edward syndrome, trisomy 18, and Patau syndrome, trisomy 13) without being imposed but, in practice screening is often so routine that women may not be fully aware that they are having tests for these conditions (bringing us back to the issue of informed consent explored in **Section 9.3.1**). Women may then be confronted with the news that their child has an increased chance of having Down syndrome, which can be associated with a number of substantial health problems, including congenital heart defects (problems with the heart's structure) and other malformations, as well as learning disabilities. At that point, the under-prepared woman may feel she is on a conveyor belt to have an amniocentesis to confirm the diagnosis and, if it is confirmed, leading her to a termination of the pregnancy. The majority of these pregnancies will be terminated, though individuals with Down syndrome and their families may contend that they lead happy, fulfilling lives and are able to contribute to society. Whilst we may suppose that we are offering personal choice to the women carrying these pregnancies, this choice is not afforded to the as yet unborn individuals, who may not be given the opportunity to live.

9.3.4 Pre-symptomatic genetic testing of children

As discussed above, informed consent for genetic and genomic testing ensures that individuals understand what the testing may tell them. However, what about when the test is for a child? Whilst older children may be able to give their informed consent if they are competent to do so, it is helpful to think generally about when genetic testing may be appropriate in children. When a child has symptoms, diagnostic genetic testing may be helpful to confirm a clinical diagnosis to inform management. However, ethical issues arise when considering pre-symptomatic genetic testing of children, as illustrated in **Case study 9.5**.

There is ongoing debate about whether parents should have the right to have their child tested for adult-onset conditions, with advocates arguing that it gives the family the

CASE STUDY 9.5 Pre-symptomatic genetic testing of children

Following a diagnosis of bowel cancer, a man aged 35 was found to have a pathogenic genetic variant that causes Lynch syndrome, an inherited predisposition to a range of different cancers. The cancer had spread too far for curative treatment to be possible, and he was receiving palliative care. He and his wife had two children aged 9 and 7, and they requested that the children be tested for the genetic variant. Lynch syndrome does not manifest in childhood, with cancer risk starting to increase in the mid to late twenties age group, so no screening or other intervention would be offered until that age and there is no medical benefit for testing the children at this age. However, the couple felt that they wanted to know whether their children would be at risk, as this was important to help them prepare themselves and their children if either or both are found to be at increased risk of developing cancer.

Q1 What do you think it would be like to grow up knowing that you are at high risk for cancer when you are older?

Q2 What are the benefits to each of the family members of testing the children at this age for Lynch syndrome?

Q3 What are the negatives of testing the children at this age?

Q4 Do you think that the parents should be able to give consent for genetic testing of their children for adult-onset conditions, or do you think that the children should decide about having a test when they are old enough to do so?

opportunity to prepare and adapt. However, others point out that less than 20% of adults at risk of developing Huntington's disease have pre-symptomatic testing, showing that many individuals do not want to know about future health risks (though this may particularly apply where there are no interventions available to counter these risks). Therefore, allowing parents to request pre-symptomatic testing for their children takes away the child's autonomy to decide for themselves about whether to find this out. The British Society for Genetic Medicine has issued guidance, indicating that this testing should usually be delayed until the 'child', by now often an adult, wishes to make their own decision about whether/when to be tested, although there may be reasons why testing could be offered sooner if the family circumstances are unusual. However, practice in other countries varies and, with the increasing availability of private genetic testing, families may be able to have their children tested for adult-onset conditions if they choose to do so. Those in patriarchal societies may argue that the decision can (and should) be made by the (male) head of the family in the interests of the family as a whole.

9.3.5 Incidental findings from whole genome testing

With advances in testing technology, it is quicker, more cost-effective, and less liable to technical failure to perform whole genome sequencing (WGS) than to sequence individual genes or even fixed (pre-determined) panels of genes, especially if attempting to identify the cause of a set of symptoms which could be due to a change in any of a range of different genes.

There has been much debate about how to manage the risk of incidental findings, balancing the potential psychological harm of identifying a risk for a different genetic condition when investigating unrelated symptoms. Findings from genomic testing which are not directly related to the symptoms being investigated can be split into three categories (though the terms used for these categories differ between authors):

- **Unexpected phenotypic information** indicates an increased risk for other health issues which are part of the same syndrome as the features being investigated. For example, carrying out a genomic test to look for the cause of degeneration of vision may identify a syndrome which causes degeneration of both vision and hearing: the risk of hearing loss is unexpected phenotypic information. The possibility of this unexpected information should be made clear to the individual when consenting for the test.
- **Additional findings** indicate a health risk which is medically actionable: there is screening and/or management to reduce morbidity and mortality from this health risk. For example, whole genome sequencing may identify a variant that increases the risk of bowel cancer, and bowel screening can be put in place, so any cancers are caught early when they are more treatable.
- **Incidental findings** are neither related to the symptoms being investigated nor medically actionable, as in **Case study 9.6**. The ethical issues around reporting incidental findings are more complex, as there is no medical reason to report them, but it could be argued that individuals should be told about their health risks. Would you wish your high risk of Huntington's disease or another form of dementia to be revealed in this way?

These categories are useful to reflect on how to manage the findings from genomic testing, though gene variants may not fall neatly into one of the latter two categories: does awareness of a health risk which encourages you to report any symptoms to your GP count as medically actionable even if there is no screening available? Does the availability of a screening programme with poor sensitivity and specificity mean that a particular condition is medically actionable, even if the chance of clarifying your risk of disease is quite low?

With genomic testing in clinical practice, the crucial question becomes which variants in the genome are to be interpreted, evaluated, and potentially reported, and which remain uninterpreted in the database that holds the sequence information. If your analysis is limited to

CASE STUDY 9.6 Incidental findings from whole genome testing

A three-year-old boy with developmental delay and unusual (dysmorphic) features was seen by the paediatrician, who felt that these features may have a genetic cause. His parents consented to whole genome sequencing, which identified a missense variant in the *MYLK* gene. This gene is associated with an increased risk of a tear in the aorta (one of the major blood vessels), which can be fatal. This usually occurs without warning so cardiac screening (such as echocardiogram) may not prevent sudden cardiac death. This variant would not explain the boy's features so is an incidental finding, resulting in the parents experiencing high anxiety about this risk to their son's health and there being no management options to reduce his risk.

Q1 Do you think it is helpful for this family to be aware of this incidental finding?

Q2 What may be the outcome for this boy and his parents?

Q3 Can you think of any reasons why these incidental findings should or should not be available for families?

those genes associated with the symptoms, you may identify unexpected phenotypic information, but not the latter two categories. However, does that limit your diagnostic ability if there are other genes associated with the symptoms which have not been identified yet? And is it a missed opportunity to identify medically actionable additional findings unrelated to the symptoms? It is worth noting that, once the whole genome has been sequenced, if the first analysis does not answer the diagnostic question being asked, a 'dry' reanalysis can be performed as a revised search ***in silico***, without the need for any further 'wet' laboratory work *in vitro*. As knowledge advances, the gene variants known to be implicated as causing a particular set of problems can be amended, with additional genes being included in a re-analysis as appropriate.

in silico
Scientific experiments or research conducted or produced by means of computer modelling or computer simulation.

Therefore, the question thrown up by the move to WGS is whether additional genes—unrelated to the diagnostic question that triggered the WGS—should be analysed, with variants being interpreted for their pathogenic potential. This might allow patients at risk of serious problems to take preventive action before they are affected by (for example) a malignancy or by cardiac disease. Such 'opportunistic' testing might also identify those at risk of rare pharmacogenetic hazards if exposed to certain drugs (e.g. flucloxacillin or halothane).

How should we approach this potential source of important information? There are potential difficulties that might result from such broad-based genetic testing, as well as potential benefits, especially if the resulting information would become part of the patient's medical records and available to others, including insurance companies. The American College of Medical Genetics & Genomics (ACMG) issued its first report on this topic, recommending that pathogenic variants in a list of genes should be disclosed to all patients and research participants, irrespective of their wishes. This approach was then revised in that the individual's wishes were acknowledged as important and that consent to receive such 'secondary' or 'incidental' findings should be sought at the time of arranging the initial genetic diagnostic test. The list of genes to which this policy is applied has now been extended by the ACMG to 73 loci, most of which are implicated in causing cancers and cardiac disease. Identifying the carrier status for recessive disorders can also be included in an even broader search for incidental findings of potential value to the patient or their relatives. However, there are real questions about whether the timing of the search for incidental findings needs to be considered very carefully. The time when the patient has just been recognized as affected by a potentially serious genetic disorder may not be the best time for the patient or their parents to be asked to consider genetic testing for a broad range of other conditions.

This brings us back to the question of testing children for adult-onset disorders—what if testing the child's genome indicates the child and one of their parents are at risk of a serious adult-onset disorder identified 'incidentally'? This would lead to the concerns around testing children for adult-onset disorders, but it may also be to the advantage of the child for the parent to know this and be able to mitigate their risk by screening or management. However, context and prior consent are crucial. If the family is already aware of a serious, adult-onset disorder in the family, and they have devised a policy of when and how to pass on the relevant information to the child, it would not be helpful for the child's genetic status to be revealed in this unplanned fashion.

SELF-CHECK 9.3

Which of the three following terms describe findings that are neither related to the symptoms being investigated nor medically actionable? Unexpected phenotypic information, additional findings, or incidental findings?

Key Points

Whenever wider sections of the genome than just known causative loci are investigated it is quite possible that genetic variations related to other conditions are found to be present. It is important that the individual whose DNA is being sequenced (and/or the individual consenting on their behalf) as well as the clinical team and those doing the testing are aware of this and that they have policies and procedures in place to manage it.

9.3.6 Ethical issues in a multicultural society

As with many healthcare interactions, cultural influences may be an important factor for patients and families being seen within medical genetics. In all cultures, patients want to maximize the benefit for their family as well as themselves, but there may be different interpretations of what this means in practice. In Western culture, the focus is often on the individual or the immediate family unit, whereas in other cultures the wider family may be the main consideration, and other family members may play a role in decision-making. Some of these cultural influences may be related to religious beliefs, and it is helpful for the healthcare professional to have some understanding of the different religions, but culture encompasses more than religion: each culture may have their own interpretation of religious doctrine and the community's lived experience over generations will also impact on cultural norms. Therefore, healthcare professionals aim to establish open relationships with their patients, to help them explore their preferences, including those influenced by culture, so that they can make an informed decision that is most appropriate for their family.

It can be difficult to distinguish strictly 'religious' considerations from broader cultural factors, and the major issue in the case below is one of these cultural factors that may sometimes be attributed to a family's religion. In most cases, these cultural and/or religious preferences do not conflict with ethical values, but there are occasions when they can lead to challenging dilemmas, such as in **Case study 9.7**.

CASE STUDY 9.7 A cultural preference for sons

A couple from South Asia attended the genetics clinic as their 3-year-old son had been diagnosed with Duchenne muscular dystrophy (DMD) (previously mentioned in **Case study 9.4**), which leads to a progressive loss of muscle strength and can also lead to educational difficulties. DMD is associated with the loss of the ability to walk, weakness of the respiratory muscles, and cardiomyopathy. It is often fatal by the third or fourth decade despite good medical care. The woman had been found to be a carrier for the condition and informs the clinicians that she is currently pregnant. The couple already has two daughters, aged 6 and 4. They are offered prenatal testing, initially by non-invasive prenatal testing (NIPT) to find out whether the foetus is male. If so, they can have invasive testing to discover whether he would be affected by the condition. The NIPT result shows that the mother is pregnant with a daughter, so would be very unlikely to be severely affected by the condition and no further testing was offered during pregnancy. When the genetic counsellor telephones the couple two weeks later, they are told that the couple have had a termination of pregnancy as they do not want to have another daughter. They plan to become pregnant again, as it is very important to them to have an unaffected son.

Q1 How do you feel about the couple using the genetic test results to terminate a pregnancy as the foetus is female?

Q2 If the couple come back to the genetics clinic when she becomes pregnant again, should prenatal testing be offered? Why, or why not?

Resolving ethical dilemmas where cultural preferences are involved can be particularly challenging, as healthcare professionals aim to respect the culture and preferences of their patients. However, they are also committed to ethical practice and may need to seek advice about how to manage these complex situations, especially as they may not have a full understanding of the relevant culture. As with all of the complex situations addressed in the case studies so far, it is helpful to use an ethical framework to assess the most appropriate course of action, as outlined in **Section 9.2**.

9.4 Ethical issues in laboratory studies and research

9.4.1 Patenting of gene sequences

During the early stages of gene discovery and sequencing, many companies who first identified a gene applied for a patent which gave them exclusive rights to that gene sequence for 20 years from the date of the patent. This enabled them to decide how the gene can be used in commercial settings, clinical practice, and research. This commonly meant that they were the only company able to test for these genes, so they were able to profit from this, and it was argued that this provided an incentive for gene discovery. More than 4300 human genes were patented, including the *BRCA1* and *BRCA2* genes which, when altered, lead to an increased risk of breast and ovarian cancer, and were patented by Myriad Genetics Inc. This patent was challenged in the Supreme Court of the United States, which ruled that human genes cannot be patented because DNA is a 'product of nature' so there is no intellectual property to protect. This decision invalidated the gene patents, so the genes became available for research and clinical testing by other companies and organizations. The normal human genome should be regarded as a discovery and not an invention, and therefore should not be subject to patenting.

9.4.2 Data sharing and data security

As genomic technologies have become faster and cheaper, increasing amounts of genomic data are generated. This data can be used to increase scientific understanding of the genome, with the aim to translate this into improved genomic medicine, improving diagnosis, developing treatments, and targeting treatments accurately. Therefore, there is a strong argument for promoting data sharing to help in this endeavour, and indeed there is a Global Network of Personal Genome Projects, where individuals can choose to make their genomic data publicly available for all to use to benefit science and society. When opting for genomic testing, some patients may prefer their data to be used for clinical purposes but not for research; however, the divide between these purposes is blurred, as outlined above: if an uncharacterized variant is found, at what point do investigations to establish its pathogenicity become research? Therefore, they may need to consent for their data to be shared in some form to help with the diagnostic process.

However, genomic data can uniquely identify an individual, which makes data security an important issue. Databases of genomic data, many of which are stored in the cloud, must be kept secure to prevent identifiable data being revealed. In England, the company Genomics England (owned by the governmental Department of Health and Social Care) is responsible for storing genomic data produced within the NHS, and then works with researchers and other companies to release de-identified data in a secure environment. Globally, there are many different repositories of genomic data, and a key issue for them all is to keep this data secure, but in a format where it can be used to benefit scientific knowledge.

Concerns about discrimination based on genomic data illustrate the importance of data security. Could individuals be denied employment or insurance due to their genomic data? To counter this concern, some countries have established legal frameworks to prevent such discrimination, such as the Genetic Information Non-discrimination Act (GINA) that has been established in the USA and the Genetic Non-discrimination Act (GNDA) in Canada. In other countries, this is covered by other legislation, such as Australia's Private Health Insurance Act (2017) which prohibits health insurers from using genetic information to discriminate. In the UK, discrimination due to genomic data is not covered by law, but there is a Code on genetic testing and insurance agreed between the UK Government and the Association of British Insurers (ABI), preventing insurers from considering predictive genetic test results except under specific circumstances. Therefore, there are differences globally in the legal framework to protect against discrimination.

Some advocates of genomic testing propose using whole genome sequencing to provide a resource for each individual to dip into throughout their life, depending on what is needed at the time. A child could have their genome sequenced at birth and the sequence data stored so that it can be used at any point when indicated by clinical need. Whilst this may sound a logical strategy, there are several arguments against this, not least the tension between being able to access the data whilst also keeping it secure, to ensure that genetic discrimination does not occur. In addition, the quality standard of data required for clinical application continues to increase so current sequencing data may become irrelevant; there are also large financial (and environmental) costs of storing all this data, and the software and hardware will need to be updated periodically over the course of a person's lifetime, which will increase cost, introduce errors, and potentially cause other problems. With the cost and speed of genomic sequencing falling rapidly, it may make more sense to simply perform a whole genome sequence each time that the data is needed at specific points during that individual's life. This may also allay some of the data security concerns.

9.4.3 Embryo selection and human germline gene editing (hGGE)

As discussed above, we are now able to carry out gene editing within the cell. Therefore, if we know enough to correct the genetic constitution of an embryo or gamete, why not carry this out at this early stage and implant only unaffected embryos?

He Jiankui, a Chinese biophysicist, announced that he had edited embryos to provide resistance to HIV infection, and these embryos had resulted in the birth of two baby girls. He has been sentenced to three years in prison for 'illegal medical practice', but it is not yet possible to understand the potential impact on the girls. It emerged soon after his announcement that the deletion that he introduced would lead to a shorter lifespan for reasons that he had not anticipated. Furthermore, the gene editing that occurred seems not to have been 100% efficient, so there is some mosaicism in the girls and also some unintended, potentially harmful sequence changes, and an enormous loss of privacy. This, as well as more recent experimental work, illustrates that CRISPR still has basic problems with both efficiency and off-target effects. At present, we have limited ability to predict the consequences of introducing sequence changes that we believe will prove advantageous. Our genomic knowledge largely comes from looking at the phenotype and then working out the genotype, which can work well and provide good assumptions. However, our ability to understand how a particular genotype will influence the phenotype of a healthy-looking embryo is very different. Finding plausible explanations for phenotypes is much more achievable than predicting what phenotype may develop. Therefore, whilst human germline gene editing is technically possible, there are still many concerns to be addressed before this can be ethically used in practice.

9.4.4 Trials of experimental therapies for rare genetic disease

For families affected by genetic conditions, there is often a hope that more research into technologies such as gene therapy and gene editing will lead to treatment and ultimately a cure. However, these are still new technologies, and achieving such a cure will certainly not be straightforward. With exciting developments related to the use of these techniques in mice models, expectations of the possible therapeutic uses in humans may be unrealistically heightened, and it is important that communication about the possibilities is more realistic to reduce the hype around these developments.

The possible implications of these therapies will depend on the genetic condition involved. For a condition such as retinal dystrophy, where the therapy can be targeted to the retinal cells and can prevent degeneration of initially healthy cells, the likelihood of unintended consequences is lower than for a condition such as Rett syndrome, which affects brain development, so it is harder to target the therapy and the possible effects of manipulating brain cells are largely unknown. However, all treatments are likely to be subject to the 'Goldilocks principle'—the amount of protein needs to be just right (not too much, not too little) in each cell. Even if this becomes possible to achieve, there may be other unintended consequences—for example, individuals with Rett syndrome are believed to have reduced sensitivity to pain. If some of this brain function is restored, will they experience high levels of pain? And what about the psychological impact and the effect on their identity? These are all issues that need to be considered, tempering the enthusiastic search for a cure. With all of these uncertainties, is it possible to gain valid informed consent for these trials, especially from parents on behalf of their children?

It seems likely that gene therapy will be more effective if administered early, before the impact on the cells is well established, which means that those who already have the condition may not benefit from these new therapies. If energy and resources are focused on developing these new therapies, this will reduce the resources for care-based therapies, to the detriment of those living with the condition.

Chapter summary

- The history of research highlights issues that have occurred due to the conflict between advancing scientific knowledge and considering the ethics of these studies.
- This has led to the development of guidelines and legal frameworks to ensure ethical practice in research.
- Some of these ethical issues relate specifically to genetics and genomics research, which has the potential to manipulate genetic material with as yet uncertain consequences.
- Ethical issues are also apparent in clinical genetics, including the importance of informed consent.
- This is particularly important with genomic testing, which may identify additional findings (for which there are treatment or management options) and incidental findings (with no management available).

- Ethical frameworks provide a structure to consider the most ethical approach to a particular situation, but there are often no straightforward answers.
- Clinicians may have a duty of care to both the individual and their family, and tensions can arise when these conflict.
- Technologies enabling couples to avoid having a child with a genetic condition promote autonomy but may devalue the lives of those living with the condition.
- Within a multicultural society, cultural preferences can sometimes clash with ethical principles, which requires the clinician to consider how to be sensitive to culture whilst maintaining their ethical practice.
- With further advances in genomic technologies, ethical issues will continue to arise, and it will be important for society to debate what is acceptable.

Discussion questions

9.1 Should/could a global moratorium on human germline gene editing (hGGE) be imposed or should we work towards an international consensus for hGGE practice instead?

9.2 Should parents/legal guardians have the right to have their child genetically tested for adult-onset conditions if there is a significant risk that the child has inherited the causative allele(s)?

9.3 Should all of us have our genome sequenced at birth?

Further reading

- Lucassen A, Hall A for the Joint Committee on Genomics in Medicine (2019) ***Consent and Confidentiality in Genomic Medicine****. Guidance on the Use of Genetic and Genomic Information in the Clinic*. 3rd Edition, Royal College of Physicians and Royal College of Pathologists, London.

  Explores consent and confidentiality in clinical genetics and genomics, and is a resource used by clinicians working in genomics.

- Miller DT, Lee K, Chung WK *et al.* (2021) ***ACMG SF v3.0 list for reporting of secondary findings in clinical exome and genome sequencing: a policy statement of the American College of Medical Genetics and Genomics (ACMG)***. *Genet Med* 23, 1381–1390.

  Contains the list of secondary findings that the American College of Medical Genetics believe should be reported if found by genomic sequencing.

- Clarke AJ, Abdala Sheikh AP (2018) **_A perspective on 'cure' for Rett syndrome_**. *Orphanet J Rare Dis*, 13, 44. https://doi-org.abc.cardiff.ac.uk/10.1186/s13023-018-0786-6

  Discusses some of the considerations about developing a cure for a genetic condition.

- Clarke A, van El C (2022) **_Genomics and justice: mitigating the potential harms and inequities that arise from the implementation of genomics in medicine_**. *Human Genetics*, 141(5), 1099–1107.

- Clarke AJ, Wallgren-Pettersson C (2019) **_Ethics in genetic counselling_**. *J Community Genet*, 10(1), 3–33.

- Cornel MC, Clarke AJ (2021) **_Costs, burdens and the prevention of genetic disorders: what role for professional influence?_** *Journal of Community Genetics*, 12, 503–5.

- Czech (2023) **_The Lancet Commission on Medicine, Nazism, and the Holocaust: historical evidence, implications for today, teaching for tomorrow_**. *Lancet*, 6736(23), 01845–7.

- deWert G, Dondorp W, Clarke A, Dequeker EMC, Cordier C, Deans Z, van El CG, Fellmann F, Hastings R, Sabine Hentze S, Howard H, Macek M, Mendes A, Patch C, Rial-Sebbag E, Stefansdottir V, Cornel MC, Forzano F (2020) **_Opportunistic genomic screening. Recommendations of the European Society of Human Genetics_**. *European Journal of Human Genetics*, 29(3), 365–77.

- Dheensa S, Fenwick A, Lucassen A (2016) '**_Is this knowledge mine and nobody else's? I don't feel that.' Patient views about consent, confidentiality and information-sharing in genetic medicine_**. *J Med Ethics*, 42, 174–9.

- European Society of Human Genetics (ESHG) (2009) **_Genetic testing in asymptomatic minors: recommendations of the European Society of Human Genetics_**. *Eur J Hum Genet*, 17(6), 720–1.

- Hamamy H, Antonarakis SE, Cavalli-Sforza LL *et al.* (2011) **_Consanguineous marriages, pearls and perils: Geneva International Consanguinity Workshop Report_**. *Genet in Med*, 13, 841–7.

- Lantos JD (2019) **_Ethical and psychosocial issues in whole genome sequencing (WGS) for newborns_**. *Pediatrics*, 143 Supplement 1, S1–S7.

- Lombardo PA (2018) **_The power of heredity and the relevance of eugenic history_**. *Genetics in Medicine*, 20(11), 1305–11.

- MacLeod R, Beach A, Henriques S, Knopp J, Nelson K, Kerzin-Storrar L (2014) **_Experiences of predictive testing in young people at risk of Huntington's disease, familial cardiomyopathy or hereditary breast and ovarian cancer_**. *Eur J Human Genetics*, 22(3), 396–401.

- MacLeod R, Tibben A, Frontali M, Evers-Kiebooms G, Jones A, Martinez-Descales A, Roos RA, Editorial Committee and Working Group 'Genetic Testing Counselling' of the European Huntington Disease Network (2013) **_Recommendations for the predictive genetic test in Huntington's disease_**. *Clin Genet*, 83, 221–31.

- Middleton A, Robson F, Burnell L, Ahmed M (2007) **_Providing a transcultural genetic counseling service in the UK_**. *Journal of Genetic Counseling*, 16(5), 567–82.

- Newson AJ (2021) **_The promise of public health ethics for precision medicine: the case of newborn preventive genomic sequencing_**. *Human Genetics*, 141(5), 1035–1043.

- Nuffield Council on Bioethics (2017) ***Noninvasive Prenatal Genetic Testing: Ethical Issues***. Nuffield Council, London.
- Parker M (2012) ***Ethical Problems and Genetics Practice***. Cambridge University Press, Cambridge.
- Shakespeare T (1998) ***Choices and rights: eugenics, genetics and disability equality***. *Disability & Society*, 13, 665–81.
- Wert D de, Pennings G, Clarke A, Eichenlaub-Ritter U, van El CG, Forzano F, Goddijn M, Heindryckx B, Howard HC, Radojkovic D, Rial-Sebbag E, Tarlatzis BC, Cornel MC, on behalf of the European Society of Human Genetics and the European Society of Human Reproduction and Embryology (2018) ***Human germline gene editing: recommendations of ESHG and ESHRE***. *European Journal of Human Genetics*, 26(4), 445–9.

10

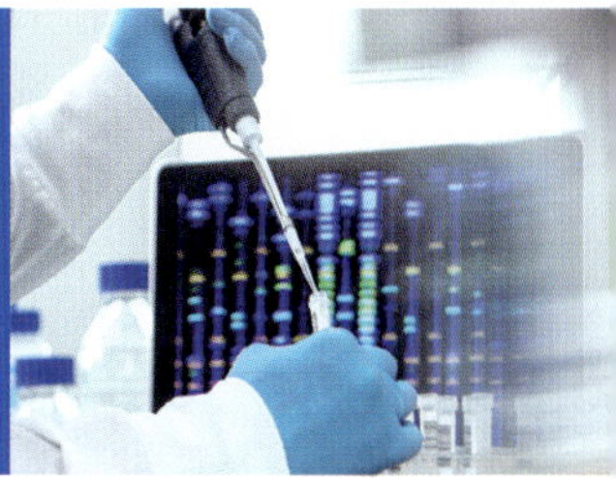

Precision Medicine

The future is here

Janine Lamb

Huge advances in science and technology in recent years have fuelled a new model of clinical medicine and led to the adoption of the term *precision medicine*. This new era provides an exciting opportunity to modify healthcare. In this chapter, we will cover what precision medicine is, and how it is applied in the fields of pharmacogenetics, lifestyle genome sequencing, gene therapy and genome editing, and the use of stem cells.

Learning Objectives

After studying this chapter, you should be able to:

- Define what precision medicine is
- Describe how current and emerging technologies are enabling precision medicine
- Provide examples of how precision medicine is being applied in fields such as pharmacogenetics, lifestyle genome sequencing, gene therapy and genome editing, and stem cells
- Describe the opportunities, and potential challenges, of precision medicine
- Suggest ways in which precision medicine might impact on the future of healthcare.

10.1 Introduction to precision medicine

Huge advances in science and technology in recent years, from biotechnology to computer power, have fuelled a new model of clinical medicine. This has led to the adoption of a new term *precision medicine*, which provides an exciting opportunity to modify healthcare, from the ways in which individuals access healthcare, to the ways in which doctors treat their patients. Precision medicine transforms healthcare from a 'one size fits all' approach developed for the average patient, to a more tailored treatment approach.

10.1.1 What is precision medicine?

In recent years, several new terms have emerged, including *personalized medicine*, *genomic*, *stratified*, and *precision medicine*. These terms are similar, but subtly different. Today, the term *personalized medicine* is not as widely used, as *personalized* implies that treatments and preventions are developed uniquely for each individual, rather than identifying approaches which will be most effective for different patients. Genomic medicine uses genomic information about an individual as part of their clinical care—this could include use in diagnosis, prognosis, prediction, prevention, or therapeutic decision-making. Stratified medicine is based on being able to identify subgroups of patients with distinct mechanisms of disease or particular responses to treatments, to identify and develop treatments that are more effective for particular groups of patients. Precision medicine is a more holistic approach: it is disease prevention and treatment that takes into account the variability in genes, environment, and lifestyle for each person (see **Figure 10.1**). In disease treatment, this means treating the right person, with the right therapy, at the right dose, at the right time. Precision medicine is defined by the four Ps: healthcare that is preventive, predictive, personalized, and participatory. Precision medicine provides an exciting new model of clinical medicine, changing the way individuals access healthcare and how doctors treat their patients.

SELF-CHECK 10.1

What is precision medicine?

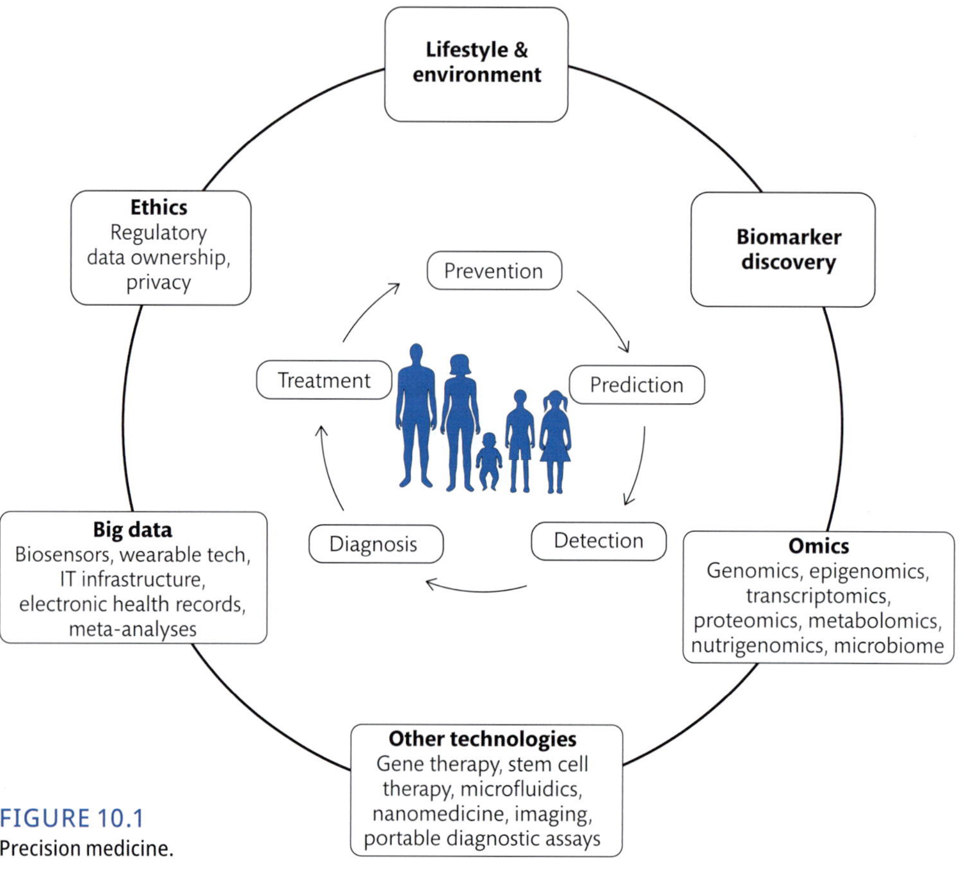

FIGURE 10.1
Precision medicine.

10.1.2 How has precision medicine come about?

Precision medicine has come about due to significant advances in science and technology that enable more personalized health and disease monitoring. Let's take an example from computer science: Moore's Law. Moore's law follows a 1965 prediction by American engineer Gordon Moore that the number of transistors packed onto a silicon microchip doubles every two years. Based on this, we should expect the speed and capability of our computers to increase, while size and costs decrease. In the decades that followed 1965, Moore's Law held true. However, in today's high-tech society, technological advance outstrips Moore's Law, with the manufacture of mobile devices, such as tablets and smartphones, with very small processors, and the development of the internet. Similar advances in biotechnology, again facilitated by advances in computing, led to completion of the full Human Genome sequence by the Human Genome Project Consortium in 2003.

You can see from **Figure 10.2** how advances in biotechnology since 2001 have led to significantly increased human genome sequencing throughput and data generation, with decreased costs. The sudden drop in the graph in 2008 reflects the time when DNA sequencing changed from Sanger sequencing to **next generation sequencing** technologies (see **Box 10.1**). These changes in DNA-sequencing capacity have revolutionized the use of genomics in medicine. This combination of changes in computing power and biotechnology have led to more exploratory data-driven approaches in biomedical research, rather than testing a very focused and specific research hypothesis. We now have the capability to ask questions and generate data at an unprecedented scale about **genomes**, **proteomes**, **metabolomes**, and to capture data from lifestyle choices, such as diet, exercise, or smoking, at both an individual and a population level.

Cross reference

Please see **Chapter 1** for further discussion on the Human Genome Project.

next generation sequencing
Also known as massively parallel or high-throughput sequencing. Millions or billions of DNA strands are sequenced in parallel.

genome
All of the genetic information in a human or other organism.

proteome
All of the proteins in an organism, tissue, or cell.

metabolome
All of the metabolites in an organism, tissue, or cell.

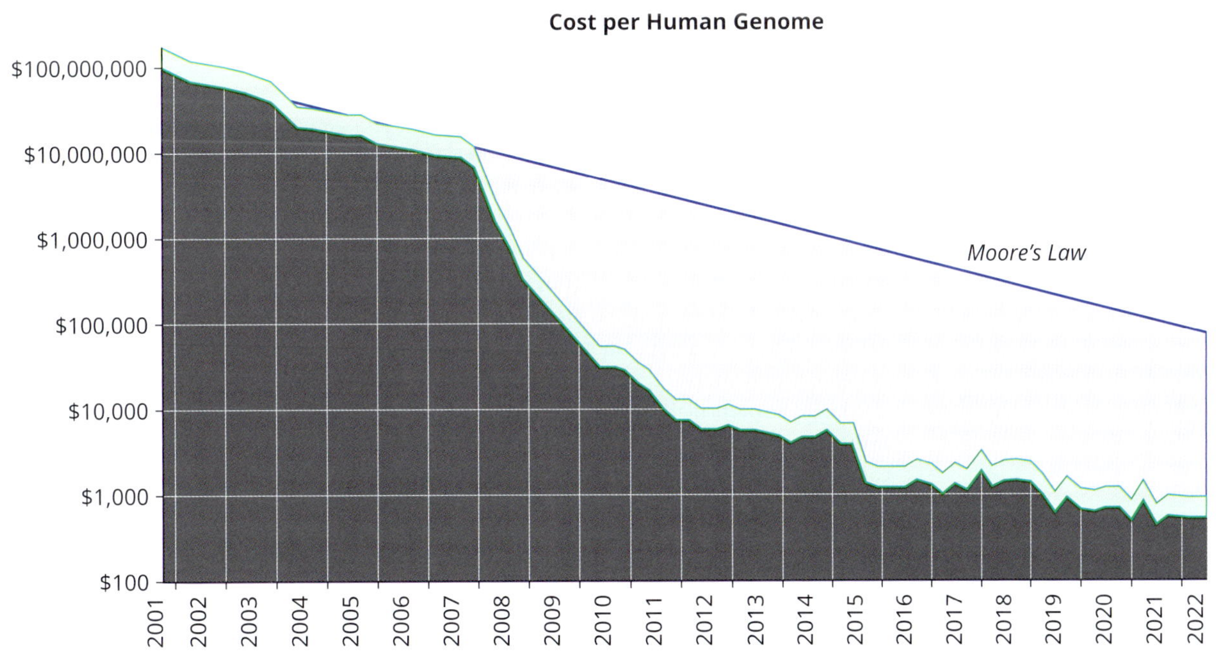

FIGURE 10.2
Change in the cost of sequencing a human-sized genome, since the completion of the draft human genome sequence.

BOX 10.1 DNA sequencing, from genes to genomes

- DNA sequencing: laboratory technique used to determine the exact sequence of bases (A, C, G, and T) in a DNA molecule.
- Next generation sequencing (NGS) (also known as massively parallel or high-throughput sequencing): millions or billions of DNA strands are sequenced in parallel; these approaches have substantially higher throughput than the traditional 'first generation' techniques.
- Targeted next generation sequencing: sequencing target regions of the genome, for example, a panel of genes known to be involved in cardiovascular disorders.
- Whole exome sequencing: sequencing the protein-coding regions of all genes in the genome.
- Whole genome sequencing: sequencing the complete genome (3.2 billion nucleotides in humans).

10.1.3 Where will precision medicine take us?

biomarker
A naturally occurring biological marker that can be measured as an indicator of a biological or pathological process, exposure, or intervention

Cross reference
For an in-depth description of how the genetics of complex disorders is explored please read **Chapter 4**.

drug repurposing
A strategy where existing drugs, including failed or abandoned drugs from clinical trials, are investigated for new therapeutic purposes.

Precision medicine aims for prevention and earlier and improved detection and diagnosis of disease. The discovery of new targets, or **biomarkers**, of health and disease will lead to a better understanding of healthy (normal) and diseased states and identify who is at risk of developing a disease or condition. In turn, this will lead to the development of new biomedical assays, improved diagnostic tests, and a precision diagnosis. The ability to classify conditions into more precise and distinct subgroups will increase our mechanistic understanding of disease, leading to more precise disease classification and a 'molecular taxonomy of disease'. This approach can be applied to both rare inherited diseases, such as muscular dystrophies, and to more common complex disorders, such as cardiovascular, inflammatory, or infectious diseases.

For example, we can now stratify people with cancer with specific genetic mutations into subgroups with a different prognosis or treatment strategy. This enables doctors to adopt a more tailored, individual treatment regimen, increasing the efficacy and cost-effectiveness of a specific treatment whilst avoiding the harmful side-effects of an iterative trial-and-error approach for those who are less likely to benefit. The development of diagnostics and therapeutics, therefore, needs to go hand-in-hand. Improved molecular characterization of disease is increasingly leading to **drug repurposing** or drug repositioning—strategies where existing drugs, including failed or abandoned drugs from clinical trials, are investigated for new therapeutic purposes. At a population level, greater stratification of populations will lead to more targeted health interventions. In this introduction, we have already seen how precision medicine might be particularly relevant for biomedical scientists working in genetics and genomics. In the following sections, we will look at how precision medicine relates to the areas of pharmacogenetics and pharmacogenomics, lifestyle genome sequencing, gene therapy and genome editing, and the use of stem cells.

SELF-CHECK 10.2

Why is precision medicine particularly relevant today?

Key Point

Precision medicine is an approach to disease prevention and treatment that takes into account individual variability in genes, environment, and lifestyle. Significant advances in science and technology enable precision medicine.

10.2 Pharmacogenetics and pharmacogenomics

Changes in genomic technologies and DNA-sequencing capacity (see **Section 10.1.2**) have revolutionized the use of genomics in medicine, and now allow the use of pharmacogenetics and pharmacogenomics in precision medicine. Pharmacogenetics and pharmacogenomics sit at the intersection of pharmacology and the study of genetic variation in the human genome. Pharmacogenetics looks at how inherited (germline) or somatic variation in single genes determines how an individual might respond to (a) particular drug(s). This includes genes involved in pharmacokinetics—drug absorption, distribution, metabolism, and excretion (ADME)—and pharmacodynamics—variability in drug action through effects on cell receptors and downstream biochemical pathways. Drug disposition is complex. As you can see from **Figure 10.3**, many drug receptors, transporters, and drug-metabolizing enzymes are involved, in organs including the liver, kidney, gastrointestinal tract, and at the blood–brain barrier. The aim of pharmacogenetics is to identify and prescribe the right drug at the right dose for that individual, predicting and maximizing efficacy and drug safety, whilst avoiding adverse drug reactions. Pharmacogenomics is a larger scale '**omics**' version of pharmacogenetics, where variation in multiple genes involved in drug response across the genome are examined simultaneously.

omics
The comprehensive or global assessment of molecules in a sample; molecular characterization can take place at a single cell, tissue, or organ level, and within an individual or a cohort of people.

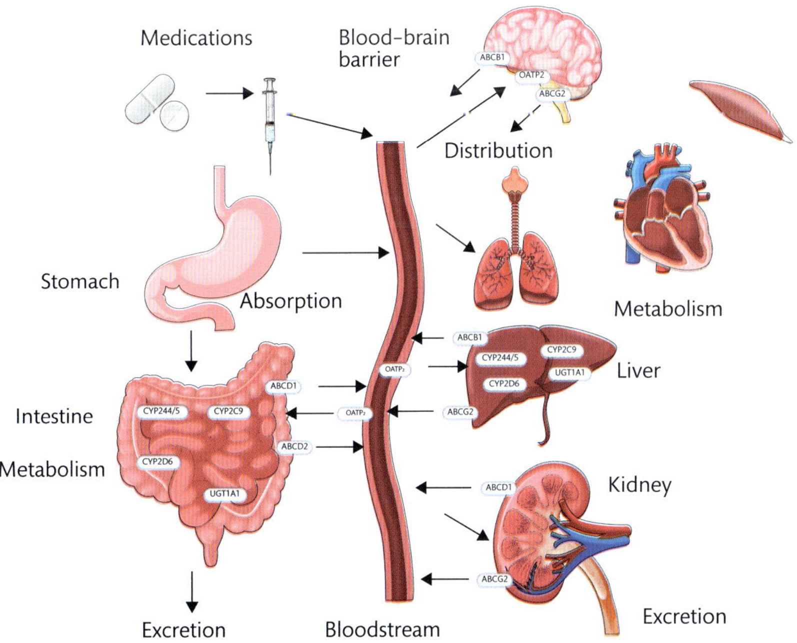

FIGURE 10.3
Genes involved in drug disposition. Many genes are involved in drug absorption, distribution, metabolism, and excretion. Phase I and Phase II drug-metabolizing enzymes include many members of the cytochrome P450 (CYP) and UDP-glucuronosyltransferase enzyme families respectively. Drug transporters include members of the organic anion transporting polypeptide (OATP) and ATP-binding cassette (ABC) transporter families.

Pharmacogenomic studies can also investigate genetic variation at a population level, to examine how drugs might work differently in different ethnic groups. In addition to genetic variation, an individual's age, gender, health, lifestyle, and environment might influence how they respond to different drugs. Other factors may include medication error, drug mis-dosing, and non-compliance. Next, we will look at some examples of the use of pharmacogenetics and pharmacogenomics in clinical practice.

SELF-CHECK 10.3

List some functions of genes relevant to pharmacogenetics.

10.2.1 Drug safety and efficacy

In the UK, adverse drug reactions are responsible for approximately 6% of hospital admissions and occur in 10 to 20% of hospital in-patients. Adverse drug reactions cost the National Health Service (NHS) approximately £1 billion annually and could be prevented by predictive pharmacogenetics testing in 20 to 30% of cases. Adverse drug reactions can range from relatively mild, such as drowsiness, fatigue, nausea and/or vomiting, diarrhoea, or skin reactions, to severe, such as drug-induced liver injury, anaphylaxis, or haemorrhage. Drug-induced toxicities have led to a significant number of drugs being withdrawn from the market. Drug efficacy or the response rate to drugs varies between individuals and between drugs, and can be disappointingly low across numerous therapeutic areas, such as rheumatoid arthritis, Alzheimer's disease, and depression. **Polypharmacy** management for individuals with **multi-morbidity** may impact on clinical utility through drug–drug interactions. Overall, if we can select the appropriate therapy, avoid adverse drug reactions, and apply more accurate dosing strategies, this will maximize drug efficacy, improve the patient's clinical experience, and reduce healthcare costs. Whole genome sequencing costs (**Figure 10.2**) may now be less than the annual prescription costs of some drugs, making it more cost-effective to determine how well someone is likely to respond to a given drug, before drug administration.

polypharmacy
The use of multiple medications (typically five or more) in a patient, commonly an older adult.

multi-morbidity
The presence of two or more long-term health conditions, which can include physical or mental health conditions.

10.2.2 Pharmacogenetics and inflammatory disease

Let's look first at an example of drug safety. Thiopurine drugs such as azathioprine and mercaptopurine are immunosuppressive drugs used in the management of autoimmune and inflammatory diseases, including inflammatory bowel disease, atopic dermatitis, rheumatological diseases, childhood leukaemia, and prevention of solid organ (such as kidney) transplant rejection. Azathioprine was first prescribed in the early 1960s, and today there are approximately 60,000 prescriptions in the UK annually. Azathioprine is effective in 55 to 70% of patients, but intolerance limits use, and side-effects can include nausea and vomiting, bone marrow suppression, lymphoma, hepatotoxicity, pancreatitis, or an allergic reaction. Azathioprine is the pro-drug for 6-mercaptopurine, which is metabolized by the enzyme thiopurine methyltransferase (TPMT). TPMT deficiency causes excess production of cytotoxic drug metabolites and bone marrow suppression, resulting in neutropenia, an abnormally low concentration of neutrophils in the blood. As you can see from **Figure 10.4**, TPMT activity has a trimodal distribution: around 90% of individuals have high or normal activity and respond to azathioprine with low risk of myelosuppression, around 10% have intermediate activity with increased risk of bone marrow suppression but a good response to azathioprine, and 0.3% of people have low or absent activity with highly increased risk of severe bone marrow toxicity and early profound

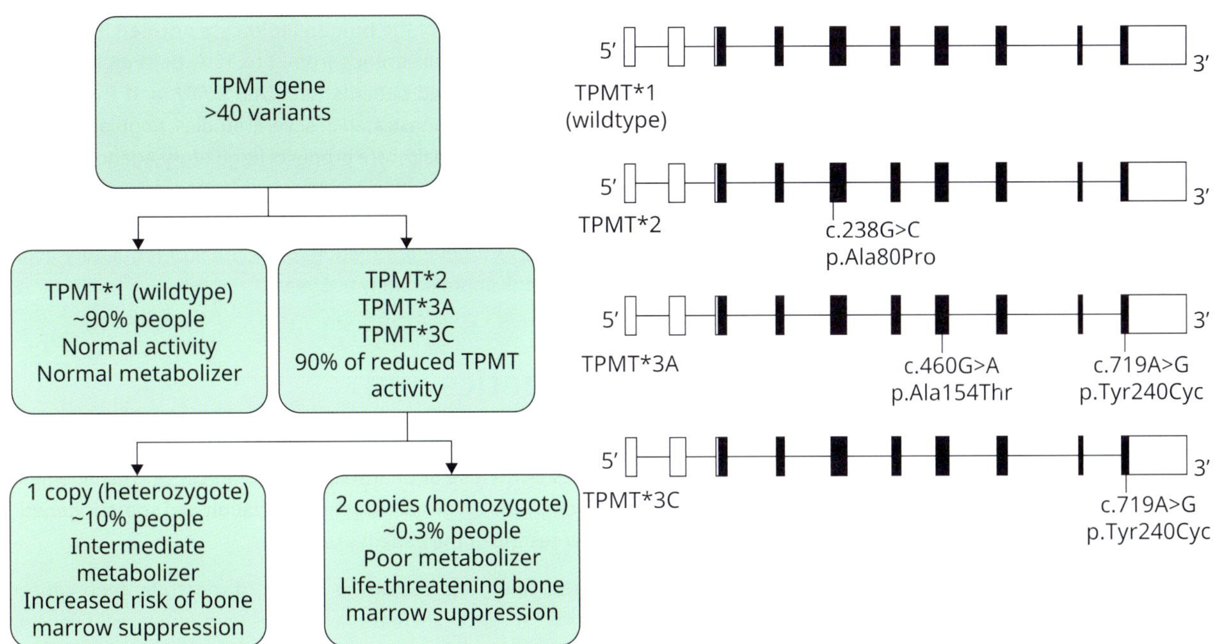

FIGURE 10.4

TPMT genetic variation and effect on azathioprine metabolism. There are more than 40 reported *TPMT* gene variants. *TPMT*1* is the most common (wild-type) allele. Three non-synonymous genetic variants account for over 90% of the reduced or absent enzyme activity. *TPMT*3A* is the most common variant allele in Caucasians and *TPMT*3C* is the most common variant allele in people with East Asian ancestry. Black rectangles represent the open reading frames and white rectangles represent the 5′ and 3′ untranslated regions. Drawing not to scale.

neutropenia at the standard dose of azathioprine. There are over 40 reported TPMT gene variants, but three common coding non-synonymous genetic variants account for over 90% of the reduced or absent enzyme activity. These variants differ in frequency in individuals of different ethnicities; for example, in people with East Asian compared to African or Caucasian ancestry. An individual's likelihood of developing azathioprine-induced neutropenia can be predicted by **genotyping** the common genetic variants in TPMT, although there is not a perfect correlation, particularly for individuals with an intermediate range of TPMT activity. TPMT enzyme activity levels can also be measured in red blood cells, although levels can be confounded by drugs or blood transfusion. Although adoption took several decades, there are now national guidelines for TPMT testing by the British National Formulary and the US Federal Drug Administration (FDA), and a significant increase in testing frequency by clinicians.

genotyping
The process of determining the DNA sequence, called a genotype, at specific positions within the genome of an individual.

10.2.3 Pharmacogenetics and infectious disease

The antiviral drug abacavir is a highly active antiretroviral therapy (HAART) introduced in 1999 as a single agent, or in combination, for treatment of people infected with the human immunodeficiency virus (HIV). However, drug hypersensitivity occurs in 1 to 10% of patients, usually within the first 6 weeks of therapy, with side-effects including fatigue, fever, vomiting, skin reactions, abdominal pain, and headache. On re-challenge, the hypersensitivity reaction occurs sooner and is more severe, causing potentially fatal severe hypotension, lactic acidosis, and severe hepatomegaly with steatosis (fatty liver disease). Pharmacogenetic studies of abacavir in people of different ethnicities from several countries showed that adverse drug reactions

human leukocyte antigen
The HLA complex is a complex of genes on chromosome 6 in humans that encode cell-surface proteins that play an important role in the immune response to foreign substances.

are associated with genetic variant *HLA-B**57:01* of the **human leukocyte antigen** (HLA) gene *HLA-B*. *HLA-B*57:01* frequency varies across ethnic groups from 0 to 10%. Before prescribing abacavir, doctors now routinely test HIV-infected patients for *HLA-B*57:01*, and the cost effectiveness of *HLA-B*57:01* testing has been demonstrated in several studies. Regulatory guidance by agencies including the Medicines and Healthcare products Regulatory Agency (MHRA), European Medicines Agency (EMA), and FDA now mandates prior genetic testing, which has reduced the frequency of adverse drug reactions. In contrast to the decades-long example of TPMT above, from the discovery of relevant genetic variation to adoption of regulatory guidance on abacavir companion testing took approximately five years.

10.2.4 Pharmacogenetics and pharmacogenomics in cancer

Today, there are multiple examples of the use of pharmacogenetics in cancer management to increase drug safety and efficacy, based on germline genetic variation and somatic genetic variation identified from molecular testing of the patient's tumour.

randomized controlled trials (RCTs)
A type of scientific experiment (e.g. a clinical trial) in which participants are randomly allocated into different groups to compare the effectiveness of different interventions or treatments.

Colorectal cancer, occurring in the colon and rectum, is the third most common cancer worldwide. The incidence of colorectal cancer increases with age, with an overall five-year survival rate of ~35% in the UK. In the UK, irinotecan is licensed for use in adults with advanced or metastatic colorectal cancer as a single or combination agent. In results from several **randomized controlled trials (RCTs)**, use of irinotecan significantly improved median overall survival and progression-free survival. Irinotecan works by inhibiting topoisomerase I, a nuclear enzyme which catalyses unwinding of DNA and is essential for cell division. Irinotecan is converted in the body to an active metabolite SN-38, which is inactivated and detoxified by a UDP-glucuronosyltransferase enzyme encoded by *UGT1A1*. Some genetic variants in *UGT1A1*, such as the common promoter variant *UGT1A1*28*, decrease enzyme activity and therefore reduce excretion and increase irinotecan metabolites in the blood. The frequency of *UGT1A1*28* varies across populations: allele frequency is ~40% in Africans, ~30% in Caucasians, and ~15% in Asian populations. Individuals with two copies of *UGT1A1*28* are more likely to develop potentially life-threatening myelosuppression and neutropenia after irinotecan treatment. Other side-effects of irinotecan include gastrointestinal effects (including severe diarrhoea), alopecia, and anorexia. In the US, the FDA recommends genetic testing before administering irinotecan, and reduction in the starting dose for patients homozygous for the *UGT1A1*28* allele, with subsequent dose adjustment based on an individual patient's tolerance to treatment. Other *UGT1A1* variants, and genetic variation in several other genes including transporter genes, may also influence irinotecan metabolism and toxicity.

Targeted therapies in cancer

Targeted therapies (also known as biological therapies) target particular proteins expressed in the cancer cells. These treatments aim to reduce damage to healthy cells, and therefore reduce side-effects. Targeted therapies include small molecule tyrosine kinase inhibitors (see **Table 10.1** for examples). Tyrosine kinases are enzymes responsible for the activation of many proteins through signal transduction cascades and play a critical role in tumour initiation and progression through regulation of cell differentiation, growth, migration, apoptosis, and death. In breast cancer, therapy with the tyrosine kinase inhibitor herceptin only works for the ~10–20% of breast cancers that are human epidermal growth factor receptor 2 (HER2) positive. HER2-positive breast cancer tends to be more aggressive than HER2-negative breast cancer. In HER2-positive cancers, somatic mutation in the tumour causes *HER2* gene amplification and an

TABLE 10.1 Cancer drugs targeting the tyrosine kinase receptor family

Protein	Drug	Cancer type
Human Epidermal Growth Factor Receptor 2 (HER2)	Herceptin	Metastatic breast
Epidermal Growth Factor Receptor (EGFR)	Erlotinib Gefitinib	Lung, pancreatic Lung
C-Kit Receptor/BCR-Abl hybrid protein	Imatinib	Philadelphia chromosome-positive acute lymphoblastic leukaemia, gastrointestinal stromal
Vascular Endothelial Growth Factor	Avastin	Colon

increased number of copies of the *HER2* gene. Increased *HER2* copy number leads to overproduction of the HER2 protein and uncontrolled breast cell growth and division. HER2 status in a breast cancer biopsy sample is most often determined by immunohistochemical staining to quantify the amount of cell surface HER2 protein expression, or by fluorescence *in situ* hybridization of HER2 proteins, to predict drug efficacy.

SELF-CHECK 10.4

What are targeted or biological therapies?

Non-small cell lung cancer (NSCLC) is the most common type of lung cancer, accounting for more than 87% of lung cancer diagnoses in the UK, and with a ten-year survival of 5%. Patients with NSCLC show a poor response to systemic chemotherapy drugs. Targeted therapies, including tyrosine kinase epidermal growth factor receptor (EGFR) inhibitors such as gefitinib and erlotinib, were developed to treat NSCLC. However, EGFR-tyrosine kinase inhibitors are associated with significant side-effects—common side-effects include skin rash, diarrhoea, loss of appetite, nausea, and fatigue. EGFR-tyrosine kinase inhibitors are therefore prescribed in NSCLC patients whose tumours have an activating mutation in the kinase domain of *EGFR* that increases drug sensitivity; in these patients efficacy is increased and progression-free survival is dramatically improved. However, it is important to note that intra- as well as inter-tumour heterogeneity exists. Individual tumours are heterogeneous collections of cancer clones, influencing cancer recurrence and acquired resistance to therapy.

Cross reference

To explore the genetics and genomics of cancer please read **Chapter 7**.

Today, clinical trials for new cancer drugs can be based on an individual's genetic profile and targeted to driver mutations. If you want to read more about how the TARGET study is stratifying patients for clinical trials in cancer, see **Box 10.2**.

10.2.5 Pharmacogenetics in warfarin sensitivity

Genetic variation in *VKORC1* (encoding the vitamin K epoxide reductase enzyme) and cytochrome P450 enzyme genes *CYP2C9* and *CYP2C19* influences response to the common anticoagulant drugs warfarin and clopidogrel. These drugs are used to prevent blood clots, for example in high numbers of patients with venous thromboembolism or atrial fibrillation. However, warfarin has a narrow **therapeutic index** (therapeutic ratio), which can lead to wide dosage variation among patients and insufficient or excessive anticoagulation (i.e. bleeding). Approximately half of the inter-individual variability in warfarin dosage is due to clinical or

therapeutic index

A ratio that compares the blood concentration at which a drug becomes toxic and the concentration at which the drug is effective. The larger the therapeutic index, the safer the drug.

BOX 10.2 A study of genetic characteristics and suitability for targeted cancer treatment, TARGET

This study looks at how genetic variation in advanced solid cancers (any type of cancer, apart from those of the blood system or lymphatic system) can help doctors decide the most suitable phase 1 clinical trial for people to take part in. Traditionally, tumours are characterized using an invasive tumour biopsy. The TARGET trial aims to develop and test whether a simple less invasive blood test (liquid biopsy) is a viable alternative to traditional biopsy. The liquid biopsy approach detects circulating tumour DNA shed via tumour cell death into the bloodstream. In TARGET, phase 1 clinical trial patients provided a blood sample at baseline. Next generation sequencing of a 641 cancer-associated gene panel was used to molecularly characterize circulating tumour DNA and a tumour biopsy from the same patient. For the first 100 patients with 22 different tumour types, the genomic profiles of the circulating tumour DNA and matched tumour biopsy showed good agreement, and results were generated within a clinically acceptable timeframe. In this first pilot phase, 40% of patients had actionable mutations for clinicians to stratify patients to the optimal phase 1 trial, and 11% of patients subsequently received a matched therapy.

lifestyle factors including age, gender, weight, concomitant medications, diet, smoking, and genetic variation. Warfarin sensitivity is caused by one copy of the *CYP2C9*2*, *CYP2C9*3*, or *VKORC1* (c.−1639G>A, rs9923231) variants in people who are warfarin 'slow metabolizers'. The rs9923231 variant is located in the promoter region of *VKORC1* and is thought to alter a CTCF transcription factor binding site, thereby influencing protein expression. The FDA-approved label on warfarin advises that *CYP2C9* and *VKORC1* genetic variation might influence response to the drug. 'Genotype-guided dosing' may therefore be used at initiation of warfarin therapy to maximize treatment time in the therapeutic range. In addition, in 2017 the Clinical Pharmacogenetics Implementation Consortium (CPIC) made recommendations for genotype-guided warfarin dosing in adults with and without self-identified African ancestry.

10.2.6 Pharmacogenetics in drug regulation

Overall, although there are still relatively few examples of clinically relevant pharmacogenetic tests, regulation has led to some mandated genetic testing and drug labelling changes. Approximately 250 drugs with pharmacogenetics label information have been approved by the FDA and EMA; you can see some examples of therapeutic areas in **Figure 10.5**. Clinically approved pharmacogenetic tests need to be sensitive and specific, clinically benign, inexpensive, technically feasible with a clinically appropriate turn-around time, and have the ability to alter prescription practice. Today, the availability of genomic data offers increasing opportunities for the application of genomic biomarkers to identify likely drug responders and non-responders, optimize drug dose, and avoid adverse events.

10.2.7 Pharmacogenomics in drug development

Historically, drug developers used to screen for chemicals with broad action against a disease. Now, genomic information is used to find or design drugs aimed at subgroups of patients with specific genetic profiles. Pharmacogenomic tools are used to identify new drug targets and search for drugs that target specific molecular and cellular pathways involved in disease. Identifying and targeting predicted drug responders increases drug safety during

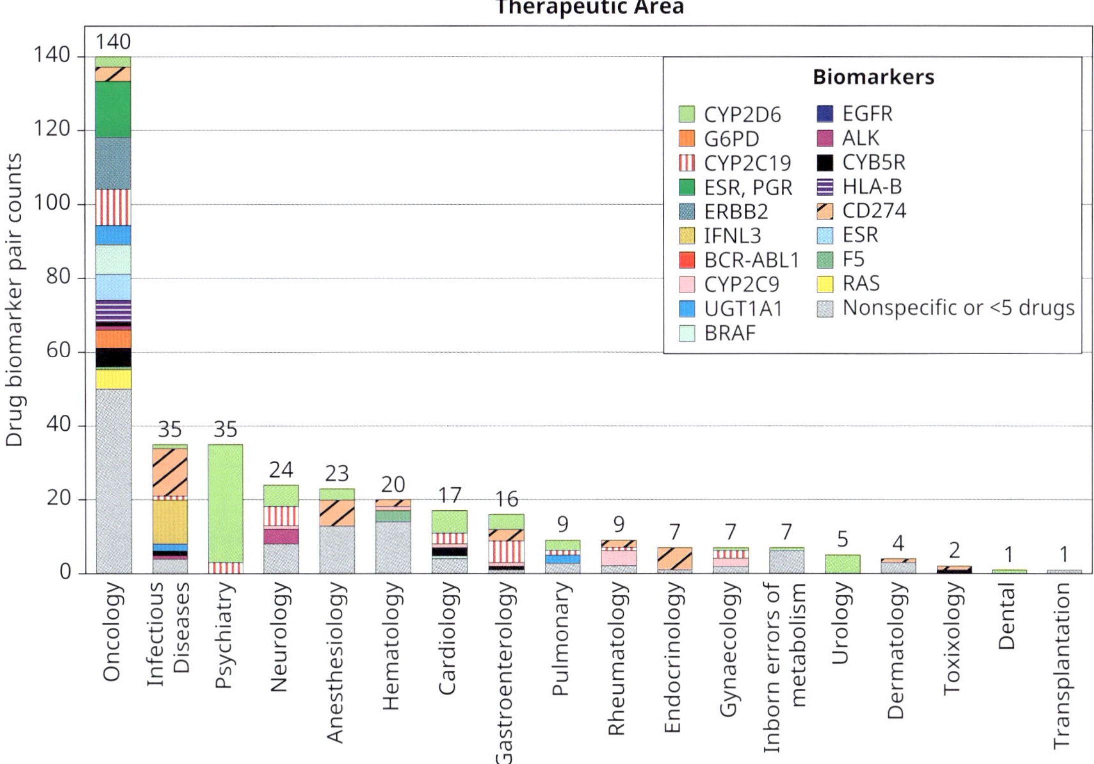

Therapeutic Area	**Biomarkers** (from top to bottom in graph)
Oncology	CYP2D6; CD274; ESR, PGR; ERBB2; CYP2C19; UGT1A1; BRAF; ESR; HLA-B; ALK; G6PD; CYB5R; F5; RAS; Nonspecific or <5 drugs
Infectious Diseases	CYP2D6; CD274; CYP2C19; IFNL3; UGT1A1; EGFR; ALK; Nonspecific or <5 drugs
Psychiatry	CYP2D6; CYP2C19
Neurobiology	CYP2D6; CYP2C19; CYP2C9; ALK; Nonspecific or <5 drugs
Anesthesiology	CYP2D6; CD274; Nonspecific or <5 drugs
Hematology	CD274; ESR, PGR; Nonspecific or <5 drugs
Cardiology	CYP2D6; CYP2C19; CYP2C9; CYB5R; BRAF; Nonspecific or <5 drugs
Gastroenterology	CYP2D6; CD274; CYP2C19; CYP2C9; CYB5R; Nonspecific or <5 drugs
Pulmonary	CYP2D6; CYP2C19; UGT1A1; Nonspecific or <5 drugs
Rheumatology	CD274; CYP2C19; CYP2C9; Nonspecific or <5 drugs
Endocrinology	CD274; Nonspecific or <5 drugs
Gynecology	CYP2D6; CYP2C19; CYP2C9; Nonspecific or <5 drugs
Inborn errors of metabolism	CYP2D6; Nonspecific or <5 drugs
Urology	CYP2D6
Dermatology	CD274; Nonspecific or <5 drugs
Toxicology	CD274; CYB5R
Dental	CYP2D6
Transplantation	Nonspecific or <5 drugs

FIGURE 10.5
FDA-approved drug labelling for different therapeutic areas.

development and post-marketing, and reduces drug attrition. Drug repurposing can be used to target failed or abandoned drugs from clinical trials to individuals with a particular genetic profile who are most likely to benefit.

SELF-CHECK 10.5

Describe pharmacogenetics and outline some examples.

Key Point

Pharmacogenetics and pharmacogenomics examine how individual genetic variation determines likely drug response, to maximize drug efficacy and safety and avoid harmful side-effects.

10.3 Lifestyle genome sequencing

In **Section 10.1.2**, we looked at how the combination of advances in biotechnology and computing power led to completion of the Human Genome sequence and have significantly increased sequencing throughput. Since completion of the Human Genome sequence in 2003, there has been a steady evolution in the number of exomes and genomes sequenced (see **Box 10.1** for a description of whole exome and whole genome sequencing), from fewer than 10 exomes in 2010 to hundreds then thousands of exomes by 2014.

In 2010, a landmark research paper incorporating personal genome information in clinical assessment was published. To investigate clinical translation of genetic risk, the authors carried out whole genome sequencing in an individual with a family history of vascular disease and early sudden death. You can explore the findings of this research in **Case study 10.1**. Since then, a number of genome sequencing initiatives have been launched in different countries, including medical initiatives and genome sequencing in healthy individuals. Public health genomics initiatives to benefit population health also provide opportunities for targeted and risk-stratified screening. In this section, we will take a closer look at some of these projects.

10.3.1 Precision medicine sequencing initiatives

In this section, we will look at some national genome sequencing initiatives.

Genomics England 100,000 Genomes Project

The flagship Genomics England 100,000 Genomes Project launched in 2012 and has now sequenced more than 100,000 genomes from over 97,000 people. This project aimed to create a new model for genomic medicine in the NHS, using and interpreting genomic data to investigate causes of disease, advance diagnosis and personalize treatment for the benefit of patients, and to kick start the UK genomics industry. The participants in this project are patients with an inherited rare disease (usually a child) plus two of their closest relatives, and patients with cancer. Cancer and rare diseases (affecting <1/2000 individuals) were selected due to the influence of genetics, and their cumulative disease burden. Approximately 7000 different rare diseases collectively affect roughly 3.5 million people in the UK, of which >80% have a genetic cause, but specific genetic testing is available for less than 1000 diseases.

Cross reference

To learn more about the processes and systems for genetic testing and screening please read **Chapter 6**.

CASE STUDY 10.1 Whole genome sequencing in a clinical context

To carry out analysis of the human genome in a clinical context, this project sequenced the genome of a 40-year-old healthy male with a family history of vascular disease and early sudden cardiac death. Clinical assessment by a cardiologist included risk prediction for coronary artery disease and screening for causes of sudden cardiac death. The individual received education and counselling from a genetic counsellor before, during, and after testing. Analysis of the whole genome sequence focused on genetic mutations known to cause inherited Mendelian disease, genetic variants involved in drug response (see **Section 10.2**), pathogenic novel mutations, and genetic variants previously associated with complex disease.

The authors analysed more than 2.6 million genetic variants. They discovered rare variants in three genes previously associated with sudden cardiac death, and a variant consistent with a family history of coronary artery disease. They found three novel variants in two genes related to development of haemochromatosis (iron overload), and a novel mutation which produces a truncated protein in a gene implicated in thyroid disorders. They also identified 64 variants relevant to drug response that suggested (amongst other things) a positive response to lipid-lowering therapy and a low initial dosing requirement for warfarin. Finally, they estimated the individual's risk across a spectrum of more than 50 complex diseases, based on age, gender, and ethnicity, and then also including genetic sequence data. They identified increased genetic risk for 8 diseases, including myocardial infarction, type 2 diabetes, and some cancers, and decreased risk for 7 diseases, including Alzheimer's disease and age-related macular degeneration.

The authors concluded that the findings of this study have future clinical and therapeutic implications for the patient, particularly considering his family history. Tools to integrate genetic data with clinical data to assist in clinical decision-making are a big step towards precision medicine. However, this will require a team approach, including medical and genetics professionals, ethicists, and healthcare delivery providers.

It is important to flag some of the limitations of this study. Interpretation of genetic variation requires up-to-date information, and the published data is not perfect; many variants of unknown importance were identified, and we have limited ability to integrate genetic information into clinical care.

Q1 What are the implications for this person in the future?

Q2 What are the possible implications for their family?

Q3 How might interpretation of this data change over time?

Q4 What would you do with this sequence data as a patient or as a healthcare professional?

The results of a pilot study generated findings with diagnostic, prognostic, or therapeutic consequences. A genetic diagnosis was provided for 16% of 7000 patients with rare disease, and ~50% of cancer cases have the potential for therapy or a clinical trial. To make this large-scale adoption of genomics in the NHS a reality, it needs to be fast and cost-effective and generate high-quality data that is easy to interpret. The challenges of this application of bioinformatics in healthcare have included automating the genetic variant calling and annotation process, data interpretation (see **Section 10.3.4**), managing the large volume of data generated (one genome occupies ~200GB), and ensuring data security.

National human genome sequencing research programmes

In the US, the 'All of Us' Research Program, launched in 2016, aims to accelerate health and medical breakthroughs by sequencing the genomes of more than one million people. All adults are eligible to participate, and the program aims to reflect the diversity of the US population, including individuals of different age, gender, race, ethnicity, regional location, socioeconomic status, education, and health. Enrolled individuals share biological samples, genetic data, lifestyle and diet information, and family health history, linked to their electronic health records, to enable individualized prevention, treatment, and care. Participants have consented to data collection on an ongoing basis throughout their lifetime. Overall, this program aims to: increase wellness and resilience and encourage healthy living; improve risk assessment and prevention strategies; provide earlier and more accurate diagnosis to reduce disease burden; improve treatment and interventions to reduce disease impact and improve health outcomes; and reduce health inequalities in historically under-represented populations in biomedical research. Addressing this lack of diversity in biomedical research is important to ensure that the genetic and genomic findings from this program are relevant and applicable across diverse populations and racial groups, so that all participants and patients benefit.

Similar genome sequencing initiatives are now taking place in many countries worldwide, including China, Australia, Brazil, Estonia, Turkey, Saudi Arabia, and Qatar. The focus of these programs varies from cancer, rare diseases, complex phenotypes (such as obesity and diabetes), population-based approaches, to infrastructure development. For the fifteen-year project announced in China in 2016, the government financial investment comparatively dwarfs the US initiative, and includes the Beijing Genomics Institute (BGI), the world's largest sequencing facility and repository of genetic material. These country-specific initiatives enable a population-focused approach to healthcare issues or diseases. In China, there is a strong focus on oncology, particularly the higher incidence of stomach and liver cancer in China, together with infectious disease, neurology, cardiology, and endocrinology. In contrast, in Saudi Arabia, the high rate of **consanguineous** marriages creates a health challenge and increases the impact of severe inherited **autosomal recessive** diseases in this country.

consanguineous
A relationship that arises from having a common ancestor.

autosomal recessive
A pattern of inheritance characteristic of some genetic diseases, where the gene is located on the non-sex chromosomes, and two copies of the mutation are needed to cause the disease.

SELF-CHECK 10.6

Outline very briefly some global examples of genome sequencing initiatives.

Key Point

A number of genome sequencing initiatives are now taking place in different countries, including medical sequencing initiatives in individuals with specific disease, and population genome sequencing in healthy individuals. These genomics projects aim to improve healthcare.

10.3.2 Direct-to-consumer genetic testing

Several commercial companies now offer lifestyle genome sequencing or genetic testing at relatively low cost, also known as direct-to-consumer genetic testing. Participants send off a sample (such as a non-invasive cheek swab or saliva sample taken at home) for DNA extraction and genotyping and answer online survey questions about their own health and lifestyle. Typically, participants might receive a report on their genetic ancestry, health reports, genetic susceptibility to various traits, and carrier status, and are able to build a pedigree diagram with other genetically related people (**Figure 10.6**).

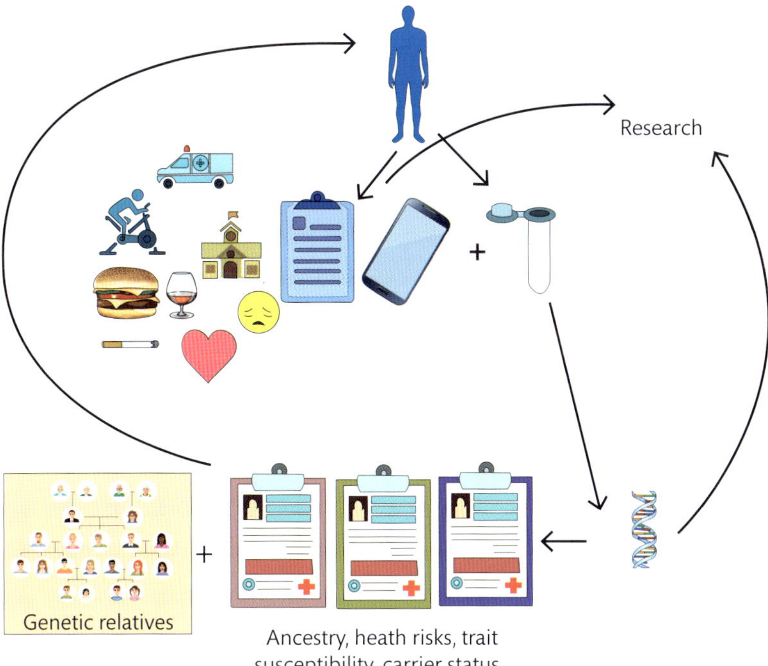

FIGURE 10.6
Overview of direct-to-consumer genetic testing. In direct-to-consumer genetic testing, a sample is used for DNA extraction and genotyping, and participants answer lifestyle and health survey questions. After testing, participants may receive reports on their genetic ancestry, health status, and susceptibility to various traits, and can build a family tree to genetically related people.

One such company, 23andMe, established in 2006, generates approximately 40 individual wellness and trait reports based on the genetic information; for example, on likelihood of coriander taste aversion, lactose intolerance, freckles, earlobe type, deep sleep, hair type, motion sickness, and muscle composition. 'Genetic Health Risks' report information includes testing for three genetic variants within the *BRCA1* and *BRCA2* genes associated with increased risk of developing breast or ovarian cancer, and variants associated with conditions such as age-related macular degeneration, coeliac disease, familial hypercholesterolaemia, hereditary thrombophilia, MUTYH-associated polyposis, late-onset Alzheimer's disease, and Parkinson's disease. Women with one of the tested *BRCA1* or *BRCA2* variants have a 45 to 85% and 46% chance of developing breast or ovarian cancer respectively, by age 70, and an increased risk for earlier-onset breast cancer. MUTYH-associated polyposis is one of the three principal hereditary colorectal cancer syndromes, and people with two variants in the *MUTYH* gene are more likely to develop colon and rectal polyps and have a 43 to 100% lifetime risk of colorectal cancer. The three variants tested in *BRCA1* and *BRCA2* are more common in people of Ashkenazi Jewish descent, and several of the genetic variants included for other conditions are more relevant to people of specific ethnicities, such as African, Lebanese, European, South Asian, and Old Order Amish. These health reports meet FDA laboratory requirements for scientific and clinical validity. Guidance provided by 23andMe recognizes the importance of other non-tested genetic variants, and non-genetic factors that might increase risk of some of these conditions, such as smoking, diet, alcohol consumption, infection, obesity, ethnicity, and family history.

TABLE 10.2 Examples of carrier status gene tests provided by 23andMe

Gene	Disease	Variants tested	Most relevant population
SLC12A6	Agenesis of the Corpus Callosum with Peripheral Neuropathy	1	French Canadian descent
PKHD1	Autosomal Recessive Polycystic Kidney Disease	3	Not specified
HBB	Beta-Thalassemia and related Haemoglobinopathies	10	Sardinian, Cypriot, Italian/Sicilian, Greek descent
CFTR	Cystic Fibrosis	29	Ashkenazi Jewish, European, Hispanic/ Latino descent
SGCB	Limb-Girdle Muscular Dystrophy Type 2E	1	Amish descent
SMPD1	Niemann–Pick Disease Type A	3	Ashkenazi Jewish
GJB2	Non-syndromic Hearing Loss and Deafness, DFNB1 (GJB2-Related)	2	Ashkenazi Jewish, European descent
PEX7	Rhizomelic Chondrodysplasia Punctata Type 1	1	Not specified
HEXA	Tay–Sachs Disease	4	Ashkenazi Jewish, Cajun descent

Cross reference
To explore the ethical issues surrounding genetic testing and counselling please read **Chapter 9**.

23andMe also tests for carrier status for variants involved in over 40 inherited autosomal recessive conditions—you can see some examples in **Table 10.2**. Carrier status may have important implications for having children. 23andMe encourages participants to speak to a genetic counsellor before and after testing, and the reports are meant for participants to have conversations with a healthcare professional, rather than to make medical decisions.

23andMe participants also answer online survey questions. Individuals consent to share their non-personally identifiable information so that this data and linked genetic data can be used to find genetic relatives and analysed by academic and pharmaceutical scientists for research and novel discoveries. This has led to >170 significant research papers published in the decade from 2010.

penetrant
How likely it is that a person who has a genetic mutation will show signs and symptoms of the disease. Not everyone who has the mutation will develop the disease. Complete penetrance means that every person who has the mutation will show signs and symptoms of the disease.

The idea behind some of these direct-to-consumer tests is that people are empowered to take ownership of their own health, and can make informed decisions about their lifestyle, based on their genetic profile. This might include, for example, eating a healthier diet, doing more exercise, or setting health-related goals. In one notable example reported in *Time* magazine in 2013, actress Angelina Jolie had an elective risk-reducing double mastectomy after receiving a positive test for a highly **penetrant** *BRCA1* genetic variant, given her family history of ovarian cancer. However, in the absence of clinical symptoms or family history, the clinical utility of data generated by genomic screening may be limited. In **Section 10.3.4** we consider some of the challenges of genetic variant interpretation, and how this might relate to personalized prevention at a population level.

SELF-CHECK 10.7

Describe direct-to-consumer genetic testing.

10.3.3 Nutrigenomics: the role of nutrients in gene expression

We have discussed in previous sections that individual lifestyle is integral to precision medicine. Non-genetic factors such as smoking, pollution, and diet all influence health and disease. Here, we examine how diet and genes interact to influence disease risk.

We know that DNA sequence variation can influence our response to nutrients, providing an opportunity for disease prevention or correction. This is the field of nutrigenetics. In contrast, nutrigenomics studies the role of nutrients on gene expression and on biomarkers such as metabolites or hormones. In nutrigenomics, gene expression is linked to measurement of food intake and clinical and behavioural data, with the aim of providing personalized dietary advice to optimize health and prevent and treat disease.

Nutrigenetics: the role of DNA sequence variation

Phenylketonuria is an autosomal recessive inborn error of metabolism where genetic mutations cause deficiency of the phenylalanine hydroxylase enzyme, leading to a build-up of phenylalanine. Phenylketonuria is treatable by a low phenylalanine diet, but if undiagnosed, can result in impaired postnatal cognitive development from the neurotoxic effect of hyperphenylalaninaemia. Glycogen storage diseases are rare disorders characterized by hypoglycaemia (liver and kidney problems) and growth retardation, caused by mutations in several genes leading to glycogen accumulation. Treatment focuses on managing diet to control blood sugar levels and prevent problems with metabolism. Other rare genetic mutations may have a significant effect on our drive to eat and response to nutrients, leading to severe obesity. A more common example is genetic variation in the lactase gene, where only 35% of people can digest lactose in milk beyond the age of 7 to 8. Combinations of common genetic variants can also alter the expression or functionality of genes. Choline is an essential macronutrient involved in neurotransmitter synthesis, cell membrane signalling, and lipid transport. Genetic variants in the choline dehydrogenase gene have been associated with choline deficiency. Choline deficiency is linked to non-alcoholic fatty liver disease, skeletal muscle atrophy, and neurodegenerative diseases. Gut microbes metabolize choline, and some choline metabolites have been linked to severity of atherosclerosis.

Cross reference

To find out more about the genetic analysis of our healthy microbiome (commensals) and disease-causing microorganisms (pathogens) please read **Chapter 8**.

The influence of vitamins on gene expression

Vitamin D is involved in mineral homeostasis, energy metabolism, immunity and inflammation, and cellular growth, differentiation, and apoptosis. Vitamin D is converted by the cytochrome P450 enzymes CYP2R1 in the liver and CYP27B1 in the kidney to a biologically active form, 1α,25-dihydroxyvitamin D3. This biologically active vitamin D metabolite is a ligand for vitamin D receptors, expressed in many cells of the body, which form a heterodimer with the retinoic acid receptor and attach to the vitamin D-response element. The vitamin D-response element then acts as a nuclear transcription factor to regulate the transcription of hundreds of different genes. Vitamins A and E also have direct effects on gene transcription.

Micronutrients such as vitamin B12 and folate may have **epigenetic** effects by affecting DNA methylation, histone modification, or chromatin remodelling, thereby switching certain genes on or off and leading to abnormal gene expression.

epigenetics
The study of how age and exposure to environmental factors, such as diet, drugs, and chemicals, may cause reversible chemical modifications of the structure and packing of DNA, influencing gene expression, without any change in the underlying DNA sequence of the organism. These changes may be heritable. The word epigenetics is of Greek origin and literally means over and above (epi) the genome.

The link between diet and cancer

population attributable fraction
The fraction of all cases of a particular disease in a population that is attributable to a specific exposure.

Diet is linked to cancer risk. Estimates suggest that genes explain 5 to 10% of cancer risk, and the environment 90 to 95%, of which diet and being overweight or obese are a significant component (see **Figure 10.7**). In the UK, the **population attributable fraction** (the fraction of all cases of a particular disease in a population that is attributable to a specific exposure) for cancer is ~5–7% for being overweight or obese, 3.5% due to insufficient dietary fibre, ~1–2% from eating processed foods and red meat, and ~3% from alcohol consumption. Alcohol is converted by the alcohol dehydrogenase (ADH) enzyme to acetaldehyde, and from acetaldehyde to acetate by aldehyde dehydrogenase enzymes (ALDH). Genetic variants in *ALDH2*—including variant *ALDH2*2*, more common in East Asian populations (a G to A change in exon 12, resulting in a glutamine to lysine amino acid substitution (E487K))—cause slow acetaldehyde metabolism, which can lead to toxic acetaldehyde build-up, DNA damage, and alcohol-related cancers.

Diet is also linked to:

- metabolic syndrome (a combination of inflammation, metabolic stress, insulin resistance, and diabetes);
- mental ill-health;
- cardiac disease (caffeine);

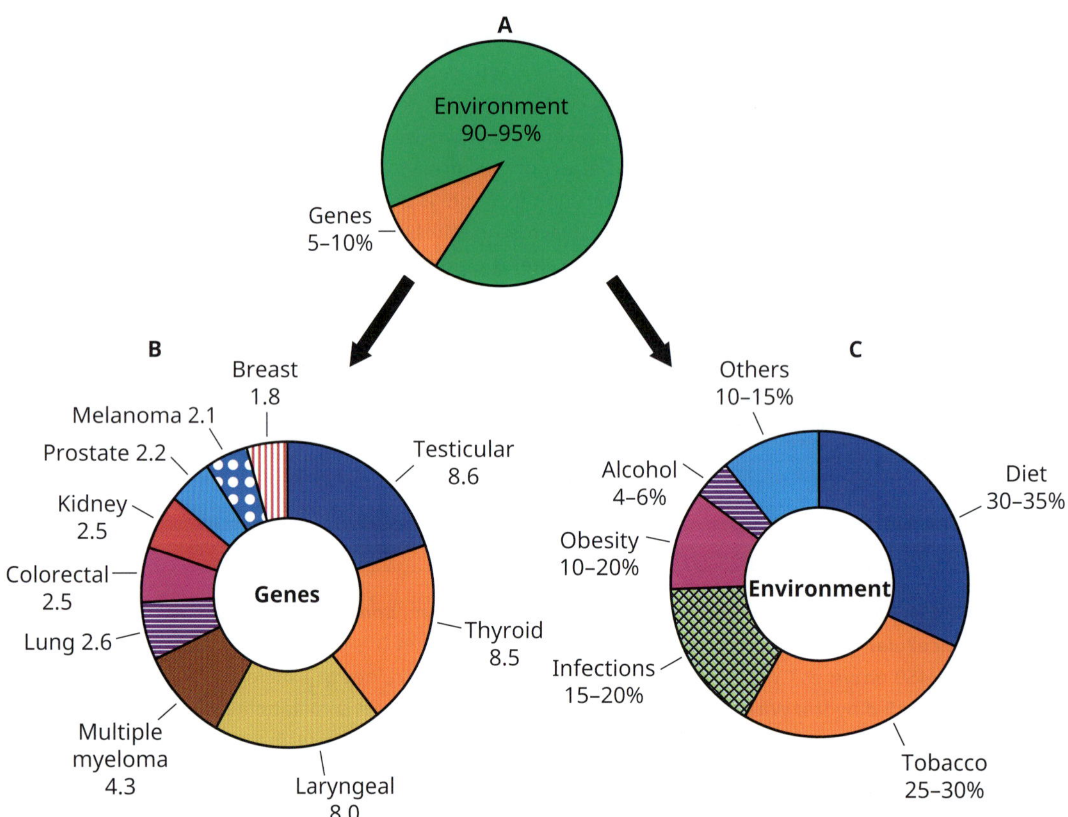

FIGURE 10.7
Genetic and environmental influences in cancer. A: Percentage contribution of genetic and environmental factors to cancer. B: Family risk ratios for selected cancers. The numbers represent familial risk ratios, defined as the risk to a given type of relative of an affected individual divided by the population prevalence. C: Percentage contribution of each environmental factor.

- malnutrition leading to obesity or weight loss;
- coeliac disease; and
- food allergies.

In nutrigenomics, it can be difficult to distinguish correlation from causation. Food diaries are subject to recall bias. Dietary intervention also needs to include the influence of age and gender, genetic differences between populations, and the cultural, economic, geographical, and social influences on diet.

SELF-CHECK 10.8

What is nutrigenomics? How might nutrients affect DNA?

10.3.4 Variant interpretation

Any successful genome sequencing initiative requires investment in data collection and analysis tools, powerful computing capabilities to make discoveries from large volumes of data at pace, and scientific research to interpret the genetic findings, understand molecular mechanisms, and implement molecular medicine. Interpretation of genetic variation requires up-to-date information that evolves over time and is not perfect, and variants of unknown significance may be identified that may be reinterpreted in later years as evidence expands. Each human genome contains 3–4 million variants when compared to the reference Human Genome sequence. Most of these variants are part of normal human variation or contribute to complex disorders. Only a minority of variants cause Mendelian disease. Indeed, each person has in the region of 15 to 30 variants in their genome that have been implicated in rare disease through databases such as ClinVar and the Human Gene Mutation Database. The genetic variants identified therefore need to have both scientific validity and clinical utility.

As we discussed in **Case study 10.1**, there are important implications for variant interpretation in precision medicine genome sequencing and direct-to-consumer genetic testing. If we take the example of people with penetrant *BRCA1* or *BRCA2* variants, not only do they need to think about a personal screening programme and/or preventative surgery, they may also need to consider the risk to their children and other biological relatives—for men, as well as women. This applies for both earlier- and later-onset conditions, such as Alzheimer's disease or Huntington's disease. We still have limited ability to integrate genetic information into clinical care. Carrier status for genetic variants leading to rare autosomal recessive conditions may have reproductive consequences. For most autosomal recessive conditions, there is no known cure, and treatment may focus on managing symptoms, for example, through physical therapy or diet, medication, preventing complications, and supportive care.

Guidelines for variant interpretation

In clinical genome sequencing, the American College of Medical Genetics and Genomics (ACMG) and the Association for Molecular Pathology and other professional bodies, such as the Association for Clinical Genomic Science (ACGS) in the UK, publish recommended standards and guidelines for interpretation of variants in clinical decision-making. ACMG recommends the use of specific standard terminology: *pathogenic*, *likely pathogenic*, *uncertain significance*, *likely benign*, and *benign* to describe variants identified in genes that cause Mendelian disorders. Variant classification criteria include population frequency data, computational prediction data, functional data from *in vitro* or *in vivo* studies of the gene or protein, and family segregation data. Publicly available resources, such as the Genome Aggregation Database (gnomAD), which collates data from different large-scale population exome and whole genome sequencing projects and provides variant frequency data for different populations, including healthy controls, help genetic variant interpretation. ACMG also publishes guidelines for reporting secondary findings unrelated to the primary reason for testing. These regularly updated guidelines are for variants detected in 'medically actionable' genes that cause conditions with accepted medical interventions. The 73 medically actionable genes (v3.0, May 2021) are mainly associated with familial cancer, cardiovascular conditions, or inborn errors of metabolism, such as adult-onset MUTYH-associated polyposis, hypertrophic cardiomyopathy, or Fabry disease.

SELF-CHECK 10.9

List some challenges of variant interpretation from genome sequence data.

10.4 Gene therapy for precision medicine

Gene therapy and genome-editing techniques come under the umbrella of regenerative medicine—treatments that aim to repair, replace, or restore function to damaged or diseased cells, tissues, or organs. Gene therapy describes cellular and molecular techniques that alter the DNA content of cells by adding, deleting, or altering DNA to change gene function, depending on the gene and disease being treated. These innovative approaches are useful in conditions where treatments are ineffective or no other treatments are available, particularly for rare genetic diseases and cancers. For recessive diseases, the aim of gene therapy is to use gene replacement to create a functional copy of a gene where no functional copies exist. Autosomal dominant genetic disorders can be caused by haploinsufficiency, dominant gain-of-function, or dominant negative mutations, where an aberrant protein interferes with normal cell function. In these diseases, the mutant protein can be silenced or knocked down. Single-use gene therapies are relatively expensive, but treat the cause, rather than the symptoms, of disease, and therefore may provide a longer-term, cheaper solution than frequent, expensive, and invasive treatments to control the condition. Gene therapies are often designed for the individual or a small group of individuals and are therefore particularly relevant for precision medicine.

There are several methods to deliver gene therapy across the cell membrane, with different safety profiles and efficacy. Non-viral and viral delivery systems are the main approaches, as viruses can infect and exert genetic or enzymatic effects on host cells by transduction. Viral gene therapies use a viral vector to deliver the synthetic nucleic acids. Common viral vectors include adenoviruses, adeno-associated viruses, retroviruses, and lentiviruses (a subtype of retrovirus). While adenoviruses normally replicate in the host nucleus without integrating their genome, retroviruses integrate into the host genome to provide stable and long-term expression, but with a risk of insertional mutagenesis. Non-viral vectors such as nanoparticles, exosomes, or liposomes can also be used to transport the DNA across membranes. The choice of the most appropriate vector depends on factors including the size of the gene to be delivered, tissue or cell specificity, host immunity, and immunogenicity. Gene therapy can take place *in vivo* by direct administration of a gene-therapy vector, or outside the body (*ex vivo*), where cells are collected, cultured, modified, and transplanted into the individual to treat the disorder. In gene therapy, the patient's own (**autologous**) cells or tissues, or donor cells (**allogeneic**) are used.

autologous
Taken from an individual's own tissues, cells, or DNA.

allogeneic
Taken from different individuals of the same species.

10.4.1 Gene therapy applications in cancer

Gene therapies applied to cancer have gained prominence in recent years. Genetically modified immunotherapies are exciting because they may be effective in late-stage cancers that have not responded to other treatments, or in patients that have relapsed after treatment. Chimeric antigen receptor T-cell (CAR-T) therapies remove T-cells from the patient's or a donor's blood, and genetically modify them to express specific antigen receptors on the cell surface. These modified T-cells are infused back into the patient's bloodstream, where the receptors allow the T-cells to detect and target tumour cell antigens that might otherwise avoid immune detection. Two CAR-T therapies are approved on the NHS for blood cancers: Kymriah (tisagenlecleucel) for acute lymphoblastic leukaemia in children and adults <25 years old, and Yescarta (axicabtagene-ciloleucel) for the treatment of diffuse large B-cell lymphoma in adults. Many other CAR-T therapies are in production, including for solid tumours. Other modified immune cell therapies use T-cell receptors and tumour-infiltrating lymphocytes. The majority of cancer immunotherapies are autologous, but increasing use of healthy donor-derived therapies would allow faster off-the-shelf products.

10.4.2 Gene therapy in rare diseases

Even though rare diseases affect few patients, collectively there are several thousand rare diseases which affect millions of people globally. Although it is hard to establish evidence for clinical effectiveness in rare diseases with low patient numbers, a small number of gene therapies are approved in the NHS to treat rare diseases. Strimvelis is an *ex vivo* viral-based gene addition immunotherapy to treat severe combined immunodeficiency (SCID) due to adenosine deaminase (ADA) deficiency. ADA-SCID is a rare life-threatening condition that presents in infancy, caused by homozygous mutation in the adenosine deaminase gene. Spinal muscular atrophies are a group of rare autosomal recessive neuromuscular disorders, characterized by degeneration of the anterior horn cells of the spinal cord, leading to symmetrical muscle weakness, atrophy, and progressive loss of movement. Spinal muscular atrophy type I is one of the most severe forms of the condition with a life expectancy of less than two years, caused by mutations in the survival motor neuron 1 (*SMN1*) gene, leading to absence of a fully functional protein. Zolgensma is an *in vivo* adeno-associated virus-mediated gene therapy approved by the National Institute for Health and Care Excellence (NICE) and the FDA that aims to restore SMN1 protein function. Clinical trials of gene therapy approaches are in progress for several genetic blood disorders such as sickle cell disease and beta-thalassaemia. In eye diseases, Luxturna is an approved *in vivo* adeno-associated virus-based gene therapy for people with Leber congenital amaurosis, a rare inherited retinal dystrophy causing vision loss due to biallelic mutations in the *RPE65* gene. The eye is a good target for novel gene therapies as the eye is relatively accessible for delivery, for example, by injection into the retina. The small, closed nature of the eye means that lower therapeutic doses are needed, and there is less risk of systemic dissemination. Moreover, the blood–retinal barrier makes the retina an immune-privileged site, so that introduction of foreign material is less likely to cause an inflammatory reaction. Importantly, the genetic basis of several eye diseases is well understood, although genetic heterogeneity, for example, in retinitis pigmentosa, causes an additional challenge for targeted treatment. To find out more about the application of genomic technologies in age-related macular degeneration, including genetic testing, polygenic risk scores, and gene therapies, see **Box 10.3**.

SELF-CHECK 10.10

What is the aim of gene therapy?

BOX 10.3 Genetic testing and gene therapies in age-related macular degeneration

Age-related macular degeneration (AMD) is an eye disease in which vision gets progressively worse over time. AMD is characterized by degeneration of cells in the macula in the retina at the back of the eye, which causes blurring and/or dark spots and loss of central vision. AMD is the leading cause of vision loss and blindness in the elderly; advanced AMD affects ~5% of those aged 65 or over, and ~12% of those older than 80. AMD is divided into dry (~90%) and wet AMD. Dry AMD is characterized by atrophy of the retina, whereas wet AMD is caused by new blood vessel formation. Repeated injections of anti-vascular endothelial growth factor into the eye are used to treat wet AMD, but there is no treatment for dry AMD. AMD is a complex disease where multiple biological and environmental factors increase risk, including age, smoking, obesity, and hypertension. Genetic variants contribute to development and progression of AMD, so risk is also linked to ethnicity.

Over 50 common and rare genetic variants contribute to AMD, including genes involved in the complement system, angiogenesis, cholesterol and lipid metabolism, extracellular matrix, and oxidative stress. Genetic testing of rare variants that confer a high risk of AMD might be useful in people with a family history of early-onset AMD. Polygenic risk scores might also be useful for risk prediction in AMD. Polygenic risk scores estimate risk based on the combined impact of multiple genetic variants; these variants individually confer a small increase in risk, but when combined can predict people at high risk of AMD or progression to severe disease, and therefore could be used to initiate a health plan in advance.

Several promising gene therapies are in clinical trials for AMD. These therapies induce anti-complement expression to inhibit inflammation, or anti-angiogenic proteins to reduce new blood vessel formation. However, there are challenges around gene delivery, and long-term effectiveness still needs to be demonstrated.

10.4.3 Genome editing and gene silencing

Genome editing approaches change the target genomic DNA sequence, by cutting the DNA at specific locations to precisely insert, delete, or edit sections of DNA underlying a disease. The aim is to permanently change the sequence in selected cells or tissues, treating the cause, rather than the symptoms of disease. Modulation of gene expression can also be carried out.

Genome editing techniques

Some genome editing techniques have existed for many years and use genetically engineered versions of naturally occurring restriction enzymes that cut the DNA at specific locations. These enzymes include zinc finger nucleases (ZFNs) and TALENS. The recent development of CRISPR/Cas9 (clustered regularly interspersed short palindromic repeats and CRISPR-associated protein 9), a bacterial defence system reprogrammed to edit genomes, has refocused attention on genome editing. This system uses a Cas9 endonuclease with a synthetic short guide RNA to target and cut the genome at any desired genomic location. The target of CRISPR/Cas9 can be altered simply by changing the guide sequence. The host cell then responds to repair the induced double-strand DNA break. You can see an overview of how CRISPR/Cas9 genome editing works in **Figure 10.8**. CRISPR/Cas9 is more flexible, cheaper, and faster than previous genome editing and gene therapy methods, and has longer-term effects than methods where DNA is inserted into cells with a limited lifespan. CRISPR/Cas9 is widely used for *in vitro* and *in vivo* research to develop new cellular

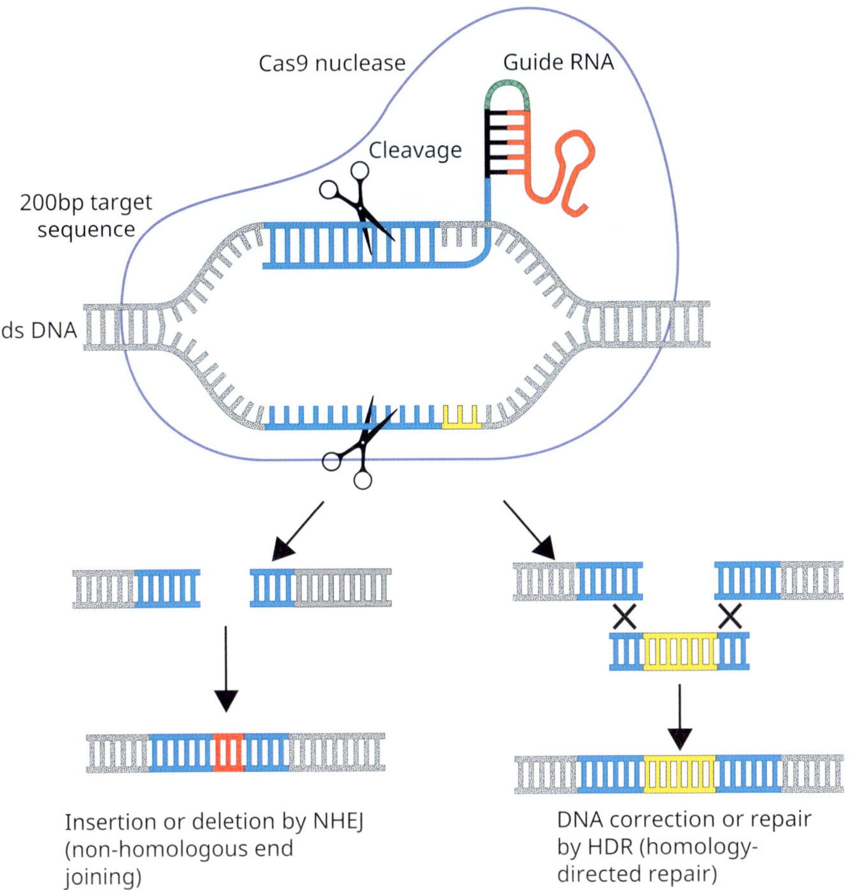

FIGURE 10.8

CRISPR/Cas9 genome editing. In CRISPR/Cas9 genome editing, the Cas9 nuclease is complexed with a synthetic guide RNA which guides the Cas9 to a specific target double-stranded DNA (dsDNA) location. Cas9 then binds to and cuts the dsDNA, creating double-strand breaks. DNA modification takes place using the cell's own repair machinery by one of two methods: DNA ligation by non-homologous end joining (NHEJ) results in small random insertions or deletions and gene disruption at the target site, or precise DNA editing by homology-directed repair (HDR) requires a homologous template to guide repair.

and animal models of human disease, for example, to allow high-throughput screening of drugs against cancer-causing mutations. For autosomal dominant genetic disorders, the aberrant protein can also be knocked down by mRNA silencing. mRNA silencing can take place using small interfering RNA molecules (siRNA), short hairpin RNA (shRNA), or microRNA (miRNA) delivered into cells to inhibit gene expression. Similarly, antisense oligonucleotides are small DNA or RNA molecules complementary to their target mRNA, which cause gene silencing or modulation.

Genome editing for healthcare takes place either *in vivo* or *ex vivo* using cells extracted from the patient or a donor. One of the primary safety concerns for genome editing is the possibility of off-target effects that could cause genome instability, or mutations in essential genes such as tumour suppressor genes.

Genome editing in healthcare

The first *in vivo* application of genome editing took place in 2017, where a small number of patients with the metabolic diseases Hunter (mucopolysaccharidosis type II) or Hurler (mucopolysaccharidosis type I) syndrome received ZFN-based treatments. These recessive diseases are caused by lysosomal enzyme deficiency and progressive glycosaminoglycan or mucopolysaccharide accumulation. These complex carbohydrates accumulate in the arteries, skeleton, eyes, joints, skin, and in other tissues including the respiratory system, liver, and central nervous system. Build-up causes progressive damage to multiple organs and a shortened lifespan. ZFNs were used to edit cells of the liver, where many metabolic enzymes are found. Initial trial results suggest the therapy is safe in the few patients that were treated, although its effectiveness has not been established.

Several other genome editing treatments are being investigated in clinical trials, primarily for rare single gene disorders. Promising examples include trials for eye disorders and for blood disorders such as beta-thalassaemia and sickle cell disease. In 2019, Victoria Gray was the first individual in the United States to undergo successful gene editing for autosomal recessive sickle cell disease, caused by mutations in the *HBB* gene. CRISPR/Cas9 was used to edit the *BCL11A* transcription factor gene in autologous bone marrow stem cells. This *BCL11A* editing reactivated foetal haemoglobin production to compensate for the defective adult beta-globin protein. Cancer immunotherapies such as CAR-T therapy (see **Section 10.4.1**) can also include genome editing. Both gene therapy and genome editing includes germline treatment of sperm or egg cells to create heritable DNA changes, or somatic treatment of dividing cells in the affected tissue to induce non-heritable changes. There is considerable ethical debate about the use of germline DNA gene therapy, from the impact of changes passed on to future generations to the implications of eugenics or 'designer babies'. Overall, important advances have recently been made in genome editing, and efforts continue to increase the efficacy and safety of genome editing tools.

Cross reference

To explore the ethical issues surrounding genome editing please read **Chapter 9**.

SELF-CHECK 10.11

List some conditions for which genome editing is beginning to be used in healthcare.

Key Points

Gene therapy and genome editing techniques alter the nucleic acid content of cells to change the function of genes underlying disease, particularly for rare genetic diseases and cancers. These techniques often target the individual or a small group of individuals, and therefore are highly relevant to precision medicine.

10.5 The use of stem cells in precision medicine

Stem cells are undifferentiated cells that have **potency**, and therefore can be differentiated to produce the specialized tissues and organs of the body. Another feature of stem cells is that they reproduce and have high self-renewal. Stem cells come from the bone marrow, and can also be isolated from embryos, umbilical cord blood, and adult tissues such as adipose tissue. Most relevant to clinical care are induced **pluripotent** cells from the blood or skin of

potency
Cell potency is the varying ability of stem cells to differentiate into specialized cell types of the body.

pluripotent
Cells that can differentiate into any cell type of the body.

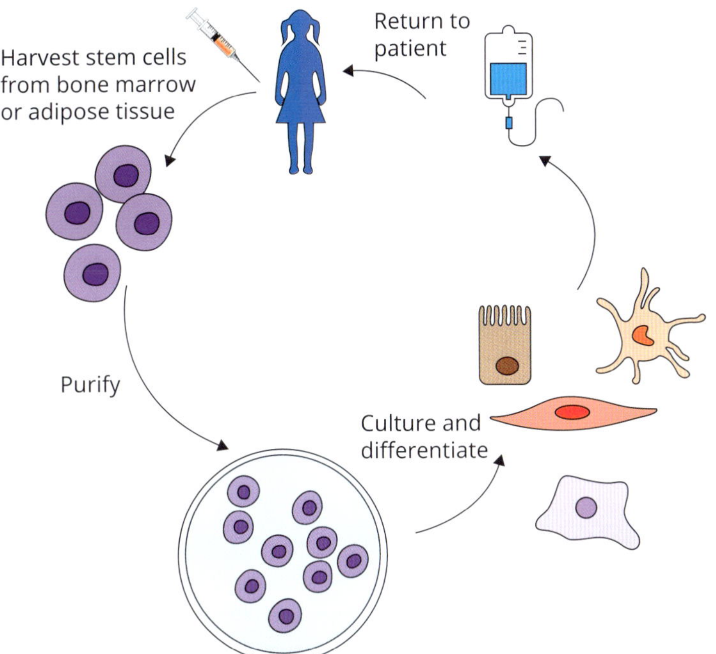

FIGURE 10.9
Overview of stem cell therapy. Stem cells can be isolated from tissues including bone marrow and adipose tissue. After stimulation and chemotherapy to release stem cells into the bloodstream, stem cells are harvested from the patient and purified. Stem cells are cultured *ex vivo* and can be differentiated into specialized cells of the body. The patient is then conditioned by high-dose chemotherapy and/or radiotherapy, before the differentiated cells are infused back into the patient.

adults that are engineered so they have the potential to develop into any cell type in the body. Mesenchymal stem cells are **multipotent** adult stem cells that can divide and differentiate into multiple skeletal tissues including bone, cartilage, muscle and fat cells, and connective tissue. As with gene therapy, stem cell therapy can use the patient's autologous cells, or allogeneic donor cells. The technically complex process of stem cell therapy (see **Figure 10.9**) involves harvesting, culturing, and transfer to the patient. Combining stem cells with gene therapy for genetic modification, as seen in the treatment of Victoria Gray's sickle cell disease (see **Section 10.4.3**), is called stem cell gene therapy.

multipotent
Cells that can differentiate into more than one cell type.

Stem cell therapies have great potential in personalized healthcare, as use of autologous organ or tissue replacement reduces the risk of rejection and the associated morbidity and mortality of mismatched transplants, and is less reliant on organ donors. However, a potential notable side-effect of stem cell therapy is uncontrolled stem cell growth leading to tumours.

SELF-CHECK 10.12

Where do stem cells come from, and what are their advantages?

10.5.1 Clinical applications of stem cell therapy

Stem cell therapy is controlled by a complex regulatory environment. Today, there are only a limited number of approved stem cell therapies, but many clinical trials are in progress, particularly in the fields of oncology (particularly lung, breast, and colorectal cancer), neurology, and gastroenterology.

In a recent clinical trial, a two-year-old patient with Sanfilippo syndrome received a new *ex vivo* gene therapy treatment using autologous bone marrow stem cells. Sanfilippo syndrome is a rare life-limiting inherited genetic condition, also known as mucopolysaccharidosis IIIA.

Children develop normally at first, but mucopolysaccharide build-up in the brain and central nervous system causes progressive developmental delays, severe behavioural problems, and premature death. In this trial, the patient's own stem cells were genetically modified *ex vivo* to correct the N-sulfoglucosamine sulfohydrolase (SGSH) gene mutation, then infused back into the patient, after conditioning, to correct the sulfamidase enzyme deficiency.

Another example of the potential of stem cell gene therapy is for Fanconi anaemia. Fanconi anaemia is a rare heterogeneous genetic syndrome characterized by progressive bone marrow failure, congenital anomalies, and a predisposition to malignancy. Approximately half of patients are diagnosed before age 10. Fanconi anaemia is caused by mutations in approximately 20 different DNA repair pathway genes leading to chromosomal instability, with ~90% of all cases caused by biallelic mutations in *FANCA*, *FANCC*, or *FANCG*. Hematopoietic stem cell transplantation is effective in Fanconi anaemia for matched, unaffected sibling donors, but graft rejection and mortality increase for matched unrelated donor transplants. Stem cell gene therapy trials are therefore ongoing in Fanconi anaemia.

Key Point

In stem cell therapy, stem cells from the patient or donor are differentiated to produce specialized tissues and organs of the body. Stem cell therapies have great potential in personalized healthcare but are not yet widely approved for routine use.

10.6 The relationship between precision medicine and other technologies

So far, in this chapter, we have focused largely on precision medicine as it relates to genes. Here, we will take a brief look at other precision medicine technologies, including those relating to gene products and emerging technologies. We will also look at the data underlying individual health management.

10.6.1 Omics technologies for precision medicine

microbiome
The community of microorganisms such as bacteria, viruses, and fungi living in and on the human body, which varies from person to person and over time.

Omics refers to the comprehensive or global assessment of molecules in a sample. Molecular characterization can take place at a single cell, tissue, or organ level, and within an individual or a cohort of people. Transcriptomics refers to measurement of all genes that are expressed in the sample at a particular time-point—both spatial and temporal expression. Epigenomics measures the chemical modifications to the structure and packaging of DNA that take place to regulate gene expression. Proteomics measures the abundance, structure, and function of proteins, whereas metabolomics quantifies the products of cellular metabolic processes. Individuals have genetically determined metabolic fingerprints, which are modulated by factors such as physical activity, diet, gut **microbiome**, and body composition. There is already a long history of using individual or combination metabolite levels in blood or urine as biomarkers for disease diagnosis and prognosis—common examples include glucose for diabetes, cholesterol or lipid levels for cardiac disease, and tests for inborn errors of metabolism in newborn babies. Metabolomics approaches can range from simple to complex data-rich technology-driven approaches. Some recent high-profile examples of using metabolomics signatures for personalized healthcare include cancer-detecting sniffer dogs, a woman who could smell Parkinson's disease, and Tanzanian rats trained to detect tuberculosis. The microbiome refers to the enormous community of microorganisms such as bacteria, viruses, and fungi living in and on the human body, which varies from person to person and over time. Increasing characterization of the microbiome is enabled by high-throughput DNA sequencing. The microbiome might influence individual response to specific diseases, such as bowel disease, obesity, diabetes, heart disease, arthritis, and cancer. The gut microbiome is of particular interest, as people could adapt their diet to manipulate their gut microbiome to promote health and combat disease.

Cross reference

To find out more about the genetic analysis of our gut bacteria please read **Chapter 8**.

Researchers are increasingly trying to move towards multidisciplinary approaches. For example, functional genomics is used to integrate multiple omics datasets from different levels of biological activity to understand the relationship between genes and their downstream effects. This integration requires joined-up data collection, analysis, and management.

SELF-CHECK 10.13

Outline some 'omics' approaches that could be applied in precision medicine.

10.6.2 Companion diagnostics and emerging technologies

We mentioned earlier that development of diagnostics and therapeutics need to go hand-in-hand. **Companion diagnostics** are tests co-developed with drugs to select patients for a particular treatment. These companion diagnostics include predictive biomarker assays to identify likely responders and non-responders, and medical devices. There are several examples from cancer pharmacogenetics implemented in the NHS, for example, the Invader® *UGT1A1* assay for irinotecan, and circulating tumour DNA testing for *EGFR* mutations in non-small cell lung cancer.

companion diagnostics
Tests co-developed with drugs to select patients for a particular treatment, including predictive biomarker assays and medical devices.

The use of nanomedicine and microfluidics to develop smaller, portable devices is increasing point-of-care and remote testing. These devices could increase diagnosis, speed up clinical decisions, and reduce costs and time to treatment. Implantable biosensors for automatic, precise, real-time measurement of validated biomarkers, such as enzymes, can be used to manage chronic conditions, or for early detection. Some examples include human chorionic gonadotropin hormone detection in urine for pregnancy, the Eversense® continuous glucose monitoring system for type 1 diabetes, AliveCor® 24-hour electrocardiogram monitoring, and the recent COVID-19 lateral flow test to detect viral RNA. These companion diagnostic and digital health devices enable continuous data gathering and remote monitoring and facilitate telemedicine. Improvements in biomedical imaging and 3D printing have increased our ability to model anatomical structures, for example, in the production of personalized implants and prosthetics and maxillofacial surgery.

10.6.3 Precision medicine and the rise of big data

As we discussed at the start of this chapter, precision medicine has come about due to significant advances in technology and digital innovation. Today, we are very familiar with the idea of digital personalized health and disease monitoring and automated data capture using wearables, sensors, and consumer mobile health apps. For example, the 'All of Us' Research Program uses Fitbits to explore the relationship between physical activity, heart rate, sleep, and other health outcomes.

Individual citizen-generated data includes deliberately produced data, and indirectly produced data from internet searches or shopping choices. Digital health systems and electronic patient records mean that the line between lifestyle and medical data is increasingly blurred. We now recognize a more complex picture of the heterogeneity of health and disease. Many individuals use digital technologies for their own health, interact with commercial and other sectors, and want support in health, as well as in disease, leading to a cultural shift in the relationship between individuals and health systems. More 'person-centred healthcare' is particularly relevant to chronic, long-term, multi-morbid disease that is becoming more prevalent in aging societies and requires a move away from specialized and fragmented clinical services.

The generation of big data enables both more stratified and more integrated data analysis. Artificial intelligence and machine learning approaches can make sense of very large and complex datasets and identify data patterns that are not obvious with less bias. These approaches can be used to detect prescribing errors or predict patients at risk of hospital readmission. Commercial companies exist that capture data from millions of individuals and use data mining to make health and wellness recommendations based on a digital, holistic view of each patient. Biobanks such as UK Biobank, a biomedical database containing in-depth genetic and health information from half a million UK participants, also allow large-scale analyses. The Accelerating Detection of Disease study will track the health of five million UK people over many years. The aim is to use artificial intelligence data analysis and develop new diagnostic tools and technologies to improve early diagnosis, intervention, and prevention in common chronic diseases such as diabetes and cancer, to allow people to live healthier lives for longer.

SELF-CHECK 10.14

What is 'big data', where does it come from, and how does it relate to precision medicine?

Key Point

Several existing and emerging technologies including digitally generated data are highly relevant to personalized healthcare.

10.6.4 Precision medicine implementation

To integrate data, we need an integrated healthcare system. We have already seen that precision medicine sits at the interface of healthcare, academic research, industry, and government. To implement precision medicine effectively, we need scientists and clinicians with biological knowledge, and computational, statistical, and analytical skills to integrate basic and clinical science. These individuals need to be able to work with a range of stakeholders, including diagnostic and technical staff in laboratories and imaging centres, bioinformaticians, and data scientists who assemble and process the data, and to consider the regulatory and ethical challenges of data ownership, privacy, and individual autonomy. Individuals need to be educated and empowered to make lifestyle changes to implement precision medicine healthcare that is truly preventive, predictive, personalized, and participatory.

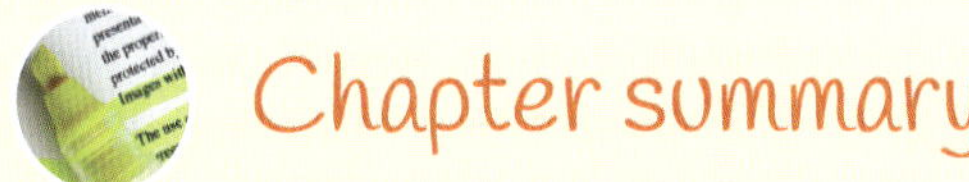

Chapter summary

- Precision medicine is an approach to disease prevention, prediction, and treatment that takes into account individual variability in genes, environment, and lifestyle. Precision medicine transforms healthcare from a 'one size fits all' approach to a more tailored approach.
- Significant advances in existing and emerging technologies, including digitally generated data, are highly relevant to more personalized healthcare.
- Greater molecular characterization of individuals combined with targeted therapies may enable earlier and/or more precise diagnosis.
- Pharmacogenetics and pharmacogenomics examine how individual genetic variation determines drug response, to maximize drug efficacy and safety. Pharmacogenetics is relevant for treatments including for cancer and for infectious and inflammatory diseases.
- There are a number of genome sequencing initiatives to improve healthcare, including in specific diseases and in healthy people. Individuals can obtain personal genetic testing from commercial companies, to make decisions about their lifestyle based on their genetic profile.
- Nutrigenetics is the influence of DNA sequence variation on our response to nutrients. Nutrigenomics studies the role of nutrients on gene expression through direct or epigenetic effects, and the relationship to health and disease.
- We have limited ability to integrate genetic information into clinical care, and genome sequencing data interpretation requires information that evolves over time.
- Gene therapy and genome editing alter the DNA of genes underlying disease using the patient's or donor cells. These approaches are currently used most often to treat rare genetic diseases and cancers.
- Stem cell therapy uses patient or donor stem cells to produce specialized tissues and organs. Stem cell therapies have great potential in personalized healthcare, but only a limited number are currently approved.
- Point of care and remote monitoring, testing, and data gathering are important for precision medicine. Other omics and emerging technologies such as nanomedicine, microfluidics, and biosensors are highly relevant, and diagnostics and therapeutics need to be co-developed.
- Mobile devices may empower individuals to take greater control of their own health.
- Individuals now generate a huge volume of digital health-related data that is being assessed using artificial intelligence. These approaches may identify disease risk and improve disease diagnosis and management.
- Effective precision medicine implementation requires an integrated healthcare system.

Discussion questions

10.1 What are the strengths and limitations of 'blockbuster' vs individually tailored biomarkers or therapeutics?

10.2 What are the implications of preventative vs therapeutic precision medicine?

10.3 What are the challenges of data generation vs data interpretation in precision medicine for individuals and for healthcare professionals?

10.4 What is the role of citizens and organized health systems in managing and optimizing their health?

10.5 Does the role of precision medicine vary globally in relation to resource access and health inequalities?

Further reading

- Blackburns L, de Villiers CB, Janus J, Johnson E, Kroese M (2020) ***Genomic innovation: technologies for personalised medicine***. The AHSN *network*. https://eprints.soton.ac.uk/449400/1/FINAL_AHSN_full_report.pdf

- Relling MV, Klein TE, Gammal RS, Whirl-Carrillo M, Hoffman JM, Caudle KE (2020) ***The Clinical Pharmacogenetics Implementation Consortium: 10 years later***. *Clin Pharmacol Ther*., 107(1), 171–175.

  Highlights the progress of the Clinical Pharmacogenetics Implementation Consortium (CPIC) over the last 10 years. The CPIC has published guidelines covering 19 genes and 46 drugs across several therapeutic areas, to implement pharmacogenetics into routine clinical practice.

- Turro E, Astle WJ, Megy K, Gräf S, Greene D, Shamardina O, Allen HL, Sanchis-Juan A, Frontini M, Thys C, Stephens J, Mapeta R, Burren OS, Downes K, Haimel M, Tuna S, Deevi SVV, Aitman TJ, Bennett DL, Calleja P, Carss K, Caulfield MJ, Chinnery PF, Dixon PH, Gale DP, James R, Koziell A, Laffan MA, Levine AP, Maher ER, Markus HS, Morales J, Morrell NW, Mumford AD, Ormondroyd E, Rankin S, Rendon A, Richardson S, Roberts I, Roy NBA, Saleem MA, Smith KGC, Stark H, Tan RYY, Themistocleous AC, Thrasher AJ, Watkins H, Webster AR, Wilkins MR, Williamson C, Whitworth J, Humphray S, Bentley DR; NIHR BioResource for the 100,000 Genomes Project, Kingston N, Walker N, Bradley JR, Ashford S, Penkett CJ, Freson K, Stirrups KE, Raymond FL, Ouwehand WH (2020) ***Whole-genome sequencing of patients with rare diseases in a national health system***. *Nature*, 583(7814), 96–102.

  Outlines the progress of the UK 100,000 Genomes Project including ~10,000 participants with rare disease, using whole genome sequencing for molecular diagnosis in routine healthcare.

- Stark Z, Dolman L, Manolio TA, Ozenberger B, Hill SL, Caulfield MJ, Levy Y, Glazer D, Wilson J, Lawler M, Boughtwood T, Braithwaite J, Goodhand P, Birney E, North KN (2019) ***Integrating genomics into healthcare: a global responsibility***. *Amer Jour Hum Gen.,* 104(1), 13–20.

  Reviews several national genomic medicine sequencing initiatives.

- Price ND, Magis AT, Earls JC, Glusman G, Levy R, Lausted C, McDonald DT, Kusebauch U, Moss CL, Zhou Y, Qin S, Moritz RL, Brogaard K, Omenn GS, Lovejoy JC, Hood L (2017) ***A wellness study of 108 individuals using personal, dense, dynamic data clouds***. *Nat Biotechnol.*, 35(8), 747–756.

  One of the first integrated studies of clinical, 'omic', and personal data to improve understanding of health and disease.

- Alfirevic A, Pirmohamed M (2017) ***Genomics of adverse drug reactions***. *Trends Pharmacol Sci*, 38(1), 100–9.
- Ashley EA, Butte AJ, Wheeler MT, Chen R, and 31 others (2010) ***Clinical assessment incorporating a personal genome***. *Lancet*, 375(9725), 1525–35.
- Ehmann F, Caneva L, Prasad K, et al. (2015) ***Pharmacogenomic information in drug labels: European Medicines Agency perspective***. *The Pharmacogenomics Journal*, 15(3), 201–10.
- Ellard S, Baple EL, Callaway A, Berry I, Forrester N, Turnbull C, Owens M, Eccles DM, Abbs S, Scott R, Deans ZC, Lester T, Campbell J, Newman WG, Ramsden S, McMullan DJ. (2020) ***ACGS Best Practice Guidelines for Variant Classification in Rare Disease***. v4.01 04/02/2020.
- Johnson E, Blackburn L (2019) ***Somatic genome editing: promise and practicalities***. *Policy Briefing. PHG Foundation.* https://www.phgfoundation.org/
- Mehta D, Uber R, Ingle T, Li C, Liu Z, Thakkar S, Ning B, Wu L, Yang J, Harris S, Zhou G, Xu J, Tong W, Lesko L, Fang H (2020) ***Study of pharmacogenomic information in FDA-approved drug labeling to facilitate application of precision medicine***. *Drug Discov Today*, 25(5), 813–20.
- Miller DT, Lee K, Gordon AS, et al. (2021) ***Recommendations for reporting of secondary findings in clinical exome and genome sequencing, 2021 update: a policy statement of the American College of Medical Genetics and Genomics (ACMG)***. *Genet Med*, 23(8), 1391–8. https://doi.org/10.1038/s41436-021-01171-4
- Richards S, Aziz N, Bale S, Bick D, Das S, Gastier-Foster J, Grody WW, Hegde M, Lyon E, Spector E, Voelkerding K, Rehm HL (2015) ***ACMG Laboratory Quality Assurance Committee. Standards and guidelines for the interpretation of sequence variants: a joint consensus recommendation of the American College of Medical Genetics and Genomics and the Association for Molecular Pathology***. *Genet Med*, 17(5), 405–24.
- Rothwell DG, Ayub M, Cook N, Thistlethwaite F, and 38 others. ***Utility of ctDNA to support patient selection for early phase clinical trials: the TARGET study***. *Nat Med*, 25(5), 738–743.
- Thomson RJ, Moshirfar M, Ronquillo Y (2021) ***Tyrosine kinase inhibitors.*** [Updated 4 May 2021]. In: ***StatPearls [Internet]***. *StatPearls Publishing, Treasure Island (FL).* Available from: https://www.ncbi.nlm.nih.gov/books/NBK563322/
- Whirl-Carrillo W, Huddart R, Gong L, Sangkuhl K, Thorn CF, Whaley R, Klein TE (2021) ***An evidence-based framework for evaluating pharmacogenomics knowledge for personalized medicine***. *Clinical Pharmacology & Therapeutics*, 110(3), 563–72. doi: 10.1002/cpt.2350
- Wishart DS, Feunang YD, Guo AC, Lo EJ, Marcu A, Grant JR, Sajed T, Johnson D, Li C, Sayeeda Z, Assempour N, Iynkkaran I, Liu Y, Maciejewski A, Gale N, Wilson A, Chin L, Cummings R, Le D, Pon A, Knox C, Wilson M (2018) ***DrugBank 5.0: a major update to the DrugBank database for 2018***. *Nucleic Acids Res*, 46(D1), D1074–D1082.

Useful websites

- HM Government. Genome UK: the future of healthcare. 2020. **https://www.gov.uk/government/publications/genome-uk-the-future-of-healthcare**.

  This strategy paper outlines the UK government vision to enable the provision of world-leading healthcare to patients in the UK and across the world.

- 23andme. **https://www.23andme.com/**
- The 100000 Genomes Project. **https://www.genomicsengland.co.uk/about-genomics-england/the-100000-genomes-project/**
- The All of Us Research program. **https://allofus.nih.gov/**
- International Agency for Research on Cancer. World Health Organization. **https://gco.iarc.fr/causes/obesity/**
- NICE. National Institute for Health and Care Excellence. Adverse drug reactions. **https://cks.nice.org.uk/topics/adverse-drug-reactions/**
- UKBiobank. **https://www.ukbiobank.ac.uk/**

Glossary

Chapter 2

Allele(s) Alternative forms of a gene that might be found at a given locus.

Anticipation A situation in which a genetic disorder becomes more severe and/or appears earlier as the causative genetic variant is passed down through the generations. Most often seen in disorders caused by trinucleotide repeat expansions.

Autosome(s) A non-sex chromosome.

Autosomal Belonging to, located on, or transmitted by an autosome.

Centromere(s) The area of a chromosome to which the spindle attaches during mitosis and meiosis.

Chromatin A complex in eukaryotic cells mostly composed of DNA, histone and other proteins involved in the regulation of gene expression, that is usually found dispersed in the interphase nucleus and is condensed during mitosis and meiosis.

Co-dominance Describes the relationship between two alleles of a gene. If the alleles are different, the dominant allele usually will be expressed, while the effect of the other allele, called recessive, is masked. In co-dominance neither allele is recessive, and the phenotypes of both alleles are expressed. An example is the ABO blood groups in humans. A and B are co-dominant alleles, O is recessive to both A and B.

Complementation Describes a situation where two parents who are homozygous for different recessive causative alleles and therefore show the phenotype for the same recessively inherited condition, albinism for example, will have no affected children because all of the children will be heterozygous at both loci.

Compound heterozygosity Describes a situation in which an individual has two different recessive causative alleles at the same locus, one on each homolog, resulting in the disease phenotype.

Consensus sequence A common nucleotide sequence or amino acid sequence found in highly conserved regions of DNA or RNA or proteins.

CpG island(s) Short stretches of DNA (typically 500–1500 bp long) with a CG:GC ratio of more than 0.6. Represented as 'CpG'. The 'p' indicates the phosphate linking the two bases to distinguish it from the hydrogen bonding of the GC base pairs in different DNA strands. CpG islands are frequently associated with vertebrate gene promoters.

Diploid Having two sets of chromosomes, one from each parent, typical of most somatic cells.

DNA polymerase An enzyme that synthesizes DNA molecules from nucleotide building blocks.

Enhancer(s) A nucleotide sequence that increases the rate of genetic transcription by preferentially increasing the activity of the nearest promoter on the same DNA molecule.

Epigenetic A heritable and stable change in gene expression that occurs through alterations in the chromosome but not in the DNA sequence. An epigenetic trait is a stably heritable phenotype resulting from epigenetic changes in a chromosome.

Epistasis (Epistatic) A situation in which the expression of one gene is affected by the expression of one or more independently inherited genes.

Euchromatin The part of chromatin that is transcriptionally active.

Exon(s) (Exonic) A polynucleotide sequence in a nucleic acid that encodes information for protein synthesis and that is copied and spliced together with other such sequences to form messenger RNA.

Exonuclease(s) An enzyme that breaks down RNA molecules by removing successive nucleotides from one end of the molecule. Exonucleases can have either 3′ to 5′ or 5′ to 3′ activity, and are involved in RNA processing, tRNA maturation, and RNA degradation.

Gamete(s) Reproductive cell of a plant or animal.

Germline The cellular lineage of a sexually reproducing organism from which eggs and sperm are derived. The genetic material contained within this cellular lineage can be passed to the next generation.

Gonosome (Gonosomal) Sex chromosomes, also referred to as allosomes, heterotypical chromosomes, gonosomes, heterochromosomes, or idiochromosomes. Gonosomes differ from autosomes in size, structure, and the way that they function within cells.

Haploid Having a single set of chromosomes, typically found in gametes (sperm and egg cells).

Hemizygous An individual who has only one member of a chromosome pair or chromosome segment. Hemizygosity is used to describe X-linked genes in males and in somatic cell genetics where cancer cell lines are often hemizygous for certain alleles or chromosomal regions.

Heterochromatin A condensed form of DNA that is transcriptionally inactive. Heterochromatin is tightly packed and inaccessible to transcription factors. Compare with Euchromatin.

Heteroplasmy A condition in which a cell or individual has more than one type of organellar genome. Typically used to describe the situation in which two or more mtDNA variants exist within the same cell.

Heterozygous Having two different alleles of a particular gene or genes.

Hexameric A polymeric protein composed of six molecules of a monomer OR a structural subunit that is part of a viral capsid composed of six subunits of similar shape.

Holandric Transmitted as part of or being a gene in the Y chromosome.

Incomplete dominance Describes the appearance in a heterozygote of a trait that is intermediate between either of the trait's homozygous phenotypes. The situation is sometimes also referred to as semi-dominance.

Incomplete penetrance Describes the situation when a dominantly inherited genetic condition is not always seen in the phenotype. It can be attributed to the effects of 'modifier genes' (epistatic interactions) or environmental differences. The level of penetrance in conditions that show incomplete penetrance is reported as a percentage or as a proportion of 1.

Insulator A long-range cis-regulatory element that can act as an enhancer blocker.

Intron(s) (Intronic) A non-coding region of a gene that is removed during RNA splicing.

Leukocyte White blood cell: any of the colourless blood cells of the immune system including basophils, eosinophils, lymphocytes, monocytes, and neutrophils.

Locus The position in a chromosome of a particular gene or allele.

Locus control region A tissue-specific, copy number-dependent long-range cis-regulatory element that enhances expression of linked genes at distal chromatin sites.

Lyonization The inactivation (through condensation and epigenetic mechanisms) of one of the X chromosomes in females to achieve dosage compensation of X-linked genes with males.

Messenger RNA (mRNA) A type of RNA that carries genetic information from DNA to the ribosome, where it is used to synthesize proteins.

Monocistronic A messenger RNA that can encode only one polypeptide per RNA molecule; compare with polycistronic.

Nucleolus A membrane-less nuclear organelle where ribosomal RNA (rRNA) is synthesized and assembled with specific proteins to form ribosomes.

Oxidative phosphorylation The metabolic pathway in which cells use enzymes to oxidize nutrients, thereby releasing chemical energy to produce adenosine triphosphate (ATP).

Promoter A region of DNA upstream of a gene where relevant proteins (such as RNA polymerase and transcription factors) bind to initiate transcription.

Replisome Multiprotein molecular machinery responsible for the replication of DNA.

Repressor A small protein that can bind to specific DNA sequences to prevent the transcription of genes and operons.

Ribonucleic acid (RNA) A nucleic acid present in all living cells that plays a role in coding, decoding, regulation, and expression of genes.

Ribosomal RNA (rRNA) Specific RNA molecules which are the major components of ribosomes.

Ribosome A cellular organelle that translates mRNA into proteins, found in both prokaryotic and eukaryotic cells.

Silencer A specific DNA sequence capable of binding repressor proteins.

Somatic Refers to the cells of the body in contrast to the germline cells.

Spliceosome A large ribonucleoprotein (RNP) complex found in the nucleus of eukaryotic cells that removes introns from pre-messenger RNAs (pre-mRNAs).

Tautomer(ic) A form of isomerism in which the isomers change into one another with great ease so that they ordinarily exist together in equilibrium.

Telomere Structure made from specific DNA sequences and proteins found at the ends of most eukaryotic chromosomes.

Terminator A specific section of nucleic acid sequence that marks the end of a gene or operon in genomic DNA during transcription.

Transcription factor A protein that regulates the transcription of a genes or set of genes.

Transfer RNA (tRNA) Small RNA molecule that carries amino acids to ribosomes, where they are covalently bound to each other to form proteins.

Variable expression Describes the situation when a genetic condition is not always seen to the same degree in the phenotype. An example is neurofibromatosis which shows almost complete penetrance (those with causative mutations almost always show the phenotype to some extent), but extreme variability of expression. The phenotypic expression can be anything from just a few faint marks on the skin (referred to as cafe au lait patches) to severe life-threatening complications.

Wildtype A phenotype, genotype, or gene that predominates in a natural population of organisms or a particular strain of organisms, in contrast to that of natural or laboratory mutant or variant forms. Adjectival form is wild-type.

Chapter 3

Acrocentric A chromosome in which the centromere is located near one of the chromosome ends.

Alpha-satellite DNA A type of non-coding DNA formed of multiple repetitions of short sequences and located at the centromere of all chromosomes.

Amplification A process leading to the gain of multiple copies of a genetic sequence.

Aneuploidy (*plural* Aneuplodies) An abnormal number of chromosomes in a cell that deviates from the haploid set or a multiple of the haploid set.

Autosome(s) Any chromosome that is not a sex chromosome.

Barr body Condensed, inactivated X chromosome that appears as a densely stained small chromatin mass at the nuclear periphery when visualized by light microscopy in mammalian female somatic cells.

Biomarker A measurable indicator of a disease state.

Bivalents Structures resulting from the pairing of homologous chromosomes during the first division of meiosis.

Centromere Constricted region of a chromosome at which the two chromatids are joined.

Centrioles Cellular organelles made of microtubules involved in spindle formation and cell division.

Chiasma (plural Chiasmata) A point of protracted contact between paired homologous chromosomes at which exchange of genetic material can happen during the first metaphase of meiosis.

Chromatids The two identical copies of a chromosome that has replicated in preparation for cell division.

Chromatin A complex of DNA and proteins found in the nucleus of eukaryotic cells.

Chromoanagenesis A catastrophic occurrence generating a complex chromosomal rearrangement.

Chromoplexy An event in which multiple chromosomes are broken and the DNA strands ligated in new configurations.

Chromosome(s) Microscopically visible macromolecular aggregates of DNA and proteins residing in the cell nucleus.

Chromosome Conformation Capture (3C) A method based on protein cross-linking that allows one to infer three-dimensional physical proximity between genomic loci within or across chromosomes.

Chromosome painting A variant of FISH in which chromosome-specific DNA libraries can be used as probes to map chromosome territories within the nuclear space.

Chromothripsis A single mutational event causing the localized shattering of one or more chromosomes.

Cohesins Specialized proteins that hold chromatids together after DNA replication in the S phase.

Condensins Proteins that play a critical role in the structural and functional organization of chromosomes.

Constitutive heterochromatin Transcriptionally silent chromatin found at centromeres and telomeres.

Crossing over The exchange of genetic material between homologous chromosomes.

Cytogenetics A branch of genetics concerned with the study of chromosomes and chromosomal abnormalities.

Cytokinesis The physical process that divides the cytoplasm of a parental cell into two daughter cells.

Deletions Physical losses of DNA.

Diploid Refers to the presence of a double set of chromosomes.

Double minutes (dmin) Small, circular fragments of extrachromosomal DNA, lacking centromeres which segregate randomly during cell division.

Duplications The presence of an extra copy of a specific genetic sequence.

Endoreduplication Replication of the nuclear genome without subsequent cell division.

Epigenetic A change in gene expression caused by chromatin modification, and not by a change in DNA sequence.

Euchromatin Transcriptionally active chromatin.

Facultative heterochromatin Transcriptionally active or potentially active chromatin that becomes heterochromatic in a developmentally regulated manner.

Fluorescence *in situ* hybridization (FISH) A combined molecular and cytological hybridization technique to quantify and evaluate genetic sequences within cellular specimens.

G-banding (or Giemsa banding) A technique that produces a longitudinal staining pattern along the length of the entire chromosome, which is unique for each chromosome pair.

Gain-of-function mutation A type of genetic mutation that leads to the acquisition of unnecessary functions.

Genetic drift A change in the frequency of a genetic variant due to random chance.

Haploid Refers to the presence of a single set of chromosomes.

Heterochromatin Transcriptionally silent chromatin.

Heteromorphisms (or chromosomal polymorphisms) Mitotically stable, natural variations of chromosomal size and morphology within the general population.

Histone proteins Positively charged proteins, rich in arginine and lysine, with a critical role in chromatin assembly and compaction.

Homogeneously stained regions (hsr) Sizeable structural anomalies that present as interstitial or terminal chromosomal segments not showing differential banding following Giemsa staining.

Homologous Refers to chromosomes within pairs.

Idiogram A diagrammatic representation of a karyotype.

Independent assortment Refers to the random alignment and parting of homologous chromosomes of maternal and parental origin in meiosis I.

Interphase The interval phase between rounds of cell division during which the cell grows and replicates its DNA.

Inversions A 180-degree rotation of the normal orientation of a DNA sequence caused by a double DNA break along the chromosome structure.

Karyogram A photograph of an entire set of chromosomes of a cell, ordered in pairs and in numbered sequence based on size and centromere position.

Karyotype A description of the whole set of chromosomes of an individual that provides genome-wide information on chromosomal variants.

Kataegis A focal hypermutation process leading to a high rate of mutations in small chromosomal regions.

Kinetochore A large protein structure that connects chromosomes to the spindle fibres.

Loss-of-function mutation A type of genetic mutation that leads to the loss of essential functions.

Meiosis A process of reductional division by which haploid gametes (sperms and eggs) are formed from precursor diploid cells.

Metacentric A chromosome with chromosomal arms of approximately equal length.

Mitosis A process of equational division by which a single cell gives rise to two daughter cells genetically identical to the parent cell.

Molecular cytogenetics The study of chromosomes and chromosomal abnormalities at the molecular level.

Monosomy An instance in which a cell presents with one less copy of a specific chromosome.

Mosaicism The presence in an individual of cell lines or clones derived from a single zygote that have evolved to contain different chromosomal complements.

Nucleolus The primary site of production and assembly of ribosomal units within the nucleus of a eukaryotic cell.

Nucleosome Basic unit of DNA packaging in eukaryotic chromosomes comprising a core of eight histone proteins around which a DNA segment of 147 bp is wrapped.

Numerical chromosomal abnormalities Changes to the expected number of chromosomes in a cell.

Paracentric inversion A chromosomal inversion that does not include the centromere.

Pericentric inversion A chromosomal inversion that includes the centromere.

Polyploidy An instance in which a cell has acquired one or more additional sets of chromosomes.

Recombination Refers to the physical swapping of genetic material that can happen between chromosomes during meiosis.

Ring chromosomes (r) Defective chromosomes with a circular shape ensuing from the fusion of their broken ends.

Robertsonian translocations Translocations that ensue from centromeric fusions of acrocentric chromosomes.

Sex chromosomes Chromosomes involved in sex determination.

Structural chromosomal abnormalities Microscopically visible changes to the content and/or order of DNA sequences on chromosomes.

Submetacentric A chromosome with chromosome arms of clearly different sizes.

Syngamy The fusion of two haploid cells.

Telomeres Structures made of DNA and proteins which protect the end of chromosomes.

Tetrad Structure resulting from the pairing of homologous chromosomes during the first division of meiosis.

Tetraploidy An instance in which a cell has acquired two extra sets of chromosomes.

Topologically associating domains (TADs) Chromosomal subdomains with well-defined and evolutionarily conserved boundaries characterized by local chromatin interactions.

Translocations Structural abnormalities that involve the exchange of genetic material between chromosomes.

Triploidy An instance in which a cell presents with an additional set of chromosomes.

Trisomy An instance in which a cell presents with one more copy of a specific chromosome.

Chapter 4

100,000 Genomes Project A UK-led project established to sequence 100,000 genomes from around 85,000 National Health Service (NHS) patients affected by a rare disease, cancer, or infectious disease (https://www.genomicsengland.co.uk/).

Allele-specific PCR Method for detecting any known mutations involving single base changes or small deletions, which is based on the use of sequence-specific PCR primers that allow amplification of test DNA only when the target allele is contained within the sample. Also referred to as amplification refractory mutation system (ARMS).

Bonferroni correction A statistical test used to reduce the instance of a false positive when performing a hypothesis test with multiple comparisons.

De novo A previously identified mutation or clinically relevant variant that arises in a family member for the first time.

Dichorionic A multiple pregnancy in which each foetus has its own placenta and amniotic sac.

Dizygotic Fraternal or non-identical (in twins).

Epistasis (Epistatic) A situation in which the expression of one gene is affected by the expression of one or more independently inherited genes.

FASTQ A text-based data file format for storing both a biological sequence (usually nucleotide sequence) and its corresponding quality scores.

Genome-wide association study (GWAS) A technique in which researchers scan genetic markers across the complete genomes of many individuals to find genetic variations that are statistically associated with a particular phenotype of interest.

Genomic Evolutionary Rate Profiling (GERP) The score measures the conservation of each nucleotide in multi-species alignment and predicts the deleteriousness of variants.

Haploinsufficiency (HI) The condition in which heterozygosity for a genetic variant leads to insufficient product and the consequence of this is observed in the phenotype.

Heel prick test Also referred to as 'Newborn blood spot screening' (NBS).

Heritable (Heritability) Capable of being inherited, or passed on from parent(s) to offspring, not necessarily genetic.

Heteroplasmy A condition in which a cell or individual has more than one type of organellar genome. Typically used to describe the situation in which two or more mitochondrial DNA (mtDNA) variants exist within the same cell.

High-resolution melting analysis A post-polymerase chain reaction analysis method in which characteristic melting temperatures are observed and used to identify genetic variation in amplicons.

Homoplasmy The condition in which all copies of the mitochondrial genome are identical. This might be either wildtype or mutated sequences.

Immunohistochemistry (IHC) A method for localizing specific antigens within fixed tissue sections based on antigen-antibody recognition.

Insertion–deletion mutations (indels) Insertion into and/or deletion of short sequences (<1kb) of nucleotides in genomic DNA.

Law of independent assortment From Mendel's observations, states that segregation of different alleles occurs independently of one another.

Law of segregation From Mendel's observations, stating that the two alleles of a single trait are separated randomly during gamete formation.

Linkage disequilibrium (LD) The non-random association of alleles at two or more loci in a general population.

Marker(s) A specific DNA sequence with a known physical location on a chromosome.

Microarray A technique that uses thousands to millions of short single-stranded nucleic acid fragments bound in specific positions to a solid surface (referred to as a 'chip'), to identify the presence of complementary molecules within samples of interest.

Monochorionic A pregnancy where multiple babies share a single placenta.

Monozygotic Identical (in twins).

Multifactorial Describes phenotypic traits that occur because of a combination of both environmental and genetic factors.

Newborn blood spot screening (NBS) UK screening programme enabling early identification, referral, and treatment of babies with an approved list of rare but serious conditions—also referred to as the 'heel prick test'.

Novel A genetic mutation or variant that has not been discovered before—contrast with *de novo*.

Our Future Health A UK initiative funded by a collaboration of the public, charity, and private sectors, to build a resource that reflects, amongst other things, the genetic diversity of the UK population (https://ourfuturehealth.org.uk/).

Pearson's chi-square test A non-parametric mathematical tool designed to analyse group differences. Used when the dependent variable is measured at a nominal level.

Polymorphism A common variation in the DNA code.

Pyrosequencing A non-gel-based DNA sequencing technique that detects inorganic pyrophosphate (light) released during DNA synthesis.

Recombination The formation of new combinations of DNA sequences, either naturally (by crossing over and independent assortment during meiosis) or, in the laboratory (by the direct manipulation of genetic material).

Single nucleotide polymorphism (SNP) Variation found at a single position in a DNA sequence when individual genomes are compared.

SNaPshot Single Nucleotide Polymorphism (SNP) genotyping technology developed by Applied Biosystems (ABI). The assay involves using fluorescently tagged dideoxynucleotides (ddNTPs) to terminate the extension of DNA primers after the addition of one nucleotide and enables the precise determination of SNPs based on emitted fluorescence.

Whole genome sequencing (WGS) A method for determining the complete DNA sequence of an organism's genome, used to identify genetic variations associated with diseases.

Chapter 5

Alternative allele (ALT) The allele of a given variant that differs from that in the reference genome.

American College of Medical Genetics (ACMG) An organization that represents the interests of medical genetics professionals including clinical geneticists, clinical laboratory geneticists, and genetic counsellors.

Ancestral allele The allele that was originally present in the genome of a shared ancestor.

Annotation The process of adding genetic information to sequence data.

Background genome The genetic sequence that an individual carries aside from a variant of interest.

Benign variant A genetic variant that is highly unlikely to directly cause a disease.

Binary alignment map (BAM) A binary equivalent of the SAM file (qv) used to save storage space.

Chromatin immunoprecipitation sequencing (ChIP-seq) A method that sequences all DNA segments bound to a specific protein to catalogue the targets of that protein.

Chromatogram A trace that represents the DNA sequence from Sanger sequencing.

Clinvar A genome database that includes information about clinically relevant genetic variation - https://www.ncbi.nlm.nih.gov/clinvar/.

Cloned contig method A method of sequencing that first involves the creation of clone libraries from small regions rather than sequencing everything all in one go.

Consensus reference A reference genome that carries the most commonly found allele at every base position.

Contig The contiguous sequence generated once DNA fragments are aligned in a massively parallel sequencing experiment.

Coverage The number of DNA fragments that capture a given base pair in a next generation sequencing experiment. Also known as read depth.

DECIPHER A genome database that includes information about clinically relevant genetic variation - https://www.deciphergenomics.org/.

Derived allele The allele generated by a mutation event in the genome of a shared ancestor.

Dideoxynucleotides (ddNTPs) Free nucleotide bases that lack a hydroxyl group at position 3 of the sugar moiety and so cannot be extended. These play an important role in Sanger sequencing.

Dosage-sensitive gene A gene that encodes a protein that must be present at a very particular level within a cell.

Emulsion PCR A PCR reaction carried out in a single drop of oil.

Ensembl A genome database. www.ensembl.org.

Epigenomics The genome-wide study of epigenetic marks.

Exome sequencing A high-throughput sequencing approach in which only the coding parts of the genome are sequenced.

FASTQ file A file format that stores the sequences of the small DNA fragments in a next generation sequencing experiment.

Gel electrophoresis A lab method used to separate DNA fragments according to size.

Genome Aggregation Database (gnomAD) A genome database that includes information about variation across populations - https://gnomad.broadinstitute.org/.

Genomics England A programme to integrate genomic medicine into clinical care in the UK.

Heterogeneity Variation between individuals carrying the same genetic variant.

Human Genome Project (HGP) A collaborative scientific effort to sequence the entire human genome.

Indel The insertion or deletion of a few base pairs in a DNA sequence.

International Genome Sample Resource (IGSR) A genome database that includes information about variation across populations - https://www.internationalgenome.org/.

Library A collection of DNA fragments from a particular part of the genome, for example the exome.

Likely benign variant A genetic variant that is unlikely to directly cause a disease.

Likely pathogenic variant A genetic variant that is likely to directly cause a disease.

Loss-of-function (LoF) variant A variant that completely prevents the function of a gene.

Major allele The allele that is most commonly found within a given population.

Massively parallel sequencing High-throughput sequencing methods where lots of small DNA fragments are sequenced all in one experiment.

Metabolomics The large-scale study of substrates and products of metabolism.

Methylated DNA immunoprecipitation sequencing (MeDIP-Seq) A method that sequences all methylated DNA segments in a genome.

Minor allele The allele that is rarest within a given population.

Next generation sequencing (NGS) Sequencing methods that allow the sequencing of a lot of DNA all in one experiment.

OMIM A genome database that includes information about clinically relevant genetic variation - https://www.omim.org/.

Pan-genome A reference genome that contains information about every possible base at every possible position.

Pathogenic variant A genetic variant that is highly likely to directly cause a disease.

Phred A logarithmic score used to indicate the quality of DNA sequencing.

Polyacrylamide A low toxicity chemical derived from acrylamide that allows us to separate DNA fragments at the base pair level.

Polymerase chain reaction (PCR) A method of DNA amplification.

Proteomics The large-scale study of proteins.

Pyrosequencing A method of next generation sequencing.

Raw DNA sequence The DNA sequence that is output from a sequencing experiment (before any analysis).

Read depth The number of DNA fragments that capture a given base pair in a next generation sequencing experiment. Also known as coverage.

Reference allele (REF) The allele found in the reference genome at a given base.

Reference genome A genome sequence against which all sequence data are positioned. The use of a single reference means that all sequence data can be compared between different experiments.

Sequence alignment map (SAM) A file format that stores the sequences of the small DNA fragments once they have been aligned against a reference genome in a next generation sequencing experiment.

Sequence motifs Characteristic patterns in the sequence that act as signals for specific biological functions.

Sequencing by ligation A method of next generation sequencing.

Sequencing by synthesis A method of next generation sequencing.

Shotgun sequencing A method of sequencing where the entire genome is sequenced randomly. In this method, the sequence fragments need to be reassembled after sequencing. This is done using a reference genome or overlapping sequences.

Single nucleotide variant (SNV) A variation that involves a single base pair of DNA.

Single-cell RNA sequencing (sc-RNAseq) The sequencing of all the RNA from a single cell.

Third-generation sequencing Sequencing methods that are capable of whole genome sequencing but do not rely upon massively parallel methods.

Transcriptomics The large-scale study of gene transcripts.

UCSC A genome database - http://genome.ucsc.edu/.

Variant call file (VCF) A file format that stores the positions at which an individual differs from the reference genome.

Variant of uncertain significance (VUS) A genetic variant for which there is not enough evidence to classify it as pathogenic or benign.

Virtual gene panels A way of analysing only the parts of the genome sequence that are believed to be relevant to the clinical presentation.

Chapter 6

Absence of heterozygosity A region of DNA where only a single allele is present.

Allele A variant form of a gene.

Amniocentesis The process of sampling amniotic fluid.

Aneuploidy An abnormal number of chromosomes (but not a whole set).

Anticipation For each generation a genetic disorder is passed on the symptoms become more severe, or appear with an earlier age.

Centromere Specialized chromosomal/DNA structure involved in linking sister chromatids, and the attachment point for spindle fibres during mitosis.

Chimerism An individual derived from two or more zygotes.

Chorionic villus Foetal-derived tissue within the placenta.

Chromosomal microarrays (CMA) A high-resolution whole-genome microarray in which the chip is designed to detect small genetic alterations, including submicroscopic abnormalities below the size resolved by conventional karyotyping or FISH (fluorescence *in situ* hybridization) analysis.

Combined test A screening test during pregnancy for Down syndrome, Edwards syndrome, and Patau syndrome performed between weeks 10 and 14.

Consanguinity Descended from the same ancestor.

Constitutional genetics Deals with inherited genetic conditions.

Diagnosis The physical manifestation or visible characteristics resulting from gene expression.

Dosage-sensitive Dosage-sensitive genes cause a phenotypic effect when deleted or duplicated.

Fluorescence *in situ* hybridization (FISH) A procedure that uses specific nucleic acid probes and fluorophores to enable researchers to locate and visualize the positions of specific DNA sequences in chromosomes.

Fluorescence microscopy Specialized microscopy using fluorescently labelled DNA probes to visualize regions of interest.

Fluorometric quantification Measuring the quantity of a chemical substance, e.g. DNA, by dyes that bind to the region of interest.

Founder mutation A genetic variant found at a higher frequency in a particular population normally due to isolation.

Functional analysis Analysis of functional effects, e.g. the effect of a genetic variant on gene expression.

G-banding The technique used to stain metaphase chromosomes to produce a unique banding pattern for each chromosome.

Gene panel Genetic sequence analysis for a given set of genes by NGS.

Genetic analyser Normally refers to capillary electrophoresis machines used to visualize Sanger sequencing and other PCR-based techniques.

Genetically heterogenous conditions Genetic conditions which can be caused by a number of genes but presenting with a similar phenotype.

Heterochromatic variation Variation in regions of condensed inactive DNA.

Heterozygous Having two different alleles for a specific locus on homologous chromosomes.

Homologous chromosomes Chromosomes exist as homologous pairs, one inherited from each parent.

Homozygous Having identical alleles for a given locus on each homologous chromosome.

Imprinting The differential expression of genes depending on maternal or paternal origin of the chromosome.

Incidental finding Findings of a test unrelated to the initial indication for testing but still of medical value.

In silico Computer simulated.

Interphase The stage of the cell cycle preparing for cell division.

Inversion A segment of DNA, or section of a chromosome, which is reversed.

Karyotyping The pairing and analysis of chromosomes.

Leukaemia Cancer of the blood cells.

Marker chromosome An abnormal small chromosome which cannot be identified by G-banding alone.

Metaphase The stage of the cell cycle where DNA is condensed into visible chromosomes and attach to the mitotic spindle.

Microsatellite DNA A type of repeat DNA sequence which can be highly variable from person to person.

Mitogen A protein that induces cell division.

Mitotic index The proportion of metaphase cells.

Mosaicism The presence of two or more cell lines derived from a single zygote.

Multidisciplinary team meetings (MDTs) Formal periodic meetings between healthcare professionals who have different, relevant expertise. In an MDT the group will discuss patients' diagnosis and condition and agree and organize their treatment plan according to the most appropriate evidence-based protocols.

Mutation A change in the DNA sequence (this should not be used to imply a pathogenic change).

Nucleic acid Complex organic molecule that contains the genetic code.

Pathogenic Disease-causing.

Phase contrast microscopy Used to visualize a specimen based on the light shift when it passes through a transparent object.

Phenotype The physical manifestation or visible characteristics resulting from gene expression.

Polyploidy Having more than 2 sets of chromosomes.

Prognosis The likely course and outcome for a medical condition.

Quadruple test A screening test during pregnancy for Down syndrome, Edwards syndrome, and Patau syndrome performed between weeks 14 and 20.

Quantitative fluorescent polymerase chain reaction (QF-PCR) A rapid method for the detection of chromosome copy number by amplification of DNA sequences at chromosome-specific loci.

Quiescent A resting or dormant phase of the cell cycle.

Recurrence risk The chance that an inherited genetic disorder will occur again in a family.

Restriction enzymes Enzymes which cleave the DNA at specific motifs.

Robertsonian translocation The fusion of the two long arms of acrocentric chromosomes.

Sanger sequencing Traditional sequencing of DNA, also known as first-generation sequencing.

SNP array A type of microarray using SNPs for detection of copy number change.

Southern blot Used for detection of specific DNA sequences by running through an electrophoretic gel and transfer to a membrane for visualization.

Spectrophotometry Measuring the quantity of a chemical substance, e.g. DNA, by light absorbance for a given wavelength.

Translocation Where chromosomes break and transfer to a different chromosome.

Triplet repeat disorders Genetic conditions caused by the expansion of triplet repeats.

Triploidy Having 3 sets of chromosomes (e.g. 69, XXX).

Variant A change in the DNA sequence, more often used in genetic reporting as this is a more neutral term than mutation.

Chapter 7

Aneuploidy The presence of an abnormal number of chromosomes in a cell.

Array comparative genomic hybridization (aCGH) A technique that analyses the entire genome for copy number aberrations by comparing the tumour DNA to a reference genome.

Chromosomal location (or locus) The specific location of a gene on a chromosome, often identified by numbers and letters that indicate its position.

Chromothripsis A phenomenon where chromosomes are shattered into many pieces and then reassembled incorrectly, leading to complex genetic rearrangements associated with cancer.

Copy number variations (CNVs) Alterations in the number of copies of a particular gene or DNA segment in the genome.

Dominant negative effect A type of genetic mutation that results in a protein that interferes with the co-expressed wild-type protein.

Ductal carcinoma *in situ* (DCIS) A non-invasive breast cancer where abnormal cells are found in the lining of a breast duct but have not spread outside the duct.

Epigenetic reprogramming The process by which gene expression is altered without changing the underlying DNA sequence.

Environmental factors External factors such as exposure to chemicals, radiation, and lifestyle choices that can influence the development of cancer.

Familial cancer genes Genes that, when mutated, can increase the risk of cancer within a family due to inheritance.

Fluorescence *in situ* hybridization (FISH) A molecular cytogenetic technique that uses fluorescent probes to detect specific DNA sequences.

Gain-of-function mutations (GOF) Mutations that result in a new or enhanced activity of a gene product, which can lead to abnormal cellular behaviour, such as uncontrolled growth.

Germline mutation Genetic alteration inherited by a cell.

Haploinsufficiency A situation in which heterozygosity for a gene mutation leads to insufficient protein product.

Heredity The passing of traits from parents to offspring, which may include genetic predispositions to certain diseases like cancer.

Heterozygous Concerning only one allelic copy.

Homozygous Concerning both allelic copies.

Immunohistochemical (IHC) staining A laboratory technique used to visualize specific proteins in tissue samples, often used to diagnose diseases like cancer.

Invasive ductal carcinoma (IDC) A common type of breast cancer that starts in the milk ducts and invades surrounding tissues.

Karyogram A diagram or a photograph of the chromosomes in a cell, arranged in homologous pairs and numbered sequence.

Karyotyping A laboratory technique that visualizes chromosomes under a microscope to detect abnormalities in chromosome number and/or structure.

Knudson's hypothesis (two-hit theory) A theory that suggests cancer is the result of accumulated variants (mutations) in a cell's genes, with a first 'hit' being a genetic predisposition variant and a second 'hit' being an acquired variant (mutation).

Loss of heterozygosity Loss of one copy of a segment of DNA.

Loss-of-function mutations (LOF) Mutations that result in the loss of a gene product.

Low penetrance variants Genetic variations that do not always result in disease but may increase the risk of disease under certain conditions.

Malignant Refers to cancerous cells that can spread and invade other parts of the body.

Microenvironment The immediate environment surrounding a cell, including other cells, molecules, and blood vessels, that influences its behaviour and function.

Microsatellite Repeated sequences of DNA (mononucleotide, dinucleotide or higher-order nucleotide repeats such as $(A)_n$ or $(CA)_n$) that are prone to mutations and are used in genetic studies, especially in the context of cancer research.

Microsatellite instability (MSI) A condition of genetic hypermutability that results from impaired DNA mismatch repair, associated with certain types of cancer like Lynch syndrome (LS).

Missense mutation A type of genetic mutation where a single nucleotide change results in a codon that codes for a different amino acid.

Next generation sequencing (NGS) A high-throughput method of DNA sequencing that allows for the rapid and comprehensive analysis of genetic material.

Nonsense mutation A type of genetic change that creates a stop signal resulting in the premature termination of the encoded protein.

Oncogenes Genes that normally help cells grow. When mutated or expressed at high levels, they can turn a normal cell into a cancerous cell.

Oncogenesis The process of tumour formation or development of cancer.

Penetrance Percentage of individuals with a certain genetic variant who present with the related trait.

Pleiotropy The phenomenon whereby a gene can affect multiple traits.

Polymerase chain reaction (PCR) A laboratory method used to make multiple copies of a specific DNA segment, enabling detailed analysis.

Polymorphisms Common variations in the DNA sequence that may or may not influence health or contribute to disease.

Proto-oncogene The unmutated version of an oncogene.

Reverse transcriptase An enzyme that synthesizes DNA from an RNA template.

Somatic mutation Genetic alteration acquired by a cell.

Translocation A type of chromosomal abnormality where a chromosome segment is moved from one location to another.

Tumour suppressor genes Genes that normally help to prevent uncontrolled cell growth. When these genes are mutated, cells can grow uncontrollably and become cancerous.

Variant (mutation) A change in the DNA sequence that can lead to changes in protein function and may contribute to diseases such as cancer.

Wild-type gene a gene in its natural, unmutated form.

Whole genome sequencing (WGS) A method for determining the complete DNA sequence of an organism's genome, used to identify genetic variations associated with diseases.

Chapter 8

Acinetobacter baumannii A Gram-negative bacterium that causes a range of infections, often resistant to antibiotics, and associated with hospital-acquired infections.

Adenovirus A group of viruses that can cause respiratory, gastrointestinal, and eye infections.

Agarose gel A gel matrix used in electrophoresis to separate DNA or RNA fragments by size.

Amplicon A piece of DNA or RNA that is the product of amplification events (such as in PCR).

Antimicrobial resistance genes Genes that provide bacteria with the ability to resist the effects of antibiotics and other antimicrobial agents.

Antimicrobial susceptibility testing Laboratory testing to determine the susceptibility of bacteria to various antibiotics.

Aspergillus A genus of fungi that includes species causing infections, particularly in immunocompromised individuals.

Astrovirus A family of viruses that cause gastroenteritis, especially in children.

Bacteria (singular bacterium) Unicellular prokaryotic microorganisms that propagate by cellular fission and are a major cause of disease.

Bacterial chromosome The main DNA molecule that contains the genetic information of a bacterium.

Bacteriophage A virus that infects and replicates within bacteria.

Campylobacter A genus of Gram-negative bacteria that is a common cause of food poisoning, leading to diarrhoea.

Candida albicans A species of yeast that can cause infections in humans, particularly in immunocompromised individuals.

Capillary electrophoresis A technique that separates ions based on their electrophoretic mobility using capillary tubes.

Cerebrospinal fluid (CSF) The clear fluid found in the brain and spinal cord, which protects and nourishes the central nervous system.

Chlamydia A genus of Gram-negative bacteria that cause chlamydia, a common sexually transmitted infection.

Clostridioides difficile A Gram-positive bacterium that causes severe diarrhoea and colitis, often associated with antibiotic use. Previously named *Clostridium difficile*.

COVID-19 The disease caused by the SARS-CoV-2 virus, characterized by respiratory illness and, in severe cases, pneumonia.

Cryptosporidium A genus of protozoa that causes cryptosporidiosis, a diarrheal disease.

Cytomegalovirus (CMV) A common virus that can cause serious illness in immunocompromised individuals and congenital infections.

Commensals Organisms that live on or within another organism (the host) without causing harm, often referring to the normal microbiota.

Conjugation A process by which one bacterium transfers genetic material to another through direct contact.

Cyclospora A genus of protozoa that causes cyclosporiasis, an intestinal infection leading to diarrhoea.

Diploid Having two sets of chromosomes, one from each parent, typical of most somatic cells.

DNA polymerase An enzyme that synthesizes DNA molecules from nucleotide building blocks.

DNase inhibitor A substance that inhibits the activity of deoxyribonucleases (enzymes that degrade DNA).

Dysbiosis An imbalance in the microbial community, often in the gut, that can be associated with disease.

Electrophoresis A technique used to separate nucleic acids or proteins based on their size and charge by applying an electric field.

Entamoeba A genus of amoebas, some species of which cause amebiasis, an intestinal illness.

Enterococcus A genus of Gram-positive bacteria that are normal inhabitants of the human intestine but can cause serious infections.

Epstein–Barr virus (EBV) A virus that causes infectious mononucleosis and is associated with certain types of cancer.

Escherichia coli A Gram-negative bacterium found in the intestines of humans and animals, some strains of which can cause various illnesses, including foodborne diseases and urinary tract infections.

Eukaryotic Referring to cells that have a nucleus and other membrane-bound organelles, characteristic of animals, plants, fungi, and protists.

Fungi A kingdom of eukaryotic organisms that includes yeasts, moulds, and mushrooms, which decompose organic material and can cause infections.

GC content The percentage of guanine (G) and cytosine (C) bases in a DNA molecule, often used as a measure of genomic composition.

Genotyping The process of determining differences in the genetic make-up (genotype), by examining their DNA sequence.

Gold standard The best available diagnostic test or benchmark against which other tests or procedures are measured.

Haemophilus influenzae A Gram-negative bacterium that can cause respiratory infections and, in severe cases, meningitis.

Haploid Having a single set of chromosomes, typically found in gametes (sperm and egg cells).

Helicobacter pylori A Gram-negative bacterium that causes chronic gastritis and peptic ulcers, and is linked to stomach cancer.

Hepatitis B virus (HBV) A virus that causes liver infection, which can lead to chronic liver disease and liver cancer.

Hepatitis C virus (HCV) A virus that causes liver infection, which can lead to chronic liver disease and liver cancer.

Herpes simplex virus (HSV) A virus that causes herpes infections, including oral and genital herpes.

Horizontal gene transfer The movement of genetic material between organisms other than by descent from parent to offspring.

Human Endogenous Retrovirus (HERV) Retroviral sequences in the human genome that originated from ancient viral infections of germ cells.

Human Immunodeficiency Virus (HIV) The virus that causes AIDS, attacking the immune system and making the body vulnerable to infections and certain cancers.

Hybridization The process of combining complementary nucleic acid strands to form a double-stranded filament.

Influenza virus The virus that causes influenza (flu), a contagious respiratory illness.

Intron A non-coding region of a gene that is removed during RNA splicing.

Karyotype The number and appearance of chromosomes in the nucleus of a eukaryotic cell.

Klebsiella A genus of Gram-negative bacteria that can cause pneumonia, bloodstream infections, and other infections.

Latent Referring to a pathogen that is present in the body but not currently causing symptoms.

Lateral flow test A simple diagnostic device used to confirm the presence or absence of a target analyte, such as a virus, in a sample.

Legionella A genus of Gram-negative bacteria that causes Legionnaires' disease, a severe form of pneumonia.

Ligase chain reaction (LCR) A variation of the PCR technique that amplifies DNA using a ligase enzyme for detecting specific sequences.

Melting curve A graph showing the denaturation of nucleic acids as temperature increases, used to analyse PCR products.

Messenger RNA (mRNA) A type of RNA that carries genetic information from DNA to the ribosome, where it is used to synthesize proteins.

Metagenomics The study of genetic material recovered directly from environmental samples, allowing the analysis of microbial communities without the need for culturing.

Microarray An assay used to simultaneously detect the expression of thousands of genes by hybridizing them to known nucleic acid sequences fixed on a solid surface.

Microbes Microscopic organisms, including bacteria, viruses, fungi, and protozoa, that can be found in various environments.

Microbiome Community of microorganisms (also called microbiota) that live in a particular environment, such as the human gut.

Mould A type of fungus that grows in multicellular filaments called hyphae, often causing food spoilage and health issues.

Multiplex real-time PCR A variant of PCR that allows simultaneous amplification of multiple targets in a single reaction.

Neisseria A genus of Gram-negative bacteria that includes species causing gonorrhoea and meningitis.

Next generation sequencing (NGS) Advanced sequencing technologies that allow rapid sequencing of large amounts of DNA.

Non-culturable bacteria Bacteria that cannot be grown in standard laboratory culture media, often detected by molecular methods.

Norovirus A highly contagious virus that causes gastroenteritis, leading to vomiting and diarrhoea.

Nucleic acid probes Short strands of DNA or RNA used to detect the presence of complementary sequences by hybridization.

Patho-adaptation The process by which a pathogen evolves to adapt to its host environment, often leading to increased virulence.

Phlebotomist A healthcare professional trained to draw blood from patients for clinical or medical testing, transfusions, donations, or research.

Phylogenetic tree A branching diagram that represents the evolutionary relationships among strains, lineages, or species based on their genetic characteristics.

Plasma The liquid component of blood that contains water, electrolytes, proteins, and other molecules, but without the cells.

Plasmid A small, circular DNA molecule found in bacteria that can replicate independently of the chromosomal DNA.

Plasmodium The genus of protozoa that causes malaria in humans.

Polycistronic mRNA An mRNA molecule that encodes multiple proteins, typically found in prokaryotes.

Polymerase chain reaction (PCR) A technique used to amplify small segments of DNA by copying them repeatedly.

Polymorphism A common variation in the DNA code.

Polyploid Having more than two sets of chromosomes, common in some plants, animals, and protozoa.

Primers Short sequences of nucleotides that provide a starting point for DNA synthesis during PCR.

Prokaryotic Referring to cells that lack a nucleus and other membrane-bound organelles, characteristic of bacteria and archaea.

Protozoa Single-celled eukaryotes that can be free-living or parasitic, causing diseases such as malaria and giardiasis.

Quantitative polymerase chain reaction (qPCR) A variant of PCR that enables quantification of the amount of DNA in a sample in real time.

Respiratory syncytial virus (RSV) A virus that causes respiratory infections, especially in young children and the elderly.

Retroviruses A family of viruses that replicate by reverse transcribing their RNA into DNA, integrating into the host genome (e.g. HIV).

Reverse transcriptase An enzyme used by retroviruses to convert their RNA genome into DNA.

Reverse transcriptase PCR (RT-PCR) A technique that combines reverse transcription of RNA into DNA and PCR amplification of the resulting DNA target.

Ribonucleic acid (RNA) A nucleic acid present in all living cells that plays a role in coding, decoding, regulation, and expression of genes.

Ribosomal RNA (rRNA) A type of RNA that makes up ribosomes and is essential for protein synthesis.

Ribosome A cellular organelle that translates mRNA into proteins, found in both prokaryotic and eukaryotic cells.

RNase inhibitor A substance that inhibits the activity of ribonucleases (enzymes that degrade RNA).

Rotavirus A virus that causes severe diarrhoea, primarily in infants and young children.

Salmonella A genus of Gram-negative bacteria that causes foodborne illnesses, such as typhoid and enteric fever.

Sanger sequencing A method of DNA sequencing that uses labelled chain-terminating nucleotides.

Serum The liquid part of blood that remains after clotting, used in diagnostic tests and research.

Severe Acute Respiratory Syndrome Coronavirus 2 (SARS-CoV-2) The virus responsible for COVID-19, a contagious respiratory illness.

Shigella A Gram-negative bacterium that causes dysentery and shigellosis, a form of severe diarrhoea.

Single nucleotide polymorphisms (SNPs) Variations at a single position in a DNA sequence among individuals.

Staphylococcus A genus of Gram-positive bacteria that includes species causing various infections, such as skin infections and pneumonia.

Stool Solid waste discharged from the intestines through the rectum, used often as a diagnostic sample.

Streptococcus A genus of Gram-positive bacteria that includes species causing strep throat, pneumonia, and other infections.

Temperature of melting (Tm) The temperature at which half of the DNA strands are in the double-helical state and half are in the 'melted' single-stranded state.

Toxoplasma gondii A protozoan parasite that causes toxoplasmosis, which can be serious for pregnant women and immunocompromised individuals.

Transduction The transfer of genetic material from one bacterium to another by a bacteriophage.

Transformation The uptake of free DNA from the environment by a bacterial cell.

Trypanosoma A genus of protozoa that causes diseases such as African sleeping sickness and Chagas disease.

Varicella zoster virus (VZV) The virus that causes chickenpox and shingles.

Vibrio cholerae A Gram-negative bacterium that causes cholera, a severe diarrhoeal disease.

Virulence factors Molecules produced by pathogens that contribute to the pathogenicity and facilitate the disease process.

Viruses Infectious agents composed of a protein coat and nucleic acid (DNA or RNA) that replicate inside living host cells.

Whole genome sequencing (WGS) A laboratory process that determines the complete DNA sequence of an organism's genome at a single time.

Yeast A type of unicellular fungus.

Chapter 9

HeLa cells A strain of continuously dividing cancer cells cultured since its isolation in 1951 from a cervical carcinoma of the patient Henrietta Lacks.

In silico Scientific experiments or research conducted or produced by means of computer modelling or computer simulation.

In vitro Scientific experiments or research conducted or produced in a test tube, culture dish, laboratory, or elsewhere outside a living organism.

Pre-implantation genetic testing A specialized technique which combines *in vitro* fertilization (IVF) technology with genetic testing.

Recombinant DNA technology Laboratory-based techniques that involve manipulating DNA fragments from different sources to create new genetic combinations.

Retinoblastoma A rare malignant tumour of the retina, affecting young children.

Tuskegee effect Describes the lower levels of participation and representation in research and in engagement with medical services by African Americans, attributed to the loss of trust which occurred as a result of the Tuskegee experiment. In the experiment, conducted between 1932 and 1972, African American men were denied treatment for syphilis in the name of research.

Chapter 10

Allogeneic Taken from different individuals of the same species.

Autologous Taken from an individual's own tissues, cells, or DNA.

Autosomal recessive A pattern of inheritance characteristic of some genetic diseases, where the gene is located on the non-sex chromosomes, and two copies of the mutation are needed to cause the disease.

Biomarker A naturally occurring biological marker that can be measured as an indicator of a biological or pathological process, exposure, or intervention.

Companion diagnostics Tests co-developed with drugs to select patients for a particular treatment, including predictive biomarker assays and medical devices.

Consanguineous A relationship that arises from having a common ancestor.

Drug repurposing A strategy where existing drugs, including failed or abandoned drugs from clinical trials, are investigated for new therapeutic purposes.

Epigenetics The study of how age and exposure to environmental factors, such as diet, drugs, and chemicals, may cause reversible chemical modifications of the structure and packing of DNA influencing gene expression, without any change in the underlying DNA sequence of the organism. These changes may be heritable. The word epigenetics is of Greek origin and literally means over and above (epi) the genome.

Genome All of the genetic information in a human or other organism.

Genotyping The process of determining the DNA sequence, called a genotype, at specific positions within the genome of an individual.

Human leukocyte antigen The HLA complex is a complex of genes on chromosome 6 in humans that encode cell-surface proteins that play an important role in the immune response to foreign substances.

Metabolome All of the metabolites in an organism, tissue, or cell.

Microbiome The community of microorganisms such as bacteria, viruses. and fungi living in and on the human body, which varies from person to person and over time.

Multi-morbidity The presence of two or more long-term health conditions, which can include physical or mental health conditions.

Multipotent Cells that can differentiate into more than one cell type.

Next generation sequencing Also known as massively parallel or high-throughput sequencing. Millions or billions of DNA strands are sequenced in parallel.

Omics The comprehensive or global assessment of molecules in a sample; molecular characterization can take place at a single cell, tissue, or organ level, and within an individual or a cohort of people.

Penetrant How likely it is that a person who has a genetic mutation will show signs and symptoms of the disease. Not everyone who has the mutation will develop the disease. Complete penetrance means that every person who has the mutation will show signs and symptoms of the disease.

Pluripotent Cells that can differentiate into any cell type of the body.

Polypharmacy The use of multiple medications (typically five or more) in a patient, commonly an older adult.

Population attributable fraction The fraction of all cases of a particular disease in a population that is attributable to a specific exposure.

Potency Cell potency is the varying ability of stem cells to differentiate into specialized cell types of the body.

Proteome All of the proteins in an organism, tissue, or cell.

Randomized controlled trials (RCTs) A type of scientific experiment (e.g. a clinical trial) in which participants are randomly allocated into different groups to compare the effectiveness of different interventions or treatments.

Therapeutic index A ratio that compares the blood concentration at which a drug becomes toxic and the concentration at which the drug is effective. The larger the therapeutic index, the safer the drug.

References

Chapter 1

Horton R, et al. (2024) Challenges of using whole genome sequencing in population newborn screening. *BMJ*, **384**, e077060.

Human Genome Project Sequencing Consortium (2004) Finishing the euchromatic sequence of the human genome. *Nature*, **431**, 931–45.

Lander E, et al. (2021) Initial sequencing and analysis of the human genome. *Nature*, **409**, 860–921.

Levy S, et al. (2007) The diploid genome sequence of an individual human. *PLOS Biology*, **5**, e254.

Lucassen A, Horton R (2024) Balancing the rights of the presymptomatic child to be found with the risk of harm to others from the screening process. *Eur J Hum Gen*.

Nurk S, et al. (2022) The complete sequence of a human genome. *Science*, **376**, 44–53.

Pairo-Castineira E, et al. (2020) Genetic mechanisms of critical illness in COVID-19. *Nature*, **591**, 92–8.

Pairo-Castineira E, et al. (2023) GWAS and meta-analysis identifies 49 genetic variants underlying critical COVID-19. *Nature*, **617**, 764–8.

Salisbury H (2024) The Generation Study. *BMJ*, **384**, q548.

Sanger F, et al. (1977) DNA sequencing with chain-terminating inhibitors. *PNAS*, **74**, 5463–7.

Sosinsky A, et al. (2024) Insights for precision oncology from the integration of genomic and clinical data of 13,880 tumors from the 100,000 Genomes Cancer Programme. *Nature Medicine*, **30**, 279–89.

The 100,000 Genomes Project Pilot Investigators (2021) 100,000 Genomes Pilot on Rare-Disease Diagnosis in Health Care—Preliminary Report. *New England Journal of Medicine*, **385**, 1868–80.

Venter JC, et al. (2001) The sequence of the human genome. *Science*, **291**, 1304–51.

Watson JD, Crick FHC (1953) Molecular structure of nucleic acids: a structure for deoxyribose nucleic acid. *Nature*, **171**, 737–8.

Chapter 3

Antonarakis SE, Skotko BG, Rafii MS, Strydom A, Pape SE, Bianchi DW, Sherman SL, Reeves RH (2020) Down syndrome. *Nat Rev Dis Primers*, **6**(1), 9. PMID: 32029743; PMCID: PMC8428796.

Augui S, Nora E, Heard E (2011) Regulation of X-chromosome inactivation by the X-inactivation centre. *Nat Rev Genet*, **12**, 429–42.

Blewitt ME (2024) Mary Lyon and the birth of X-inactivation research. *Nat Rev Genet*, **25**(1), 6. PMID: 37704716.

Bolzer A, Kreth G, Solovei I, Koehler D, Saracoglu K, et al. (2005) Three-dimensional maps of all chromosomes in human male fibroblast nuclei and prometaphase rosettes. *PLOS Biology*, **3**(5), e157.

Branco MR, Branco T, Ramirez F, et al. (2008) Changes in chromosome organization during PHA-activation of resting human lymphocytes measured by cryo-FISH. *Chromosome Res*, **16**, 413–26.

Brooker AS, Berkowitz KM (2014) The roles of cohesins in mitosis, meiosis, and human health and disease. *Methods Mol Biol*, **1170**, 229–66. PMID: 24906316; PMCID: PMC4495907.

Cremer T, Cremer C (2001) Chromosome territories, nuclear architecture and gene regulation in mammalian cells. *Nat Rev Genet*, **2**(4), 292–301. PMID: 11283701.

Croft JA, Bridger JM, Boyle S, Perry P, Teague P, Bickmore WA (1999) Differences in the localization and morphology of chromosomes in the human nucleus. *J Cell Biol*, **145**(6), 1119–31. PMID: 10366586; PMCID: PMC2133153.

da Costa-Nunes, JA, Noordermeer D (2023) TADs: Dynamic structures to create stable regulatory functions. *Current Opinion in Structural Biology*, **81**, 102622. ISSN 0959-440X.

Dhital B, Rodriguez-Bravo V (2023) Mechanisms of chromosomal instability (CIN) tolerance in aggressive tumors: surviving the genomic chaos. *Chromosome Res*, **31**.

Dixon JR, Gorkin DU, Ren B (2016) Chromatin domains: the unit of chromosome organization. *Mol Cell*, **62**(5), 668–80. PMID: 27259200; PMCID: PMC5371509.

Finlan LE, Sproul D, Thomson I, Boyle S, Kerr E, et al. (2008) Recruitment to the nuclear periphery can alter expression of genes in human cells. *PLOS Genetics*, **4**(3), e1000039.

Finn EH, Misteli T (2021) A high-throughput DNA FISH protocol to visualize genome regions in human cells. *STAR Protocols*, **2**(3), 100741. ISSN 2666-1667.

Gerber A, van Otterdijk S, Bruggeman FJ, Tutucci E (2023) Understanding spatiotemporal coupling of gene expression using single molecule RNA imaging technologies. *Transcription*, **14**(3–5), 105–126.

Howe B, Umrigar A, Tsien F (2014) Chromosome preparation from cultured cells. *J Vis Exp.*, **83**, e50203. PMID: 24513647; PMCID: PMC4091199.

Lenormand T, Engelstädter J, Johnston SE, Wijnker E, Haag CR (2016) Evolutionary mysteries in meiosis. *Phil Trans R Soc Lond B Biol Sci*, **371**(1706), 20160001. PMID: 27619705; PMCID: PMC5031626.

Maderspacher F. (2008) Theodor Boveri and the natural experiment. *Curr Biol*, **18**(7), R279–R286. PMID: 18397731.

Mandahl et al. (2024) Gene amplification in neoplasia: A cytogenetic survey of 80131 cases. *Genes Chromosomes Cancer*, **63**(1), e23214.

Sati S, Cavalli G (2017) Chromosome conformation capture technologies and their impact in understanding genome function. *Chromosoma*, **126**, 33–44.

Shah A (2024) Rethinking cancer initiation: The role of large-scale mutational events. *Genes Chromosomes Cancer*, **63**(1), e23213.

Shorokhova M, Nikolsky N, Grinchuk T. (2021) Chromothripsis: Explosion in Genetic Science. *Cells*, **10**(5), 1102. PMID: 34064429; PMCID: PMC8147837.

Skibbens RV (2019) Condensins and cohesins—one of these things is not like the other! *J Cell Sci*, **132**(3), jcs220491. PMID: 30733374; PMCID: PMC6382015.

Solovei I, Cremer M (2010) 3D-FISH on cultured cells combined with immunostaining. *Methods Mol Biol*, **659**, 117–26. PMID: 20809307.

Spicer MFD, Gerlich DW (2023) The material properties of mitotic chromosomes. *Curr Opin Struct Biol,* **81**, 102617. Epub 2023 Jun 6. PMID: 37279615; PMCID: PMC10448380.

Stephens PJ, et al. (2011) Massive genomic rearrangement acquired in a single catastrophic event during cancer development. Cell, **144**, 27–40.

Tanabe H, Habermann FA, Solovei I, Cremer M, Cremer T (2002) Non-random radial arrangements of interphase chromosome territories: evolutionary considerations and functional implications. *Mutat Res*, **504**(1–2), 37–45. PMID: 12106644.

Thompson LL, Jeusset LM, Lepage CC, McManus KJ (2017) Evolving therapeutic strategies to exploit chromosome instability in cancer. *Cancers (Basel),* **9**(11), 151. PMID: 29104272; PMCID: PMC5704169.

Volpi EV, Chevret E, Jones T, Vatcheva R, Williamson J, Beck S, Campbell RD, Goldsworthy M, Powis SH, Ragoussis J, Trowsdale J, Sheer D (2000) Large-scale chromatin organization of the major histocompatibility complex and other regions of human chromosome 6 and its response to interferon in interphase nuclei. *J Cell Sci*, **113**(9), 1565–76.

Chapter 4

Abdellah Z, Ahmadi A, Ahmed S, et al. (2004) Finishing the euchromatic sequence of the human genome. *Nature,* **431,** 931–45.

Abdellaoui A, Ehli EA, Hottenga JJ, et al. (2015) CNV concordance in 1,097 MZ twin pairs. *Twin Res Hum Genet.*

Acuna-Hidalgo R, Bo T, Kwint MP, et al. (2015) Post-zygotic point mutations are an underrecognized source of de novo genomic variation. *Am J Hum Genet,* **97**, 67–74.

Acuna-Hidalgo R, Veltman JA, Hoischen A (2016) New insights into the generation and role of de novo mutations in health and disease. *Genome Biol,* **17**, 1–19.

Asbury K, Dunn JF, Pike A, et al. (2003) Nonshared environmental influences on individual differences in early behavioral development: a monozygotic twin differences study. *Child Dev,* **74**, 933–43.

Bateson W, Saunders ER, Punnett RC (1909) Experimental studies in the physiology of heredity. *Z Indukt Abstamm Vererbungsl,* **2**, 17–9.

Bell JT, Loomis AK, Butcher LM, et al. (2014) Differential methylation of the TRPA1 promoter in pain sensitivity. *Nat Commun,* **5**, 12.

Bell JT, Spector TD (2011) A twin approach to unraveling epigenetics. *Trends Genet,* **27**, 116–25.

Bird TD (2008) Genetic aspects of Alzheimer disease. *Genet Med,* **10**, 231–9.

Bouhlal Y, Martinez S, Gong H, et al. (2013) Twin mitochondrial sequence analysis. *Mol Genet Genomic Med,* **1**, 174–86.

Brosens E, Marsch F, De Jong EM, et al. (2016) Copy number variations in 375 patients with oesophageal atresia and/or tracheoesophageal fistula. *Eur J Hum Genet,* **24**, 1715–23.

Bruder CEG, Piotrowski A, Gijsbers AACJ, et al. (2008) Phenotypically concordant and discordant monozygotic twins display different DNA copy-number-variation profiles. *Am J Hum Genet,* **82**, 763–71.

Cacheiro P, Muñoz-Fuentes V, Murray SA, et al. (2020) Human and mouse essentiality screens as a resource for disease gene discovery. *Nat Commun*, **11**, 1–16.

Campbell CD, Chong JX, Malig M, et al. (2012) Estimating the human mutation rate using autozygosity in a founder population. *Nat Genet,* **44**, 1277–81.

Campbell CD, Eichler EE (2013) Properties and rates of germline mutations in humans. *Trends Genet*, **29**, 575–84.

Campbell IM, Shaw CA, Stankiewicz P, et al. (2015) Somatic mosaicism: Implications for disease and transmission genetics. *Trends Genet,* **31**, 382–92.

Castellani CA, Melka MG, Wishart AE, et al. (2014) Biological relevance of CNV calling methods using familial relatedness including monozygotic twins. *BMC Bioinformatics*, **15**, 114.

Ciregia F, Giusti L, Da Valle Y, et al. (2013) A multidisciplinary approach to study a couple of monozygotic twins discordant for the chronic fatigue syndrome: A focus on potential salivary biomarkers. *J Transl Med,* **11**.

Civelek M, Wu Y, Pan C, et al. (2017) Genetic regulation of adipose gene expression and cardio-metabolic traits. *Am J Hum Genet*, **100**, 428–43.

Colvert E, Tick B, McEwen F, et al. (2015) Heritability of autism spectrum disorder in a UK population-based twin sample. *JAMA Psychiatry*, **72**, 415–23.

Czyz W, Morahan JM, Ebers GC, et al. (2012) Genetic, environmental and stochastic factors in monozygotic twin discordance with a focus on epigenetic differences. *BMC Med*, **10**, 93.

Dal GM, Ergüner B, Sağiroğlu MS, et al. (2014) Early postzygotic mutations contribute to de novo variation in a healthy monozygotic twin pair. *J Med Genet,* **51**, 455–9.

Davydov EV, Goode DL, Sirota M, et al. (2010) Identifying a high fraction of the human genome to be under selective constraint using GERP++. *PLoS Comput Biol,* **6**, e1001025.

Dichgans M, Pulit SL, Rosand J (2019) Stroke genetics: discovery, biology, and clinical applications. *Lancet Neurol*, **18**, 587–99.

DiMauro S, Schon EA (2003) Mitochondrial respiratory-chain diseases. *N Engl J Med*, **348**, 2656–68.

Dunham I, Kundaje A, Aldred SF, et al. (2012) An integrated encyclopedia of DNA elements in the human genome. *Nature,* **489**, 57–74.

Firth HV, Wright CF (2011) The Deciphering Developmental Disorders (DDD) study. *Dev Med Child Neurol*, **53**, 702–3.

Fizelova M, Jauhiainen R, Stančáková A, et al. (2016) Finnish diabetes risk score is associated with impaired insulin secretion and insulin sensitivity, drug-treated hypertension and cardiovascular disease: A follow-up study of the METSIM cohort. *PLoS One*, **11**.

Forsberg LA, Rasi C, Razzaghian HR, et al. (2012) Age-related somatic structural changes in the nuclear genome of human blood cells. *Am J Hum Genet*, **90**, 217–28.

Franke A, McGovern DPB, Barrett JC, et al. (2010) Genome-wide meta-analysis increases to 71 the number of confirmed Crohn's disease susceptibility loci. *Nat Genet*, **42**, 1118–25.

Freeman JL, Perry GH, Feuk L, et al. (2006) Copy number variation: New insights in genome diversity. *Genome Res*, **16**, 949–61.

Gervin K, Vigeland MD, Mattingsdal M, et al. (2012) DNA methylation and gene expression changes in monozygotic twins discordant for psoriasis: identification of epigenetically dysregulated genes. *PLoS Genet*, **8**, e1002454

González JR, Rodríguez-Santiago B, Cáceres A, et al. (2011) A fast and accurate method to detect allelic genomic imbalances underlying mosaic rearrangements using SNP array data. *BMC Bioinformatics*, **12**, 166.

Hales CN, Barker DJP (2001) The thrifty phenotype hypothesis. *Br Med Bull*, **60**, 5–20.

Handunnetthi L, Handel AE, Ramagopalan SV (2010) Contribution of genetic, epigenetic and transcriptomic differences to twin discordance in multiple sclerosis. *Expert Rev Neurother*, **10**, 1379–81.

Heinig M, Petretto E, Wallace C, et al. (2010) A trans-acting locus regulates an anti-viral expression network and type 1 diabetes risk. *Nature*, **467**, 460–4.

Heyn H, Carmona FJ, Gomez A, et al. (2013) DNA methylation profiling in breast cancer discordant identical twins identifies DOK7 as novel epigenetic biomarker. *Carcinogenesis*, **34**, 102–8.

Ho DSW, Schierding W, Wake M, et al. (2019) Machine learning SNP based prediction for precision medicine. *Front Genet,* **10**, 267.

Huang N, Lee I, Marcotte EM, et al. (2010) Characterising and predicting haploinsufficiency in the human genome. *PLoS Genet*, **6**:e1001154.

IJzerman RG, Stehouwer CDA, Boomsma DI (2000) Evidence for genetic factors explaining the birth weight–blood pressure relation. *Hypertension*, **36**, 1008–12.

Ionita-Laza I, Rogers AJ, Lange C, et al. (2009) Genetic association analysis of copy-number variation (CNV) in human disease pathogenesis. *Genomics*, **93**, 22–6.

Kazuno A, Ohtawa K, Otsuki K, et al. (2013) Proteomic analysis of lymphoblastoid cells derived from monozygotic twins discordant for bipolar disorder: a preliminary study. *PLoS One*, **8**, 53855.

Kelley JT (2012) The City of God (De Civitate Dei): Books 1–10. By Saint Augustine, Bishop of Hippo. Introduced and translated by WS Babcock. New City Press, Hyde Park, NY.

Kendler KS, Neale MC, Kessler RC, et al. (1992) A population-based twin study of major depression in women: the impact of varying definitions of illness. *Arch Gen Psychiatry*, **49**, 257–66.

Ketelaar ME, Hofstra EMW, Hayden MR (2012) What monozygotic twins discordant for phenotype illustrate about mechanisms influencing genetic forms of neurodegeneration. *Clin Genet*, **81**, 325–33.

King DA, Sifrim A, Fitzgerald TW, et al. (2017) Detection of structural mosaicism from targeted and whole-genome sequencing data. *Genome Res*, **27**, 1704–14.

Kloosterman WP, Francioli LC, Hormozdiari F, et al. (2015) Characteristics of de novo structural changes in the human genome. *Genome Res*, **25**, 792–801.

Kloss-Brandstätter A, Weissensteiner H, Erhart G, et al. (2015) Validation of next-generation sequencing of entire mitochondrial genomes and the diversity of mitochondrial DNA mutations in oral squamous cell carcinoma. *PLoS One*, **10**(8), e0135643.

Kong A, Frigge ML, Masson G, et al. (2012) Rate of de novo mutations and the importance of father's age to disease risk. *Nature*, **488**, 471–5.

Laakso M, Kuusisto J, Stančáková A, et al. (2017) The Metabolic Syndrome in Men study: a resource for studies of metabolic and cardiovascular diseases. *J Lipid Res,* **58**(3), 481–93.

Landstrom AP, Adekola BA, Bos JM, et al. (2011) PLN-encoded phospholamban mutation in a large cohort of hypertrophic cardiomyopathy cases: Summary of the literature and implications for genetic testing. *Am Heart J*, **161**, 165–71.

Laplana M, Royo JL, Aluja A, et al. (2014) Absence of substantial copy number differences in a pair of monozygotic twins discordant for features of autism spectrum disorder. *Case Rep Genet.*

Legault MA, Girard S, Perreault LPL, et al. (2015) Comparison of sequencing based CNV discovery methods using monozygotic twin quartets. *PLoS One*, **10**.

Lejeune J, Gauthier M, Turpin R. (1959) Les chromosomes humaines en culture de tissues. *C R Hebd Seances Acad Sci*, **248**, 602–3.

Lewis CM, Vassos E. (2020) Polygenic risk scores: From research tools to clinical instruments. *Genome Med*, **12**, 44.

Lindhurst MJ, Sapp JC, Teer JK, et al. (2011) A mosaic activating mutation in AKT1 associated with the Proteus syndrome. *N Engl J Med*, **365**, 611–19.

Loke YJ, Hannan AJ, Craig JM (2015) The role of epigenetic change in autism spectrum disorders. *Front Neurol.*

Magaard Koldby K, Nygaard M, Christensen K, et al. (2016) Somatically acquired structural genetic differences: A longitudinal study of elderly Danish twins. *Eur J Hum Genet.*

Magnusson PKE, Lee D, Chen X, et al. (2016) One CNV discordance in NRXN1 observed upon genome-wide screening in 38 pairs of adult healthy monozygotic twins. *Twin Res Hum Genet.*

Manolio TA, Collins FS, Cox NJ, et al. (2009) Finding the missing heritability of complex diseases. *Nature*, **461**, 747–53.

Manuck SB, McCaffery JM (2014) Gene-environment interaction. *Annu Rev Psychol.*

Martin NG, Carr AB, Oakeshott JG, et al. (1982) Co-twin control studies: vitamin C and the common cold. *Prog Clin Biol Res*, **103** Pt A, 365–73.

Mattson MP, Gleichmann M, Cheng A. (2008) Mitochondria in neuroplasticity and neurological disorders. *Neuron*, **60**, 748–66.

McCarroll SA, Kuruvilla FG, Korn JM, et al. (2008) Integrated detection and population-genetic analysis of SNPs and copy number variation. *Nat Genet*, **40**, 1166–74.

Mitchell KJ (2012) What is complex about complex disorders? *Genome Biol*, **13**, 1–11.

Morrow EM (2020) Paternal sperm DNA mosaicism and recurrence risk of autism in families. *Nat Med*, **26**, 26–8.

Myers RH, Macdonald ME, Koroshetz WJ, et al. (1993) De novo expansion of a (CAG)n repeat in sporadic Huntington's disease. *Nat Genet*, **5**, 168–73.

O'Hanlon TP, Li Z, Gan L, et al. (2011) Plasma proteomic profiles from disease-discordant monozygotic twins suggest that molecular pathways are shared in multiple systemic autoimmune diseases. *Arthritis Res Ther*, **13**.

O'Rawe J, Jiang T, Sun G, et al. (2013) Low concordance of multiple variant-calling pipelines: practical implications for exome and genome sequencing. *Genome Med.*

Petronis A, Gottesman II, Kan P, et al. (2003) Monozygotic twins exhibit numerous epigenetic differences: Clues to twin discordance? *Schizophr Bull*, **29**(1), 169–78.

Pirooznia M, Kramer M, Parla J, et al. (2014) Validation and assessment of variant calling pipelines for next-generation sequencing. *Hum Genomics*, **8**, 14.

Polderman TJC, Benyamin B, De Leeuw CA, et al. (2015) Meta-analysis of the heritability of human traits based on fifty years of twin studies. *Nat Genet*, **47**, 702–9.

Poulsen P, Esteller M, Vaag A, et al. (2007) The epigenetic basis of twin discordance in age-related diseases. *Pediatr Res.*

Raffel LJ, Mohandas T, Rimoin DL (1986) Chromosomal mosaicism in the Killian Teschler-Nicola syndrome. *Am J Med Genet*, **24**, 607–11.

Redon R, Ishikawa S, Fitch KR, et al. (2006) Global variation in copy number in the human genome. *Nature.*

Rodríguez-Santiago B, Malats N, Rothman N, et al. (2010) Mosaic uniparental disomies and aneuploidies as large structural variants of the human genome. *Am J Hum Genet.*

Rollins B, Martin MV, Sequeira PA, et al. (2009) Mitochondrial variants in schizophrenia, bipolar disorder, and major depressive disorder. *PLoS One.*

Rygiel KA, Miller J, Grady JP, et al. (2015) Mitochondrial and inflammatory changes in sporadic inclusion body myositis. *Neuropathol Appl Neurobiol.*

Schaid DJ, Chen W, Larson NB (2018) From genome-wide associations to candidate causal variants by statistical fine-mapping. *Nat Rev Genet*, **19**, 491–504.

Schuster SC (2008) Next-generation sequencing transforms today's biology. *Nat Methods.*

Siemens HW (1924) Die Zwillingspathologie. *Z Indukt Abstamm Vererbungsl.*

Song S, Wheeler LJ, Mathews CK (2003) Deoxyribonucleotide pool imbalance stimulates deletions in HeLa cell mitochondrial DNA. *J Biol Chem.*

Stamouli S, Anderlid BM, Willfors C, et al. (2018) Copy number variation analysis of 100 twin pairs enriched for neurodevelopmental disorders. *Twin Res Hum Genet.*

Stefan M, Zhang W, Concepcion E, et al. (2014) DNA methylation profiles in type 1 diabetes twins point to strong epigenetic effects on etiology. *J Autoimmun*, **50**, 33–7.

Stewart JB, Chinnery PF (2015) The dynamics of mitochondrial DNA heteroplasmy: implications for human health and disease. *Nat Rev Genet*, **16**, 530–42.

Sudlow C, Gallacher J, Allen N, et al. (2015) UK Biobank: an open access resource for identifying the causes of a wide range of complex diseases of middle and old age. *PLOS Med*, **12**, e1001779.

Turro E, Astle WJ, Megy K, et al. (2020) Whole-genome sequencing of patients with rare diseases in a national health system. *Nature*, **583**, 96–102.

Vadgama N, Gaze D, Ranson J, et al. (2015a) Elevated γ-glutamyltransferase and erythrocyte sedimentation rate in ischemic stroke in discordant monozygotic twin study. *Int J Stroke*, **10**, E32–3.

Vadgama N, Lamont D, Hardy J, et al. (2019a) Distinct proteomic profiles in monozygotic twins discordant for ischaemic stroke. *Mol Cell Biochem*, **456**(1–2), 157–165.

Vadgama N, Nirmalananthan N, Sadiq M, et al. (2015b) Identical non-identical twins and non-identical identical twins. *BMJ*, **351**.

Vadgama N, Pittman A, Simpson M, et al. (2019b) De novo single-nucleotide and copy number variation in discordant monozygotic twins reveals disease-related genes. *Eur J Hum Genet.*

van Dongen J, Slagboom P, Draisma H et al. (2012) The continuing value of twin studies in the omics era. *Nat Rev Genet*, **13**, 640–53.

Veltman JA, Brunner HG (2012) De novo mutations in human genetic disease. *Nat Rev Genet.*

Weischenfeldt J, Symmons O, Spitz F, et al. (2013) Phenotypic impact of genomic structural variation: insights from and for human disease. *Nat Rev Genet.*

Wilhelm M, Schlegl J, Hahne H, et al. (2014) Mass-spectrometry-based draft of the human proteome. *Nature*.

Wishart DS, Jewison T, Guo AC, et al. (2012) HMDB 3.0—The Human Metabolome Database in 2013. *Nucleic Acids Res*, **41**, D801–7.

Wolff DJ, Miller AP, Van Dyke DL, et al. (1996) Molecular definition of breakpoints associated with human Xq isochromosomes: implications for mechanisms of formation. *Am J Hum Genet*, **58**, 154–60.

Wrede JE, Mengel-From J, Buchwald D, et al. (2015) Mitochondrial DNA copy number in sleep duration discordant monozygotic twins. *Sleep*, **38**, 1655–1658A.

Xu H, DiCarlo J, Satya RV et al. (2014) Comparison of somatic mutation calling methods in amplicon and whole exome sequence data. *BMC Genomics.*

Yao Y-G, Kajigaya S, Young NS (2015) Mitochondrial DNA mutations in single human blood cells. *Mutat Res*, **779**, 68–77.

Youngson NA, Whitelaw E (2008) Transgenerational epigenetic effects. *Annu Rev Genomics Hum Genet*, **9**, 233–57.

Zierer J, Menni C, Kastenmüller G, et al. (2015) Integration of 'omics' data in aging research: from biomarkers to systems biology. *Aging Cell*, **14**, 933–44.

Chapter 5

JASPAR: Rauluseviciute I, Riudavets-Puig R, Blanc-Mathieu R, Castro-Mondragon JA, Ferenc K, Kumar V, Lemma RB, Lucas J, Chèneby J, Baranasic D, Khan A, Fornes O, Gundersen S, Johansen M, Hovig E, Lenhard B, Sandelin A, Wasserman WW, Parcy F, Mathelier A. JASPAR 2024: 20th anniversary of the open-access database of transcription factor binding profiles. *Nucleic Acids Res*, **52**(D1), D174–D182.

Karczewski KJ, Francioli LC, Tiao G, Cummings BB, and 61; Genome Aggregation Database Consortium (2020) The mutational constraint spectrum quantified from variation in 141,456 humans. *Nature*, **581**, 434–43.

Lander ES, Linton LM, Birren B, Nusbaum C, and 254 others. International Human Genome Sequencing Consortium (2001) Initial sequencing and analysis of the human genome. *Nature*, **409**, 860–921.

O'Neal WK, Knowles MR (2018) Cystic fibrosis disease modifiers: complex genetics defines the phenotypic diversity in a monogenic disease. *Annu Rev Genomics Hum Genet*, **19**, 201–22.

Richards S, Aziz N, Bale S, Bick D, and 8 others; ACMG Laboratory Quality Assurance Committee (2015) Standards and guidelines for the interpretation of sequence variants: a joint consensus recommendation of the American College of Medical Genetics and Genomics and the Association for Molecular Pathology. *Genet Med*, **17**, 405–24.

The 1000 Genomes Project Consortium (2015) A global reference for human genetic variation *Nature*, **526**, 68.

Chapter 7

Fearon ER, Vogelstein B (1990) A genetic model for colorectal tumorigenesis. *Cell*, **61**(5), 759–67. PMID: 2188735.

Hanahan D (2022) Hallmarks of cancer: new dimensions. *Cancer Discov*, **12**(1), 31–46.

Hanahan D, Weinberg RA (2011) Hallmarks of cancer: the next generation. *Cell*, **144**(5), 646–74.

Lane DP, Crawford LV (1979) T antigen is bound to at host protein in SV40-transformed cells. *Nature*, **278**, 261–3.

Linzer DI, Levine AJ (1979) Characterization of a 54K Dalton cellular SV40 tumor antigen present in SV40-transformed cells and uninfected embryonal carcinoma cells. *Cell*, **17**, 43–52. 10.1016/0092-8674(79)90293-9

McClellan J, King MC (2010) Genetic heterogeneity in human disease. *Cell*, **141**, 210–17.

Menter DG, Davis JS, Broom BM, Overman MJ, Morris J, Kopetz S (2019) Back to the colorectal cancer consensus molecular subtype future. *Curr Gastroenterol Rep*, **21**(2), 5. PMID: 30701321; PMCID: PMC6622456.

Mullis KB, Faloona FA (1987) Specific synthesis of DNA in vitro via a polymerase-catalyzed chain reaction. *Methods Enzymol.*, **155**, 335–50.

Rous P (1983) Landmark article (JAMA 1911, 56, 198). Transmission of a malignant new growth by means of a cell-free filtrate. By Peyton Rous. *JAMA*, **250**(11), 1445–9.

Chapter 9

Beauchamp TL, Childress JF (2019) *Principles of biomedical ethics* (8th ed.). Oxford University Press.

Kaan T, Xafis V, Schaefer GO, Zhu Y, Labude MK, Chadwick R (2021) Germline genome editing: Moratorium, hard law, or an informed adaptive consensus? *PLoS Genet*, **17**(9), e1009742. PMID: 34499642; PMCID: PMC8428541.

Index

Note: Tables, figures, and boxes are indicated by an italic *t*, *f*, and *b* following the page/paragraph number.

D

H

M

N

Q

R

S

T

U

V

W

X

Y

Z